Learning Data Science

Data Wrangling, Exploration, Visualization, and Modeling with Python

Sam Lau, Joseph Gonzalez, and Deborah Nolan

Beijing • Boston • Farnham • Sebastopol • Tokyo O'REILLY®

Learning Data Science
by Sam Lau, Joseph Gonzalez, and Deborah Nolan

Published by O'Reilly Media, Inc., 1005 Gravenstein Highway North, Sebastopol, CA 95472.

O'Reilly books may be purchased for educational, business, or sales promotional use. Online editions are also available for most titles (*http://oreilly.com*). For more information, contact our corporate/institutional sales department: 800-998-9938 or *corporate@oreilly.com*.

Acquisitions Editor: Aaron Black
Development Editor: Melissa Potter
Production Editor: Katherine Tozer
Copyeditor: Audrey Doyle
Proofreader: J.M. Olejarz
Indexer: Potomac Indexing, LLC
Interior Designer: David Futato
Cover Designer: Karen Montgomery
Illustrator: Kate Dullea

September 2023: First Edition

Revision History for the First Release
2023-09-15: First Release

See *http://oreilly.com/catalog/errata.csp?isbn=9781098113001* for release details.

978-1-098-11300-1

[LSI]

Table of Contents

Part III. Understanding The Data

Part VI. Classification

Preface

Data science is exciting work. The ability to draw insights from messy data is valuable for all kinds of decision making across business, medicine, policy, and more. This book, *Learning Data Science*, aims to prepare readers to do data science. To achieve this, we've designed this book with the following special features:

Focus on the fundamentals
: Technologies come and go. While we work with specific technologies in this book, our goal is to equip readers with the fundamental building blocks of data science. We do this by revealing how to think about data science problems and challenges, and by covering the fundamentals behind the individual technologies. Our aim is to serve readers even as technologies change.

Cover the entire data science lifecycle
: Instead of just focusing on a single topic, like how to work with data tables or how to apply machine learning techniques, we cover the entire data science lifecycle—the process of asking a question, obtaining data, understanding the data, and understanding the world. Working through the entire lifecycle can often be the hardest part of being a data scientist.

Use real data
: To be prepared for working on real problems, we consider it essential to learn from examples that use real data, with their warts and all. We chose the datasets presented in this book by carefully picking from actual data analyses that have made an impact, rather than using overly refined or synthetic data.

Apply concepts through case studies
: We've included extended case studies throughout the book that follow or extend analyses from other data scientists. These case studies show readers how to navigate the data science lifecycle in real settings.

Combine both computational and inferential thinking
: On the job, data scientists need to foresee how the decisions they make when writing code and how the size of a dataset might affect statistical analysis. To prepare readers for their future work, *Learning Data Science* integrates computational and statistical thinking. We also motivate statistical concepts through simulation studies rather than mathematical proofs.

The text and code for this book are open source and available on GitHub (*https://github.com/DS-100/textbook/*).

Expected Background Knowledge

We expect readers to be proficient in Python and understand how to use built-in data structures like lists, dictionaries, and sets; import and use functions and classes from other packages; and write functions from scratch. We also use the `numpy` Python package without introduction but don't expect readers to have much prior experience using it.

Readers will get more from this book if they also know a bit of probability, calculus, and linear algebra, but we aim to explain mathematical ideas intuitively.

Organization of the Book

This book has 21 chapters, divided into six parts:

Part I (Chapters 1–5)
: Part I describes what the lifecycle is, makes one full pass through the lifecycle at a basic level, and introduces terminology that we use throughout the book. The part concludes with a short case study about bus arrival times.

Part II (Chapters 6–7)
: Part II introduces dataframes and relations and how to write code to manipulate data using `pandas` and SQL.

Part III (Chapters 8–12)
: Part III is all about obtaining data, discovering its traits, and spotting issues. After understanding these concepts, a reader can take a datafile and describe the dataset's interesting features to someone else. This part ends with a case study about air quality.

Part IV (Chapters 13–14)
: Part IV looks at widely used alternative sources of data, like text, binary, and data from the web.

Part V (Chapters 15–18)
: Part V focuses on understanding the world using data. It covers inferential topics like confidence intervals and hypothesis testing in addition to model fitting, feature engineering, and model selection. This part ends with a case study about predicting donkey weights for veterinarians in Kenya.

Part VI (Chapters 19–21)
: Part VI completes our study of supervised learning with logistic regression and optimization. It ends with a case study on predicting whether news articles make real or fake statements.

At the end of the book, we included resources to learn more about many of the topics this book introduces, and we provided the complete list of datasets used throughout the book.

Conventions Used in This Book

The following typographical conventions are used in this book:

Italic
: Indicates new terms, URLs, email addresses, filenames, and file extensions.

`Constant width`
: Used for program listings, as well as within paragraphs to refer to program elements such as variable or function names, databases, data types, environment variables, statements, and keywords.

`Constant width bold`
: Shows commands or other text that should be typed literally by the user.

`Constant width italic`
: Shows text that should be replaced with user-supplied values or by values determined by context.

This element signifies a general note.

This element indicates a tip.

This element indicates a warning or caution.

Using Code Examples

Supplemental material (code examples, exercises, etc.) is available for download at *https://learningds.org*.

If you have a technical question or a problem using the code examples, please email *bookquestions@oreilly.com*.

This book is here to help you get your job done. In general, if example code is offered with this book, you may use it in your programs and documentation. You do not need to contact us for permission unless you're reproducing a significant portion of the code. For example, writing a program that uses several chunks of code from this book does not require permission. Selling or distributing examples from O'Reilly books does require permission. Answering a question by citing this book and quoting example code does not require permission. Incorporating a significant amount of example code from this book into your product's documentation does require permission.

We appreciate attribution. An attribution usually includes the title, author, publisher, and ISBN. For example: "*Learning Data Science* by Sam Lau, Joseph Gonzalez, and Deborah Nolan (O'Reilly). Copyright 2023 Sam Lau, Joseph Gonzalez, and Deborah Nolan, 978-1-098-11300-1."

If you feel your use of code examples falls outside fair use or the permission given above, feel free to contact us at *bookquestions@oreilly.com*.

O'Reilly Online Learning

For more than 40 years, *O'Reilly Media* has provided technology and business training, knowledge, and insight to help companies succeed.

Our unique network of experts and innovators share their knowledge and expertise through books, articles, and our online learning platform. O'Reilly's online learning platform gives you on-demand access to live training courses, in-depth learning paths, interactive coding environments, and a vast collection of text and video from O'Reilly and 200+ other publishers. For more information, visit *https://oreilly.com*.

How to Contact Us

Please address comments and questions concerning this book to the publisher:

O'Reilly Media, Inc.
1005 Gravenstein Highway North
Sebastopol, CA 95472
800-889-8969 (in the United States or Canada)
707-829-7019 (international or local)
707-829-0104 (fax)
support@oreilly.com
https://www.oreilly.com/about/contact.html

We have a web page for this book, where we list errata, examples, and any additional information. You can access this page at *https://oreil.ly/learning-data-science*.

For news and information about our books and courses, visit *https://oreilly.com*.

Find us on LinkedIn: *https://linkedin.com/company/oreilly-media*.

Follow us on Twitter: *https://twitter.com/oreillymedia*.

Watch us on YouTube: *https://youtube.com/oreillymedia*.

Acknowledgments

This book has come about from our joint experience designing and teaching "Principles and Techniques of Data Science," an undergraduate course at the University of California, Berkeley. We first taught "Data 100" in spring 2017 in response to student demand for a second course in data science; they wanted a course that would prepare them for advanced study in data science and for the workforce.

The thousands of students we have taught since that first offering have been an inspiration for us. We've also benefited from co-teaching with other instructors, including Ani Adhikari, Andrew Bray, John DeNero, Sandrine Dudoit, Will Fithian, Joe Hellerstein, Josh Hug, Anthony Joseph, Scott Lee, Fernando Perez, Alvin Wan, Lisa Yan, and Bin Yu. We especially thank Joe Hellerstein for insights around data wrangling, Fernando Perez for encouraging us to include more complex data structures like NetCDF, Josh Hug for the idea of the PurpleAir case study, and Duncan Temple Lang for collaboration on an earlier version of the course. We also thank the Berkeley students who have been our teaching assistants, and especially mention those who have contributed to previous versions of the book: Ananth Agarwal, Ashley Chien, Andrew Do, Tiffany Jann, Sona Jeswani, Andrew Kim, Jun Seo Park, Allen Shen, Katherine Yen, and Daniel Zhu.

A core part of this book is the many datasets that we wrangle and analyze, and we are immensely thankful to the individuals and organizations that made their data open and available to us. At the end of this book, we list these contributors along with the original data sources, and related research papers, blog posts, and reports.

Lastly, we are grateful to the O'Reilly team for their work to bring this book from class notes to publication, especially Melissa Potter, Jess Haberman, Aaron Black, Danny Elfanbaum, and Mike Loukides. We'd also like to thank the technical reviewers whose comments have improved the book: Sona Jeswani, Thomas Nield, Siddharth Yadav, and Abhijit Dasgupta.

PART I
The Data Science Lifecycle

CHAPTER 1
The Data Science Lifecycle

Data science is a rapidly evolving field. At the time of this writing, people are still trying to pin down exactly what data science is, what data scientists do, and what skills data scientists should have. What we do know, though, is that data science uses a combination of methods and principles from statistics and computer science to work with and draw insights from data. And learning computer science and statistics in combination makes us better data scientists. We also know that any insights we glean need to be interpreted in the context of the problem that we are working on.

This book covers fundamental principles and skills that data scientists need to help make all sorts of important decisions. With both technical skills and conceptual understanding we can work on data-centric problems to, say, assess whether a vaccine works, filter out fake news automatically, calibrate air quality sensors, and advise analysts on policy changes.

To help you keep track of the bigger picture, we've organized topics around a workflow that we call the *data science lifecycle*. In this chapter, we introduce this lifecycle. Unlike other data science books, which tend to focus on one part of the lifecycle or address only computational or statistical topics, we cover the entire cycle from start to finish and consider both statistical and computational aspects together.

The Stages of the Lifecycle

Figure 1-1 shows the data science lifecycle, which is divided into four stages: Ask a Question, Obtain Data, Understand the Data, and Understand the World. We've purposefully made these stages broad. In our experience, the mechanics of the lifecycle change frequently. Computer scientists and statisticians continue to build new software packages and programming languages for working with data, and they develop new methodologies that are more specialized.

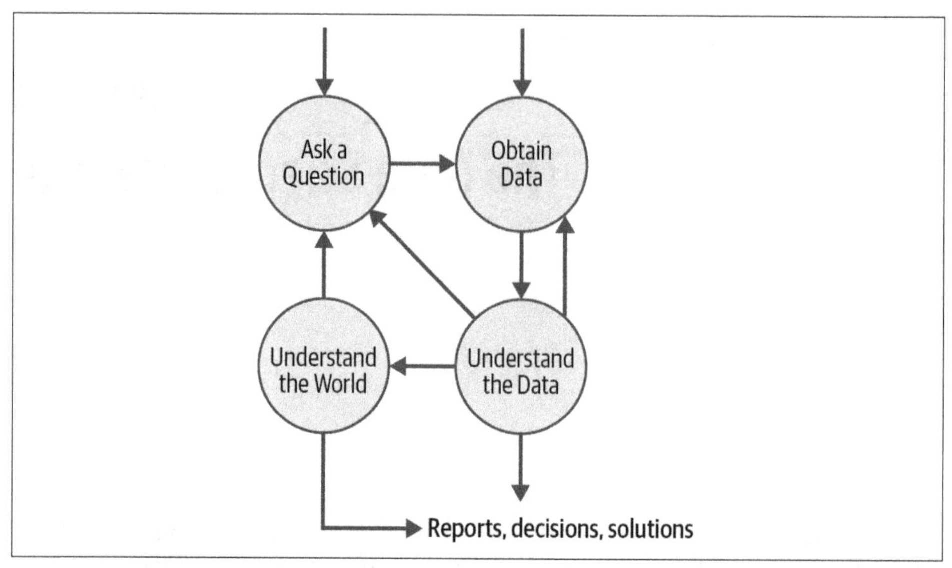

Figure 1-1. The four high-level stages of the data science lifecycle with arrows indicating how the stages can lead into one another

Despite these changes, we've found that almost every data project consists of these four stages:

Ask a Question
: Asking good questions is at the heart of data science, and recognizing different kinds of questions guides us in our analyses. We cover four categories of questions: descriptive, exploratory, inferential, and predictive. For example, "How have house prices changed over time?" is descriptive in nature, whereas "Which aspects of houses are related to sale price?" is exploratory. Narrowing down a broad question into one that can be answered with data is a key element of this first stage in the lifecycle. It can involve consulting the people participating in a study, figuring out how to measure something, and designing data collection protocols. A clear and focused research question helps us determine the data we need, the patterns to look for, and how to interpret results. It can also help us refine our question, recognize the type of question being asked, and plan the data collection phase of the lifecycle.

Obtain Data
: When data are expensive and hard to gather and when our goal is to generalize from the data to the world, we aim to define precise protocols for collecting the data. Other times, data are cheap and easily accessed. This is especially true for online data sources. For example, Twitter (*https://oreil.ly/WvUhe*) lets people quickly download millions of data points. When data are plentiful, we can start an analysis by obtaining and exploring the data, and then honing a research

question. In both situations, most data have missing or unusual values and other anomalies that we need to account for. No matter the source, we need to check the data quality. Considering the scope of the data is equally important; for example, we identify how representative the data are and look for potential sources of bias in the collection process. These considerations help us determine how much faith we can place in our findings. And, typically, we must manipulate the data before we can analyze it more formally. We may need to modify structure, clean data values, and transform measurements to prepare for analysis.

Understand the Data
: After obtaining and preparing data, we want to carefully examine them, and *exploratory data analysis* is often key. In our explorations, we make plots to uncover interesting patterns and summarize the data visually. We also continue to look for problems with the data. As we search for patterns and trends, we use summary statistics and build statistical models, like linear and logistic regression. In our experience, this stage of the lifecycle is highly iterative. Understanding the data can also lead us back to earlier stages in the data science lifecycle. We may find that we need to modify or redo the data cleaning and manipulation, acquire more data to supplement our analysis, or refine our research question given the limitations of the data. The descriptive and exploratory analyses that we carry out in this stage may adequately answer our question, or we may need to go on to the next stage in order to make generalizations beyond our data.

Understand the World
: When our goals are purely descriptive or exploratory, the analysis ends at the Understand the Data stage of the lifecycle. At other times, we aim to quantify how well the trends we find generalize beyond our data. We may want to use a model that we have fit to our data to make inferences about the world or give predictions for future observations. To draw inferences from a sample to a population, we use statistical techniques like A/B testing and confidence intervals. And to make predictions for future observations, we create prediction intervals and use train-test splits of the data.

For each stage of the lifecycle, we explain theoretical concepts, introduce data technologies and statistical methodologies, and show how they work in practical examples. Throughout, we rely on authentic data and analyses by other data scientists, not made-up data, so you can learn how to perform your own data acquisition, cleaning, exploration, and formal analyses, and draw sound conclusions. Each chapter in this book tends to focus on one stage of the data science lifecycle, but we also include chapters with case studies that demonstrate the full lifecycle.

Understanding the differences between exploration, inference, prediction, and causation can be a challenge. We can easily slip into confusing a correlation found in data with a causal relationship. For example, an exploratory or inferential analysis might look for correlations in response to the question "Do people who have a greater exposure to air pollution have a higher rate of lung disease?" Whereas a causal question might ask "Does giving an award to a Wikipedia contributor increase productivity?" We typically cannot answer causal questions unless we have a randomized experiment (or approximate one). We point out these important distinctions throughout the book.

Examples of the Lifecycle

Several case studies that address the entire data science lifecycle are placed throughout this book. These cases serve double duty. They focus on one stage in the lifecycle to provide a specific example of the topics in the part of the book where they are located, and they also demonstrate the entire cycle.

The focus of Chapter 5 is on the interplay between a question of interest and how data can be used to answer the question. The simple question "Why is my bus always late?" provides a rich case study that is basic enough for the beginning data scientist to track the stages of the lifecycle, and yet nuanced enough to demonstrate how we apply both statistical and computational thinking to answer the question. In this case study, we build a simulation study to inform us about the distribution of wait times for riders. And we fit a simple model to summarize the wait times with a statistic. This case study also demonstrates how, as a data scientist, you can collect your own data to answer questions that interest you.

Chapter 12 studies the accuracy of mass-market air sensors that are used across the United States. We devise a way to leverage data from highly accurate sensors maintained by the Environmental Protection Agency to improve readings from less expensive sensors. This case study shows how crowdsourced, open data can be improved with data from rigorously maintained, precise, government-monitored equipment. In the process, we focus on cleaning and merging data from multiple sources, but we also fit models to adjust and improve air quality measurements.

In Chapter 18 our focus is on model building and prediction. But we cover the full lifecycle and see how the question of interest impacts the model that we build. Our aim is to enable veterinarians in rural Kenya, who have no access to a scale to weigh a donkey, to prescribe medication for a sick animal. As we learn about the design of the study, clean the data, and balance simplicity with accuracy, we assess the predictive capabilities of our model and show how scientists can partner with people facing practical problems and assist them with solutions.

Finally, in Chapter 21 we examine hand-classified news stories in an effort to algorithmically differentiate fake news from real news. In this case study, we again see how readily accessible information creates amazing opportunities for data scientists to develop new technologies and investigate today's important problems. These data have been scraped from news stories on the web and classified as fake or real news by people reading the stories. We also see how data scientists thinking creatively can take general information, such as the content of a news article, and transform it into analyzable data to address topical questions.

Summary

The data science lifecycle provides an organizing structure for this book. We keep the lifecycle in mind as we work with many datasets from a wide range of sources, including science, medicine, politics, social media, and government. The first time we use a dataset, we provide the context in which the data were collected, the question of interest in examining the data, and descriptions needed to understand the data. In this way, we aim to practice good data science throughout the book.

The first stage of the lifecycle—asking a question—is often seen in books as a question that requires an application of a technique to get a number, such as "What's the *p*-value for this A/B test?" Or a vague question that is often seen in practice, like "Can we restore the American Dream?" Answering the first sort of question gives little practice in developing a research question. Answering the second is hard to do without guidance on how to turn a general area of interest into a question that can be answered with data. The interplay between asking a question and understanding the limitations of data to answer it is the topic of the next chapter.

CHAPTER 2

Questions and Data Scope

As data scientists, we use data to answer questions, and the quality of the data collection process can significantly impact the validity and accuracy of the data, the strength of the conclusions we draw from an analysis, and the decisions we make. In this chapter, we describe a general approach for understanding data collection and evaluating the usefulness of the data in addressing the question of interest. Ideally, we aim for data to be representative of the phenomenon that we are studying, whether that phenomenon is a population characteristic, a physical model, or some type of social behavior. Typically, our data do not contain complete information (the scope is restricted in some way), yet we want to use the data to accurately describe a population, estimate a scientific quantity, infer the form of a relationship between features, or predict future outcomes. In all of these situations, if our data are not representative of the object of our study, then our conclusions can be limited, possibly misleading, or even wrong.

To motivate the need to think about these issues, we begin with an example of the power of big data and what can go wrong. We then provide a framework that can help you connect the goal of your study (your question) with the data collection process. We refer to this as the *data scope*,[1] and we provide terminology to help describe data scope, along with examples from surveys, government data, scientific instruments, and online resources. Later in this chapter, we consider what it means for data to be accurate. There, we introduce different forms of bias and variation, and describe situations where they can arise. Throughout, the examples cover the spectrum of the sorts of data that you may be using as a data scientist; these examples are from science, politics, public health, and online communities.

1 The notion of "scope" has been adapted from Joseph Hellerstein's course notes (*https://oreil.ly/VrByF*) on scope, temporality, and faithfulness.

Big Data and New Opportunities

The tremendous increase in openly available data has created new roles and opportunities in data science. For example, data journalists look for interesting stories in data much like how traditional beat reporters hunt for news stories. The lifecycle for the data journalist begins with the search for existing data that might have an interesting story, rather than beginning with a research question and figuring out how to collect new or use existing data to address the question.

Citizen science projects are another example. They engage many people (and instruments) in data collection. Collectively, these data are made available to researchers who organize the project, and often they are made available in repositories for the general public to further investigate.

The availability of administrative and organizational data creates other opportunities. Researchers can link data collected from scientific studies with, say, medical data that have been collected for health-care purposes; these administrative data have been collected for reasons that don't directly stem from the question of interest, but they can be useful in other settings. Such linkages can help data scientists expand the possibilities of their analyses and cross-check the quality of their data. In addition, found data can include digital traces, such as your web-browsing activity, your posts on social media, and your online network of friends and acquaintances, and they can be quite complex.

When we have large amounts of administrative data or expansive digital traces, it can be tempting to treat them as more definitive than data collected from traditional, smaller research studies. We might even consider these large datasets to be a replacement for scientific studies and essentially a census. This overreach is referred to as "big data hubris" (*https://doi.org/10.1126/science.1248506*). Data with a large scope does not mean that we can ignore foundational issues of how representative the data are, nor can we ignore issues with measurement, dependency, and reliability. (And it can be easy to discover meaningless or nonsensical relationships just by coincidence.) One well-known example is the Google Flu Trends tracking system.

Example: Google Flu Trends

Digital epidemiology (*https://oreil.ly/i2PVM*), a new subfield of epidemiology, leverages data generated outside the public health system to study patterns of disease and health dynamics in populations. The Google Flu Trends (GFT) tracking system was one of the earliest examples of digital epidemiology. In 2007, researchers found that counting the searches people made for flu-related terms could accurately estimate the number of flu cases. This apparent success made headlines, and many researchers became excited about the possibilities of big data. However, GFT did not live up to expectations and was abandoned in 2015.

What went wrong? After all, GFT used millions of digital traces from online queries for terms related to influenza to predict flu activity. Despite initial success, in the 2011–2012 flu season, Google's data scientists found that GFT was not a substitute for the more traditional surveillance reports of three-week-old counts collected by the US Centers for Disease Control and Prevention (CDC) from laboratories across the country. In comparison, GFT overestimated the CDC numbers for 100 out of 108 weeks. Week after week, GFT came in too high for the cases of influenza, even though it was based on big data:

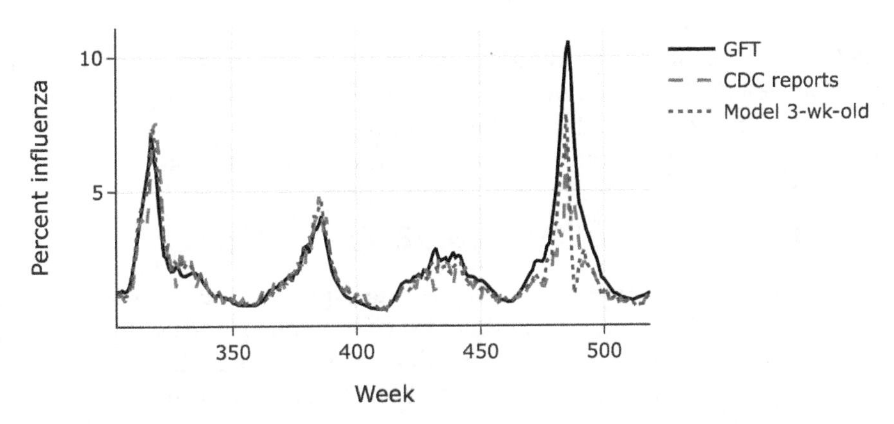

From weeks 412 to 519 in this plot, GFT (solid line) overestimated the actual CDC reports (dashed line) 100 times. Also plotted here are predictions from a model based on three-week-old CDC data and seasonal trends (dotted line), which follows the actuals more closely than GFT.

Data scientists found that a simple model built from past CDC reports that used three-week-old CDC data and seasonal trends did a better job of predicting flu prevalence than GFT. GFT overlooked considerable information that can be extracted by basic statistical methods. This does not mean that big data captured from online activity is useless. In fact, researchers have shown (*https://oreil.ly/Qlw6u*) that the combination of GFT data with CDC data can substantially improve both GFT predictions and the CDC-based model. It is often the case that combining different approaches leads to improvements over individual methods.

The GFT example shows us that even when we have tremendous amounts of information, the connections between the data and the question being asked are paramount. Understanding this framework can help us avoid answering the wrong question, applying inappropriate methods to the data, and overstating our findings.

In the age of big data, we are tempted to collect more and more data to answer a question precisely. After all, a census gives us perfect information, so shouldn't big data be nearly perfect? Unfortunately, this is often not the case, especially with administrative data and digital traces. The inaccessibility of a small fraction of the people you want to study (see the 2016 election upset in Chapter 3) or the measurement process itself (as in this GFT example) can lead to poor predictions. It is important to consider the scope of the data as it relates to the question under investigation.

A key factor to keep in mind is the scope of the data. Scope includes considering the population we want to study, how to access information about that population, and what we are actually measuring. Thinking through these points can help us see potential gaps in our approach. We investigate this in the next section.

Target Population, Access Frame, and Sample

An important initial step in the data lifecycle is to express the question of interest in the context of the subject area and consider the connection between the question and the data collected to answer that question. It's a good practice to do this before even thinking about the analysis or modeling steps because it may uncover a disconnect where the question of interest cannot be directly addressed by the data. As part of making the connection between the data collection process and the topic of investigation, we identify the population, the means to access the population, instruments of measurement, and additional protocols used in the collection process. These concepts—the target population, the access frame, and the sample—help us understand the scope of the data, whether we aim to gain knowledge about a population, scientific quantity, physical model, social behavior, or something else:

Target population
: The *target population* consists of the collection of elements comprising the population that you ultimately intend to describe and draw conclusions about. The element may be a person in a group of people, a voter in an election, a tweet from a collection of tweets, or a county in a state. We sometimes call an element a *unit* or an *atom*.

Access frame
: The *access frame* is the collection of elements that are accessible to you for measurement and observation. These are the units through which you can study the target population. Ideally, the access frame and population are perfectly aligned, meaning they consist of the exact same elements. However, the units in an access frame may be only a subset of the target population; additionally, the frame may include units that don't belong to the population. For example, to find out how a voter intends to vote in an election, you might call people by phone. Someone

you call may not be a voter, so they are in your frame but not in the population. On the other hand, a voter who never answers a call from an unknown number can't be reached, so they are in the population but not in your frame.

Sample
: The *sample* is the subset of units taken from the access frame to observe and measure. The sample gives you the data to analyze in order to make predictions or generalizations about the population of interest. When resources have been put into following up with nonrespondents and tracking down hard-to-find units, a small sample can be more effective than a large sample or an attempt at a census where subsets of the population have been overlooked.

The contents of the access frame, in comparison to the target population, and the method used to select units from the frame to be in the sample are important factors in determining whether or not the data can be considered representative of the target population. If the access frame is not representative of the target population, then the data from the sample is most likely not representative either. And if the units are sampled in a biased manner, problems with representativeness also arise.

You will also want to consider time and place in the data scope. For example, the effectiveness of a drug trial tested in one part of the world where a disease is raging might not compare favorably with a trial in a different part of the world where background infection rates are lower (see Chapter 3). Additionally, data collected for the purpose of studying changes over time, like with the monthly measurements of carbon dioxide (CO_2) in the atmosphere (see Chapter 9) and the weekly reporting of Google searches for predicting flu trends, have a *temporal* structure that we need to be mindful of as we examine the data. At other times, there might be *spatial* patterns in the data. For example, the environmental health data, described later in this section, are reported for each census tract in the State of California, and we might make maps to look for spatial correlations.

And if you didn't collect the data, you will want to consider who did and for what purpose. This is especially relevant now since more data is passively collected instead of collected with a specific goal in mind. Taking a hard look at found data and asking yourself whether and how these data might be used to address your question can save you from making a fruitless analysis or drawing inappropriate conclusions.

For the examples in the following subsections, we begin with a general question, narrow it to one that can be answered with data, and in doing so, identify the target population, access frame, and sample. These concepts are represented by circles and rectangles in diagrams, and the configuration of the overlap of these shapes helps reveal key aspects of the scope. Also in each example, we describe relevant temporal and spatial features of the data scope.

Example: What Makes Members of an Online Community Active?

Content on Wikipedia is written and edited by volunteers who belong to the Wikipedia community. This online community is crucial to the success and vitality of Wikipedia. In trying to understand how to incentivize members of online communities, researchers (*https://oreil.ly/j74rl*) carried out an experiment with Wikipedia contributors as subjects. A narrowed version of the general question asks: do awards increase the activity of Wikipedia contributors? For this experiment, the target population is the collection of top, active contributors—the 1% most active contributors to Wikipedia in the month before the start of the study. The access frame eliminated anyone in the population who had received an incentive (award) that month. The access frame purposely excluded some of the contributors in the population because the researchers wanted to measure the impact of an incentive, and those who had already received one incentive might behave differently (see Figure 2-1).

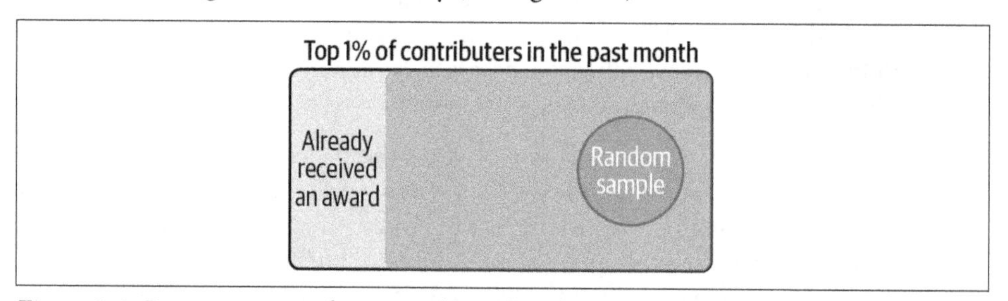

Figure 2-1. Representation of scope in the Wikipedia experiment

The sample is a randomly selected set of 200 contributors from the frame. The contributors were observed for 90 days, and digital traces of their activities on Wikipedia were collected. Notice that the contributor population is not static; there is regular turnover. In the month prior to the start of the study, more than 144,000 volunteers produced content for Wikipedia. Selecting top contributors from among this group limits the generalizability of the findings, but given the size of the group of top contributors, if they can be influenced by an informal reward to maintain or increase their contributions, this is still a valuable finding.

In many experiments and studies, we don't have the ability to include all population units in the frame. It is often the case that the access frame consists of volunteers who are willing to join the study/experiment.

Example: Who Will Win the Election?

The outcome of the US presidential election in 2016 took many people and many pollsters by surprise. Most preelection polls predicted Hillary Clinton would beat Donald Trump. Political polling is a type of public opinion survey held prior to an election that attempts to gauge whom people will vote for. Since opinions change over time, the focus is reduced to a "horse race" question, where respondents are asked whom they would vote for in a head-to-head race if the election were tomorrow: candidate A or candidate B.

Polls are conducted regularly throughout the presidential campaign, and as election day approaches, we expect the polls to get better at predicting the outcome as preferences stabilize. Polls are also typically conducted statewide and later combined to make predictions for the overall winner. For these reasons, the timing and location of a poll matters. The pollster matters too (*https://oreil.ly/iHApH*); some have consistently been closer to the mark than others.

In these preelection surveys, the target population consists of those who will vote in the election, which in this example was the 2016 US presidential election. However, pollsters can only guess at whether someone will vote in the election, so the access frame consists of those deemed to be likely voters (this is usually based on past voting records, but other factors may also be used). And since people are contacted by phone, the access frame is limited to those who have a landline or mobile phone. The sample consists of those people in the frame who are chosen according to a random dialing scheme (see Figure 2-2).

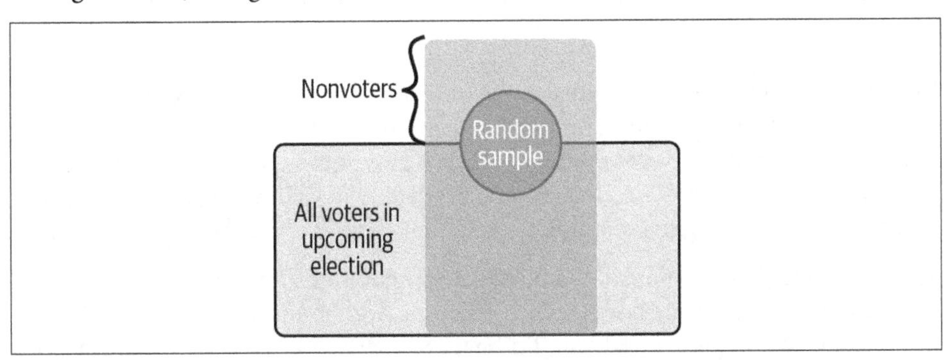

Figure 2-2. Representation of scope in the 2016 presidential election survey

In Chapter 3, we discuss the impact on the election predictions of people's unwillingness to answer their phone or participate in the poll.

Example: How Do Environmental Hazards Relate to an Individual's Health?

To address this question, the California Environmental Protection Agency (CalEPA), the California Office of Environmental Health Hazard Assessment (OEHHA), and the public developed the CalEnviroScreen (*https://oreil.ly/qeVD0*) project. The project studies connections between population health and environmental pollution in California communities using data collected from several sources that include demographic summaries from the US census, health statistics from the California

Department of Health Care Access and Information, and pollution measurements from air monitoring stations around the state maintained by the California Air Resources Board.

Ideally, we want to study the people of California and assess the impact of these environmental hazards on an individual's health. However, in this situation, the data can only be obtained at the level of a census tract. The access frame consists of groups of residents living in the same census tract. So the units in the frame are census tracts and the sample is a census—all of the tracts—since data are provided for all of the tracts in the state (see Figure 2-3).

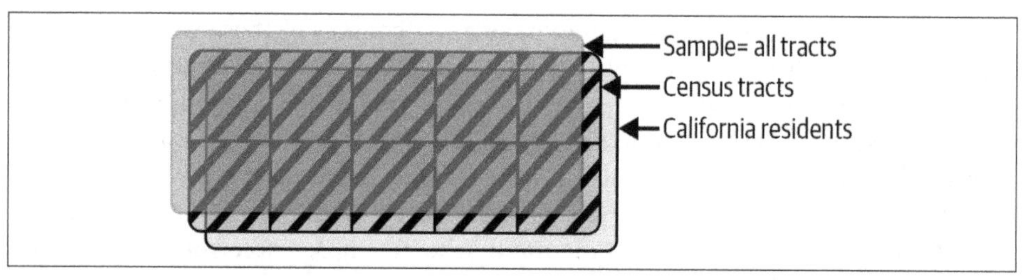

Figure 2-3. Scope of the CalEnviroScreen project; the grid in the access frame represents the census tracts

Unfortunately, we cannot disaggregate the information in a tract to examine an individual person. This aggregation impacts the questions we can address and the conclusions that we can draw. For example, we can ask questions about the relation between rates of hospitalizations due to asthma and air quality in California communities. But we can't answer the original question posed about an individual's health.

These examples have demonstrated possible configurations for a target, access frame, and sample. When a frame doesn't reach everyone, we should consider how this missing information might impact our findings. Similarly, we ask what might happen when a frame includes those not in the population. Additionally, the techniques for drawing the sample can affect how representative the sample is of the population. When you think about generalizing your data findings, you also want to consider the quality of the instruments and procedures used to collect the data. If your sample is a census that matches your target, but the information is poorly collected, then your findings will be of little value. This is the topic of the next section.

Instruments and Protocols

When we consider the scope of the data, we also consider the instrument being used to take the measurements and the procedure for taking measurements, which we call the *protocol*. For a survey, the instrument is typically a questionnaire that an individual in the sample answers. The protocol for a survey includes how the sample is chosen, how nonrespondents are followed up on, interviewer training, protections for confidentiality, and so on.

Good instruments and protocols are important to all kinds of data collection. If we want to measure a natural phenomenon, such as carbon dioxide in the atmosphere, we need to quantify the accuracy of the instrument. The protocol for calibrating the instrument and taking measurements is vital to obtaining accurate measurements. Instruments can go out of alignment and measurements can drift over time, leading to poor, highly inaccurate measurements.

Protocols are also critical in experiments. Ideally, any factor that can influence the outcome of the experiment is controlled. For example, temperature, time of day, confidentiality of a medical record, and even the order in which measurements are taken need to be consistent to rule out potential effects from these factors getting in the way.

With digital traces, the algorithms used to support online activity are dynamic and continually reengineered. For example, Google's search algorithms are continually tweaked to improve user service and advertising revenue. Changes to the search algorithms can impact the data generated from the searches, which in turn impact systems built from these data, such as the Google Flu Trends tracking system. This changing environment can make it untenable to maintain data collection protocols and difficult to replicate findings.

Many data science projects involve linking data together from multiple sources. Each source should be examined through this data-scope construct, and any difference across sources should be considered. Additionally, matching algorithms used to combine data from multiple sources need to be clearly understood so that populations and frames from the sources can be compared.

Measurements from an instrument taken to study a natural phenomenon can be cast in the scope diagram of a target, access frame, and sample. This approach is helpful in understanding the instrument's accuracy.

Measuring Natural Phenomena

The scope diagram introduced for observing a target population can be extended to the situation where we want to measure a quantity, such as a particle count in the air, the age of a fossil, or the speed of light. In these cases, we consider the quantity we want to measure as an unknown exact value. (This unknown value is often referred to as a *parameter*.) We can adapt our scope diagram to this setting: we shrink the target to a point that represents the unknown; the instrument's accuracy acts as the access frame; and the sample consists of the measurements taken by the instrument. You might think of the frame as a dartboard, where the instrument is the person throwing the darts, and the darts land within the circle, scattered around the bullseye. The scatter of darts corresponds to the measurements taken by the instrument. The target point is not seen by the dart thrower, but ideally it coincides with the bullseye.

To illustrate the concept of measurement error and its connection to sampling error, we examine the problem of measuring CO_2 levels in the air.

Example: What Is the Level of CO_2 in the Air?

CO_2 is an important signal of global warming because it traps heat in the Earth's atmosphere. Without CO_2, the Earth would be impossibly cold, but it's a delicate balance. An increase in CO_2 drives global warming and threatens our planet's climate. To address this question, CO_2 concentrations have been monitored at Mauna Loa Observatory (*https://oreil.ly/HpqFr*) since 1958. These data offer a crucial benchmark for understanding the threat of global warming.

When thinking about the scope of the data, we consider the location and time of data collection. Scientists chose to measure CO_2 on the Mauna Loa volcano because they wanted a place where they could measure the background level of CO_2 in the air. Mauna Loa is in the Pacific Ocean, far away from pollution sources, and the observatory is high up on a mountain surrounded by bare lava, away from plants that remove CO_2 from the air.

It's important that the instrument measuring CO_2 is as accurate as possible. Rigorous protocols (*https://oreil.ly/r_Da9*) are in place to keep the instrument in top condition. For example, samples of air are routinely measured at Mauna Loa by different types of equipment, and other samples are sent off-site to a laboratory for more accurate measurement. These measurements help determine the accuracy of the instrument. In addition, a reference gas is measured for 5 minutes every hour, and two other reference gases are measured for 15 minutes every day. These reference gases have known CO_2 levels. A comparison of the measured concentrations against the known values helps identify bias in the instrument.

While the CO_2 in background air is relatively steady at Mauna Loa, the five-minute average concentrations that are measured in any hour deviate from the hourly average. These deviations reflect the accuracy of the instrument and variation in airflow.

The scope for data collection can be summarized as follows: at this particular location (high up on Mauna Loa) during a particular one-hour period, there is a true background concentration of CO_2; this is our target (see Figure 2-4). The instrument takes measurements and reports five-minute averages. These readings form a sample contained in the access frame, the dartboard. If the instrument is working properly, the bullseye coincides with the target (the one-hour average concentration of CO_2) and the measurements are centered on the bullseye, with deviations of about 0.30 parts per million (ppm). The measurement of CO_2 is the number of CO_2 molecules per 1 million molecules of dry air, so the unit of measurement is ppm.

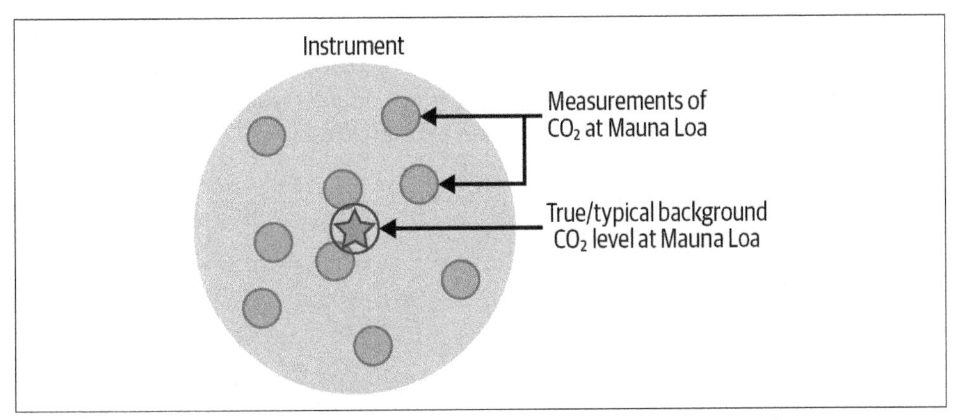

Figure 2-4. The access frame represents the accuracy of the instrument; the star represents the true value of interest

We continue the dartboard analogy in the next section to introduce the concepts of bias and variation, describe common ways in which a sample might not be representative of the population, and draw connections between accuracy and protocol.

Accuracy

In a census, the access frame matches the population, and the sample captures the entire population. In this situation, if we administer a well-designed questionnaire, then we have complete and accurate information on the population, and the scope is perfect. Similarly, in measuring CO_2 concentrations in the atmosphere, if our instrument has perfect accuracy and is properly used, then we can measure the exact value of the CO_2 concentration (ignoring air fluctuations). These situations are rare, if not impossible. In most settings, we need to quantify the accuracy of our measurements in order to generalize our findings to the unobserved. For example, we often use the sample to estimate an average value for a population, infer the value of a scientific unknown from measurements, or predict the behavior of a new individual. In each of these settings, we also want a quantifiable degree of accuracy. We want to know how close our estimates, inferences, and predictions are to the truth.

The analogy of darts thrown at a dartboard that was introduced earlier can be useful in understanding accuracy. We divide *accuracy* into two basic parts: *bias* and *precision* (also known as *variation*). Our goal is for the darts to hit the bullseye on the dartboard and for the bullseye to line up with the unseen target. The spray of the darts on the board represents the precision in our measurements, and the gap from the bullseye to the unknown value that we are targeting represents the bias.

Figure 2-5 shows combinations of low and high bias and precision. In each of these diagrams, the dots represent the measurements taken, and the star represents the true, unknown parameter value. The dots form a scattershot within the access frame, represented by the dartboard. When the bullseye of the access frame is roughly centered on the star (top row), the measurements are scattered around the value of interest and bias is low. The larger dartboards (on the right) indicate a wider spread (lower precision) in the measurements.

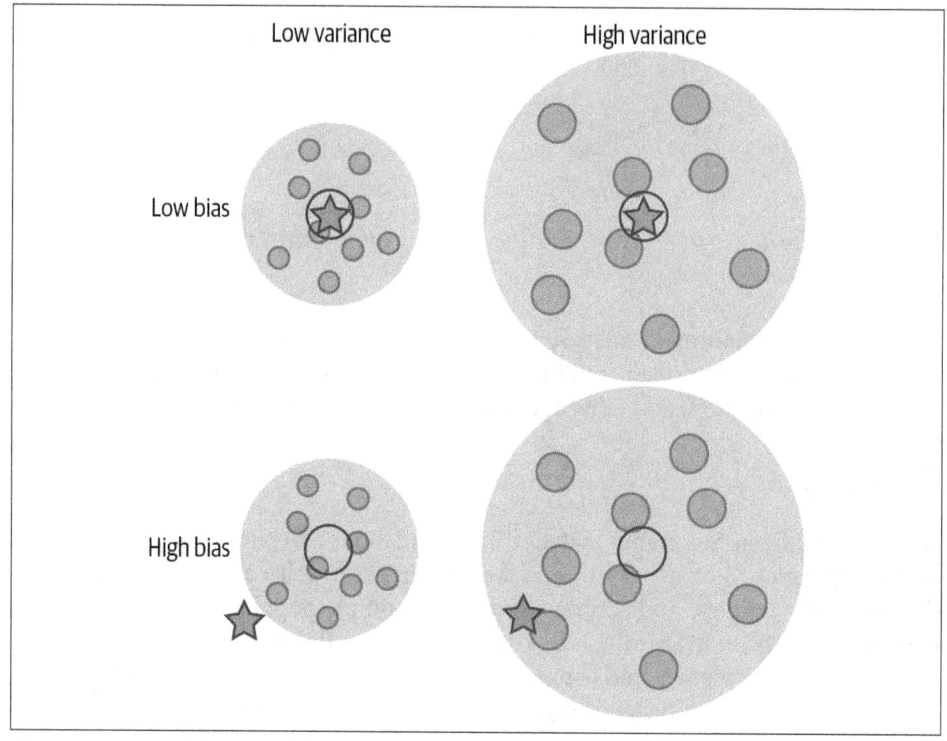

Figure 2-5. Combinations of low and high measurement bias and precision

Representative data puts us in the top row of the diagram, where there is low bias, meaning that the unknown target aligns with the bullseye. Ideally, our instruments and protocols put us in the upper-left part of the diagram, where the variation is also low. The pattern of points in the bottom row systematically misses the targeted value. Taking larger samples will not correct this bias.

Types of Bias

Bias comes in many forms. We describe some classic types here and connect them to our target-access-sample framework:

Coverage bias
: Occurs when the access frame does not include everyone in the target population. For example, a survey based on phone calls cannot reach those without a

phone. In this situation, those who cannot be reached may differ in important ways from those in the access frame.

Selection bias
: Arises when the mechanism used to choose units for the sample tends to select certain units more often than they should be selected. As an example, a convenience sample chooses the units that are most easily available. Problems can arise when those who are easy to reach differ in important ways from those who are harder to reach. As another example, observational studies and experiments often rely on volunteers (people who choose to participate), and this self-selection has the potential for bias if the volunteers differ from the target population in important ways.

Nonresponse bias
: Comes in two forms: unit and item. Unit nonresponse happens when someone selected for a sample is unwilling to participate (they may never answer a phone call from an unknown caller). Item nonresponse occurs when, say, someone answers the phone but refuses to respond to a particular survey question. Nonresponse can lead to bias if those who choose not to respond are systematically different from those who choose to respond.

Measurement bias
: Happens when an instrument systematically misses the target in one direction. For example, low humidity can systematically give us incorrectly high measurements of air pollution. In addition, measurement devices can become unstable and drift over time and so produce systematic errors. In surveys, measurement bias can arise when questions are confusingly worded or leading, or when respondents may not be comfortable answering honestly.

Each of these types of bias can lead to situations where the data are not centered on the unknown targeted value. Often, we cannot assess the potential magnitude of the bias, since little to no information is available on those who are outside the access frame, less likely to be selected for the sample, or disinclined to respond. Protocols are key to reducing these sources of bias. Chance mechanisms to select a sample from the frame or to assign units to experimental conditions can eliminate selection bias. A nonresponse follow-up protocol to encourage participation can reduce nonresponse bias. A pilot survey can improve question wording and so reduce measurement bias. Procedures to calibrate instruments and protocols to take measurements in, say, random order can reduce measurement bias.

In the 2016 US presidential election, nonresponse bias and measurement bias were key factors in the inaccurate predictions of the winner. Nearly all voter polls leading up to the election predicted Clinton a winner over Trump. Trump's upset victory came as a surprise. After the election, many polling experts attempted to diagnose where things went wrong in the polls. The American Association for Public Opinion

Research (*https://oreil.ly/uPDlR*) found that the predictions were flawed for two key reasons:

- College-educated voters were overrepresented. College-educated voters are more likely to participate in surveys than those with less education (*https://oreil.ly/K4BvY*), and in 2016 they were more likely to support Clinton. Higher response rates from more highly educated voters biased the sample and overestimated support for Clinton.
- Voters were undecided or changed their preferences a few days before the election. Since a poll is static and can only directly measure current beliefs, it cannot reflect a shift in attitudes.

It's difficult to figure out whether people held back their preference or changed their preference and how large a bias this created. However, exit polls have helped polling experts understand what happened after the fact. They indicate that in battleground states, such as Michigan, many voters made their choice in the final week of the campaign, and that group went for Trump by a wide margin.

Bias does not need to be avoided under all circumstances. If an instrument is highly precise (low variance) and has a small bias, then that instrument might be preferable to another with higher variance and no bias. As an example, biased studies are potentially useful to pilot a survey instrument or to capture useful information for the design of a larger study. Many times we can at best recruit volunteers for a study. Given this limitation, it can still be useful to enroll these volunteers in the study and use random assignment to split them into treatment groups. That's the idea behind randomized controlled experiments.

Whether or not bias is present, data typically also exhibit variation. Variation can be introduced purposely by using a chance mechanism to select a sample, and it can occur naturally through an instrument's precision. In the next section, we identify three common sources of variation.

Types of Variation

The following types of variation results from a chance mechanism and have the advantage of being quantifiable:

Sampling variation
: Results from using chance to select a sample. In this case, we can, in principle, compute the chance that a particular collection of elements is selected for the sample.

Assignment variation
: Occurs in a controlled experiment when we assign units at random to treatment groups. In this situation, if we split the units up differently, then we can get

different results from the experiment. This assignment process allows us to compute the chance of a particular group assignment.

Measurement error
  Results from the measurement process. If the instrument used for measurement has no drift or bias and a reliable distribution of errors, then when we take multiple measurements on the same object, we get random variations in measurements that are centered on the truth.

The *urn model* is a simple abstraction that can be helpful for understanding variation. This model sets up a container (an urn, which is like a vase or a bucket) full of identical marbles that have been labeled, and we use the simple action of drawing marbles from the urn to reason about sampling schemes, randomized controlled experiments, and measurement error. For each of these types of variation, the urn model helps us estimate the size of the variation using either probability or simulation (see Chapter 3). The example of selecting Wikipedia contributors to receive an informal award provides two examples of the urn model.

Recall the Wikipedia experiment, where 200 contributors were selected at random from 1,440 top contributors. These 200 contributors were then split, again at random, into two groups of 100 each. One group received an informal award and the other didn't. Here's how we use the urn model to characterize this process of selection and splitting:

1. Imagine an urn filled with 1,440 marbles that are identical in shape and size, and written on each marble is one of the 1,440 Wikipedia usernames. (This is the access frame.)
2. Mix the marbles in the urn really well, select one marble, and set it aside.
3. Repeat the mixing and selecting of the marbles to obtain 200 marbles.

The marbles drawn form the sample. Next, to determine which of the 200 contributors receive awards, we work with another urn:

1. In a second urn, put in the 200 marbles from the preceding sample.
2. Mix these marbles well, select one marble, and set it aside.
3. Repeat, choosing 100 marbles. That is, choose marbles one at a time, mixing in between, and setting the chosen marble aside.

The 100 drawn marbles are assigned to the treatment group and correspond to the contributors who receive an award. The 100 left in the urn form the control group and receive no award.

Both the selection of the sample and the choice of award recipients use a chance mechanism. If we were to repeat the first sampling activity again, returning all 1,440 marbles to the original urn, then we would most likely get a different sample. This variation is the source of *sampling variation*. Likewise, if we were to repeat the random assignment process again (keeping the sample of 200 unchanged), then we would get a different treatment group. *Assignment variation* arises from this second chance process.

The Wikipedia experiment provided an example of both sampling and assignment variation. In both cases, the researcher imposed a chance mechanism on the data collection process. Measurement error can at times also be considered a chance process that follows an urn model. For example, we can characterize the measurement error of the CO_2 monitor at Mauna Loa in this way.

If we can draw an accurate analogy between variation in the data and the urn model, the urn model provides us the tools to estimate the size of the variation (see Chapter 3). This is highly desirable because we can give concrete values for the variation in our data. However, it's vital to confirm that the urn model is a reasonable depiction of the source of variation. Otherwise, our claims of accuracy can be seriously flawed. We need to know as much as possible about data scope, including instruments and protocols and chance mechanisms used in data collection, to apply these urn models.

Summary

No matter the kind of data you are working with, before diving into cleaning, exploration, and analysis, take a moment to look into the data's source. If you didn't collect the data, ask yourself:

- Who collected the data?
- Why were the data collected?

Answers to these questions can help determine whether these found data can be used to address the question of interest to you.

Consider the scope of the data. Questions about the temporal and spatial aspects of data collection can provide valuable insights:

- When were the data collected?
- Where were the data collected?

Answers to these questions help you determine whether your findings are relevant to the situation that interests you, or whether your situation may not be comparable to this other place and time.

Core to the notion of scope are answers to the following questions:

- What is the target population (or unknown parameter value)?
- How was the target accessed?
- What methods were used to select samples/take measurements?
- What instruments were used and how were they calibrated?

Answering as many of these questions as possible can give you valuable insights as to how much trust you can place in your findings and whether you can generalize from them.

This chapter provided you with terminology and a framework for thinking about and answering these questions. The chapter also outlined ways to identify possible sources of bias and variance that can impact the accuracy of your findings. To help you reason about bias and variance, we introduced the following diagrams and notions:

- Scope diagram to indicate the overlap between target population, access frame, and sample
- Dartboard to describe an instrument's bias and variance
- Urn model for situations when a chance mechanism has been used to select a sample from an access frame, divide a group into experimental treatment groups, or take measurements from a well-calibrated instrument

These diagrams and models attempt to boil down key concepts that are required to understand how to identify limitations and judge the usefulness of your data in answering your question. Chapter 3 continues the development of the urn model to more formally quantify accuracy and design simulation studies.

CHAPTER 3

Simulation and Data Design

In this chapter, we develop the basic theoretical foundation needed to reason about how data is sampled and the implications on bias and variance. We build this foundation not on the dry equations of classic statistics but on the story of an urn filled with marbles. We use the computational tools of simulation to reason about the properties of selecting marbles from the urn and what they tell us about data collection in the real world. We connect the simulation process to common statistical distributions (the dry equations), but the basic tools of simulation enable us to go beyond what can be directly modeled using equations.

As an example, we study how the pollsters failed to predict the outcome of the US presidential election in 2016. Our simulation study uses the actual votes cast in Pennsylvania. We simulate the sampling variation for a poll of these six million voters to uncover how response bias can skew polls and see how simply collecting more data would not have helped.

In a second simulation study, we examine a controlled experiment that demonstrated the efficacy of a COVID-19 vaccine but also launched a heated debate on the relative efficacy of vaccines. Abstracting the experiment to an urn model gives us a tool for studying assignment variation in randomized controlled experiments. Through simulation, we find the expected outcome of the clinical trial. Our simulation, along with careful examination of the data scope, debunks claims of vaccine ineffectiveness.

A third example uses simulation to imitate a measurement process. When we compare the fluctuations in our artificial measurements of air quality to real measurements, we can evaluate the appropriateness of the urn to model fluctuations in air quality measurements. This comparison creates the backdrop against which we calibrate PurpleAir monitors so that they can more accurately measure air quality in times of low humidity, like during fire season.

However, before we tackle some of the most significant data debates of our time, we first start small, very small, with the story of a few marbles sitting in an urn.

The Urn Model

The urn model was developed by Jacob Bernoulli in the early 1700s as a way to model the process of selecting items from a population. The urn model shown in Figure 3-1 gives a visual depiction of the process of randomly sampling marbles from an urn. Five marbles were originally in the urn: three black and two white. The diagram shows that two draws were made: first a white marble was drawn and then a black marble.

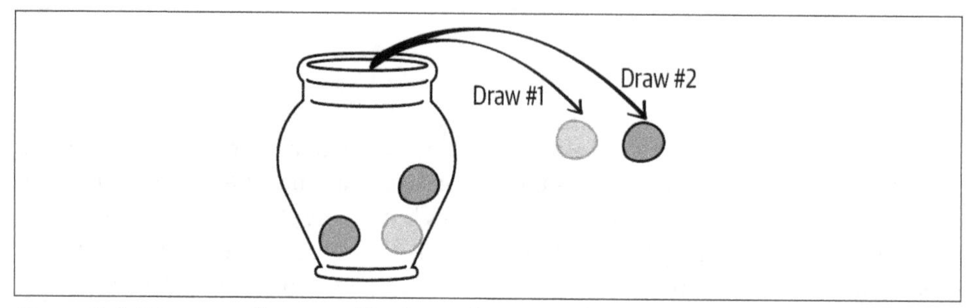

Figure 3-1. Diagram of two marbles being drawn, without replacement, from an urn

To set up an urn model, we first need to make a few decisions:

- The number of marbles in the urn
- The color (or label) on each marble
- The number of marbles to draw from the urn

Finally, we also need to decide on the sampling process. For our process, we mix the marbles in the urn, and as we select a marble for our sample, we can choose to record the color and return the marble to the urn (with replacement), or set aside the marble so that it cannot be drawn again (without replacement).

These decisions make up the parameters of our model. We can adapt the urn model to describe many real-world situations by our choice for these parameters. To illustrate, consider the example in Figure 3-1. We can *simulate* the draw of two marbles from our urn without replacement between draws using `numpy`'s `random.choice` method. The `numpy` library supports functions for arrays, which can be particularly useful for data science:

```
import numpy as np

urn = ["b", "b", "b", "w", "w"]
```

```
print("Sample 1:", np.random.choice(urn, size=2, replace=False))
print("Sample 2:", np.random.choice(urn, size=2, replace=False))
```

```
Sample 1: ['b' 'w']
Sample 2: ['w' 'b']
```

Notice that we set the `replace` argument to `False` to indicate that once we sample a marble, we don't return it to the urn.

With this basic setup, we can get approximate answers to questions about the kinds of samples we would expect to see. What is the chance that our sample contains marbles of only one color? Does the chance change if we return each marble after selecting it? What if we changed the number of marbles in the urn? What if we draw more marbles from the urn? What happens if we repeat the process many times?

The answers to these questions are fundamental to our understanding of data collection. We can build from these basic skills to simulate the urn and apply simulation techniques to real-world problems that can't be easily solved with classic probability equations.

For example, we can use simulation to easily estimate the fraction of samples where both marbles that we draw match in color. In the following code, we run 10,000 rounds of sampling two marbles from our urn. Using these samples, we can directly compute the proportion of samples with matching marbles:

```
n = 10_000
samples = [np.random.choice(urn, size=2, replace=False) for _ in range(n)]
is_matching = [marble1 == marble2 for marble1, marble2 in samples]
print(f"Proportion of samples with matching marbles: {np.mean(is_matching)}")
```

```
Proportion of samples with matching marbles: 0.4032
```

We just carried out a *simulation study*. Our call to `np.random.choice` imitates the chance process of drawing two marbles from the urn without replacement. Each call to `np.random.choice` gives us one possible sample. In a simulation study, we repeat this chance process many times (`10_000` in this case) to get a whole bunch of samples. Then we use the typical behavior of these samples to reason about what we might expect to get from the chance process. While this might seem like a contrived example (it is), consider if we replaced the marbles with people on a dating service, replaced the colors with more complex attributes, and perhaps used a neural network to score a match, and you can start to see the foundation of much more sophisticated analysis.

So far we have focused on the sample, but we are often interested in the relationship between the sample we might observe and what it can tell us about the "population" of marbles that were originally in the urn.

We can draw an analogy to data scope from Chapter 2: a set of marbles drawn from the urn is a *sample*, and the collection of all marbles placed in the urn is the *access frame*, which in this situation we take to be the same as the *population*. This blurring of the difference between the access frame and the population points to the gap between simulation and reality. Simulations tend to simplify models. Nonetheless, they can give helpful insights to real-world phenomena.

The urn model, where we do not replace the marbles between draws, is a common selection method called the *simple random sample*. We describe this method and other sampling techniques based on it next.

Sampling Designs

The process of drawing marbles without replacement from an urn is equivalent to a simple random sample. *In a simple random sample, every sample has the same chance of being selected.* While the method name has the word *simple* in it, constructing a simple random sample is often anything, but simple and in many cases is also the best sampling procedure. Plus, if we are being honest, it can also be a little confusing.

To better understand this sampling method, we return to the urn model. Consider an urn with seven marbles. Instead of coloring the marbles, we label each uniquely with a letter `A` through `G`. Since each marble has a different label, we can more clearly identify all possible samples that we might get. Let's select three marbles from the urn without replacement, and use the `itertools` library to generate the list of all combinations:

```
from itertools import combinations

all_samples = ["".join(sample) for sample in combinations("ABCDEFG", 3)]
print(all_samples)
print("Number of Samples:", len(all_samples))
```

```
['ABC', 'ABD', 'ABE', 'ABF', 'ABG', 'ACD', 'ACE', 'ACF', 'ACG', 'ADE', 'ADF',
'ADG', 'AEF', 'AEG', 'AFG', 'BCD', 'BCE', 'BCF', 'BCG', 'BDE', 'BDF', 'BDG',
'BEF', 'BEG', 'BFG', 'CDE', 'CDF', 'CDG', 'CEF', 'CEG', 'CFG', 'DEF', 'DEG',
'DFG', 'EFG']
Number of Samples: 35
```

Our list shows that there are 35 unique sets of three marbles. We could have drawn each of these sets six different ways. For example, the set $\{A, B, C\}$ can be sampled:

```
from itertools import permutations

print(["".join(sample) for sample in permutations("ABC")])
```

```
['ABC', 'ACB', 'BAC', 'BCA', 'CAB', 'CBA']
```

In this small example, we can get a complete picture of all the ways in which we can draw any three marbles from the urn.

Since each set of three marbles from the population of seven is equally likely to occur, the chance of any one particular sample must be $1/35$:

$$\mathbb{P}(ABC) = \mathbb{P}(\text{ABD}) = \cdots = \mathbb{P}(\text{EFG}) = \frac{1}{35}$$

We use the special symbol $\mathbb{P}$ to stand for "probability" or "chance," and we read the statement $\mathbb{P}(ABC)$ as "the chance the sample contains the marbles labeled A, B, and C in any order."

We can use the enumeration of all of the possible samples from the urn to answer additional questions about this chance process. For example, to find the chance that marble A is in the sample, we can add up the chance of all samples that contain A. There are 15 of them, so the chance is:

$$\mathbb{P}(A \textit{ is in the sample}) = \frac{15}{35} = \frac{3}{7}$$

When it's too difficult to list and count all of the possible samples, we can use simulation to help understand this chance process.

Many people mistakenly think that the defining property of a simple random sample is that every unit has an equal chance of being in the sample. However, this is not the case. A simple random sample of n units from a population of N means that every possible collection of n of the N units has the same chance of being selected. A slight variant of this is the *simple random sample with replacement*, where the units/marbles are returned to the urn after each draw. This method also has the property that every sample of n units from a population of N is equally likely to be selected. The difference, though, is that there are more possible sets of n units because the same marble can appear more than once in the sample.

The simple random sample (and its corresponding urn) is the main building block for more complex survey designs. We briefly describe two of the more widely used designs:

Stratified sampling
: Divide the population into nonoverlapping groups, called *strata* (one group is called a *stratum* and more than one are strata), and then take a simple random sample from each. This is like having a separate urn for each stratum and drawing marbles from each urn, independently. The strata do not have to be the same size, and we need not take the same number of marbles from each.

Cluster sampling
: Divide the population into nonoverlapping subgroups, called *clusters*, take a simple random sample of the clusters, and include all of the units in a cluster in the

sample. We can think of this as a simple random sample from one urn that contains large marbles that are themselves containers of small marbles. (The large marbles need not have the same number of marbles in them.) When opened, the sample of large marbles turns into the sample of small marbles. (Clusters tend to be smaller than strata.)

As an example, we might organize our seven marbles, labeled A–G, into three clusters, (A, B), (C, D), and (E, F, G). Then, a cluster sample of size one has an equal chance of drawing any of the three clusters. In this scenario, each marble has the same chance of being in the sample:

$$\begin{aligned}
\mathbb{P}(A \text{ in sample}) &= \mathbb{P}(\text{cluster } (A, B) \text{ chosen}) &&= \frac{1}{3} \\
\mathbb{P}(B \text{ in sample}) &= \mathbb{P}(\text{cluster } (A, B) \text{ chosen}) &&= \frac{1}{3} \\
&\vdots \\
\mathbb{P}(G \text{ in sample}) &= \mathbb{P}(\text{cluster } (E, F, G) \text{ chosen}) &&= \frac{1}{3}
\end{aligned}$$

But every combination of elements is not equally likely: it is not possible for the sample to include both A and C, because they are in different clusters.

Often, we are interested in a summary of the sample; in other words, we are interested in a *statistic*. For any sample, we can calculate the statistic, and the urn model helps us find the distribution of possible values that statistic may take on. Next, we examine the distribution of a statistic for our simple example.

Sampling Distribution of a Statistic

Suppose we are interested in testing the failure pressure of a new fuel tank design for a rocket. It's expensive to carry out the pressure tests since we need to destroy the fuel tank, and we may need to test more than one fuel tank to address variations in manufacturing.

We can use the urn model to choose the prototypes to be tested, and we can summarize our test results by the proportion of prototypes that fail the test. The urn model provides us the knowledge that each of the samples has the same chance of being selected, and so the pressure test results are representative of the population.

To keep the example simple, let's say we have seven fuel tanks that are labeled like the marbles from before. Let's see what happens when tanks A, B, D, and F fail the pressure test, if chosen, and tanks C, E, and G pass.

For each sample of three marbles, we can find the proportion of failures according to how many of these four defective prototypes are in the sample. We give a few examples of this calculation:

Sample	ABC	BCE	BDF	CEG
Proportion	2/3	1/3	1	0

Since we are drawing three marbles from the urn, the only possible sample proportions are 0, 1/3, 2/3, and 1, and for each triple, we can calculate its corresponding proportion. There are four samples that give us all failed tests (a sample proportion of 1). These are *ABD*, *ABF*, *ADF*, and *BDF*, so the chance of observing a sample proportion of 1 is 4/35. We can summarize the distribution of values for the sample proportion into a table, which we call the *sampling distribution* of the proportion:

Proportion of failures	No. of samples	Fraction of samples
0	1	$1/35 \approx 0.03$
1/3	12	$12/35 \approx 0.34$
2/3	18	$18/35 \approx 0.51$
1	4	$4/35 \approx 0.11$
Total	35	1

While these calculations are relatively straightforward, we can approximate them through a simulation study. To do this, we take samples of three from our population over and over—say 10,000 times. For each sample, we calculate the proportion of failures. That gives us 10,000 simulated sample proportions. The table of the simulated proportions should come close to the sampling distribution. We confirm this with a simulation study.

Simulating the Sampling Distribution

Simulation can be a powerful tool to understand complex random processes. In our example of seven fuel tanks, we are able to consider all possible samples from the corresponding urn model. However, in situations with large populations and samples and more complex sampling processes, it may not be tractable to directly compute the chance of certain outcomes. In these situations, we often turn to simulation to provide accurate estimates of the quantities we can't compute directly.

Let's set up the problem of finding the sampling distribution of the proportion of failures in a simple random sample of three fuel tanks as an urn model. Since we are interested in whether or not the tank fails, we use 1 to indicate a failure and 0 to indicate a pass, giving us an urn with marbles labeled as follows:

```
urn = [1, 1, 0, 1, 0, 1, 0]
```

We have encoded the tanks *A* through *G* using 1 for fail and 0 for pass, so we can take the mean of the sample to get the proportion of failures in a sample:

```
sample = np.random.choice(urn, size=3, replace=False)
print(f"Sample: {sample}")
print(f"Prop Failures: {sample.mean()}")
```

```
Sample: [1 0 0]
Prop Failures: 0.3333333333333333
```

In a simulation study, we repeat the sampling process thousands of times to get thousands of proportions, and then we estimate the sampling distribution of the proportion from what we get in our simulation. Here, we construct 10,000 samples (and so 10,000 proportions):

```
samples = [np.random.choice(urn, size=3, replace=False) for _ in range(10_000)]
prop_failures = [s.mean() for s in samples]
```

We can study these 10,000 sample proportions and match our findings against what we calculated already using the complete enumeration of all 35 possible samples. We expect the simulation results to be close to our earlier calculations because we have repeated the sampling process many, many times. That is, we want to compare the fraction of the 10,000-sample proportion that is 0, 1/3, 2/3, and 1 to those we computed exactly; those fractions are 1/35, 12/35, 18/35, and 4/35, or about 0.03, 0.34, 0.51, and 0.11:

```
unique_els, counts_els = np.unique(prop_failures, return_counts=True)
pd.DataFrame({
    "Proportion of failures": unique_els,
    "Fraction of samples": counts_els / 10_000,
})
```

	Proportion of failures	Fraction of samples
0	0.00	0.03
1	0.33	0.35
2	0.67	0.51
3	1.00	0.11

The simulation results are very close to the exact chances that we calculated earlier.

Simulation studies leverage random number generators to sample many outcomes from a random process. In a sense, simulation studies convert complex random processes into data that we can readily analyze using the broad set of computational tools we cover in this book. While simulation studies typically do not provide definitive proof of a particular hypothesis, they can provide important evidence. In many situations, simulation is the most accurate estimation process we have.

Drawing marbles from an urn with 0s and 1s is such a popular framework for understanding randomness that this chance process has been given a formal name, *hypergeometric distribution*, and most software provides functionality to rapidly carry out simulations of this process. In the next section, we simulate the hypergeometric distribution of the fuel tank example.

Simulation with the Hypergeometric Distribution

Instead of using `random.choice`, we can use `numpy`'s `random.hypergeometric` to simulate drawing marbles from the urn and counting the number of failures. The `random.hypergeometric` method is optimized for the 0-1 urn and allows us to ask for 10,000 simulations in one call. For completeness, we repeat our simulation study and calculate the empirical proportions:

```
simulations_fast = np.random.hypergeometric(
    ngood=4, nbad=3, nsample=3, size=10_000
)
print(simulations_fast)
```

```
[1 1 2 ... 1 2 2]
```

(We don't think that a pass is "bad"; it's just a naming convention to call the type you want to count "good" and the other "bad.")

We tally the fraction of the 10,000 samples with 0, 1, 2, or 3 failures:

```
unique_els, counts_els = np.unique(simulations_fast, return_counts=True)
pd.DataFrame({
    "Number of failures": unique_els,
    "Fraction of samples": counts_els / 10_000,
})
```

	Number of failures	Fraction of samples
0	0	0.03
1	1	0.34
2	2	0.52
3	3	0.11

You might have asked yourself already: since the hypergeometric is so popular, why not provide the exact distribution of the possible values? In fact, we can calculate these exactly:

```
from scipy.stats import hypergeom

num_failures = [0, 1, 2, 3]
pd.DataFrame({
    "Number of failures": num_failures,
    "Fraction of samples": hypergeom.pmf(num_failures, 7, 4, 3),
})
```

	Number of failures	Fraction of samples
0	0	0.03
1	1	0.34
2	2	0.51
3	3	0.11

Whenever possible, it's a good idea to use the functionality provided in a third-party package for simulating from a named distribution, such as the random number generators offered in `numpy`, rather than writing your own function. It's best to take advantage of efficient and accurate code that others have developed. That said, building from scratch on occasion can help you gain an understanding of an algorithm, so we recommend trying it.

Perhaps the two most common chance processes are those that arise from counting the number of 1s drawn from a 0-1 urn: drawing without replacement is the *hypergeometric* distribution and drawing with replacement is the *binomial* distribution.

While this simulation was so simple that we could have used `hypergeom.pmf` to directly compute our distribution, we wanted to demonstrate the intuition that a simulation study can reveal. The approach we take in this book is to develop an understanding of chance processes based on simulation studies. However, we do formalize the notion of a probability distribution of a statistic (like the proportion of fails in a sample) in Chapter 17.

Now that we have simulation as a tool for understanding accuracy, we can revisit the election example from Chapter 2 and carry out a post-election study of what might have gone wrong with the voter polls. This simulation study imitates drawing more than a thousand marbles (voters who participate in the poll) from an urn of six million. We can examine potential sources of bias and the variation in the polling results, and we can carry out a what-if analysis where we examine how the predictions might have gone if a larger number of draws from the urn were taken.

Example: Simulating Election Poll Bias and Variance

In 2016, nearly every prediction for the outcome of the US presidential election was wrong. This was a historic level of prediction error that shocked the statistics and data science communities. Here, we examine why nearly every political poll was so confident and yet also so wrong. This story both illustrates the power of simulation and reveals the hubris of data and the challenge of bias.

The president of the United States is chosen by the Electoral College, not by popular vote. Each state is allotted a certain number of votes to cast in the Electoral College according to the size of its population. Typically, whomever wins the popular vote in a state receives all of the Electoral College votes for that state. With the aid of polls conducted in advance of the election, pundits identify "battleground" states where the election is expected to be close and the Electoral College votes might swing the election.

In 2016, pollsters correctly predicted the election outcome in 46 of the 50 states. Not bad! After all, for those 46 states, Donald Trump received 231 and Hillary Clinton received 232 Electoral College votes—nearly a tie, with Clinton having a very narrow lead. Unfortunately, the remaining four states, Florida, Michigan, Pennsylvania, and Wisconsin, were identified as battleground states and accounted for a total of 75 votes. The margins of the popular vote in these four states were narrow. For example, in Pennsylvania, Trump received 48.18% and Clinton received 47.46% of the 6,165,478 votes cast. Such narrow margins can make it hard to predict the outcome given the sample sizes that the polls used. But there was an even greater challenge hidden in the survey process itself.

Many experts have studied the 2016 election results to dissect and identify what went wrong. According to the American Association for Public Opinion Research (*https://oreil.ly/4FWW2*), one online, opt-in poll adjusted its polling results for the education of the respondents but used only three broad categories (high school or less, some college, and college graduate). The pollsters found that if they had separated out respondents with advanced degrees from those with college degrees, then they would have reduced Clinton's estimated percentage by 0.5 points. In other words, after the fact, they were able to identify an education bias where highly educated voters tended to be more willing to participate in polls. This bias matters because these voters also tended to prefer Clinton over Trump.

Now that we know how people actually voted, we can carry out a simulation study like Manfred te Grotenhuis et al.'s (*https://oreil.ly/hOSC2*), which imitates election polling under different scenarios to help develop intuition for accuracy, bias, and variance.[1] We can simulate and compare the polls for Pennsylvania under two scenarios:

- People surveyed didn't change their minds, didn't hide who they voted for, and were representative of those who voted on election day.
- People with a higher education were more likely to respond, which led to a bias for Clinton.

1 Manfred te Grotenhuis et al., "Better Poll Sampling Would Have Cast More Doubt on the Potential for Hillary Clinton to Win the 2016 Election" *London School of Economics*, February 1, 2018.

Our ultimate goal is to understand the frequency with which a poll incorrectly calls the election for Hillary Clinton when a sample is collected with absolutely no bias and when there is a small amount of nonresponse bias. We begin by setting up the urn model for the first scenario.

The Pennsylvania Urn Model

Our urn model for carrying out a poll of Pennsylvania voters is an after-the-fact situation where we use the outcome of the election. The urn has 6,165,478 marbles in it, one for each voter. Like with our tiny population, we write on each marble the candidate that they voted for, draw 1,500 marbles from the urn (1,500 is a typical size for these polls), and tally the votes for Trump, Clinton, and any other candidate. From the tally, we can calculate Trump's lead over Clinton.

Since we care only about Trump's lead over Clinton, we can lump together all votes for other candidates. This way, each marble has one of three possible votes: Trump, Clinton, or Other. We can't ignore the "Other" category, because it impacts the size of the lead. Let's divvy up the voter counts between these three groups:

```
proportions = np.array([0.4818, 0.4746, 1 - (0.4818 + 0.4746)])
n = 1_500
N = 6_165_478
votes = np.trunc(N * proportions).astype(int)
votes
```

```
array([2970527, 2926135,  268814])
```

This version of the urn model has three types of marbles in it. It is a bit more complex than the hypergeometric distribution, but it is still common enough to have a named distribution: the *multivariate hypergeometric*. In Python, the urn model with more than two types of marbles is implemented by the `scipy.stats.multivariate_hypergeom.rvs` method. The function returns the number of each type of marble drawn from the urn. We call the function as follows:

```
from scipy.stats import multivariate_hypergeom

multivariate_hypergeom.rvs(votes, n)
```

```
array([727, 703,  70])
```

As before, each time we call `multivariate_hypergeom.rvs` we get a different sample and counts:

```
multivariate_hypergeom.rvs(votes, n)
```

```
array([711, 721,  68])
```

We need to compute Trump's lead for each sample: $(n_T - n_C)/n$, where n_T is the number of Trump votes in the sample and n_C the number for Clinton. If the lead is positive, then the sample shows a win for Trump.

We know the actual lead was 0.4818 – 0.4746 = 0.0072. To get a sense of the variation in the poll, we can simulate the chance process of drawing from the urn over and over and examine the values that we get in return. Now we can simulate 100,000 polls of 1,500 voters from the votes cast in Pennsylvania:

```
def trump_advantage(votes, n):
    sample_votes = multivariate_hypergeom.rvs(votes, n)
    return (sample_votes[0] - sample_votes[1]) / n

simulations = [trump_advantage(votes, n) for _ in range(100_000)]
```

On average, the polling results show Trump with close to a 0.7% lead, as expected given the composition of the more than six million votes cast:

```
np.mean(simulations)
```

```
0.007177066666666666
```

However, many times the lead in a sample was negative, meaning Clinton was the winner for that sample of voters. The following histogram shows the sampling distribution of Trump's advantage in Pennsylvania for a sample of 1,500 voters. The vertical dashed line at 0 shows that more often than not, Trump is called, but there are many times when the poll of 1,500 shows Clinton in the lead:

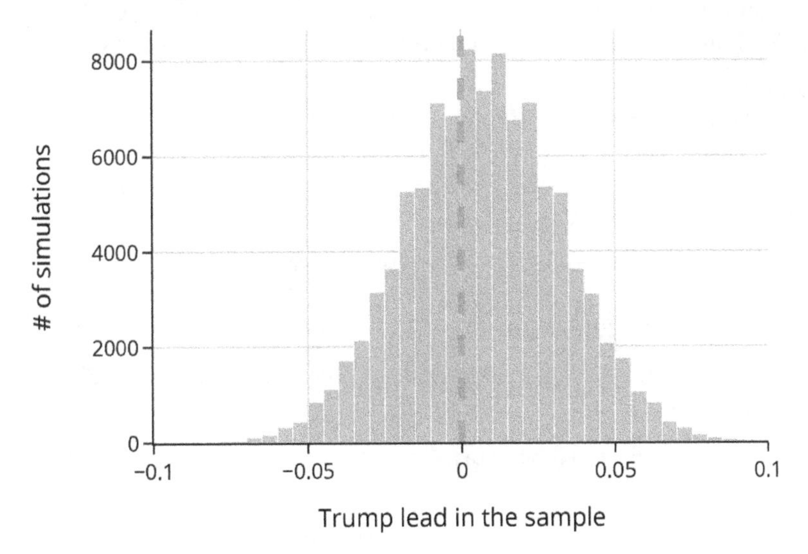

In the 100,0000 simulated polls, we find Trump a victor about 60% of the time:

```
np.mean(np.array(simulations) > 0)
```

```
0.60613
```

In other words, a sample will correctly predict Trump's victory *even if the sample was collected with absolutely no bias* about 60% of the time. And this unbiased sample will be wrong about 40% of the time.

We have used the urn model to study the variation in a simple poll, and we found how a poll's prediction might look if there was no bias in our selection process (the marbles are indistinguishable, and every possible collection of 1,500 marbles out of the more than six million marbles is equally likely). Next, we see what happens when a little bias enters the mix.

An Urn Model with Bias

According to Grotenhuis, "In a perfect world, polls sample from the population of voters, who would state their political preference perfectly clearly and then vote accordingly."[2] That's the simulation study that we just performed. In reality, it is often difficult to control for every source of bias.

We investigate here the effect of a small education bias on the polling results. Specifically, we examine the impacts of a 0.5% bias in favor of Clinton. This bias essentially means that we see a distorted picture of voter preferences in our poll. Instead of 47.46% votes for Clinton, we have 47.96%, and we have 48.18 – 0.5 = 47.68% for Trump. We adjust the proportions of marbles in the urn to reflect this change:

```
bias = 0.005
proportions_bias = np.array([0.4818 - bias, 0.4747 + bias,
                             1 - (0.4818 + 0.4746)])
proportions_bias
```

```
array([0.48, 0.48, 0.04])
```

```
votes_bias = np.trunc(N * proportions_bias).astype(int)
votes_bias
```

```
array([2939699, 2957579,  268814])
```

When we carry out the simulation study again, this time with the biased urn, we find a quite different result:

```
simulations_bias = [trump_advantage(votes_bias, n) for _ in range(100_000)]
```

2 Grotenhuis et al., "Better Poll Sampling Would Have Cast More Doubt on the Potential for Hillary Clinton to Win the 2016 Election."

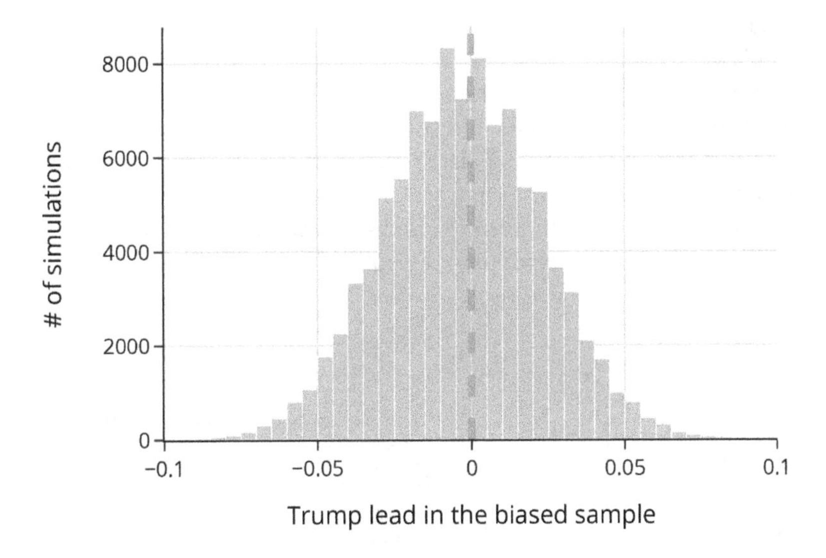


```
np.mean(np.array(simulations_bias) > 0)
```

```
0.44967
```

Now, Trump would have a positive lead in about 45% of the polls. Notice that the histograms from the two simulations are similar in shape. They are symmetric with tails of reasonable length. That is, they appear to roughly follow the normal curve. The second histogram is shifted slightly to the left, which reflects the nonresponse bias we introduced. Would increasing the sample size have helped? We investigate this topic next.

Conducting Larger Polls

With our simulation study we can gain insight on the impact of a larger poll on the sample lead. For example, we can try a sample size of 12,000, eight times the size of the actual poll, and run 100,000 simulations for both the unbiased and biased scenarios:

```
simulations_big = [trump_advantage(votes, 12_000) for _ in range(100_000)]
simulations_bias_big = [trump_advantage(votes_bias, 12_000)
                        for _ in range(100_000)]
```

```
scenario_no_bias = np.mean(np.array(simulations_big) > 0)
scenario_bias = np.mean(np.array(simulations_bias_big) > 0)
print(scenario_no_bias, scenario_bias)
```

```
0.78968 0.36935
```

The simulation shows that Trump's lead is detected in only about one-third of the simulated biased scenario. The spread of the histogram of these results is narrower than the spread when only 1,500 voters were polled. Unfortunately, it has narrowed in on the wrong value. We haven't overcome the bias; we just have a more accurate picture of the biased situation. Big data has not come to the rescue. Additionally, larger polls have other problems. They are often harder to conduct because pollsters are working with limited resources, and efforts that could go into improving the data scope are being redirected to expanding the poll:

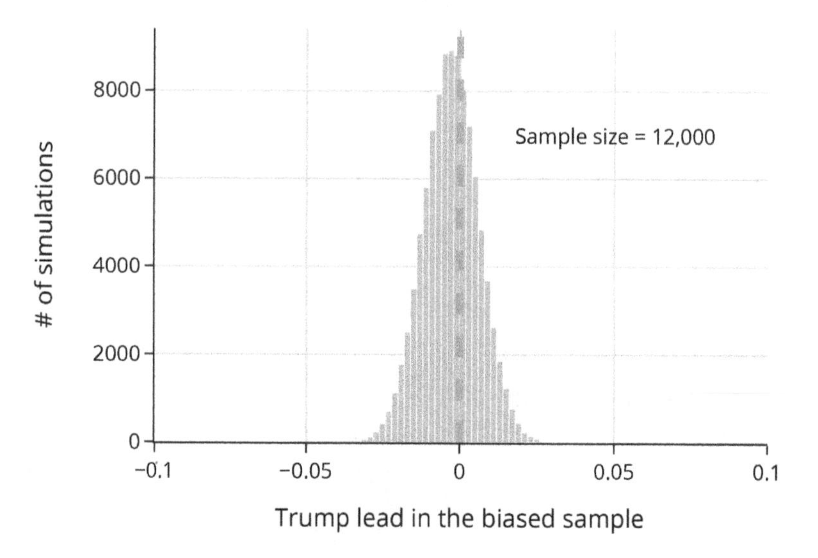

After the fact, with multiple polls for the same election, we can detect bias. In a post-election analysis (*http://dx.doi.org/10.1080/01621459.2018.1448823*) of over 4,000 polls for 600 state-level, gubernatorial, senatorial, and presidential elections, researchers found that, on average, election polls exhibit a bias of about 1.5 percentage points, which helps explain why so many polls got it wrong.

When the margin of victory is relatively small, as it was in 2016, a larger sample size reduces the sampling error, but unfortunately, if there is bias, then the predictions are close to the biased estimate. If the bias pushes the prediction from one candidate (Trump) to another (Clinton), then we have a "surprise" upset. Pollsters develop voter selection schemes that attempt to reduce bias, like the separation of voters' preference by education level. But, as in this case, it can be difficult, even impossible, to account for new, unexpected sources of bias. Polls are still useful, but we need to acknowledge the issues with bias and do a better job at reducing it.

In this example, we used the urn model to study a simple random sample in polling. Another common use of the urn is in randomized controlled experiments.

Example: Simulating a Randomized Trial for a Vaccine

In a drug trial, volunteers for the trial receive either the new treatment or a placebo (a fake treatment), and researchers control the assignment of volunteers to the treatment and placebo groups. In a *randomized controlled experiment*, they use a chance process to make this assignment. Scientists essentially use an urn model to select the subjects for the treatment and control (those given the placebo groups). We can simulate the chance mechanism of the urn to better understand variation in the outcome of an experiment and the meaning of efficacy in clinical trials.

In March 2021, Detroit Mayor Mike Duggan made national news (*https://oreil.ly/kB757*) when he turned down a shipment of over 6,000 Johnson & Johnson (J&J) vaccine doses, stating that the citizens of his city should "get the best." The mayor was referring to the efficacy rate of the vaccine, which was reported to be about 66%. In comparison, Moderna and Pfizer both reported efficacy rates of about 95% for their vaccines.

On the surface, Duggan's reasoning seems valid, but the scopes of the three clinical trials are not comparable, meaning direct comparisons of the experimental results are problematic. Moreover, the CDC (*https://oreil.ly/25Pok*) considers a 66% efficacy rate quite good, which is why it was given emergency approval.

Let's consider the points of scope and efficacy in turn.

Scope

Recall that when we evaluate the scope of the data, we consider the who, when, and where of the study. For the Johnson & Johnson clinical trial, the participants:

- Included adults age 18 and over, where roughly 40% had preexisting conditions associated with an increased risk for getting severe COVID-19
- Enrolled in the study from October to November 2020
- Came from eight countries across three continents, including the US and South Africa

The participants in the Moderna and Pfizer trials were primarily from the US, roughly 40% had preexisting conditions, and the trial took place earlier, over summer 2020. The timing and location of the trials make them difficult to compare. Cases of COVID-19 were at a low point in the summer in the US, but they rose rapidly in the late fall. Also, a variant of the virus that was more contagious was spreading rapidly in South Africa at the time of the J&J trial.

Each clinical trial was designed to test a vaccine against the situation of no vaccine under similar circumstances through the random assignment of subjects to treatment and control groups. While the scope from one trial to the next is quite different, the randomization within a trial keeps the scope of the treatment and control groups roughly the same. This enables meaningful comparisons between groups in the same trial. The scope was different enough across the three vaccine trials to make direct comparisons of the three trials problematic.

In the trial carried out for the J&J vaccine (*https://oreil.ly/epz0T*), 43,738 people were enrolled. These participants were split into two groups at random. Half received the new vaccine, and the other half received a placebo, such as a saline solution. Then everyone was followed for 28 days to see whether they contracted COVID-19.

A lot of information was recorded on each patient, such as their age, race, and sex, and in addition whether they caught COVID-19, including the severity of the disease. At the end of 28 days, the researchers found 468 cases of COVID-19, with 117 of these in the treatment group and 351 in the control group.

The random assignment of patients to treatment and control gives the scientists a framework to assess the effectiveness of the vaccine. The typical reasoning goes as follows:

1. Begin with the assumption that the vaccine is ineffective.
2. So the 468 who caught COVID-19 would have caught it whether or not they received the vaccine.
3. And the remaining 43,270 people in the trial who did not get sick would have remained healthy whether or not they received the vaccine.
4. The split of 117 sick people in treatment and 351 in control was solely due to the chance process in assigning participants to treatment or control.

We can set up an urn model that reflects this scenario and then study, via simulation, the behavior of the experimental results.

The Urn Model for Random Assignment

Our urn has 43,738 marbles, one for each person in the clinical trial. Since there were 468 cases of COVID-19 among them, we label 468 marbles with a 1 and the remaining 43,270 with a 0. We draw half the marbles (21,869) from the urn to receive the treatment, and the remaining half receive the placebo. The key result of the experiment is simply the count of the number of marbles marked 1 that were randomly drawn from the urn.

We can simulate this process to get a sense of how likely it would be under these assumptions to draw at most 117 marbles marked 1 from the urn. Since we draw half of the marbles from the urn, we would expect about half of the 468, or 234, to be drawn. The simulation study gives us a sense of the variation that might result from the random assignment process. That is, the simulation can give us an approximate chance that the trials would result in so few cases of the virus in the treatment group.

Several key assumptions enter into this urn model, such as the assumption that the vaccine is ineffective. It's important to keep track of the reliance on these assumptions because our simulation study gives us an approximation of the rarity of an outcome like the one observed only under these key assumptions.

As before, we can simulate the urn model using the hypergeometric probability distribution, rather than having to program the chance process from scratch:

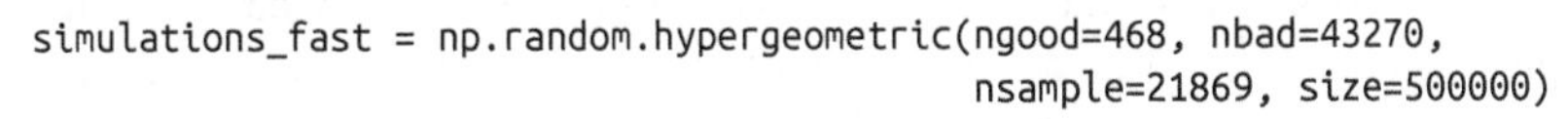

```
simulations_fast = np.random.hypergeometric(ngood=468, nbad=43270,
                                            nsample=21869, size=500000)
```

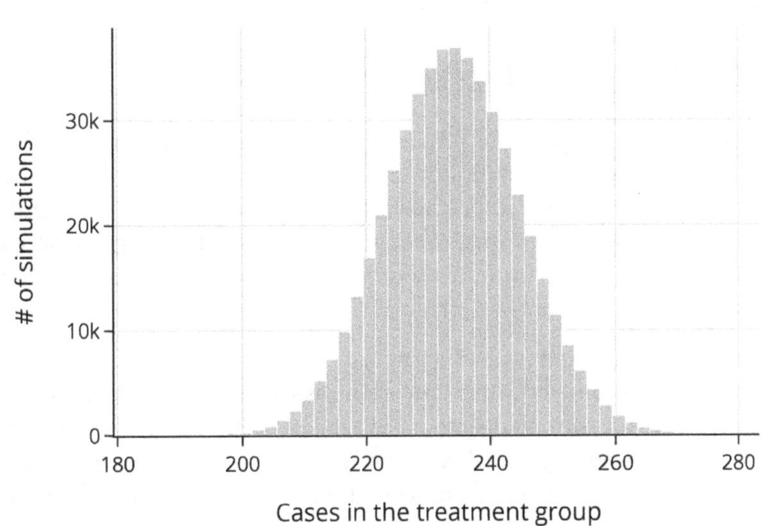

In our simulation, we repeated the process of random assignment to the treatment group 500,000 times. Indeed, we found that not one of the 500,000 simulations had 117 or fewer cases. It would be an extremely rare event to see so few cases of COVID-19 if in fact the vaccine was not effective.

After the problems with comparing drug trials that have different scopes and the efficacy for preventing severe cases of COVID-19 was explained, Mayor Duggan retracted his original statement, saying, "I have full confidence that the Johnson & Johnson vaccine is both safe and effective."[3]

3 Unfortunately, despite the vaccine's efficacy, the US Food and Drug Administration limited the use of the J&J vaccine in May 2022 due to a heightened risk of developing rare and potentially life-threatening blood clots.

This example has shown that:

- Using a chance process in the assignment of subjects to treatments in clinical trials can help us answer what-if scenarios.
- Considering data scope can help us determine whether it is reasonable to compare figures from different datasets.

Simulating the draw of marbles from an urn is a useful abstraction for studying the possible outcomes from survey samples and controlled experiments. The simulation works because it imitates the chance mechanism used to select a sample or to assign people to a treatment. In settings where we measure natural phenomena, our measurements tend to follow a similar chance process. As described in Chapter 2, instruments typically have an error associated with them, and we can use an urn to represent the variability in measuring an object.

Example: Measuring Air Quality

Across the US, sensors to measure air pollution are widely used (*https://oreil.ly/t6JzZ*) by individuals, community groups, and state and local air monitoring agencies. For example, on two days in September 2020, approximately 600,000 Californians and 500,000 Oregonians viewed PurpleAir's map as fire spread through their states and evacuations were planned. (PurpleAir (*https://www2.purpleair.com*) creates air quality maps from crowdsourced data that streams in from its sensors.)

The sensors measure the amount of particulate matter in the air that has a diameter smaller than 2.5 micrometers (the unit of measurement is micrograms per cubic meter: $\mu g/m^3$). The measurements recorded are the average concentrations over two minutes. While the level of particulate matter changes over the course of a day as, for example, people commute to and from work, there are certain times of the day, like at midnight, when we expect the two-minute averages to change little in a half hour. If we examine the measurements taken during these times of the day, we can get a sense of the combined variability in the instrument recordings and the mixing of particles in the air.

Anyone can access sensor measurements from PurpleAir's site. The site provides a download tool, and data are available for any sensor that appears on PurpleAir's map. We downloaded data from one sensor over a 24-hour period and selected three half-hour time intervals spread throughout the day where the readings were roughly constant over the 30-minute period. This gave us three sets of 15 two-minute averages, for a total of 45 measurements:

	aq2.5	time	hour	meds	diff30
0	6.14	2022-04-01 00:01:10 UTC	0	5.38	0.59
1	5.00	2022-04-01 00:03:10 UTC	0	5.38	-0.55
2	5.29	2022-04-01 00:05:10 UTC	0	5.38	-0.26
...	...	...	...	...	...
42	7.55	2022-04-01 19:27:20 UTC	19	8.55	-1.29
43	9.47	2022-04-01 19:29:20 UTC	19	8.55	0.63
44	8.55	2022-04-01 19:31:20 UTC	19	8.55	-0.29

```
45 rows × 5 columns
```

Line plots can give us a sense of variation in the measurements. In one 30-minute period, we expect the measurements to be roughly the same, with the exception of minor variations from the particles moving in the air and the measurement error of the instrument:

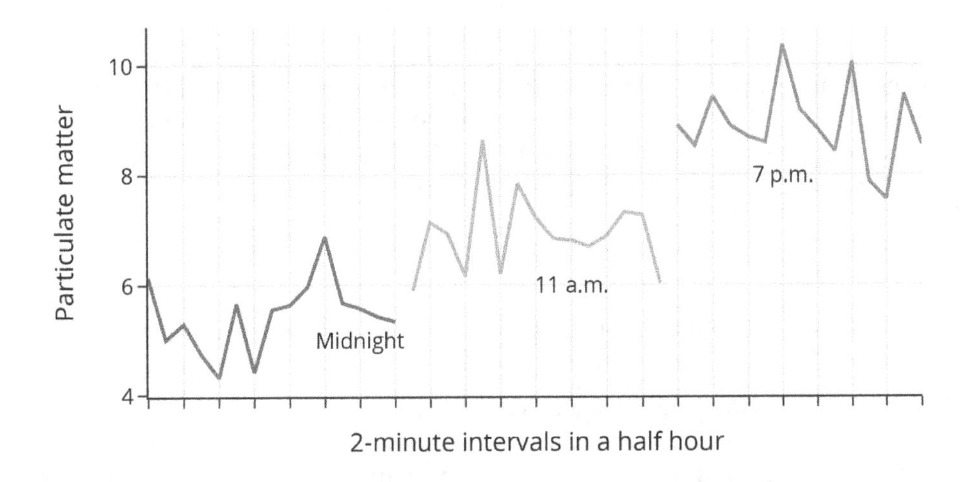

The plot shows us how the air quality worsens throughout the day, but in each of these half-hour intervals, the air quality is roughly constant at 5.4, 6.6, and 8.6 μg/m³ at midnight, 11 a.m., and 7 p.m., respectively. We can think of the data scope as follows: at this particular location in a specific half-hour time interval, there is an average particle concentration in the air surrounding the sensor. This concentration is our target, and our instrument, the sensor, takes many measurements that form a sample from the access frame. (See Chapter 2 for the dartboard analogy of this process.) If the instrument is working properly, the measurements are centered on the target: the 30-minute average.

To get a better sense of the variation in a half-hour interval, we can examine the differences of the measurements from the median for the corresponding half hour. The distribution of these "errors" is as follows:

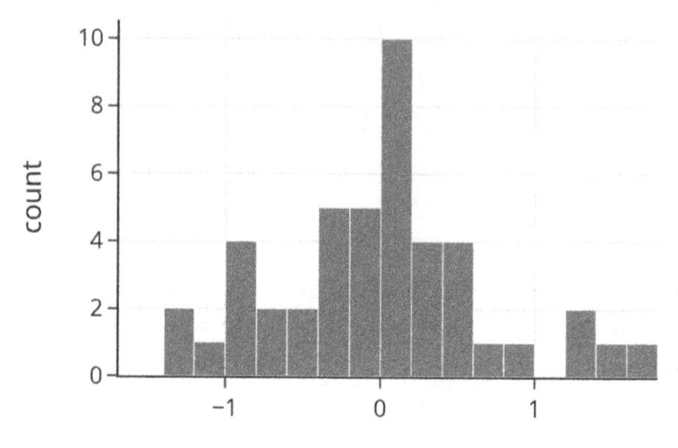

The histogram shows us that the typical fluctuations in measurements are often less than 0.5 μg/m³ and rarely greater than 1 μg/m³. With instruments, we often consider their *relative standard error*, which is the standard deviation as a percentage of the mean. The standard deviation of these 45 deviations is:

```
np.std(pm['diff30'])
```

```
0.6870817156282193
```

Given that the hourly measurements range from 5 to 9 μg/m³, the relative error is 8% to 12%, which is reasonably accurate.

We can use the urn model to simulate the variability in this measurement process. We place in the urn the deviations of the measurements from their 30-minute medians for all 45 readings, and we simulate a 30-minute air quality sequence of measurements by drawing 15 times *with replacement* from the urn and adding the deviations drawn to a hypothetical 30-minute average:

```
urn = pm["diff30"]
```

```
np.random.seed(221212)
sample_err = np.random.choice(urn, size=15, replace=True)
aq_imitate = 11 + sample_err
```

We can add a line plot for this artificial set of measurements to our earlier line plots, and compare it to the three real ones:

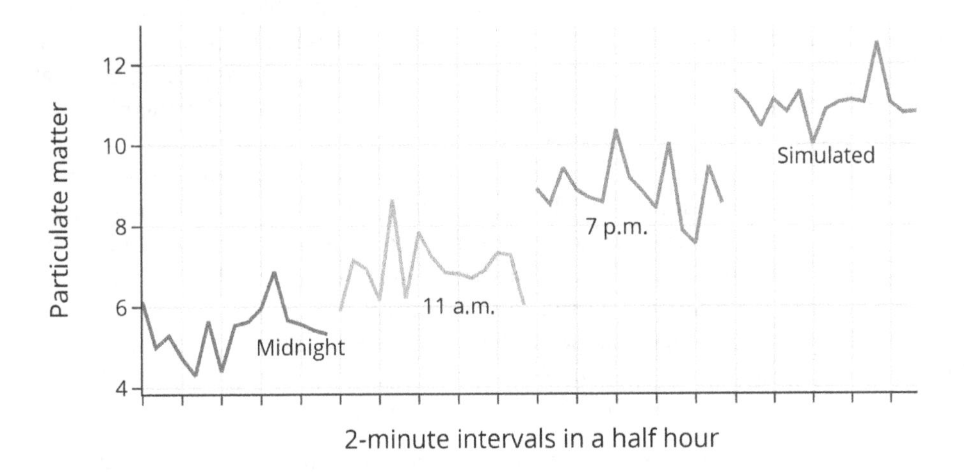

The shape of the line plot from the simulated data is similar to the others, which indicates that our model for the measurement process is reasonable. Unfortunately, what we don't know is whether the measurements are close to the true air quality. To detect bias in the instrument, we need to make comparisons against a more accurate instrument or take measurements in a protected environment where the air has a known quantity of particulate matter. In fact, researchers (*https://oreil.ly/XkvhO*) have found that low humidity can distort the readings so that they are too high. In Chapter 12, we carry out a more comprehensive analysis of the PurpleAir sensor data and calibrate the instruments to improve their accuracy.

Summary

In this chapter, we used the analogy of drawing marbles from an urn to model random sampling from populations and random assignment of subjects to treatments in experiments. This framework enables us to run simulation studies for hypothetical surveys, experiments, or other chance processes in order to study their behavior. We found the chance of observing particular results from a clinical trial under the assumption that the treatment was not effective, and we studied the support for Clinton and Trump with samples based on actual votes cast in the election. These simulation studies enabled us to quantify the typical deviations in the chance process and to approximate the distribution of summary statistics, like Trump's lead over Clinton. These simulation studies revealed the sampling distribution of a statistic and helped us answer questions about the likelihood of observing results like ours under the urn model.

The urn model reduces to a few basics: the number of marbles in the urn, what is written on each marble, the number of marbles to draw from the urn, and whether or not they are replaced between draws. From there, we can simulate increasingly complex data designs. However, the crux of the urn's usefulness is the mapping from the data design to the urn. If samples are not randomly drawn, subjects are not randomly assigned to treatments, or measurements are not made on well-calibrated equipment, then this framework falls short in helping us understand our data and make decisions. On the other hand, we also need to remember that the urn is a simplification of the actual data collection process. If in reality there is bias in data collection, then the randomness we observe in the simulation doesn't capture the complete picture. Too often, data scientists wave these annoyances aside and address only the variability described by the urn model. That was one of the main problems in the surveys predicting the outcome of the 2016 US presidential election.

In each of these examples, the summary statistics that we have studied were given to us as part of the example. In the next chapter, we address the question of how to choose a summary statistic to represent the data.

CHAPTER 4

Modeling with Summary Statistics

We saw in Chapter 2 the importance of data scope and in Chapter 3 the importance of data generation mechanisms, such as one that can be represented by an urn model. Urn models address one aspect of modeling: they describe chance variation and ensure that the data are representative of the target. Good scope and representative data lay the groundwork for extracting useful information from data, which is the other part of modeling. This information is often referred to as the *signal* in the data. We use models to approximate the signal, with the simplest of these being the constant model, where the signal is approximated by a single number, like the mean or median. Other, more complex models summarize relationships between features in the data, such as humidity and particulate matter in air quality (Chapter 12), upward mobility and commute time in communities (Chapter 15), and height and weight of animals (Chapter 18). These more complex models are also approximations built from data. When a model fits the data well, it can provide a useful approximation to the world or simply a helpful description of the data.

In this chapter, we introduce the basics of model fitting through a *loss* formulation. We demonstrate how to model patterns in the data by considering the loss that arises from using a simple summary to describe the data, the constant model. We delve deeper into the connections between the urn model and the fitted model in Chapter 16, where we examine the balance between signal and noise when fitting models, and in Chapter 17, where we tackle the topics of inference, prediction, and hypothesis testing.

The constant model lets us introduce model fitting from the perspective of *loss minimization* in a simple context, and it helps us connect summary statistics, like the mean and median, to more complex modeling scenarios in later chapters. We begin with an example that uses data about the late arrival of a bus to introduce the constant model.

The Constant Model

A transit rider, Jake, often takes the northbound C bus at the 3rd & Pike bus stop in downtown Seattle.[1] The bus is supposed to arrive every 10 minutes, but Jake notices that he sometimes waits a long time for the bus. He wants to know how late the bus usually is. Jake was able to acquire the scheduled arrival and actual arrival times for his bus from the Washington State Transportation Center. From these data, he can calculate the number of minutes that each bus is late to arrive at his stop:

```
times = pd.read_csv('data/seattle_bus_times_NC.csv')
times
```

	route	direction	scheduled	actual	minutes_late
0	C	northbound	2016-03-26 06:30:28	2016-03-26 06:26:04	-4.40
1	C	northbound	2016-03-26 01:05:25	2016-03-26 01:10:15	4.83
2	C	northbound	2016-03-26 21:00:25	2016-03-26 21:05:00	4.58
...	...	...	...	...	...
1431	C	northbound	2016-04-10 06:15:28	2016-04-10 06:11:37	-3.85
1432	C	northbound	2016-04-10 17:00:28	2016-04-10 16:56:54	-3.57
1433	C	northbound	2016-04-10 20:15:25	2016-04-10 20:18:21	2.93

```
1434 rows × 5 columns
```

The `minutes_late` column in the data table records how late each bus was. Notice that some of the times are negative, meaning that the bus arrived early. Let's examine a histogram of the number of minutes each bus is late:

```
fig = px.histogram(times, x='minutes_late', width=450, height=250)
fig.update_xaxes(range=[-12, 60], title_text='Minutes late')
fig
```

1 We (the authors) first learned of the bus arrival time data from an analysis by a data scientist named Jake VanderPlas. We've named the protagonist of this section in his honor.

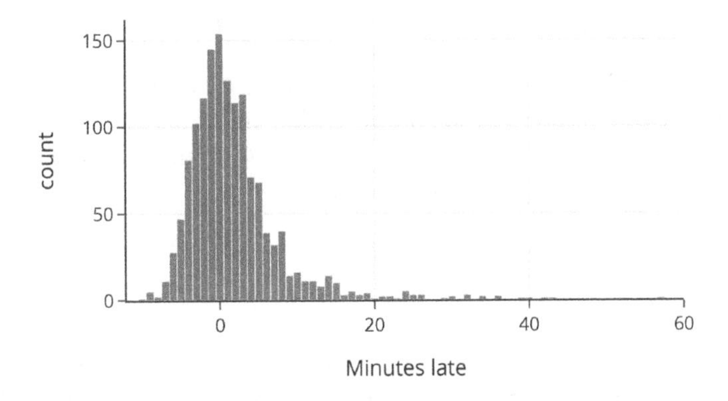

We can already see some interesting patterns in the data. For example, many buses arrive earlier than scheduled, but some are well over 20 minutes late. We also see a clear mode (high point) at 0, meaning many buses arrive roughly on time.

To understand how late a bus on this route typically is, we'd like to summarize lateness by a constant—this is a statistic, a single number, like the mean, median, or mode. Let's find each of these summary statistics for the `minutes_late` column in the data table.

From the histogram, we estimate the mode of the data to be 0, and we use Python to compute the mean and median:

```
mean:    1.92 mins late
median:  0.74 mins late
mode:    0.00 mins late
```

Naturally, we want to know which of these numbers best represents a summary of lateness. Rather than relying on rules of thumb, we take a more formal approach. We make a constant model for bus lateness. Let's call this constant θ (in modeling, θ is often referred to as a *parameter*). For example, if we consider $\theta = 5$, then our model approximates the bus to typically be five minutes late.

Now, $\theta = 5$ isn't a particularly good guess. From the histogram of minutes late, we saw that there are many more points closer to 0 than 5. But it isn't clear that $\theta = 0$ (the mode) is a better choice than $\theta = 0.74$ (the median), $\theta = 1.92$ (the mean), or something else entirely. To make choices between different values of θ, we would like to assign any value of θ a score that measures how well that constant fits the data. That is, we want to assess the loss involved in approximating the data by a constant, like $\theta = 5$. And ideally, we want to pick the constant that best fits our data, meaning the constant that has the smallest loss. In the next section, we describe more formally what we mean by loss and show how to use it to fit a model.

Minimizing Loss

We want to model how late the northbound C bus is by a constant, which we call θ, and we want to use the data of actual number of minutes each bus is late to figure out a good value for θ. To do this, we use a *loss function*—a function that measures how far away our constant, θ, is from the actual data.

A loss function is a mathematical function that takes in θ and a data value y. It outputs a single number, the *loss*, that measures how far away θ is from y. We write the loss function as $\ell(\theta, y)$.

By convention, the loss function outputs lower values for better values of θ and larger values for worse θ. To fit a constant to our data, we select the particular θ that produces the lowest average loss across all choices for θ. In other words, we find the θ that *minimizes the average loss* for our data, $y_1, \ldots, y_n$. More formally, we write the average loss as $L(\theta, y_1, y_2, \ldots, y_n)$, where:

$$
\begin{aligned}
L(\theta, y_1, y_2, \ldots, y_n) &= \text{mean}\{\ell(\theta, y_1), \ell(\theta, y_2), \ldots, \ell(\theta, y_n)\} \\
&= \frac{1}{n} \sum_{i=1}^{n} \ell(\theta, y_i)
\end{aligned}
$$

As a shorthand, we often use the vector $\mathbf{y} = [y_1, y_2, \ldots, y_n]$. Then we can write the average loss as:

$$
L(\theta, \mathbf{y}) = \frac{1}{n} \sum_{i=1}^{n} \ell(\theta, y_i)
$$

Notice that $\ell(\theta, y)$ tells us the model's loss for a single data point while $L(\theta, \mathbf{y})$ gives the model's average loss for all the data points. The capital L helps us remember that the average loss combines multiple smaller ℓ values.

Once we define a loss function, we can find the value of θ that produces the smallest average loss. We call this minimizing value $\hat{\theta}$. In other words, of all the possible θ values, $\hat{\theta}$ is the one that produces the smallest average loss for our data. We call this optimization process *model fitting*; it finds the best constant model for our data.

Next, we look at two particular loss functions: absolute error and squared error. Our goal is to fit the model and find $\hat{\theta}$ for each of these loss functions.

Mean Absolute Error

We start with the *absolute error* loss function. Here's the idea behind absolute loss. For some value of θ and data value y:

1. Find the error, $y - \theta$.
2. Take the absolute value of the error, $|y - \theta|$.

So the loss function is $\ell(\theta, y) = |y - \theta|$.

Taking the absolute value of the error is a simple way to convert negative errors into positive ones. For instance, the point $y = 4$ is equally far away from $\theta = 2$ and $\theta = 6$, so the errors are equally "bad."

The average of the absolute errors is called the *mean absolute error* (MAE). The MAE is the average of each of the individual absolute errors:

$$L(\theta, \mathbf{y}) = \frac{1}{n} \sum_{i=1}^{n} |y_i - \theta|$$

Notice that the name MAE tells you how to compute it: take the Mean of the Absolute value of the Errors, $\{y_i - \theta\}$.

We can write a simple Python function to compute this loss:

```
def mae_loss(theta, y_vals):
    return np.mean(np.abs(y_vals - theta))
```

Let's see how this loss function behaves when we have just five data points $[-1, 0, 2, 5, 10]$. We can try different values of θ and see what the MAE outputs for each value:

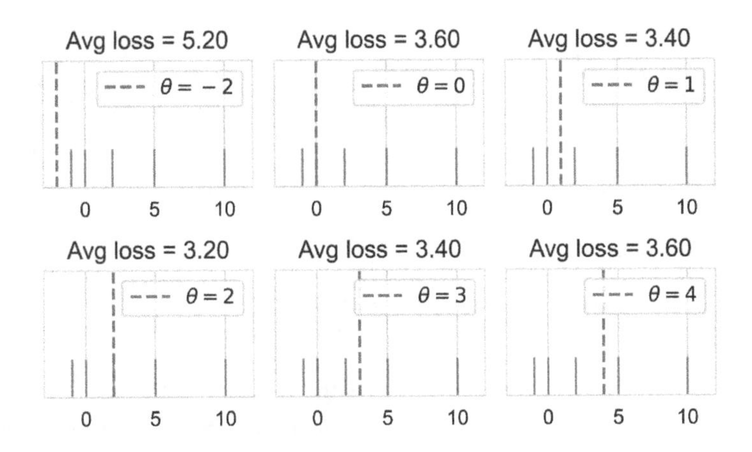

We suggest verifying some of these loss values by hand to check that you understand how the MAE is computed.

Of the values of θ that we tried, we found that $\theta = 2$ has the lowest mean absolute error. For this simple example, 2 is the median of the data values. This isn't a coincidence. Let's now check what the average loss is for the original dataset of bus late times. We find the MAE when we set θ to the mode, median, and mean of the minutes late, respectively:

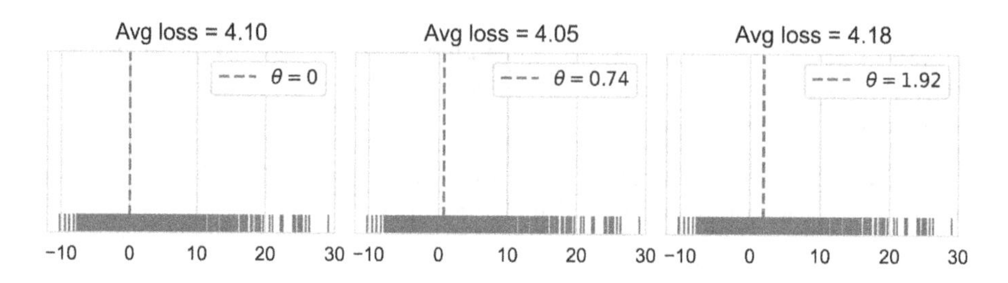

We see again that the median (middle plot) gives a smaller loss than the mode and mean (left and right plots). In fact, for absolute loss, the minimizing $\hat{\theta}$ is the $\text{median}\{y_1, y_2, \ldots, y_n\}$.

So far, we have found the best value of θ by simply trying out a few values and then picking the one with the smallest loss. To get a better sense of the MAE as a function of θ, we can try many more values of θ and plot a curve that shows how $L(\theta, \mathbf{y})$ changes as θ changes. We draw the curve for the preceding example with the five data values $[-1, 0, 2, 5, 10]$:

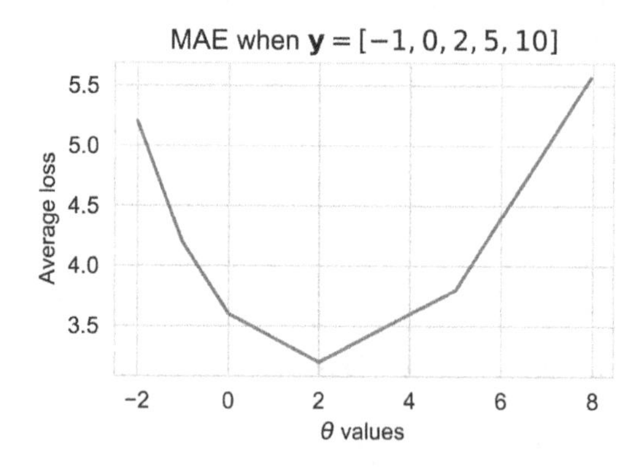

The preceding plot shows that in fact, $\theta = 2$ is the best choice for this small dataset of five values. Notice the shape of the curve. It is piecewise linear, where the line segments connect at the location of the data values (–1, 0, 2, and 5). This is a property of the absolute value function. With a lot of data, the flat pieces are less obvious. Our bus data have over 1,400 points and the MAE curve appears smoother:

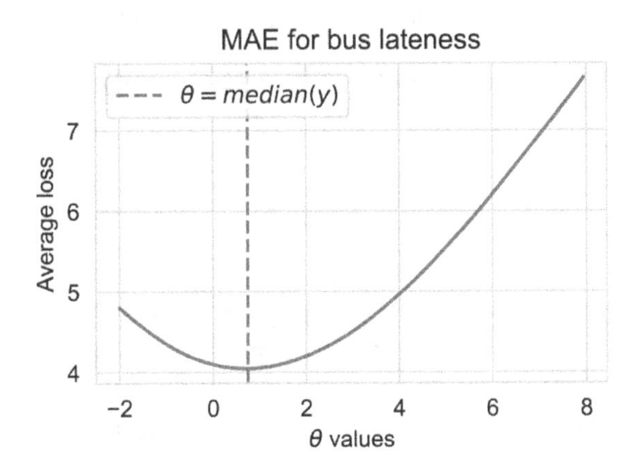

We can use this plot to help confirm that the median of the data is the minimizing value; in other words, $\hat{\theta} = 0.74$. This plot is not really a proof, but hopefully it's convincing enough for you.

Next, let's look at another loss function that squares error.

Mean Squared Error

We have fitted a constant model to our data and found that with mean absolute error, the minimizer is the median. Now we'll keep our model the same but switch to a different loss function: squared error. Instead of taking the absolute difference between each data value y and the constant θ, we'll square the error. That is, for some value of θ and data value y:

1. Find the error, $y - \theta$.
2. Take the square of the error, $(y - \theta)^2$.

This gives the loss function $\ell(\theta, y) = (y - \theta)^2$.

As before, we want to use all of our data to find the best θ, so we compute the mean squared error, or MSE for short:

$$L(\theta, \mathbf{y}) = L(\theta, y_1, y_2, \ldots, y_n) = \frac{1}{n} \sum_{i=1}^{n} (y_i - \theta)^2$$

We can write a simple Python function to compute the MSE:

```
def mse_loss(theta, y_vals):
    return np.mean((y_vals - theta) ** 2)
```

Let's again try the mean, median, and mode as potential minimizers of the MSE:

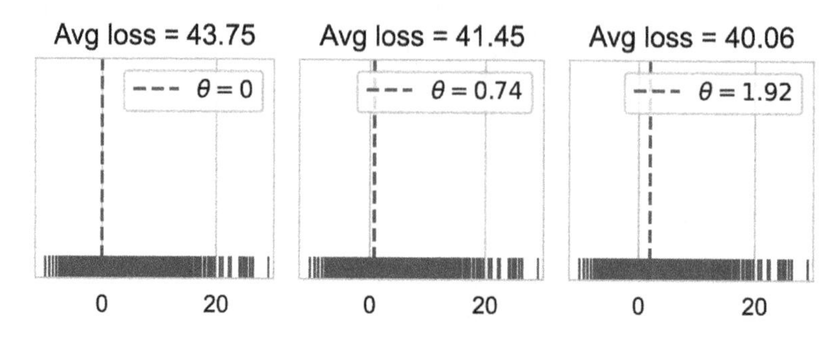

Now when we fit the constant model using MSE loss, we find that the mean (right plot) has a smaller loss than the mode and the median (left and middle plots).

Let's plot the MSE curve for different values of θ given our data. The curve shows that the minimizing value $\hat{\theta}$ is close to 2:

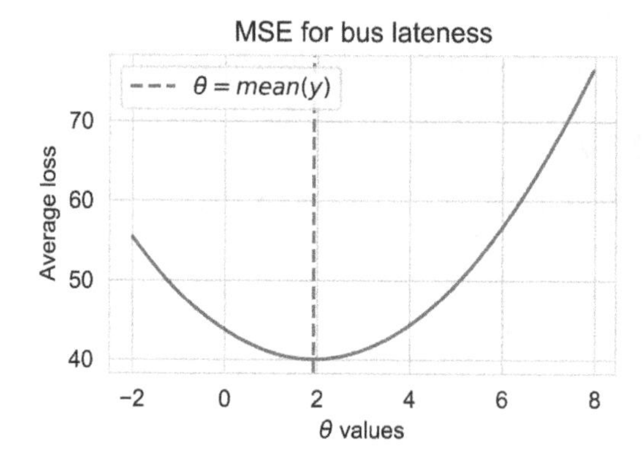

One feature of this curve that is quite noticeable is how rapidly the MSE grows compared to the MAE (note the range on the vertical axis). This growth has to do with the nature of squaring errors; it places a much higher loss on data values further away from θ. If $\theta = 10$ and $y = 110$, the squared loss is $(10 - 110)^2 = 10,000$ whereas the absolute loss is $|10 - 110| = 100$. For this reason, the MSE is more sensitive to unusually large data values than the MAE.

From the MSE curve, it appears that the minimizing $\hat{\theta}$ is the mean of $\mathbf{y}$. Again, this is no mere coincidence; the mean of the data always coincides with $\hat{\theta}$ for squared error. We show how this comes about from the quadratic nature of the MSE. Along the way, we demonstrate a common representation of squared loss as a sum of variance and bias terms, which is at the heart of model fitting with squared loss. To begin, we add and subtract $\bar{y}$ in the loss function and expand the square as follows:

$$
\begin{aligned}
L(\theta, \mathbf{y}) &= \frac{1}{n}\sum_{i=1}^{n}(y_i - \theta)^2 \\
&= \frac{1}{n}\sum_{i=1}^{n}[(y_i - \bar{y}) + (\bar{y} - \theta)]^2 \\
&= \frac{1}{n}\sum_{i=1}^{n}[(y_i - \bar{y})^2 + 2(y_i - \bar{y})(\bar{y} - \theta) + (\bar{y} - \theta)^2]
\end{aligned}
$$

Next, we split the MSE into the sum of these three terms and note that the middle term is 0, due to the simple property of the average: $\sum(y_i - \bar{y}) = 0$:

$$
\begin{aligned}
&\frac{1}{n}\sum_{i=1}^{n}(y_i - \bar{y})^2 + \frac{1}{n}\sum_{i=1}^{n}2(y_i - \bar{y})(\bar{y} - \theta) + \frac{1}{n}\sum_{i=1}^{n}(\bar{y} - \theta)^2 \\
&\quad = \frac{1}{n}\sum_{i=1}^{n}(y_i - \bar{y})^2 + 2(\bar{y} - \theta)\frac{1}{n}\sum_{i=1}^{n}(y_i - \bar{y}) + \frac{1}{n}\sum_{i=1}^{n}(\bar{y} - \theta)^2 \\
&\quad = \frac{1}{n}\sum_{i=1}^{n}(y_i - \bar{y})^2 + (\bar{y} - \theta)^2
\end{aligned}
$$

Of the remaining two terms, the first does not involve θ. You probably recognize it as the variance of the data. The second term is always non-negative. It is called the *bias squared*. This second term, the bias squared, is 0 when θ is $\bar{y}$, so $\hat{\theta} = \bar{y}$ gives the smallest MSE for any dataset.

We have seen that for absolute loss, the best constant model is the median, but for squared error, it's the mean. The choice of the loss function is an important aspect of model fitting.

Choosing Loss Functions

Now that we've worked with two loss functions, we can return to our original question: how do we choose whether to use the median, mean, or mode? Since these statistics minimize different loss functions,[2] we can equivalently ask: what is the most appropriate loss function for our problem? To answer this question, we look at the context of our problem.

Compared to the MAE, the MSE gives especially large losses when the bus is much later (or earlier) than expected. A bus rider who wants to understand the typical late times would use the MAE and the median (0.74 minutes late), but a rider who despises unexpected large late times might summarize the data using the MSE and the mean (1.92 minutes late).

If we want to refine the model even more, we can use a more specialized loss function. For example, suppose that when a bus arrives early, it waits at the stop until the scheduled time of departure; then we might want to assign an early arrival 0 loss. And if a really late bus is a larger aggravation than a moderately late one, we might choose an *asymmetric loss function* that gives a larger penalty to super-late arrivals.

In essence, context matters when choosing a loss function. By thinking carefully about how we plan to use the model, we can pick a loss function that helps us make good data-driven decisions.

Summary

We introduced the constant model: a model that summarizes the data by a single value. To fit the constant model, we chose a loss function that measured how well a given constant fits a data value, and we computed the average loss over all of the data values. We saw that depending on the choice of loss function, we get a different minimizing value: we found that the mean minimizes the average squared error (MSE), and the median minimizes the average absolute error (MAE). We also discussed how we can incorporate context and knowledge of our problem to pick a loss function.

The idea of fitting models through loss minimization ties simple summary statistics—like the mean, median, and mode—to more complex modeling situations. The steps we took to model our data apply to many modeling scenarios:

1. Select the form of a model (such as the constant model).
2. Select a loss function (such as absolute error).
3. Fit the model by minimizing the loss over all the data (such as average loss).

For the rest of this book, our modeling techniques expand upon one or more of these steps. We introduce new models, new loss functions, and new techniques for

2 The mode minimizes a loss function called 0-1 loss. Although we haven't covered this specific loss, the procedure is identical: pick the loss function, then find what minimizes the loss.

minimizing loss. Chapter 5 revisits the study of a bus arriving late at its stop. This time, we present the problem as a case study and visit all stages of the data science lifecycle. By going through these stages, we make some unusual discoveries; when we augment our analysis by considering data scope and using an urn to simulate a rider arriving at the bus stop, we find that modeling bus lateness is not the same as modeling the rider's experience waiting for a bus.

CHAPTER 5

Case Study: Why Is My Bus Always Late?

Jake VanderPlas's blog, Pythonic Perambulations (*http://jakevdp.github.io*), offers a great example of what it's like to be a modern data scientist. As data scientists, we see data in our work, daily routines, and personal lives, and we tend to be curious about what insights these data might bring to our understanding of the world. In this first case study, we borrow from one of the posts on Pythonic Perambulations, "The Waiting Time Paradox, or, Why Is My Bus Always Late?" (*https://oreil.ly/W8Ih5*), to model waiting for a bus on a street corner in Seattle. We touch on each stage of the data lifecycle, but in this first case study, our focus is on the process of thinking about the question, data, and model, rather than on data structures and modeling techniques. A constant model and simulation study get us a long way toward understanding the issues.

VanderPlas's post was inspired by his experience waiting for the bus. The wait always seemed longer than expected. This experience did not match the reasoning that if a bus comes every 10 minutes and you arrive at the stop at a random time, then, on average, the wait should be about 5 minutes. Armed with data provided by the Washington State Transportation Center, the author was able to investigate this phenomenon. We do the same.

We apply concepts introduced in earlier chapters, beginning with the general question, Why is my bus always late?, and refining this question to one that is closer to our goal and that we can investigate with data. We then consider the data scope, such as how these data were collected and potential sources of biases, and we prepare the data for analysis. Our understanding of the data scope helps us design a model for waiting at a bus stop, which we simulate to study this phenomenon.

Question and Scope

Our original question comes from the experience of a regular bus rider wondering why their bus is always late. We are not looking for actual reasons for its lateness, like a traffic jam or maintenance delay. Instead, we want to study patterns in the actual arrival times of buses at a stop, compared to their scheduled times. This information will help us better understand what it's like to wait for the bus.

Bus lines differ across the world and even across a city, so we narrow our investigation to one bus stop in the city of Seattle. The data we have are for the stops of Seattle's Rapid Ride lines C, D, and E at 3rd Avenue and Pike Street. The Washington State Transportation Center has provided times for all of the actual and scheduled stop times of these three bus lines between March 26 and May 27, 2016.

Considering our narrowed scope to buses at one particular stop over a two-month period and our access to all of the administrative data collected in this window of time, the population, access frame, and sample are one and the same. Yet, we can imagine that our analysis might prove useful for other locations in and beyond Seattle and for other times of the year. If we are lucky, the ideas that we uncover, or the approach that we take, can be useful to others. For now, we keep a narrowed focus.

Let's take a look at these data to better understand their structure.

Data Wrangling

Before we start our analysis, we check the quality of the data, simplify the structure where possible, and derive new measurements that might help us in our analysis. We cover these types of operations in Chapter 9, so don't worry about the details of the code for now. Instead, focus on the differences between the data tables as we clean the data. We start by loading the data into Python.

The first few rows in the data table are shown here:

```
bus.head(3)
```

	OPD_DATE	VEHICLE_ID	RTE	DIR	...	STOP_ID	STOP_NAME	SCH_STOP_TM	ACT_STOP_TM
0	2016-03-26	6201	673	S	...	431	3RD AVE & PIKE ST (431)	01:11:57	01:13:19
1	2016-03-26	6201	673	S	...	431	3RD AVE & PIKE ST (431)	23:19:57	23:16:13
2	2016-03-26	6201	673	S	...	431	3RD AVE & PIKE ST (431)	21:19:57	21:18:46

```
3 rows × 9 columns
```

(The raw data are available as comma-separated values in a file, which we have loaded into this table; see Chapter 8 for details on this process.)

It looks like some of the columns in the table might be redundant, like the columns labeled `STOP_ID` and `STOP_NAME`. We can find the number of unique values and their counts to confirm this:

```
bus[['STOP_ID','STOP_NAME']].value_counts()
```

```
STOP_ID  STOP_NAME
578      3RD AVE & PIKE ST (578)    19599
431      3RD AVE & PIKE ST (431)    19318
dtype: int64
```

There are two `3RD AVE & PIKE ST` names for the stop. We wonder whether they are related to the direction of the bus, which we can check against the possible combinations of direction, stop ID, and stop name:

```
bus[['DIR','STOP_ID','STOP_NAME']].value_counts()
```

```
DIR  STOP_ID  STOP_NAME
N    578      3RD AVE & PIKE ST (578)    19599
S    431      3RD AVE & PIKE ST (431)    19318
dtype: int64
```

Indeed, the northern direction corresponds to stop ID 578 and the southern direction corresponds to stop ID 431. Since we are looking at only one stop in our analysis, we don't really need anything more than the direction.

We can also check the number of unique route names:

```
673    13228
674    13179
675    12510
Name: RTE, dtype: int64
```

These routes are numbered and don't match the names C, D, and E from the original description of the problem. This issue involves another aspect of data wrangling: we need to dig up information that connects the route letters and numbers. We can get this info from the Seattle transit site. Yet another part of wrangling is to translate data values into ones that are easier to understand, so we replace the route numbers with their letters:

```
def clean_stops(bus):
    return bus.assign(
        route=bus["RTE"].replace({673: "C", 674: "D", 675: "E"}),
        direction=bus["DIR"].replace({"N": "northbound", "S": "southbound"}),
    )
```

We can also create new columns in the table that help us in our investigations. For example, we can use the scheduled and actual arrival times to calculate how late a bus is. Doing this requires some work with date and time formats, which is covered in Chapter 9.

Let's examine the values of this new quantity to make sure that our calculations are correct:

```
smallest amount late: -12.87 minutes
greatest amount late:  150.28 minutes
median amount late:    0.52 minutes
```

It's a bit surprising that there are negative values for how late a bus is, but this just means the bus arrived earlier than scheduled. While the median lateness is only about half a minute, some of the buses are 2.5 hours late! Let's take a look at the histogram of how many minutes late the buses are:

```
px.histogram(bus, x="minutes_late", nbins=120, width=450, height=300,
             labels={'minutes_late':'Minutes late'})
```

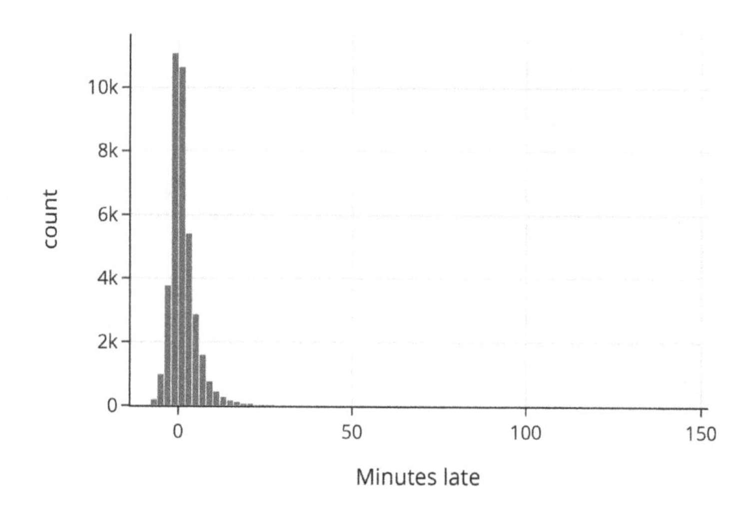

We saw a similarly shaped histogram in Chapter 4. The distribution of how late the buses are is highly skewed to the right, but many arrive close to on time.

Finally, we conclude our wrangling by creating a simplified version of the data table. Since we only need to keep track of the route, direction, scheduled and actual arrival time, and how late the bus is, we create a smaller table and give the columns names that are a bit easier to read:

```
bus = bus[["route", "direction", "scheduled", "actual", "minutes_late"]]
bus.head()
```

	route	direction	scheduled	actual	minutes_late
0	C	southbound	2016-03-26 01:11:57	2016-03-26 01:13:19	1.37
1	C	southbound	2016-03-26 23:19:57	2016-03-26 23:16:13	-3.73
2	C	southbound	2016-03-26 21:19:57	2016-03-26 21:18:46	-1.18
3	C	southbound	2016-03-26 19:04:57	2016-03-26 19:01:49	-3.13
4	C	southbound	2016-03-26 16:42:57	2016-03-26 16:42:39	-0.30

These table manipulations are covered in Chapter 6.

Before we begin to model bus lateness, we want to explore and learn more about these data. We do that next.

Exploring Bus Times

We learned a lot about the data as we cleaned and simplified it, but before we begin to model wait time, we want to dig deeper to better understand the phenomenon of bus lateness. We narrowed our focus to the bus activity at one stop (3rd Avenue and Pike Street) over a two-month period. And we saw that the distribution of the lateness of a bus is skewed to the right, with some buses being very late indeed. In this exploratory phase, we might ask:

- Does the distribution of lateness look the same for all three bus lines?
- Does it matter whether the bus is traveling north or south?
- How does the time of day relate to how late the bus is?
- Are the buses scheduled to arrive at regular intervals throughout the day?

Answering these questions helps us better determine how to model.

Recall from Chapter 4 that we found the median time a bus was late was 3/4 of a minute. But this doesn't match the median we calculated for all bus routes and directions (1/2 a minute). Let's check whether that could be due to the focus on northbound line C buses in that chapter. Let's create histograms of lateness for each of the six combinations of bus line and direction to address this question and the first two questions on our list:

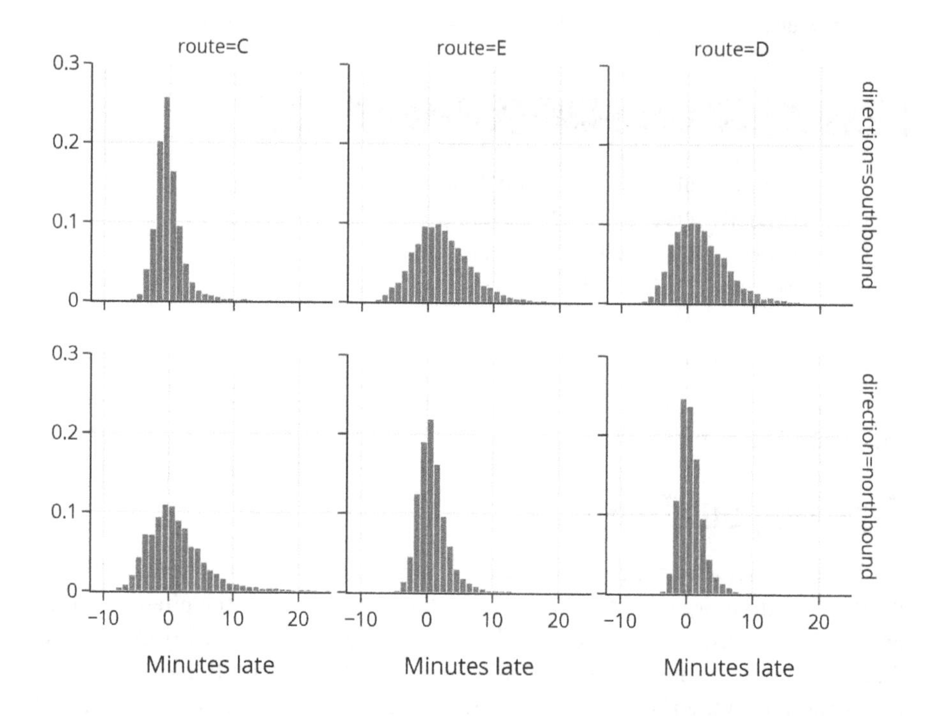

The scale on the y-axis is proportion (or density). This scale makes it easier to compare the histograms since we are not misled by different counts in the groups. The range on the x-axis is the same across the six plots, making it easier to detect the different center and spread of the distributions. (These notions are described in Chapter 11.)

The northbound and southbound distributions are different for each line. When we dig deeper into the context, we learn that line C originates in the north and the other two lines originate in the south. The histograms imply there is greater variability in arrival times in the second half of the bus routes, which makes sense to us since delays get compounded as the day progresses.

Next, to explore lateness by time of day, we need to derive a new quantity: the hour of the day that the bus is scheduled to arrive. Given the variation in route and direction that we just saw in bus lateness, we again create separate plots for each route and direction:

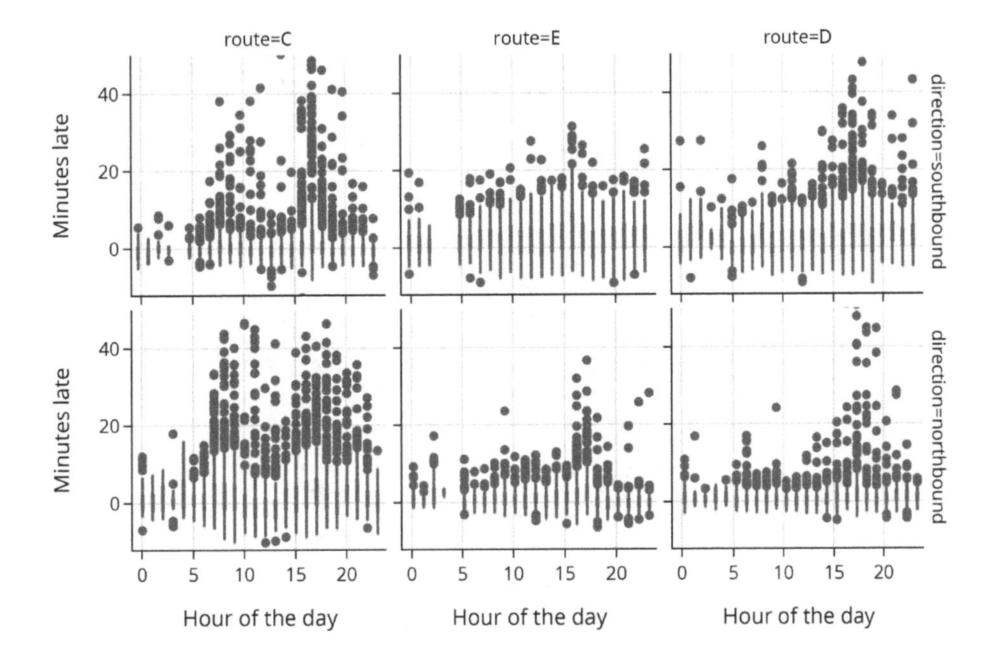

Indeed, there does appear to be a rush-hour effect, and it seems worse for the evening rush hour compared to the morning. The northbound C line looks to be the most impacted.

Lastly, to examine the scheduled frequency of the buses, we need to compute the intervals between scheduled bus times. We create a new column in our table that contains the time between the scheduled arrival times for the northbound C buses:

```
minute = pd.Timedelta('1 minute')
bus_c_n = (
    bus[(bus['route'] == 'C') & (bus['direction'] == 'northbound')]
    .sort_values('scheduled')
    .assign(sched_inter=lambda x: x['scheduled'].diff() / minute)
)
bus_c_n.head(3)
```

	route	direction	scheduled	actual	minutes_late	sched_inter
19512	C	northbound	2016-03-26 00:00:25	2016-03-26 00:05:01	4.60	NaN
19471	C	northbound	2016-03-26 00:30:25	2016-03-26 00:30:19	-0.10	30.0
19487	C	northbound	2016-03-26 01:05:25	2016-03-26 01:10:15	4.83	35.0

Let's examine a histogram of the distribution of inter-arrival times of these buses:

```
fig = px.histogram(bus_c_n, x='sched_inter',
                   title="Bus line C, northbound",
                   width=450, height=300)
```

```
fig.update_xaxes(range=[0, 40], title="Time between consecutive buses")
fig.update_layout(margin=dict(t=40))
```

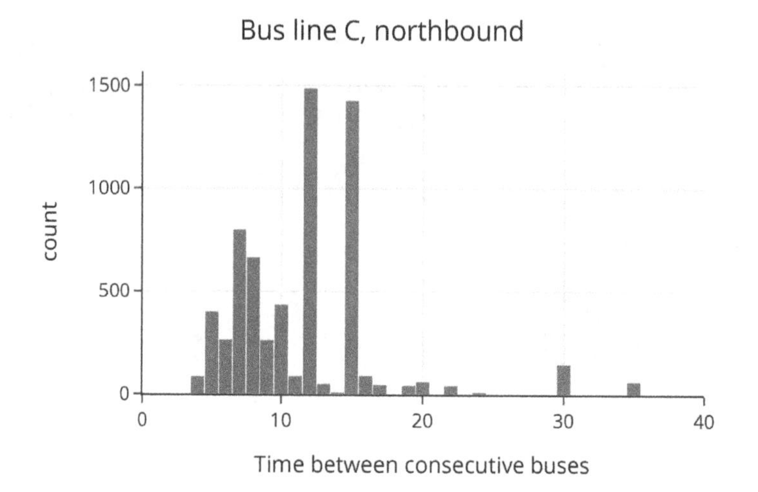

We see that the buses are scheduled to arrive at different intervals throughout the day. In this two-month period, about 1,500 of the buses are scheduled to arrive 12 minutes apart and about 1,400 are supposed to arrive 15 minutes after the previous bus.

We have learned a lot in our exploration of the data and are in a better position to fit a model. Most notably, if we want to get a clear picture of the experience of waiting for a bus, we need to take into account the scheduled interval between buses, as well as the bus line and direction.

Modeling Wait Times

We are interested in modeling the experience of someone waiting at a bus stop. We could develop a complex model that involves the intervals between scheduled arrivals, the bus line, and direction. Instead, we take a simpler approach and narrow the focus to one line, one direction, and one scheduled interval. We examine the northbound C line stops that are scheduled to arrive 12 minutes apart:

```
bus_c_n_12 = bus_c_n[bus_c_n['sched_inter'] == 12]
```

Both the complex and the narrow approaches are legitimate, but we do not yet have the tools to approach the complex model (see Chapter 15 for more details on modeling).

So far, we have examined the distribution of the number of minutes the bus is late. We create another histogram of this delay for the subset of data that we are analyzing (northbound C line stops that are scheduled to arrive 12 minutes after the previous bus):

```
fig = px.histogram(bus_c_n_12, x='minutes_late',
                   labels={'minutes_late':'Minutes late'},
                   nbins=120, width=450, height=300)

fig.add_annotation(x=20, y=150,  showarrow=False,
  text="Line C, northbound<br>Scheduled arrivals: 12 minutes apart" )
fig.update_xaxes(range=[-13, 40])
fig.show()
```

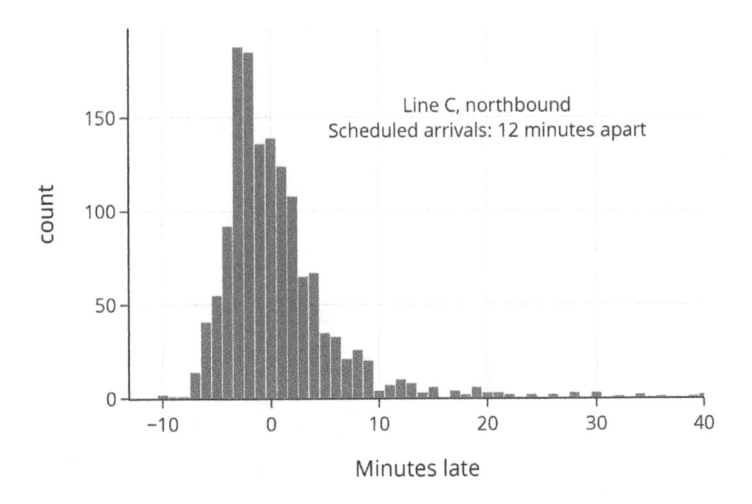

And let's calculate the minimum, maximum, and median lateness:

```
smallest amount late:  -10.20 minutes
greatest amount late:  57.00 minutes
median amount late:    -0.50 minutes
```

Interestingly, the northbound buses on the C line that are 12 minutes apart are more often early than not!

Now let's revisit our question to confirm that we are on track for answering it. A summary of how late the buses are does not quite address the experience of the person waiting for the bus. When someone arrives at a bus stop, they need to wait for the next bus to arrive. Figure 5-1 shows an idealization of time passing as passengers and buses arrive at the bus stop. If people are arriving at the bus stop at random times, notice that they are more likely to arrive in a time interval where the bus is delayed because there's a longer interval between buses. This arrival pattern is an example of size-biased sampling. So to answer the question of what people experience when waiting for a bus, we need to do more than summarize how late the bus is.

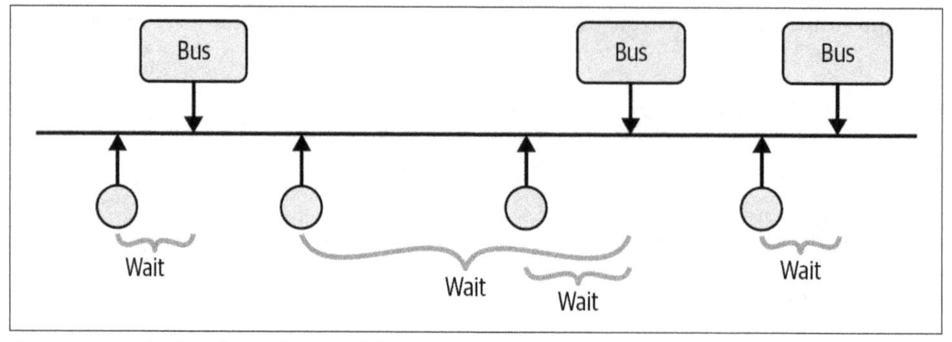

Figure 5-1. Idealized timeline with buses arriving (rectangles), passengers arriving (circles), and time the rider waits for the next bus to arrive (curly brackets)

We can design a simulation that mimics waiting for a bus over the course of one day, using the ideas from Chapter 3. To do this, we set up a string of scheduled bus arrivals that are 12 minutes apart from 6 a.m. to midnight:

```
scheduled = 12 * np.arange(91)
scheduled
```

```
array([   0,   12,   24, ..., 1056, 1068, 1080])
```

Then, for each scheduled arrival, we simulate its actual arrival time by adding a random number of minutes each bus is late. To do this, we choose the minutes late from the distribution of observed lateness of the actual buses. Notice how we have incorporated the real data in our simulation study by using the distribution of actual delays of the buses that are 12 minutes apart:

```
minutes_late = bus_c_n_12['minutes_late']
actual = scheduled + np.random.choice(minutes_late, size=91, replace=True)
```

We need to sort these arrival times because when a bus is super late, another may well come along before it:

```
actual.sort()
actual
```

```
array([  -1.2 ,   25.37,   32.2 , ..., 1051.02, 1077.  , 1089.43])
```

We also need to simulate the arrival of people at the bus stop at random times throughout the day. We can use another, different urn model for the passenger arrivals. For the passengers, we put a marble in the urn with a time on it. These run from time 0, which stands for 6 a.m., to the arrival of the last bus at midnight, which is 1,068 minutes past 6 a.m. To match the way the bus times are measured in our data, we make the times 1/100th of a minute apart:

```
pass_arrival_times = np.arange(100*1068)
pass_arrival_times / 100
```

```
array([   0.  ,    0.01,    0.02, ..., 1067.97, 1067.98, 1067.99])
```

Now we can simulate the arrival of, say, five hundred people at the bus stop throughout the day. We draw five hundred times from this urn, replacing the marbles between draws:

```
sim_arrival_times = (
    np.random.choice(pass_arrival_times, size=500, replace=True) / 100
)
sim_arrival_times.sort()
sim_arrival_times
```

```
array([   2.06,    3.01,    8.54, ..., 1064.  , 1064.77, 1066.42])
```

To find out how long each individual waits, we look for the soonest bus to arrive after their sampled time. The difference between these two times (the sampled time of the person and the soonest bus arrival after that) is how long the person waits:

```
i = np.searchsorted(actual, sim_arrival_times, side='right')
sim_wait_times = actual[i] - sim_arrival_times
sim_wait_times
```

```
array([23.31, 22.36, 16.83, ..., 13.  , 12.23, 10.58])
```

We can set up a complete simulation where we simulate, say, two hundred days of bus arrivals, and for each day, we simulate five hundred people arriving at the bus stop at random times throughout the day. In total, that's 100,000 simulated wait times:

```
sim_wait_times = []

for day in np.arange(0, 200, 1):
    bus_late = np.random.choice(minutes_late, size=91, replace=True)
    actual = scheduled + bus_late
    actual.sort()
    sim_arrival_times = (
        np.random.choice(pass_arrival_times, size=500, replace=True) / 100
    )
    sim_arrival_times.sort()
    i = np.searchsorted(actual, sim_arrival_times, side="right")
    sim_wait_times = np.append(sim_wait_times, actual[i] - sim_arrival_times)
```

Let's make a histogram of these simulated wait times to examine the distribution:

```
fig = px.histogram(x=sim_wait_times, nbins=40,
                   histnorm='probability density',
                   width=450, height=300)

fig.update_xaxes(title="Simulated wait times for 100,000 passengers")
fig.update_yaxes(title="proportion")
fig.show()
```

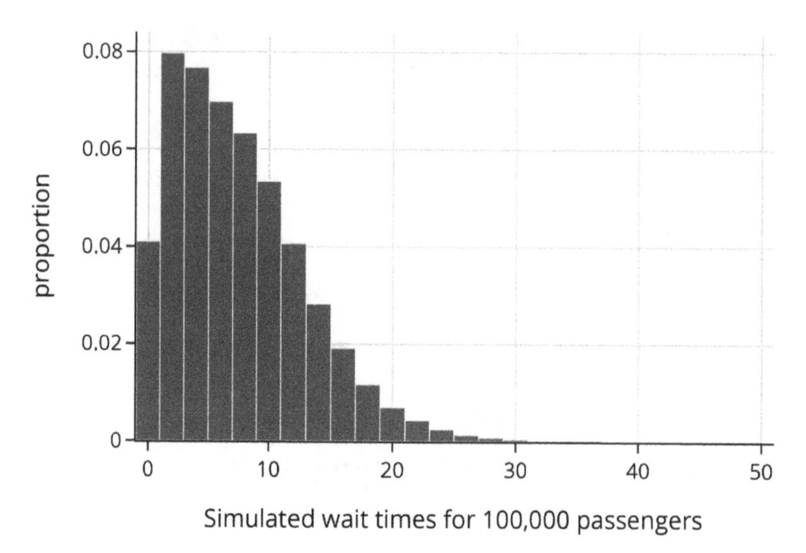

As we expect, we find a skewed distribution. We can model this with a constant where we use absolute loss to select the best constant. We saw in Chapter 4 that absolute loss gives us the median wait time:

```
print(f"Median wait time: {np.median(sim_wait_times):.2f} minutes")
```

```
Median wait time: 6.49 minutes
```

The median of about six and a half minutes doesn't seem too long. While our model captures the typical wait time, we also want to provide an estimate of the variability in the process. This topic is covered in Chapter 17. We can compute the upper quartile of wait times to give us a sense of variability:

```
print(f"Upper quartile: {np.quantile(sim_wait_times, 0.75):.2f} minutes")
```

```
Upper quartile: 10.62 minutes
```

The upper quartile is quite large. It's undoubtedly memorable when you have to wait more than 10 minutes for a bus that is supposed to arrive every 12 minutes, and this happens one in four times you take the bus!

Summary

In our first case study, we have traversed the full lifecycle of data modeling. It might strike you that such a simple question is not immediately answerable with the data collected. We needed to combine the data of scheduled and actual arrival times of buses with a simulation study of riders arriving at the bus stop at random times to uncover the riders' waiting experience.

This simulation simplified many of the real patterns in bus riding. We focused on one bus line traveling in one direction with buses arriving at 12-minute intervals. Further, the exploration of the data revealed that the patterns in lateness correlated with the time of day, which we have not accounted for in our analysis. Nonetheless, our findings can still be useful. For example, they confirm that the typical wait time is longer than half the scheduled interval. And the distribution of wait times has a long right tail, meaning a rider's experience may well be impacted by the variability in the process.

We also saw how deriving new quantities, such as how late a bus is and the time between buses, and exploring the data can be useful in modeling. Our histograms showed that the particular line and direction of the bus matter and they need to be accounted for. We also discovered that the schedules change throughout the day, with many buses arriving 10, 12, and 15 minutes after another, and some arriving more frequently or more separated. This observation further informed the modeling stage.

Finally, we used data tools, such as the `pandas` and `plotly` libraries, that will be covered in later chapters. Our focus here was not on how to manipulate tables or how to create a plot. Instead, we focused on the lifecycle, connecting questions to data to modeling to conclusions. In the next chapter, we turn to the practicalities of working with data tables.

PART II

Rectangular Data

CHAPTER 6

Working with Dataframes Using pandas

Data scientists work with data stored in tables. This chapter introduces *dataframes*, one of the most widely used ways to represent data tables. We also introduce pandas, the standard Python package for working with dataframes. Here is an example of a dataframe that holds information about popular dog breeds:

breed	grooming	food_cost	kids	size
Labrador Retriever	weekly	466.0	high	medium
German Shepherd	weekly	466.0	medium	large
Beagle	daily	324.0	high	small
Golden Retriever	weekly	466.0	high	medium
Yorkshire Terrier	daily	324.0	low	small
Bulldog	weekly	466.0	medium	medium
Boxer	weekly	466.0	high	medium

In a dataframe, each row represents a single record—in this case, a single dog breed. Each column represents a feature about the record—for example, the grooming column represents how often each dog breed needs to be groomed.

Dataframes have labels for both columns and rows. For instance, this dataframe has a column labeled grooming and a row labeled German Shepherd. The columns and rows of a dataframe are ordered—we can refer to the Labrador Retriever row as the first row of the dataframe.

Within a column, data have the same type. For instance, the cost of food contains numbers, and the size of the dog consists of categories. But data types can be different within a row.

Because of these properties, dataframes enable all sorts of useful operations.

Data scientists often find themselves working with people from different backgrounds who use different terms. For instance, computer scientists say that the columns of a dataframe represent *features* of the data, while statisticians call them *variables* instead.

Other times, people use the same term to refer to slightly different ideas. *Data types* in a programming sense refers to how a computer stores data internally. For instance, the `size` column has a string data type in Python. But from a statistical point of view, the type of the `size` column is ordered categorical data (ordinal data). We talk more about this specific distinction in Chapter 10.

In this chapter, we introduce common dataframe operations. Data scientists use the `pandas` library when working with dataframes in Python. First, we explain the main objects that `pandas` provides: the `DataFrame` and `Series` classes. Then we show how to use `pandas` to perform common data manipulation tasks, like slicing, filtering, sorting, grouping, and joining.

Subsetting

This section introduces operations for taking subsets of dataframes. When data scientists first read in a dataframe, they often want to subset the specific data that they plan to use. For example, a data scientist can *slice* out the 10 relevant features from a dataframe with hundreds of columns. Or they can *filter* a dataframe to remove rows with incomplete data. For the rest of this chapter, we demonstrate dataframe operations using a dataframe of baby names.

Data Scope and Question

There's a 2021 *New York Times* article (*https://oreil.ly/qL1dt*) that talks about Prince Harry and Meghan Markle's unique choice for their new baby daughter's name: Lilibet. The article has an interview with Pamela Redmond, an expert on baby names, who talks about interesting trends in how people name their kids. For example, she says that names that start with the letter L have become very popular in recent years, while names that start with the letter J were most popular in the 1970s and 1980s. Are these claims reflected in data? We can use `pandas` to find out.

First, we import the package as `pd`, the canonical abbreviation:

```
import pandas as pd
```

We have a dataset of baby names stored in a comma-separated values (CSV) file called *babynames.csv*. We use the `pd.read_csv` function to read the file as a `pandas.DataFrame` object:

```
baby = pd.read_csv('babynames.csv')
baby
```

	Name	Sex	Count	Year
0	Liam	M	19659	2020
1	Noah	M	18252	2020
2	Oliver	M	14147	2020
...	...	...	...	...
2020719	Verona	F	5	1880
2020720	Vertie	F	5	1880
2020721	Wilma	F	5	1880

```
2020722 rows × 4 columns
```

The data in the baby table comes from the US Social Security Administration (SSA) (*https://oreil.ly/EhTlP*), which records the baby name and birth sex for birth certificate purposes. The SSA makes the baby names data available on its website. We've loaded this data into the baby table.

The SSA website has a page (*https://oreil.ly/jzCVF*) that describes the data in more detail. We won't go in depth in this chapter about the data's limitations, but we'll point out this relevant information from the website:

> All names are from Social Security card applications for births that occurred in the United States after 1879. Note that many people born before 1937 never applied for a Social Security card, so their names are not included in our data. For others who did apply, our records may not show the place of birth, and again their names are not included in our data.
>
> All data are from a 100% sample of our records on Social Security card applications as of March 2021.

It's also important to point out that at the time of this writing, the SSA dataset only provides the binary options of male and female. We hope that in the future, national datasets like this one will provide more inclusive options.

Dataframes and Indices

Let's examine the baby dataframe in more detail. A dataframe has rows and columns. Every row and column has a label, as highlighted in Figure 6-1.

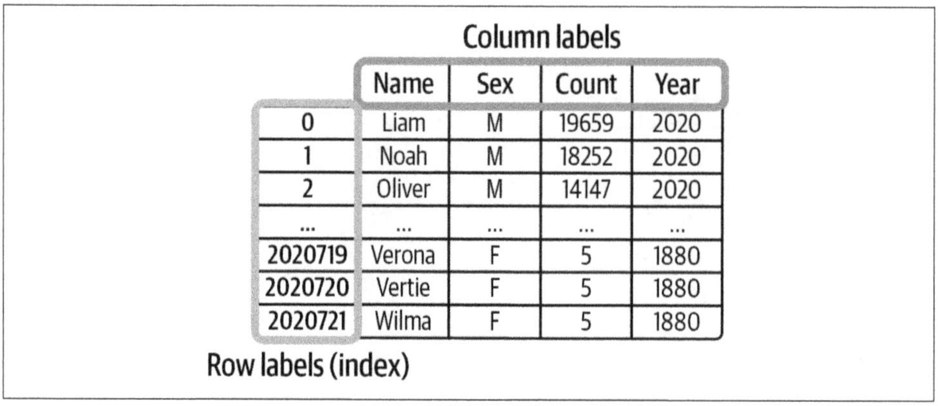

	Name	Sex	Count	Year
0	Liam	M	19659	2020
1	Noah	M	18252	2020
2	Oliver	M	14147	2020
...	...	...	...	...
2020719	Verona	F	5	1880
2020720	Vertie	F	5	1880
2020721	Wilma	F	5	1880

Column labels

Row labels (index)

Figure 6-1. The baby dataframe has labels for both rows and columns (boxed)

By default, `pandas` assigns row labels as incrementing numbers starting from 0. In this case, the data at the row labeled `0` and column labeled `Name` has the data `'Liam'`.

Dataframes can also have strings as row labels. Figure 6-2 shows a dataframe of dog data where the row labels are strings.

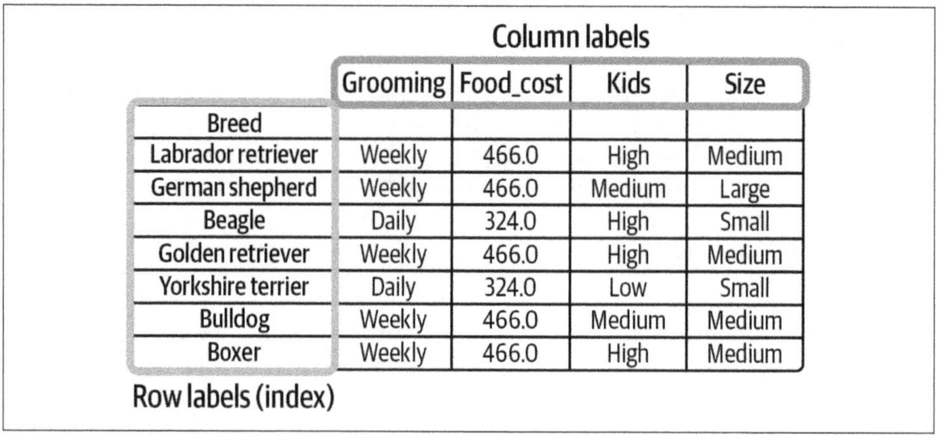

	Grooming	Food_cost	Kids	Size
Breed				
Labrador retriever	Weekly	466.0	High	Medium
German shepherd	Weekly	466.0	Medium	Large
Beagle	Daily	324.0	High	Small
Golden retriever	Weekly	466.0	High	Medium
Yorkshire terrier	Daily	324.0	Low	Small
Bulldog	Weekly	466.0	Medium	Medium
Boxer	Weekly	466.0	High	Medium

Column labels

Row labels (index)

Figure 6-2. Row labels in dataframes can also be strings, as in this example, in which each row is labeled using the dog breed name

The row labels have a special name. We call them the *index* of a dataframe, and `pandas` stores the row labels in a special `pd.Index` object. We won't discuss the `pd.Index` object since it's less common to manipulate the index itself. For now, it's important to remember that even though the index looks like a column of data, the index really represents row labels, not data. For instance, the dataframe of dog breeds has four columns of data, not five, since the index doesn't count as a column.

Slicing

Slicing is an operation that creates a new dataframe by taking a subset of rows or columns out of another dataframe. Think about slicing a tomato—slices can go both vertically and horizontally. To take slices of a dataframe in `pandas`, we use the `.loc` and `.iloc` properties. Let's start with `.loc`.

Here's the full `baby` dataframe:

```
baby
```

	Name	Sex	Count	Year
0	Liam	M	19659	2020
1	Noah	M	18252	2020
2	Oliver	M	14147	2020
...	...	...	...	...
2020719	Verona	F	5	1880
2020720	Vertie	F	5	1880
2020721	Wilma	F	5	1880

```
2020722 rows × 4 columns
```

`.loc` lets us select rows and columns using their labels. For example, to get the data in the row labeled `1` and column labeled `Name`:

```
#       The first argument is the row label
#       ↓
baby.loc[1, 'Name']
#           ↑
#           The second argument is the column label
```

```
'Noah'
```

Notice that `.loc` needs square brackets; running `baby.loc(1, 'Name')` will result in an error.

To slice out multiple rows or columns, we can use Python slice syntax instead of individual values:

```
baby.loc[0:3, 'Name':'Count']
```

	Name	Sex	Count
0	Liam	M	19659
1	Noah	M	18252

	Name	Sex	Count
2	Oliver	M	14147
3	Elijah	M	13034

To get an entire column of data, we can pass an empty slice as the first argument:

```
baby.loc[:, 'Count']
```

```
0          19659
1          18252
2          14147
           ...
2020719        5
2020720        5
2020721        5
Name: Count, Length: 2020722, dtype: int64
```

Notice that the output of this doesn't look like a dataframe, and it's not. Selecting out a single row or column of a dataframe produces a `pd.Series` object:

```
counts = baby.loc[:, 'Count']
counts.__class__.__name__
```

```
'Series'
```

What's the difference between a `pd.Series` object and a `pd.DataFrame` object? Essentially, a `pd.DataFrame` is two-dimensional—it has rows and columns and represents a table of data. A `pd.Series` is one-dimensional—it represents a list of data. `pd.Series` and `pd.DataFrame` objects have many methods in common, but they really represent two different things. Confusing the two can cause bugs and confusion.

To select specific columns of a dataframe, pass a list into `.loc`. Here's the original dataframe:

```
baby
```

	Name	Sex	Count	Year
0	Liam	M	19659	2020
1	Noah	M	18252	2020
2	Oliver	M	14147	2020
...	...	...	...	...
2020719	Verona	F	5	1880
2020720	Vertie	F	5	1880
2020721	Wilma	F	5	1880

```
2020722 rows × 4 columns
```

```
# And here's the dataframe with only Name and Year columns
baby.loc[:, ['Name', 'Year']]
```

```
#          └──────────┬──────────┘
#              list of column labels
```

	Name	Year
0	Liam	2020
1	Noah	2020
2	Oliver	2020
...	...	...
2020719	Verona	1880
2020720	Vertie	1880
2020721	Wilma	1880

```
2020722 rows × 2 columns
```

Selecting columns is very common, so there's a shorthand:

```
# Shorthand for baby.loc[:, 'Name']
baby['Name']
```

```
0            Liam
1            Noah
2          Oliver
            ...
2020719    Verona
2020720    Vertie
2020721     Wilma
Name: Name, Length: 2020722, dtype: object
```

```
# Shorthand for baby.loc[:, ['Name', 'Count']]
baby[['Name', 'Count']]
```

	Name	Count
0	Liam	19659
1	Noah	18252
2	Oliver	14147
...	...	...
2020719	Verona	5
2020720	Vertie	5
2020721	Wilma	5

```
2020722 rows × 2 columns
```

Slicing using `.iloc` works similarly to `.loc`, except that `.iloc` uses the *positions* of rows and columns rather than labels. It's easiest to show the difference between `.iloc` and `.loc` when the dataframe index has strings, so for demonstration purposes, let's look at a dataframe with information on dog breeds:

```
dogs = pd.read_csv('dogs.csv', index_col='breed')
dogs
```

breed	grooming	food_cost	kids	size
Labrador Retriever	weekly	466.0	high	medium
German Shepherd	weekly	466.0	medium	large
Beagle	daily	324.0	high	small
Golden Retriever	weekly	466.0	high	medium
Yorkshire Terrier	daily	324.0	low	small
Bulldog	weekly	466.0	medium	medium
Boxer	weekly	466.0	high	medium

To get the first three rows and the first two columns by position, use `.iloc`:

```
dogs.iloc[0:3, 0:2]
```

breed	grooming	food_cost
Labrador Retriever	weekly	466.0
German Shepherd	weekly	466.0
Beagle	daily	324.0

The same operation using `.loc` requires us to use the dataframe labels:

```
dogs.loc['Labrador Retriever':'Beagle', 'grooming':'food_cost']
```

breed	grooming	food_cost
Labrador Retriever	weekly	466.0
German Shepherd	weekly	466.0
Beagle	daily	324.0

Next, we'll look at filtering rows.

Filtering Rows

So far, we've shown how to use `.loc` and `.iloc` to slice a dataframe using labels and positions.

However, data scientists often want to *filter* rows—they want to take subsets of rows using some criteria. Let's say we want to find the most popular baby names in 2020. To do this, we can filter rows to keep only the rows where the `Year` is 2020.

To filter, we'd like to check whether each value in the `Year` column is equal to 1970 and then keep only those rows.

To compare each value in `Year`, we slice out the column and make a boolean comparison (this is similar to what we'd do with a `numpy` array). Here's the dataframe for reference:

```
baby
```

	Name	Sex	Count	Year
0	Liam	M	19659	2020
1	Noah	M	18252	2020
2	Oliver	M	14147	2020
...	...	...	...	...
2020719	Verona	F	5	1880
2020720	Vertie	F	5	1880
2020721	Wilma	F	5	1880

```
2020722 rows × 4 columns
```

```
# Get a Series with the Year data
baby['Year']
```

```
0          2020
1          2020
2          2020
           ...
2020719    1880
2020720    1880
2020721    1880
Name: Year, Length: 2020722, dtype: int64
```

```
# Compare with 2020
baby['Year'] == 2020
```

```
0           True
1           True
2           True
           ...
2020719    False
2020720    False
2020721    False
Name: Year, Length: 2020722, dtype: bool
```

Notice that a boolean comparison on a `Series` gives a `Series` of booleans. This is nearly equivalent to writing:

```
is_2020 = []
for value in baby['Year']:
    is_2020.append(value == 2020)
```

But the boolean comparison is easier to write and much faster to execute than a for loop.

Now we tell pandas to keep only the rows where the comparison evaluated to True:

```
baby.loc[baby['Year'] == 2020, :]
```

	Name	Sex	Count	Year
0	Liam	M	19659	2020
1	Noah	M	18252	2020
2	Oliver	M	14147	2020
...	...	...	...	...
31267	Zylynn	F	5	2020
31268	Zynique	F	5	2020
31269	Zynlee	F	5	2020

```
31270 rows × 4 columns
```

Passing a Series of booleans into .loc only keeps rows where the Series has a True value.

Filtering has a shorthand. This computes the same table as the preceding snippet without using .loc:

```
baby[baby['Year'] == 2020]
```

	Name	Sex	Count	Year
0	Liam	M	19659	2020
1	Noah	M	18252	2020
2	Oliver	M	14147	2020
...	...	...	...	...
31267	Zylynn	F	5	2020
31268	Zynique	F	5	2020
31269	Zynlee	F	5	2020

```
31270 rows × 4 columns
```

Finally, to find the most common names in 2020, sort the dataframe by Count in descending order. Wrapping a long expression in parentheses lets us easily add line breaks to make it more readable:

```
(baby[baby['Year'] == 2020]
 .sort_values('Count', ascending=False)
```

```
    .head(7) # take the first seven rows
)
```

	Name	Sex	Count	Year
0	Liam	M	19659	2020
1	Noah	M	18252	2020
13911	Emma	F	15581	2020
2	Oliver	M	14147	2020
13912	Ava	F	13084	2020
3	Elijah	M	13034	2020
13913	Charlotte	F	13003	2020

We see that Liam, Noah, and Emma were the most popular baby names in 2020.

Example: How Recently Has Luna Become a Popular Name?

The *New York Times* article mentions that the name Luna was almost nonexistent before 2000 but has since grown to become a very popular name for girls. When exactly did Luna become popular? We can check this using slicing and filtering. When approaching a data manipulation task, we recommend breaking the problem down into smaller steps. For example, we could think:

1. Filter: keep only rows with `'Luna'` in the `Name` column.
2. Filter: keep only rows with `'F'` in the `Sex` column.
3. Slice: keep the `Count` and `Year` columns.

Now it's a matter of translating each step into code:

```
luna = baby[baby['Name'] == 'Luna'] # [1]
luna = luna[luna['Sex'] == 'F']     # [2]
luna = luna[['Count', 'Year']]      # [3]
luna
```

	Count	Year
13923	7770	2020
45366	7772	2019
77393	6929	2018
...	...	...
2014083	17	1883
2018187	18	1881
2020223	15	1880

```
128 rows × 2 columns
```

In this book, we use a library called `plotly` for plotting. We won't cover plotting in depth here since we talk more about it in Chapter 11. For now, we use `px.line()` to make a simple line plot:

```
px.line(luna, x='Year', y='Count', width=350, height=250)
```

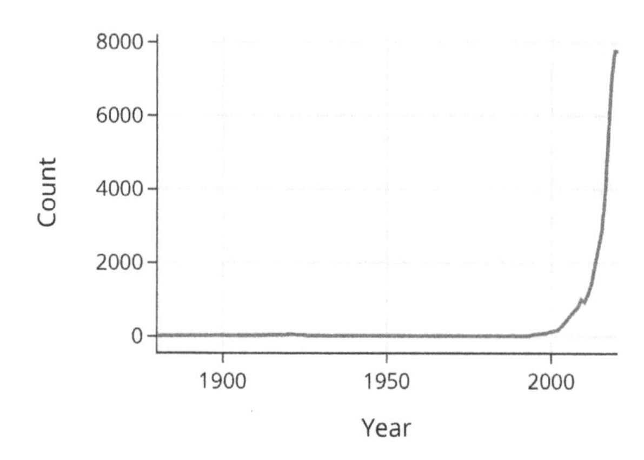

It's just as the article says. Luna wasn't popular at all until the year 2000 or so. In other words, if someone tells you that their name is Luna, you can take a pretty good guess at their age even without any other information about them!

Just for fun, here's the same plot for the name Siri:

```
siri = (baby.query('Name == "Siri"')
        .query('Sex == "F"'))
px.line(siri, x='Year', y='Count', width=350, height=250)
```

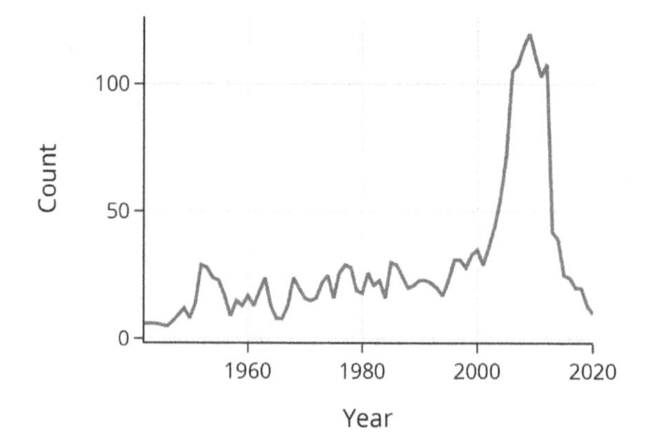

Using `.query` is similar to using `.loc` with a boolean series. `query()` has more restrictions on filtering but can be convenient as a shorthand.

Why might the popularity have dropped so suddenly after 2010? Well, Siri happens to be the name of Apple's voice assistant and was introduced in 2011. Let's draw a line for the year 2011 and take a look:

```
fig = px.line(siri, x="Year", y="Count", width=350, height=250)
fig.add_vline(
    x=2011, line_color="red", line_dash="dashdot", line_width=4, opacity=0.7
)
```

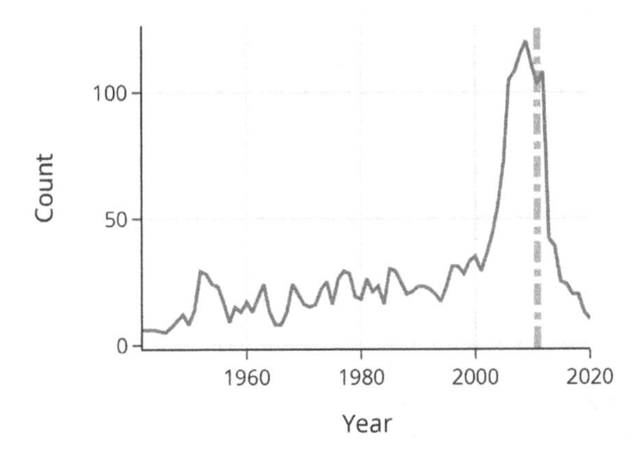

It looks like parents don't want their kids to be confused when other people say "Hey Siri" to their phones.

In this section, we introduced dataframes in `pandas`. We covered the common ways that data scientists subset dataframes—slicing with labels and filtering using a boolean condition. In the next section, we explain how to aggregate rows together.

Aggregating

This section introduces operations for aggregating rows in a dataframe. Data scientists aggregate rows together to make summaries of data. For instance, a dataset containing daily sales can be aggregated to show monthly sales instead. This section introduces *grouping* and *pivoting*, two common operations for aggregating data.

We work with the baby names data, as introduced in the previous section:

```
baby = pd.read_csv('babynames.csv')
baby
```

	Name	Sex	Count	Year
0	Liam	M	19659	2020
1	Noah	M	18252	2020
2	Oliver	M	14147	2020
...	...	...	...	...
2020719	Verona	F	5	1880
2020720	Vertie	F	5	1880
2020721	Wilma	F	5	1880

```
2020722 rows × 4 columns
```

Basic Group-Aggregate

Let's say we want to find out the total number of babies born as recorded in this data. This is simply the sum of the `Count` column:

```
baby['Count'].sum()
```

```
352554503
```

Summing up the name counts is one simple way to aggregate the data—it combines data from multiple rows.

But let's say we instead want to answer a more interesting question: are US births trending upward over time? To answer this question, we can sum the `Count` column within each year rather than taking the sum over the entire dataset. In other words, we split the data into groups based on `Year`, then sum the `Count` values within each group. This process is depicted in Figure 6-3.

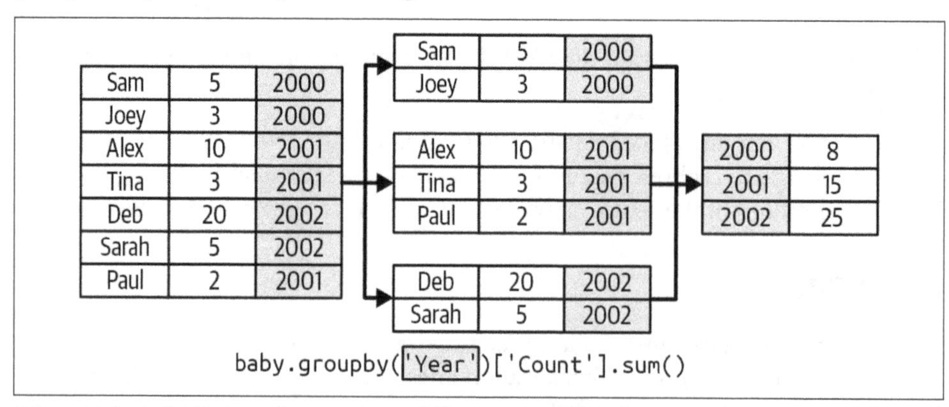

Figure 6-3. A depiction of grouping and then aggregating for example data

We call this operation *grouping* followed by *aggregating*. In pandas, we write:

```
baby.groupby('Year')['Count'].sum()
```

```
Year
1880     194419
1881     185772
1882     213385
          ...
2018    3487193
2019    3437438
2020    3287724
Name: Count, Length: 141, dtype: int64
```

Notice that the code is nearly the same as the nongrouped version, except that it starts with a call to `.groupby('Year')`.

The result is a `pd.Series` with the total number of babies born for each year in the data. Notice that the index of this series contains the unique `Year` values. Now we can plot the counts over time:

```
counts_by_year = baby.groupby('Year')['Count'].sum().reset_index()
px.line(counts_by_year, x='Year', y='Count', width=350, height=250)
```

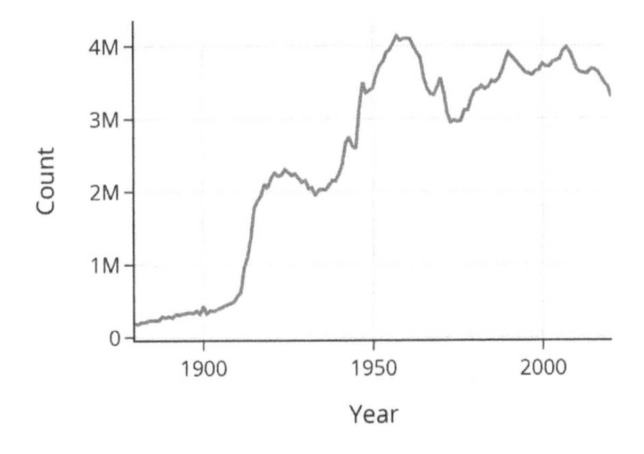

What do we see in this plot? First, we notice that there seem to be suspiciously few babies born before 1920. One likely explanation is that the SSA was created in 1935, so its data for prior births could be less complete.

We also notice the dip when World War II began in 1939, and the postwar baby boomer era from 1946 to 1964.

Here's the basic recipe for grouping in pandas:

```
(baby                # the dataframe
 .groupby('Year')    # column(s) to group
 ['Count']           # column(s) to aggregate
```

```
 .sum()              # how to aggregate
)
```

Example: Using .value_counts()

One of the more common dataframe tasks is to count the number of times every unique item in a column appears. For example, we might be interested in the number of times each name appears in the following `classroom` dataframe:

```
classroom
```

	name
0	Eden
1	Sachit
2	Eden
3	Sachit
4	Sachit
5	Luke

One way to do this is to use our grouping recipe with the `.size()` aggregation function:

```
(classroom
 .groupby('name')
 ['name']
 .size()
)
```

```
name
Eden      2
Luke      1
Sachit    3
Name: name, dtype: int64
```

This operation is so common that `pandas` provides a shorthand—the `.value_counts()` method for `pd.Series` objects:

```
classroom['name'].value_counts()
```

```
name
Sachit    3
Eden      2
Luke      1
Name: count, dtype: int64
```

By default, the `.value_counts()` method will sort the resulting series from highest to lowest number, making it convenient to see the most and least common values. We point out this method because we use it often in other chapters of the book.

Grouping on Multiple Columns

We pass multiple columns into `.groupby` as a list to group by multiple columns at once. This is useful when we need to further subdivide our groups. For example, we can group by both year and sex to see how many male and female babies were born over time:

```
counts_by_year_and_sex = (baby
 .groupby(['Year', 'Sex']) # Arg to groupby is a list of column names
 ['Count']
 .sum()
)
counts_by_year_and_sex
```

```
Year  Sex
1880  F        83929
      M       110490
1881  F        85034
               ...
2019  M      1785527
2020  F      1581301
      M      1706423
Name: Count, Length: 282, dtype: int64
```

Notice how the code closely follows the grouping recipe.

The `counts_by_year_and_sex` series has what we call a multilevel index with two levels, one for each column that was grouped. It's a bit easier to see if we convert the series to a dataframe. The result only has one column:

```
counts_by_year_and_sex.to_frame()
```

Year	Sex	Count
1880	F	83929
	M	110490
1881	F	85034
...	...	...
2019	M	1785527
2020	F	1581301
	M	1706423

```
282 rows × 1 columns
```

There are two levels to the index because we grouped by two columns. It can be a bit tricky to work with multilevel indices, so we can reset the index to go back to a dataframe with a single index:

```
counts_by_year_and_sex.reset_index()
```

	Year	Sex	Count
0	1880	F	83929
1	1880	M	110490
2	1881	F	85034
...	...	...	...
279	2019	M	1785527
280	2020	F	1581301
281	2020	M	1706423

```
282 rows × 3 columns
```

Custom Aggregation Functions

After grouping, pandas gives us flexible ways to aggregate the data. So far, we've seen how to use .sum() after grouping:

```
(baby
 .groupby('Year')
 ['Count']
 .sum() # aggregate by summing
)
```

```
Year
1880     194419
1881     185772
1882     213385
          ...
2018    3487193
2019    3437438
2020    3287724
Name: Count, Length: 141, dtype: int64
```

pandas also supplies other aggregation functions, like .mean(), .size(), and .first(). Here's the same grouping using .max():

```
(baby
 .groupby('Year')
 ['Count']
 .max() # aggregate by taking the max within each group
)
```

```
Year
1880     9655
1881     8769
1882     9557
         ...
2018    19924
2019    20555
2020    19659
Name: Count, Length: 141, dtype: int64
```

But sometimes pandas doesn't have the exact aggregation function we want to use. In these cases, we can define and use a custom aggregation function. pandas lets us do this through .agg(fn), where fn is a function that we define.

For instance, if we want to find the difference between the largest and smallest values within each group (the range of the data), we could first define a function called data_range, then pass that function into .agg(). The input to this function is a pd.Series object containing a single column of data. It gets called once for each group:

```
def data_range(counts):
    return counts.max() - counts.min()

(baby
 .groupby('Year')
 ['Count']
 .agg(data_range) # aggregate using custom function
)
```

```
Year
1880     9650
1881     8764
1882     9552
         ...
2018    19919
2019    20550
2020    19654
Name: Count, Length: 141, dtype: int64
```

We start by defining a count_unique function that counts the number of unique values in a series. Then we pass that function into .agg(). Since this function is short, we could use a lambda instead:

```
def count_unique(s):
    return len(s.unique())

unique_names_by_year = (baby
 .groupby('Year')
 ['Name']
 .agg(count_unique) # aggregate using the custom count_unique function
)
unique_names_by_year
```

```
Year
1880     1889
1881     1829
1882     2012
         ...
2018    29619
2019    29417
2020    28613
Name: Name, Length: 141, dtype: int64
```

```
px.line(unique_names_by_year.reset_index(),
        x='Year', y='Name',
        labels={'Name': '# unique names'},
        width=350, height=250)
```

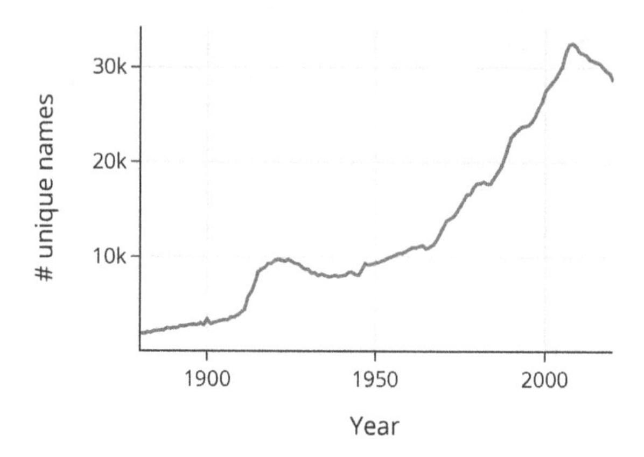

We see that the number of unique names has generally increased over time, even though the number of babies born annually has plateaued since the 1960s.

Pivoting

Pivoting is essentially a convenient way to arrange the results of a group and aggregation when grouping with two columns. Earlier in this section we grouped the baby names data by year and sex:

```
counts_by_year_and_sex = (baby
 .groupby(['Year', 'Sex'])
 ['Count']
 .sum()
)
counts_by_year_and_sex.to_frame()
```

		Count
Year	Sex	
1880	F	83929
	M	110490
1881	F	85034
...	...	...
2019	M	1785527
2020	F	1581301
	M	1706423

```
282 rows × 1 columns
```

This produces a `pd.Series` with the counts. We can also imagine the same data with the `Sex` index level "pivoted" to the columns of a dataframe. It's easier to see with an example:

```
mf_pivot = pd.pivot_table(
    baby,
    index='Year',    # Column to turn into new index
    columns='Sex',   # Column to turn into new columns
    values='Count',  # Column to aggregate for values
    aggfunc=sum)     # Aggregation function
mf_pivot
```

Sex	F	M
Year		
1880	83929	110490
1881	85034	100738
1882	99699	113686
...	...	...
2018	1676884	1810309
2019	1651911	1785527
2020	1581301	1706423

```
141 rows × 2 columns
```

As we can see in `mf_pivot` table, dataframe indexes can also be named. To read the output, it's important to notice that the dataframe has two columns, M and F, stored in an index named Sex. Likewise, the dataframe has 141 rows, each with their own label. These labels are stored in an index named Year. Here, Sex and Year are the names of the dataframe indexes, and are not row or column labels themselves.

Notice that the data values are identical in the pivot table and the table produced with `.groupby()`; the values are just arranged differently. Pivot tables are useful for quickly summarizing data using two attributes and are often seen in articles and papers.

The `px.line()` function also happens to work well with pivot tables, since the function draws one line for each column of data in the table:

```
fig = px.line(mf_pivot, width=350, height=250)
fig.update_traces(selector=1, line_dash='dashdot')
fig.update_yaxes(title='Value')
```

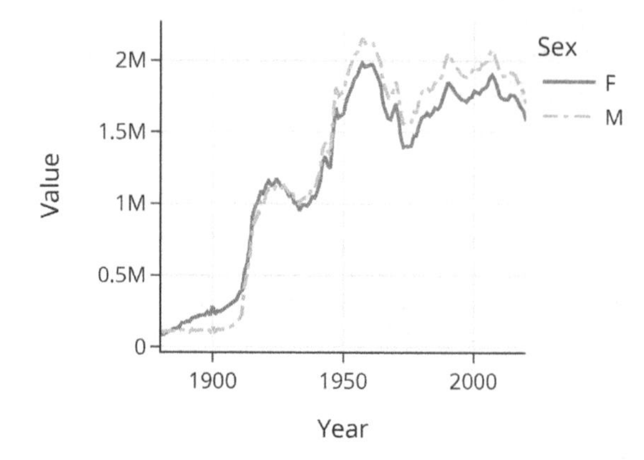

This section covered common ways to aggregate data in `pandas` using the `.groupby()` function with one or more columns, or using the `pd.pivot_table()` function. In the next section, we'll explain how to join dataframes together.

Joining

Data scientists very frequently want to *join* two or more dataframes together in order to connect data values across dataframes. For instance, an online bookstore might have one dataframe with the books each user has ordered and a second dataframe with the genres of each book. By joining the two dataframes together, the data scientist can see what genres each user prefers.

We'll continue looking at the baby names data. We'll use joins to check some trends mentioned in the *New York Times* article about baby names (*https://oreil.ly/qL1dt*). The article talks about how certain categories of names have become more or less popular over time. For instance, it mentions that mythological names like Julius and Cassius have become popular, while baby boomer names like Susan and Debbie have become less popular. How has the popularity of these categories changed over time?

We've taken the names and categories in the *NYT* article and put them in a small dataframe:

```
nyt = pd.read_csv('nyt_names.csv')
nyt
```

	nyt_name	category
0	Lucifer	forbidden
1	Lilith	forbidden
2	Danger	forbidden
...	...	...

	nyt_name	category
20	Venus	celestial
21	Celestia	celestial
22	Skye	celestial

```
23 rows × 2 columns
```

To see how popular the categories of names are, we join the `nyt` dataframe with the `baby` dataframe to get the name counts from `baby`:

```
baby = pd.read_csv('babynames.csv')
baby
```

	Name	Sex	Count	Year
0	Liam	M	19659	2020
1	Noah	M	18252	2020
2	Oliver	M	14147	2020
...	...	...	...	...
2020719	Verona	F	5	1880
2020720	Vertie	F	5	1880
2020721	Wilma	F	5	1880

```
2020722 rows × 4 columns
```

For intuition, we can imagine going down each row in `baby` and asking: is this name in the `nyt` table? If so, then add the value in the `category` column to the row. This is the basic idea behind a join. Let's look at a few examples on smaller dataframes first.

Inner Joins

We start by making smaller versions of the `baby` and `nyt` tables so that it's easier to see what happens when we join tables together:

```
nyt_small
```

	nyt_name	category
0	Karen	boomer
1	Julius	mythology
2	Freya	mythology

```
baby_small
```

	Name	Sex	Count	Year
0	Noah	M	18252	2020
1	Julius	M	960	2020
2	Karen	M	6	2020
3	Karen	F	325	2020
4	Noah	F	305	2020

To join tables in pandas, we'll use the `.merge()` method:

```
baby_small.merge(nyt_small,
                 left_on='Name',        # column in left table to match
                 right_on='nyt_name')   # column in right table to match
```

	Name	Sex	Count	Year	nyt_name	category
0	Julius	M	960	2020	Julius	mythology
1	Karen	M	6	2020	Karen	boomer
2	Karen	F	325	2020	Karen	boomer

Notice that the new table has the columns of both the `baby_small` and `nyt_small` tables. The rows with the name `Noah` are gone. And the remaining rows have their matching `category` from `nyt_small`.

Readers should also be aware that pandas has a `.join()` method for joining two dataframes together. However, the `.merge()` method has more flexibility for how the dataframes are joined, which is why we focus on `.merge()`. We encourage readers to consult the pandas documentation for the exact difference between the two.

When we join two tables together, we tell pandas the column(s) from each table that we want to use to make the join (the `left_on` and `right_on` arguments). pandas matches rows together when the values in the joining columns match, as shown in Figure 6-4.

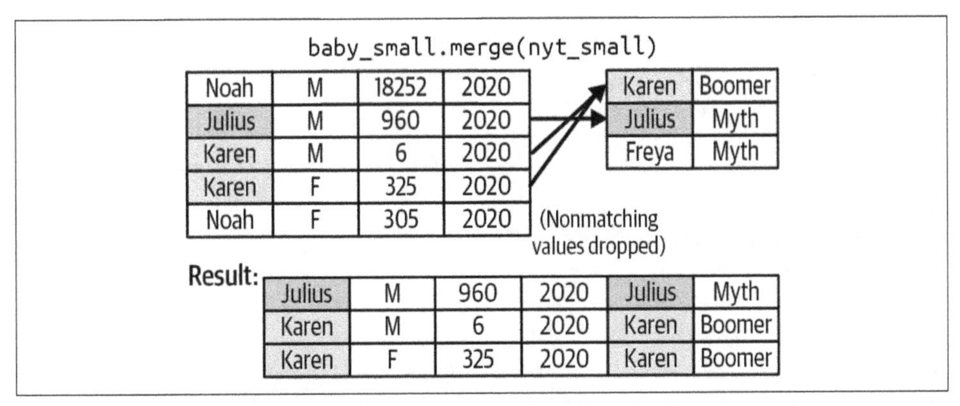

Figure 6-4. To join, `pandas` matches rows using the values in the `Name` and `nyt_name` columns, dropping rows that don't have matching values

By default, `pandas` does an *inner join*. If either table has rows that don't have matches in the other table, `pandas` drops those rows from the result. In this case, the `Noah` rows in `baby_small` don't have matches in `nyt_small`, so they are dropped. Also, the `Freya` row in `nyt_small` doesn't have matches in `baby_small`, so it's dropped as well. Only the rows with a match in both tables stay in the final result.

Left, Right, and Outer Joins

We sometimes want to keep rows without a match instead of dropping them entirely. There are other types of joins—left, right, and outer—that keep rows even when they don't have a match.

In a *left join*, rows in the left table without a match are kept in the final result, as shown in Figure 6-5.

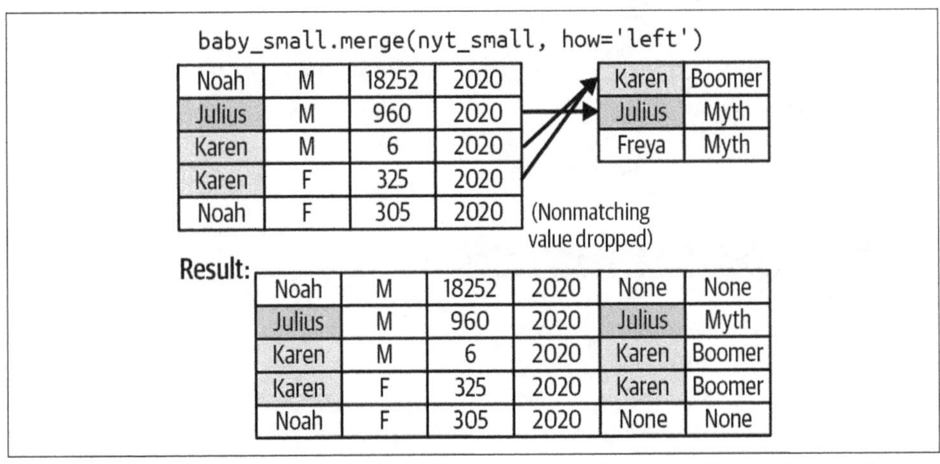

Figure 6-5. In a left join, rows in the left table that don't have matching values are kept

To do a left join in pandas, use how='left' in the call to .merge():

```
baby_small.merge(nyt_small,
                 left_on='Name',
                 right_on='nyt_name',
                 how='left')         # left join instead of inner
```

	Name	Sex	Count	Year	nyt_name	category
0	Noah	M	18252	2020	NaN	NaN
1	Julius	M	960	2020	Julius	mythology
2	Karen	M	6	2020	Karen	boomer
3	Karen	F	325	2020	Karen	boomer
4	Noah	F	305	2020	NaN	NaN

Notice that the Noah rows are kept in the final table. Since those rows didn't have a match in the nyt_small dataframe, the join leaves NaN values in the nyt_name and category columns. Also, notice that the Freya row in nyt_small is still dropped.

A *right join* works similarly to the left join, except that nonmatching rows in the right table are kept instead of the left table:

```
baby_small.merge(nyt_small,
                 left_on='Name',
                 right_on='nyt_name',
                 how='right')
```

	Name	Sex	Count	Year	nyt_name	category
0	Karen	M	6.0	2020.0	Karen	boomer
1	Karen	F	325.0	2020.0	Karen	boomer
2	Julius	M	960.0	2020.0	Julius	mythology
3	NaN	NaN	NaN	NaN	Freya	mythology

Finally, an *outer join* keeps rows from both tables even when they don't have a match:

```
baby_small.merge(nyt_small,
                 left_on='Name',
                 right_on='nyt_name',
                 how='outer')
```

	Name	Sex	Count	Year	nyt_name	category
0	Noah	M	18252.0	2020.0	NaN	NaN
1	Noah	F	305.0	2020.0	NaN	NaN
2	Julius	M	960.0	2020.0	Julius	mythology
3	Karen	M	6.0	2020.0	Karen	boomer
4	Karen	F	325.0	2020.0	Karen	boomer
5	NaN	NaN	NaN	NaN	Freya	mythology

Example: Popularity of NYT Name Categories

Now let's return to the full dataframes baby and nyt. .head() slices out the first few rows, which is convenient for saving space:

```
baby.head(2)
```

	Name	Sex	Count	Year
0	Liam	M	19659	2020
1	Noah	M	18252	2020

```
nyt.head(2)
```

	nyt_name	category
0	Lucifer	forbidden
1	Lilith	forbidden

We want to know how the popularity of name categories in nyt has changed over time. To answer this question:

1. Inner join baby with nyt.
2. Group the table by category and Year.
3. Aggregate the counts using a sum:

```
cate_counts = (
    baby.merge(nyt, left_on='Name', right_on='nyt_name') # [1]
    .groupby(['category', 'Year'])                       # [2]
    ['Count']                                            # [3]
    .sum()                                               # [3]
    .reset_index()
)
cate_counts
```

	category	Year	Count
0	boomer	1880	292
1	boomer	1881	298
2	boomer	1882	326
...	...	...	...
647	mythology	2018	2944
648	mythology	2019	3320
649	mythology	2020	3489

```
650 rows × 3 columns
```

Now we can plot the popularity of `boomer` names and `mythology` names:

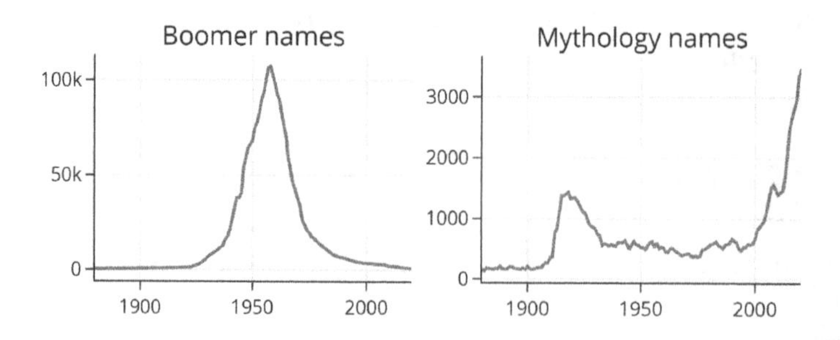

As the *NYT* article claims, baby boomer names have become less popular since 2000, while mythology names have become more popular.

We can also plot the popularities of all the categories at once. Take a look at the following plots and see whether they support the claims made in the *New York Times* article:

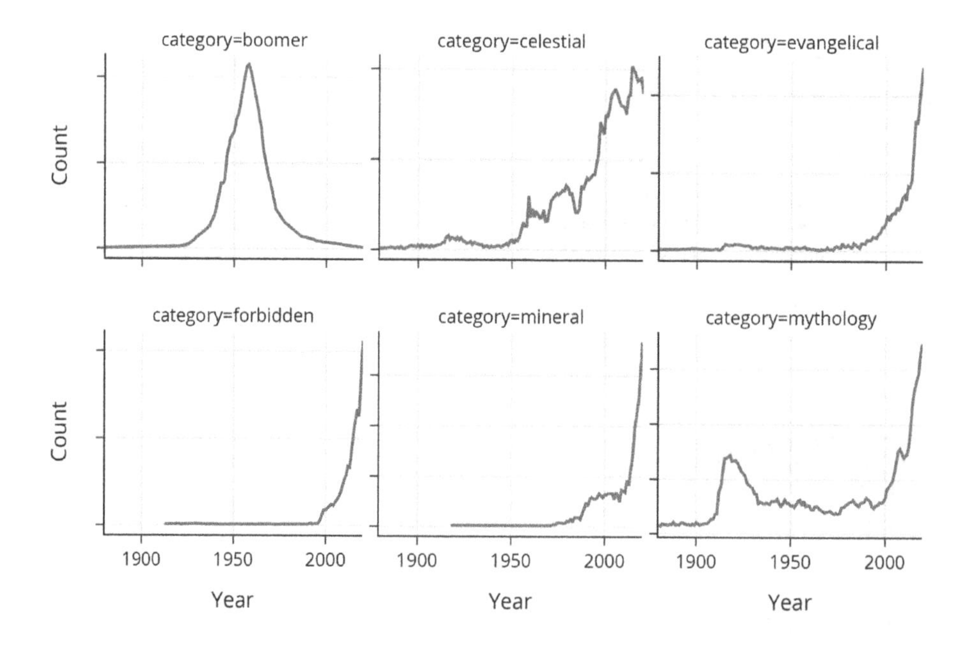

In this section, we introduced joins for dataframes. When joining dataframes together, we match rows using the `.merge()` function. It's important to consider the type of join (inner, left, right, or outer) when joining dataframes. In the next section, we'll explain how to transform values in a dataframe.

Transforming

Data scientists transform dataframe columns when they need to change each value in a feature in the same way. For example, if a feature contains heights of people in feet, a data scientist might want to transform the heights to centimeters. In this section, we'll introduce *apply*, an operation that transforms columns of data using a user-defined function:

```
baby = pd.read_csv('babynames.csv')
baby
```

	Name	Sex	Count	Year
0	Liam	M	19659	2020
1	Noah	M	18252	2020
2	Oliver	M	14147	2020
...	...	...	...	...
2020719	Verona	F	5	1880
2020720	Vertie	F	5	1880
2020721	Wilma	F	5	1880

```
2020722 rows × 4 columns
```

In the baby names *New York Times* article, Pamela mentions that names starting with the letter L or K became popular after 2000. On the other hand, names starting with the letter J peaked in popularity in the 1970s and 1980s and dropped off in popularity since. We can verify these claims using the `baby` dataset.

We approach this problem using the following steps:

1. Transform the `Name` column into a new column that contains the first letters of each value in `Name`.
2. Group the dataframe by the first letter and year.
3. Aggregate the name counts by summing.

To complete the first step, we'll *apply* a function to the `Name` column.

Apply

`pd.Series` objects contain an `.apply()` method that takes in a function and applies it to each value in the series. For instance, to find the lengths of each name, we apply the `len` function:

```
names = baby['Name']
names.apply(len)
```

```
0          4
1          4
2          6
          ..
2020719    6
2020720    6
2020721    5
Name: Name, Length: 2020722, dtype: int64
```

To extract the first letter of each name, we define a custom function and pass it into `.apply()`. The argument to the function is an individual value in the series:

```
def first_letter(string):
    return string[0]

names.apply(first_letter)
```

```
0          L
1          N
2          O
          ..
2020719    V
2020720    V
2020721    W
Name: Name, Length: 2020722, dtype: object
```

Using `.apply()` is similar to using a `for` loop. The preceding code is roughly equivalent to writing:

```
result = []
for name in names:
    result.append(first_letter(name))
```

Now we can assign the first letters to a new column in the dataframe:

```
letters = baby.assign(Firsts=names.apply(first_letter))
letters
```

	Name	Sex	Count	Year	Firsts
0	Liam	M	19659	2020	L
1	Noah	M	18252	2020	N
2	Oliver	M	14147	2020	O
...	...	...	...	...	...
2020719	Verona	F	5	1880	V
2020720	Vertie	F	5	1880	V
2020721	Wilma	F	5	1880	W

```
2020722 rows × 5 columns
```

To create a new column in a dataframe, you might also encounter this syntax:

```
baby['Firsts'] = names.apply(first_letter)
```

This mutates the baby table by adding a new column called `Firsts`. In the preceding code, we use `.assign()`, which doesn't mutate the baby table itself; it creates a new dataframe instead. Mutating dataframes isn't wrong but can be a common source of bugs. Because of this, we'll mostly use `.assign()` in this book.

Example: Popularity of "L" Names

Now we can use the `letters` dataframe to see the popularity of first letters over time:

```
letter_counts = (letters
 .groupby(['Firsts', 'Year'])
 ['Count']
 .sum()
 .reset_index()
)
letter_counts
```

	Firsts	Year	Count
0	A	1880	16740
1	A	1881	16257
2	A	1882	18790
...	...	...	...
3638	Z	2018	55996
3639	Z	2019	55293
3640	Z	2020	54011

```
3641 rows × 3 columns
```

```
fig = px.line(letter_counts.loc[letter_counts['Firsts'] == 'L'],
              x='Year', y='Count', title='Popularity of "L" names',
              width=350, height=250)
fig.update_layout(margin=dict(t=30))
```

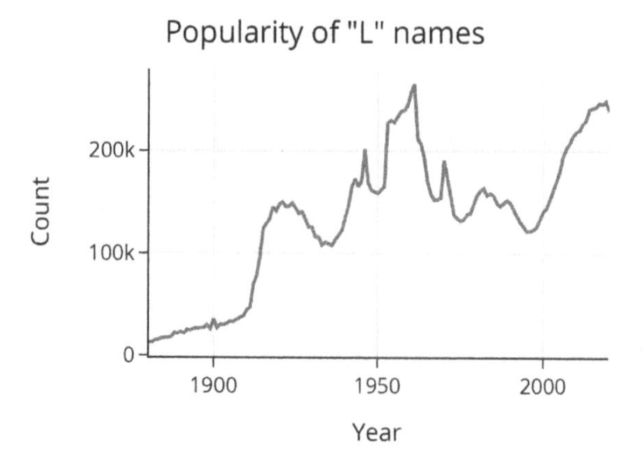

The plot shows that L names were popular in the 1960s, dipped in the decades after, but have indeed resurged in popularity since 2000.

What about J names?

```
fig = px.line(letter_counts.loc[letter_counts['Firsts'] == 'J'],
              x='Year', y='Count', title='Popularity of "J" names',
              width=350, height=250)
fig.update_layout(margin=dict(t=30))
```

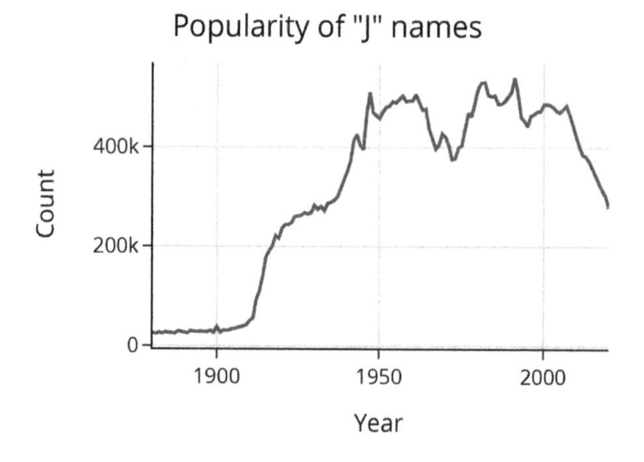

The *NYT* article says that J names were popular in the 1970s and '80s. The plot agrees and shows that they have become less popular since 2000.

The Price of Apply

The power of `.apply()` is its flexibility—you can call it with any function that takes in a single data value and outputs a single data value.

Its flexibility has a price, though. Using `.apply()` can be slow, since `pandas` can't optimize arbitrary functions. For example, using `.apply()` for numeric calculations is much slower than using vectorized operations directly on `pd.Series` objects:

```
%%timeit

# Calculate the decade using vectorized operators
baby['Year'] // 10 * 10
```

```
9.66 ms ± 755 µs per loop (mean ± std. dev. of 7 runs, 100 loops each)
```

```
%%timeit

def decade(yr):
    return yr // 10 * 10

# Calculate the decade using apply
baby['Year'].apply(decade)
```

```
658 ms ± 49.6 ms per loop (mean ± std. dev. of 7 runs, 1 loop each)
```

The version using `.apply()` is 30 times slower! For numeric operations in particular, we recommend operating on `pd.Series` objects directly.

In this section, we introduced data transformations. To transform values in a dataframe, we commonly use the `.apply()` and `.assign()` functions. In the next section, we'll compare dataframes with other ways to represent and manipulate data tables.

How Are Dataframes Different from Other Data Representations?

Dataframes are just one way to represent data stored in a table. In practice, data scientists encounter many other types of data tables, like spreadsheets, matrices, and relations. In this section, we'll compare and contrast the dataframe with other representations to explain why dataframes have become so widely used for data analysis. We'll also point out scenarios where other representations might be more appropriate.

Dataframes and Spreadsheets

Spreadsheets are computer applications in which users can enter data in a grid and use formulas to perform calculations. One well-known example today is Microsoft Excel, although spreadsheets date back to at least 1979 with VisiCalc (*https://doi.org/10.1109/MAHC.2007.4338439*). Spreadsheets make it easy to see and directly manipulate data since spreadsheet formulas can automatically recalculate results whenever the data change. In contrast, dataframe code typically needs to be manually rerun when datasets are updated. These properties make spreadsheets highly popular—by a

2005 estimate (*https://doi.org/10.1109/VLHCC.2005.34*), there are over 55 million spreadsheet users compared to 3 million professional programmers in industry.

Dataframes have several key advantages over spreadsheets. Writing dataframe code in a computational notebook like Jupyter naturally produces a data lineage. Someone who opens the notebook can see the input files for the notebook and how the data were changed. Spreadsheets do not make a data lineage visible; if a person manually edits data values in a cell, it can be difficult for future users to see which values were manually edited and how they were edited. Dataframes can handle larger datasets than spreadsheets, and users can also use distributed programming tools to work with huge datasets that would be very hard to load into a spreadsheet.

Dataframes and Matrices

A matrix is a two-dimensional array of data used primarily for linear algebra operations. In this next example, **X** is a matrix with three rows and two columns:

$$\mathbf{X} = \begin{bmatrix} 1 & 0 \\ 0 & 4 \\ 0 & 0 \end{bmatrix}$$

Matrices are mathematical objects defined by the operators that they allow. For instance, matrices can be added or multiplied together. Matrices also have a transpose. These operators have very useful properties that data scientists rely on for statistical modeling.

One important difference between a matrix and a dataframe is that when matrices are treated as mathematical objects, they can only contain numbers. Dataframes, on the other hand, can also have other types of data, like text. This makes dataframes more useful for loading and processing raw data that may contain all kinds of data types. In practice, data scientists often load data into dataframes, then manipulate the data into matrix form. In this book, we'll generally use dataframes for exploratory data analysis and data cleaning, then process the data into matrices for machine learning models.

Data scientists refer to matrices not only as mathematical objects but also as program objects. For instance, the R programming language has a matrix object, while in Python we could represent a matrix using a two-dimensional `numpy` array. Matrices as implemented in Python and R can contain other data types besides numbers, but they lose mathematical properties when doing so. This is yet another example of how domains can refer to different things with the same term.

Dataframes and Relations

A relation is a data table representation used in database systems, especially SQL systems like SQLite and PostgreSQL. (We cover relations and SQL in Chapter 7.) Relations share many similarities with dataframes; both use rows to represent records and columns to represent features. Both have column names, and data within a column have the same type.

One key advantage of dataframes is that they don't *require* rows to represent records and columns to represent features. Many times, raw data don't come in a convenient format that can directly be put into a relation. In these scenarios, data scientists use the dataframe to load and process data since dataframes are more flexible in this regard. Often, data scientists will load raw data into a dataframe, then process the data into a format that can easily be stored in a relation.

One key advantage that relations have over dataframes is that relations are used by relational database systems like PostgreSQL (*https://oreil.ly/3zXyH*), which have highly useful features for data storage and management. Consider a data scientist at a company that runs a large social media website. The database might hold data that is far too large to read into a `pandas` dataframe all at once; instead, data scientists use SQL queries to subset and aggregate data since database systems are more capable of handling large datasets. Also, website users constantly update their data by making posts, uploading pictures, and editing their profiles. Here, database systems let data scientists reuse their existing SQL queries to update their analyses with the latest data rather than having to repeatedly download large CSV files.

Summary

In this chapter, we explained what dataframes are, why they're useful, and how to work with them using `pandas` code. Subsetting, aggregating, joining, and transforming are useful in nearly every data analysis. We'll rely on these operations often in the rest of the book, especially in Chapters 8, 9, and 10.

CHAPTER 7
Working with Relations Using SQL

In Chapter 6, we used dataframes to represent tables of data. This chapter introduces *relations*, another widely used way to represent data tables. We also introduce SQL, the standard programming language for working with relations. Here's an example of a relation that holds information about popular dog breeds.

Like dataframes, each row in a relation represents a single record—in this case, a single dog breed. Each column represents a feature about the record—for example, the `grooming` column represents how often each dog breed needs to be groomed.

Both relations and dataframes have labels for each column in the table. However, one key difference is that the rows in a relation don't have labels, while rows in a dataframe do.

In this chapter, we demonstrate common relation operations using SQL. We start by explaining the structure of SQL queries. Then we show how to use SQL to perform common data manipulation tasks, like slicing, filtering, sorting, grouping, and joining.

This chapter replicates the data analyses in Chapter 6 using relations and SQL instead of dataframes and Python. The datasets, data manipulations, and conclusions are nearly identical across the two chapters for ease of comparison between performing data manipulations in `pandas` and SQL.

Subsetting

To work with relations, we'll introduce a domain-specific programming language called *SQL* (Structured Query Language). We commonly pronounce "SQL" like "sequel" instead of spelling out the acronym. SQL is a specialized language for working with relations—as such, SQL has a different syntax than Python for writing programs that operate on relational data.

In this chapter, we'll use SQL queries within Python programs. This illustrates a common workflow—data scientists often process and subset data in SQL before loading the data into Python for further analysis. SQL databases make it easier to work with large amounts of data compared to pandas programs. However, loading data into pandas makes it easier to visualize the data and build statistical models.

Why do SQL systems tend to work better with larger datasets? In short, SQL systems have sophisticated algorithms for managing data stored on disk. For example, when working with a large dataset, SQL systems will transparently load and manipulate small portions of data at a time; doing this in pandas can be quite difficult in comparison. We cover this topic in more detail in Chapter 8.

SQL Basics: SELECT and FROM

We'll use the pd.read_sql function, which runs a SQL query and stores the output in a pandas dataframe. Using this function requires some setup. We start by importing the pandas and sqlalchemy Python packages:

```
import pandas as pd
import sqlalchemy
```

Our database is stored in a file called *babynames.db*. This file is a SQLite (*https://oreil.ly/sGYWE*) database, so we'll set up a sqlalchemy object that can process this format:

```
db = sqlalchemy.create_engine('sqlite:///babynames.db')
```

In this book, we use SQLite, an extremely useful database system for working with data stored locally. Other systems make different trade-offs that are useful for different domains. For instance, PostgreSQL and MySQL are more complex systems that are useful for large web applications where many end users are writing data at the same time. Although each SQL system has slight differences, they provide the same core SQL functionality. Readers may also be aware that Python provides SQLite support in its standard sqlite3 library. We choose to use sqlalchemy because it's easier to reuse the code for other SQL systems beyond SQLite.

Now we can use pd.read_sql to run SQL queries on this database. This database has two relations: baby and nyt. Here's a simple example that reads in the entire baby relation. We write a SQL query as a Python string and pass it into pd.read_sql:

```
query = '''
SELECT *
FROM baby;
'''

pd.read_sql(query, db)
```

	Name	Sex	Count	Year
0	Liam	M	19659	2020
1	Noah	M	18252	2020
2	Oliver	M	14147	2020
...	...	...	...	...
2020719	Verona	F	5	1880
2020720	Vertie	F	5	1880
2020721	Wilma	F	5	1880

```
2020722 rows × 4 columns
```

The text inside the `query` variable contains SQL code. `SELECT` and `FROM` are SQL keywords. We read the preceding query like this:

```
SELECT *      -- Get all the columns...
FROM baby;    -- ...from the baby relation
```

The `baby` relation contains the same data as the `baby` dataframe in Chapter 6: the names of all babies registered by the US Social Security Administration.

What's a Relation?

Let's examine the `baby` relation in more detail. A relation has rows and columns. Every column has a label, as illustrated in Figure 7-1. Unlike dataframes, however, individual rows in a relation don't have labels. Also unlike dataframes, rows of a relation aren't ordered.

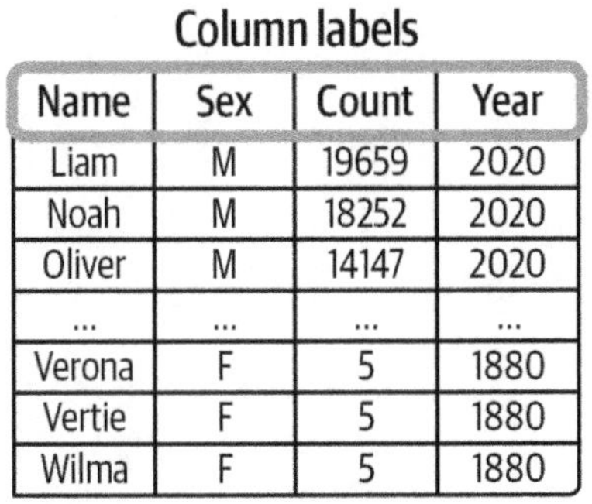

Name	Sex	Count	Year
Liam	M	19659	2020
Noah	M	18252	2020
Oliver	M	14147	2020
...	...	...	...
Verona	F	5	1880
Vertie	F	5	1880
Wilma	F	5	1880

Figure 7-1. The `baby` relation has labels for columns (boxed)

Relations have a long history. More formal treatments of relations use the term *tuple* to refer to the rows of a relation, and *attribute* to refer to the columns. There is also a rigorous way to define data operations using relational algebra, which is derived from mathematical set algebra.

Slicing

Slicing is an operation that creates a new relation by taking a subset of rows or columns out of another relation. Think about slicing a tomato—slices can go both vertically and horizontally. To slice columns of a relation, we give the `SELECT` statement the columns we want:

```
query = '''
SELECT Name
FROM baby;
'''

pd.read_sql(query, db)
```

	Name
0	Liam
1	Noah
2	Oliver
...	...
2020719	Verona
2020720	Vertie
2020721	Wilma

```
2020722 rows × 1 columns
```

```
query = '''
SELECT Name, Count
FROM baby;
'''

pd.read_sql(query, db)
```

	Name	Count
0	Liam	19659
1	Noah	18252
2	Oliver	14147
...	...	...
2020719	Verona	5
2020720	Vertie	5
2020721	Wilma	5

```
2020722 rows × 2 columns
```

To slice out a specific number of rows, use the `LIMIT` keyword:

```
query = '''
SELECT Name
FROM baby
LIMIT 10;
'''

pd.read_sql(query, db)
```

	Name
0	Liam
1	Noah
2	Oliver
...	...
7	Lucas
8	Henry
9	Alexander

```
10 rows × 1 columns
```

In sum, we use the `SELECT` and `LIMIT` keywords to slice columns and rows of a relation.

Filtering Rows

Now we turn to *filtering* rows—taking subsets of rows using one or more criteria. In `pandas`, we slice dataframes using Boolean series objects. In SQL, we instead use the `WHERE` keyword with a predicate. The following query filters the `baby` relation to have only the baby names in 2020:

```
query = '''
SELECT *
FROM baby
WHERE Year = 2020;
'''

pd.read_sql(query, db)
```

	Name	Sex	Count	Year
0	Liam	M	19659	2020
1	Noah	M	18252	2020
2	Oliver	M	14147	2020
...	...	...	...	...
31267	Zylynn	F	5	2020

	Name	Sex	Count	Year
31268	Zynique	F	5	2020
31269	Zynlee	F	5	2020

```
31270 rows × 4 columns
```

Note that when comparing for equality, SQL uses a single equals sign:

```
SELECT *
FROM baby
WHERE Year = 2020;
--         ↑
--         Single equals sign
```

In Python, however, single equals signs are used for variable assignment. The statement `Year = 2020` will assign the value `2020` to the variable `Year`. To compare for equality, Python code uses double equals signs:

```
# Assignment
my_year = 2021

# Comparison, which evaluates to False
my_year == 2020
```

To add more predicates to the filter, use the `AND` and `OR` keywords. For instance, to find the names that have more than 10,000 babies in either 2020 or 2019, we write:

```
query = '''
SELECT *
FROM baby
WHERE Count > 10000
  AND (Year = 2020
       OR Year = 2019);
-- Notice that we use parentheses to enforce evaluation order
'''

pd.read_sql(query, db)
```

	Name	Sex	Count	Year
0	Liam	M	19659	2020
1	Noah	M	18252	2020
2	Oliver	M	14147	2020
...	...	...	...	...
41	Mia	F	12452	2019

	Name	Sex	Count	Year
42	Harper	F	10464	2019
43	Evelyn	F	10412	2019

```
44 rows × 4 columns
```

Finally, to find the 10 most common names in 2020, we can sort the dataframe by `Count` in descending order using the `ORDER BY` keyword with the `DESC` option (short for DESCending):

```
query = '''
SELECT *
FROM baby
WHERE Year = 2020
ORDER BY Count DESC
LIMIT 10;
'''

pd.read_sql(query, db)
```

	Name	Sex	Count	Year
0	Liam	M	19659	2020
1	Noah	M	18252	2020
2	Emma	F	15581	2020
...	...	...	...	...
7	Sophia	F	12976	2020
8	Amelia	F	12704	2020
9	William	M	12541	2020

```
10 rows × 4 columns
```

We see that Liam, Noah, and Emma were the most popular baby names in 2020.

Example: How Recently Has Luna Become a Popular Name?

As we mentioned in Chapter 6, a *New York Times* article mentions that the name Luna was almost nonexistent before 2000 but has since grown to become a very popular name for girls. When exactly did Luna become popular? We can check this in SQL using slicing and filtering:

```
query = '''
SELECT *
FROM baby
WHERE Name = "Luna"
  AND Sex = "F";
'''
```

```
luna = pd.read_sql(query, db)
luna
```

	Name	Sex	Count	Year
0	Luna	F	7770	2020
1	Luna	F	7772	2019
2	Luna	F	6929	2018
...	...	...	...	...
125	Luna	F	17	1883
126	Luna	F	18	1881
127	Luna	F	15	1880

```
128 rows × 4 columns
```

`pd.read_sql` returns a `pandas.DataFrame` object, which we can use to make a plot. This illustrates a common workflow—process the data using SQL, load it into a `pan das` dataframe, then visualize the results:

```
px.line(luna, x='Year', y='Count', width=350, height=250)
```

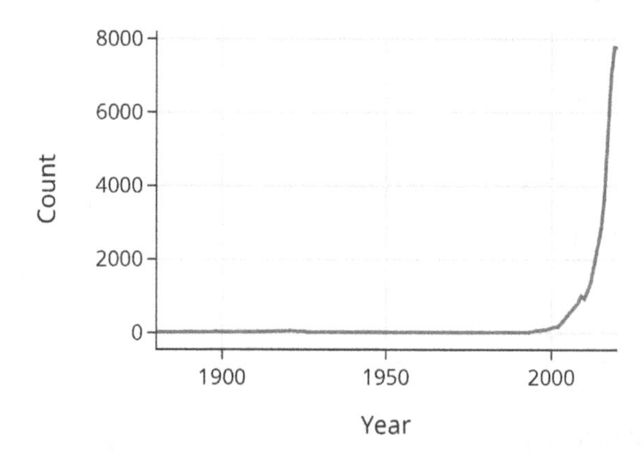

In this section, we introduced the common ways that data scientists subset relations—slicing with column labels and filtering using a boolean condition. In the next section, we explain how to aggregate rows together.

Aggregating

This section introduces grouping and aggregating in SQL. We'll work with the baby names data, as in the previous section:

```
import sqlalchemy
db = sqlalchemy.create_engine('sqlite:///babynames.db')
```

```
query = '''
SELECT *
FROM baby
LIMIT 10
'''

pd.read_sql(query, db)
```

	Name	Sex	Count	Year
0	Liam	M	19659	2020
1	Noah	M	18252	2020
2	Oliver	M	14147	2020
...	...	...	...	...
7	Lucas	M	11281	2020
8	Henry	M	10705	2020
9	Alexander	M	10151	2020

```
10 rows × 4 columns
```

Basic Group-Aggregate Using GROUP BY

Let's say we want to find out the total number of babies born as recorded in this data. This is simply the sum of the `Count` column. SQL provides functions that we use in the `SELECT` statement, like `SUM`:

```
query = '''
SELECT SUM(Count)
FROM baby
'''

pd.read_sql(query, db)
```

	SUM(Count)
0	352554503

In Chapter 6, we used grouping and aggregation to figure out whether US births are trending upward over time. We grouped the dataset by year using `.groupby()`, then summed the counts within each group using `.sum()`.

In SQL, we instead group using the `GROUP BY` clause, then call aggregation functions in `SELECT`:

```
query = '''
SELECT Year, SUM(Count)
FROM baby
GROUP BY Year
'''
```

```
pd.read_sql(query, db)
```

	Year	SUM(Count)
0	1880	194419
1	1881	185772
2	1882	213385
...	...	...
138	2018	3487193
139	2019	3437438
140	2020	3287724

```
141 rows × 2 columns
```

As with dataframe grouping, notice that the `Year` column contains the unique `Year` values—there are no duplicate `Year` values anymore since we grouped them together. When grouping in `pandas`, the grouping columns become the index of the resulting dataframe. However, relations don't have row labels, so the `Year` values are just a column in the resulting relation.

Here's the basic recipe for grouping in SQL:

```
SELECT
  col1,            -- column used for grouping
  SUM(col2)        -- aggregation of another column
FROM table_name    -- relation to use
GROUP BY col1      -- the column(s) to group by
```

Note that the order of clauses in a SQL statement is important. To avoid a syntax error, `SELECT` needs to appear first, then `FROM`, then `WHERE`, then `GROUP BY`.

When using `GROUP BY` we need to be careful about the columns given to `SELECT`. In general, we can only include columns without an aggregation when we use those columns to group. For instance, in the preceding example we grouped by the `Year` column, so we can include `Year` in the `SELECT` clause. All other columns included in `SELECT` should be aggregated, as we did earlier with `SUM(Count)`. If we included a "bare" column like `Name` that wasn't used for grouping, it's ambiguous which name within the group should be returned. Although bare columns won't cause an error for SQLite, they cause other SQL engines to error, so we recommend avoiding them.

Grouping on Multiple Columns

We pass multiple columns into `GROUP BY` to group by multiple columns at once. This is useful when we need to further subdivide our groups. For example, we can group by both year and sex to see how many male and female babies were born over time:

```
query = '''
SELECT Year, Sex, SUM(Count)
FROM baby
GROUP BY Year, Sex
'''

pd.read_sql(query, db)
```

	Year	Sex	SUM(Count)
0	1880	F	83929
1	1880	M	110490
2	1881	F	85034
...	...	...	...
279	2019	M	1785527
280	2020	F	1581301
281	2020	M	1706423

```
282 rows × 3 columns
```

Notice that the preceding code is very similar to grouping by a single column, except that it gives multiple columns to `GROUP BY` to group by both `Year` and `Sex`.

Unlike `pandas`, SQLite doesn't provide a simple way to pivot a relation. Instead, we can use `GROUP BY` on two columns in SQL, read the result into a dataframe, and then use the `unstack()` dataframe method.

Other Aggregation Functions

SQLite has several other built-in aggregation functions besides `SUM`, such as `COUNT`, `AVG`, `MIN`, and `MAX`. For the full list of functions, consult the SQLite website (*https://oreil.ly/ALtjb*).

To use another aggregation function, we call it in the `SELECT` clause. For instance, we can use `MAX` instead of `SUM`:

```
query = '''
SELECT Year, MAX(Count)
FROM baby
GROUP BY Year
'''

pd.read_sql(query, db)
```

	Year	MAX(Count)
0	1880	9655
1	1881	8769
2	1882	9557
...	...	...
138	2018	19924
139	2019	20555
140	2020	19659

```
141 rows × 2 columns
```

The built-in aggregation functions are one of the first places a data scientist may encounter differences in SQL implementations. For instance, SQLite has a relatively minimal set of aggregation functions while PostgreSQL has many more (*https://oreil.ly/gqYoK*). That said, almost all SQL implementations provide `SUM`, `COUNT`, `MIN`, `MAX`, and `AVG`.

This section covered common ways to aggregate data in SQL using the `GROUP BY` keyword with one or more columns. In the next section, we'll explain how to join relations together.

Joining

To connect records between two data tables, SQL relations can be joined together similar to dataframes. In this section, we introduce SQL joins to replicate our analysis of the baby names data. Recall that Chapter 6 mentions a *New York Times* article that talks about how certain name categories, like mythological and baby boomer names, have become more or less popular over time.

We've taken the names and categories in the *NYT* article and put them in a small relation named `nyt`. First, the code sets up a connection to a database, then runs a SQL query to display the `nyt` relation:

```
import sqlalchemy
db = sqlalchemy.create_engine('sqlite:///babynames.db')

query = '''
SELECT *
FROM nyt;
'''

pd.read_sql(query, db)
```

	nyt_name	category
0	Lucifer	forbidden
1	Lilith	forbidden
2	Danger	forbidden
...	...	...
20	Venus	celestial
21	Celestia	celestial
22	Skye	celestial

```
23 rows × 2 columns
```

Notice that the preceding code runs a query on babynames.db, the same database that contains the larger baby relation from the previous sections. SQL databases can hold more than one relation, making them very useful when we need to work with many data tables at once. CSV files, on the other hand, typically contain one data table each—if we perform a data analysis that uses 20 data tables, we might need to keep track of the names, locations, and versions of 20 CSV files. Instead, it could be simpler to store all the data tables in a SQLite database stored in a single file.

To see how popular the categories of names are, we join the nyt relation with the baby relation to get the name counts from baby.

Inner Joins

As in Chapter 6, we've made smaller versions of the baby and nyt tables so that it's easier to see what happens when we join tables together. The relations are called baby_small and nyt_small:

```
query = '''
SELECT *
FROM baby_small;
'''

pd.read_sql(query, db)
```

	Name	Sex	Count	Year
0	Noah	M	18252	2020
1	Julius	M	960	2020
2	Karen	M	6	2020
3	Karen	F	325	2020
4	Noah	F	305	2020

```
query = '''
SELECT *
FROM nyt_small;
'''

pd.read_sql(query, db)
```

	nyt_name	category
0	Karen	boomer
1	Julius	mythology
2	Freya	mythology

To join relations in SQL, we use the `INNER JOIN` clause to say which tables we want to join and the `ON` clause to specify a predicate for joining the tables. Here's an example:

```
query = '''
SELECT *
FROM baby_small INNER JOIN nyt_small
  ON baby_small.Name = nyt_small.nyt_name
'''

pd.read_sql(query, db)
```

	Name	Sex	Count	Year	nyt_name	category
0	Julius	M	960	2020	Julius	mythology
1	Karen	M	6	2020	Karen	boomer
2	Karen	F	325	2020	Karen	boomer

Notice that this result is the same as doing an inner join in `pandas`: the new table has the columns of both the `baby_small` and `nyt_small` tables. The rows with the name Noah are gone, and the remaining rows have their matching `category` from `nyt_small`.

To join two tables together, we tell SQL the column(s) from each table that we want to do the join with, using a predicate with the `ON` keyword. SQL matches rows together when the values in the joining columns fulfill the predicate, as shown in Figure 7-2.

Unlike `pandas`, SQL gives more flexibility on how rows are joined. The `pd.merge()` method can only join using simple equality, but the predicate in the `ON` clause can be arbitrarily complex. As an example, we take advantage of this extra versatility in "Finding Collocated Sensors" on page 284.

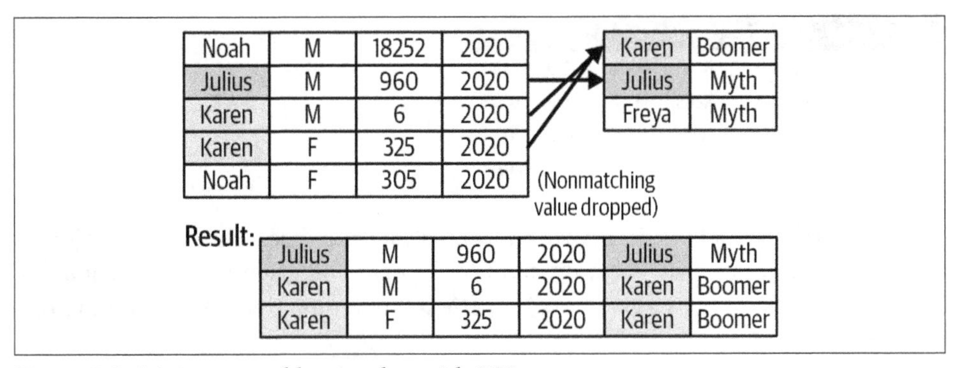

Figure 7-2. Joining two tables together with SQL

Left and Right Joins

Like pandas, SQL also supports left joins. Instead of saying `INNER JOIN`, we use `LEFT JOIN`:

```
query = '''
SELECT *
FROM baby_small LEFT JOIN nyt_small
  ON baby_small.Name = nyt_small.nyt_name
'''

pd.read_sql(query, db)
```

	Name	Sex	Count	Year	nyt_name	category
0	Noah	M	18252	2020	None	None
1	Julius	M	960	2020	Julius	mythology
2	Karen	M	6	2020	Karen	boomer
3	Karen	F	325	2020	Karen	boomer
4	Noah	F	305	2020	None	None

As we might expect, the "left" side of the join refers to the table that appears on the left side of the `LEFT JOIN` keyword. We can see the `Noah` rows are kept in the resulting relation even when they don't have a match in the righthand relation.

Note that SQLite doesn't support right joins directly, but we can perform the same join by swapping the order of relations, then using `LEFT JOIN`:

```
query = '''
SELECT *
FROM nyt_small LEFT JOIN baby_small
  ON baby_small.Name = nyt_small.nyt_name
'''

pd.read_sql(query, db)
```

	nyt_name	category	Name	Sex	Count	Year
0	Karen	boomer	Karen	F	325.0	2020.0
1	Karen	boomer	Karen	M	6.0	2020.0
2	Julius	mythology	Julius	M	960.0	2020.0
3	Freya	mythology	None	None	NaN	NaN

SQLite doesn't have a built-in keyword for outer joins. In cases where an outer join is needed, we have to either use a different SQL engine or perform an outer join via `pandas`. However, in our (the authors') experience, outer joins are rarely used in practice compared to inner and left joins.

Example: Popularity of NYT Name Categories

Now let's return to the full `baby` and `nyt` relations.

We want to know how the popularity of name categories in `nyt` has changed over time. To answer this question, we should:

1. Inner join `baby` with `nyt`, matching rows where the names are equal.
2. Group the table by `category` and `Year`.
3. Aggregate the counts using a sum:

```
query = '''
SELECT
  category,
  Year,
  SUM(Count) AS count            -- [3]
FROM baby INNER JOIN nyt         -- [1]
  ON baby.Name = nyt.nyt_name    -- [1]
GROUP BY category, Year          -- [2]
'''

cate_counts = pd.read_sql(query, db)
cate_counts
```

	category	Year	count
0	boomer	1880	292
1	boomer	1881	298
2	boomer	1882	326
...	...	...	...
647	mythology	2018	2944
648	mythology	2019	3320
649	mythology	2020	3489

```
650 rows × 3 columns
```

The numbers in square brackets ([1], [2], [3]) in the preceding query show how each step in our plan maps to the parts of the SQL query. The code re-creates the dataframe from Chapter 6, where we created plots to verify the claims of the *New York Times* article. For brevity, we omit duplicating the plots here.

Notice that in the SQL code in this example, the numbers appear out of order—[3], [1], then [2]. As a rule of thumb for first-time SQL learners, we can often think of the SELECT statement as the *last* piece of the query to execute even though it appears first.

In this section, we introduced joins for relations. When joining relations together, we match rows using the INNER JOIN or LEFT JOIN keyword and a boolean predicate. In the next section, we'll explain how to transform values in a relation.

Transforming and Common Table Expressions

In this section, we show how to call functions to transform columns of data using built-in SQL functions. We also demonstrate how to use common table expressions to build up complex queries from simpler ones. As usual, we start by loading the database:

```
# Set up connection to database
import sqlalchemy
db = sqlalchemy.create_engine('sqlite:///babynames.db')
```

SQL Functions

SQLite provides a variety of *scalar functions*, or functions that transform single data values. When called on a column of data, SQLite will apply these functions on each value in the column. In contrast, aggregation functions like SUM and COUNT take a column of values as input and compute a single value as output.

SQLite provides a comprehensive list of the built-in scalar functions in its online documentation (*https://oreil.ly/kznBO*). For instance, to find the number of characters in each name, we use the LENGTH function:

```
query = '''
SELECT Name, LENGTH(Name)
FROM baby
LIMIT 10;
'''

pd.read_sql(query, db)
```

	Name	LENGTH(Name)
0	Liam	4
1	Noah	4
2	Oliver	6
...	...	...
7	Lucas	5
8	Henry	5
9	Alexander	9

```
10 rows × 2 columns
```

Notice that the `LENGTH` function is applied to each value within the `Name` column.

Like aggregation functions, each implementation of SQL provides a different set of scalar functions. SQLite has a relatively minimal set of functions, while PostgreSQL has many more (*https://oreil.ly/i2KIA*). That said, almost all SQL implementations provide some equivalent to SQLite's `LENGTH`, `ROUND`, `SUBSTR`, and `LIKE` functions.

Although scalar functions use the same syntax as an aggregation function, they behave differently. This can result in confusing output if the two are mixed together in a single query:

```
query = '''
SELECT Name, LENGTH(Name), AVG(Count)
FROM baby
LIMIT 10;
'''

pd.read_sql(query, db)
```

	Name	LENGTH(Name)	AVG(Count)
0	Liam	4	174.47

Here, the `AVG(Name)` computes the average of the entire `Count` column, but the output is confusing—a reader could easily think the average is related to the name Liam. For this reason, we must be careful when scalar and aggregation functions appear together within a `SELECT` statement.

To extract the first letter of each name, we can use the `SUBSTR` function (short for *substring*). As described in the documentation, the `SUBSTR` function takes three arguments. The first is the input string, the second is the position to begin the substring (1-indexed), and the third is the length of the substring:

```
query = '''
SELECT Name, SUBSTR(Name, 1, 1)
```

```
FROM baby
LIMIT 10;
'''

pd.read_sql(query, db)
```

	Name	SUBSTR(Name, 1, 1)
0	Liam	L
1	Noah	N
2	Oliver	O
...	...	...
7	Lucas	L
8	Henry	H
9	Alexander	A

```
10 rows × 2 columns
```

We can use the AS keyword to rename the column:

```
query = '''
SELECT *, SUBSTR(Name, 1, 1) AS Firsts
FROM baby
LIMIT 10;
'''

pd.read_sql(query, db)
```

	Name	Sex	Count	Year	Firsts
0	Liam	M	19659	2020	L
1	Noah	M	18252	2020	N
2	Oliver	M	14147	2020	O
...	...	...	...	...	...
7	Lucas	M	11281	2020	L
8	Henry	M	10705	2020	H
9	Alexander	M	10151	2020	A

```
10 rows × 5 columns
```

After calculating the first letter of each name, our analysis aims to understand the popularity of first letters over time. To do this, we want to take the output of this SQL query and use it as a single step within a longer chain of operations.

SQL provides several options to break queries into smaller steps, which is helpful in more complex analyses like this one. The most common options for doing this are to create a new relation using a CREATE TABLE statement, create a new view using CREATE VIEW, or create a temporary relation using WITH. Each of these methods has different use-cases. For simplicity, we only describe the WITH statement in this section and suggest that readers look over the SQLite documentation for details.

Multistep Queries Using a WITH Clause

The WITH clause lets us assign a name to any SELECT query. Then we can treat that query as though it exists as a relation in the database just for the duration of the query. SQLite calls these temporary relations *common table expressions*. For instance, we can take the earlier query that calculates the first letter of each name and call it letters:

```
query = '''
-- Create a temporary relation called letters by calculating
-- the first letter for each name in baby
WITH letters AS (
  SELECT *, SUBSTR(Name, 1, 1) AS Firsts
  FROM baby
)
-- Then, select the first ten rows from letters
SELECT *
FROM letters
LIMIT 10;
'''

pd.read_sql(query, db)
```

	Name	Sex	Count	Year	Firsts
0	Liam	M	19659	2020	L
1	Noah	M	18252	2020	N
2	Oliver	M	14147	2020	O
...	...	...	...	...	...
7	Lucas	M	11281	2020	L
8	Henry	M	10705	2020	H
9	Alexander	M	10151	2020	A

```
10 rows × 5 columns
```

WITH statements are very useful since they can be chained together. We can create multiple temporary relations in a WITH statement that each perform a bit of work on the previous result, which lets us gradually build complicated queries a step at a time.

Example: Popularity of "L" Names

We can use WITH statements to look at the popularity of names that start with the letter L over time. We'll group the temporary letters relation by the first letter and year, then aggregate the Count column using a sum, then filter to get only names with the letter L:

```
query = '''
WITH letters AS (
  SELECT *, SUBSTR(Name, 1, 1) AS Firsts
  FROM baby
)
SELECT Firsts, Year, SUM(Count) AS Count
FROM letters
WHERE Firsts = "L"
GROUP BY Firsts, Year;
'''

letter_counts = pd.read_sql(query, db)
letter_counts
```

	Firsts	Year	Count
0	L	1880	12799
1	L	1881	12770
2	L	1882	14923
...	...	...	...
138	L	2018	246251
139	L	2019	249315
140	L	2020	239760

```
141 rows × 3 columns
```

This relation contains the same data as the one from Chapter 6. In that chapter, we make a plot of the Count column over time, which we omit here for brevity.

In this section, we introduced data transformations. To transform values in a relation, we commonly use SQL functions like LENGTH() or SUBSTR(). We also explained how to build up complex queries using the WITH clause.

Summary

In this chapter, we explained what relations are, why they're useful, and how to work with them using SQL code. SQL databases are useful for many real-world settings. For example, SQL databases typically have robust data recovery mechanisms—if the computer crashes while in the middle of a SQL operation, the database system can recover as much data as possible without corruption. As mentioned earlier, SQL databases can also handle larger scale; organizations use SQL databases to store and query databases that are far too large to analyze in memory using pandas code. These are just a few reasons why SQL is an important part of the data science toolbox, and we expect that many readers will soon encounter SQL code as part of their work.

PART III

Understanding The Data

CHAPTER 8
Wrangling Files

Before you can work with data in Python, it helps to understand the files that store the source of the data. You want answers to a couple of basic questions:

- How much data do you have?
- How is the source file formatted?

Answers to these questions can be very helpful. For example, if your file is too large or is not formatted the way you expect, you might not be able to properly load it into a dataframe.

Although many types of structures can represent data, in this book we primarily work with data tables, such as Pandas DataFrames and SQL relations. (But do note that Chapter 13 examines less-structured text data, and Chapter 14 introduces hierarchical formats and binary files.) We focus on data tables for several reasons. Research on how to store and manipulate data tables has resulted in stable and efficient tools for working with tables. Plus, data in a tabular format are close cousins of matrices, the mathematical objects of the immensely rich field of linear algebra. And of course, data tables are quite common.

In this chapter, we introduce typical file formats and encodings for plain text, describe measures of file size, and use Python tools to examine source files. Later in the chapter, we introduce an alternative approach for working with files: the shell interpreter. Shell commands give us a programmatic way to get information about a file outside the Python environment, and the shell can be very useful with big data. Finally, we check the data table's shape (the number of rows and columns) and granularity (what a row represents). These simple checks are the starting point for cleaning and analyzing our data.

We first provide brief descriptions of the datasets that we use as examples throughout this chapter.

Data Source Examples

We have selected two examples to demonstrate file wrangling concepts: a government survey about drug abuse, and administrative data from the San Francisco Department of Public Health about restaurant inspections. Before we start wrangling, we give an overview of the data scope for these examples (see Chapter 2).

Drug Abuse Warning Network (DAWN) Survey

DAWN is a national health-care survey that monitors trends in drug abuse. The survey aims to estimate the impact of drug abuse on the country's health-care system and improve how emergency departments monitor substance abuse crises. DAWN was administered annually from 1998 through 2011 by the Substance Abuse and Mental Health Services Administration (SAMHSA) (*https://www.samhsa.gov*). In 2018, due in part to the opioid epidemic, the DAWN survey was restarted. In this example, we look at the 2011 data, which have been made available through the SAMHSA Data Archive (*https://oreil.ly/Y2SKG*).

The target population consists of all drug-related emergency room visits in the US. These visits are accessed through a frame of emergency rooms in hospitals (and their records). Hospitals are selected for the survey through probability sampling (see Chapter 3), and all drug-related visits to the sampled hospital's emergency room are included in the survey. All types of drug-related visits are included, such as drug misuse, abuse, accidental ingestion, suicide attempts, malicious poisonings, and adverse reactions. For each visit, the record may contain up to 16 different drugs, including illegal drugs, prescription drugs, and over-the-counter medications.

The source file for this dataset is an example of fixed-width formatting that requires external documentation, like a codebook, to decipher. Also, it is a reasonably large file and so motivates the topic of how to find a file's size. And the granularity is unusual because an ER visit, not a person, is the subject of investigation.

The San Francisco restaurant files have other characteristics that make them a good example for this chapter.

San Francisco Restaurant Food Safety

The San Francisco Department of Public Health (*https://oreil.ly/kG1PN*) routinely makes unannounced visits to restaurants and inspects them for food safety. The inspector calculates a score based on the violations found and provides descriptions of the violations. The target population here is all restaurants in San Francisco. These restaurants are accessed through a frame of restaurant inspections that were conducted between 2013 and 2016. Some restaurants have multiple inspections in a year, and not all of the 7,000+ restaurants are inspected annually.

Food safety scores are available through the city's Open Data initiative (*https://oreil.ly/kwh-F*), called DataSF (*https://datasf.org*). DataSF is one example a city government making their data publicly available; the DataSF mission is to "empower the use of data in decision making and service delivery" with the goal of improving the quality of life and work for residents, employers, employees, and visitors.

San Francisco requires restaurants to publicly display their scores (see Figure 8-1 for an example placard).[1] These data offer an example of multiple files with different structures, fields, and granularity. One dataset contains summary results of inspections, another provides details about the violations found, and a third contains general information about the restaurants. The violations include both serious problems related to the transmission of foodborne illnesses and minor issues such as not properly displaying the inspection placard.

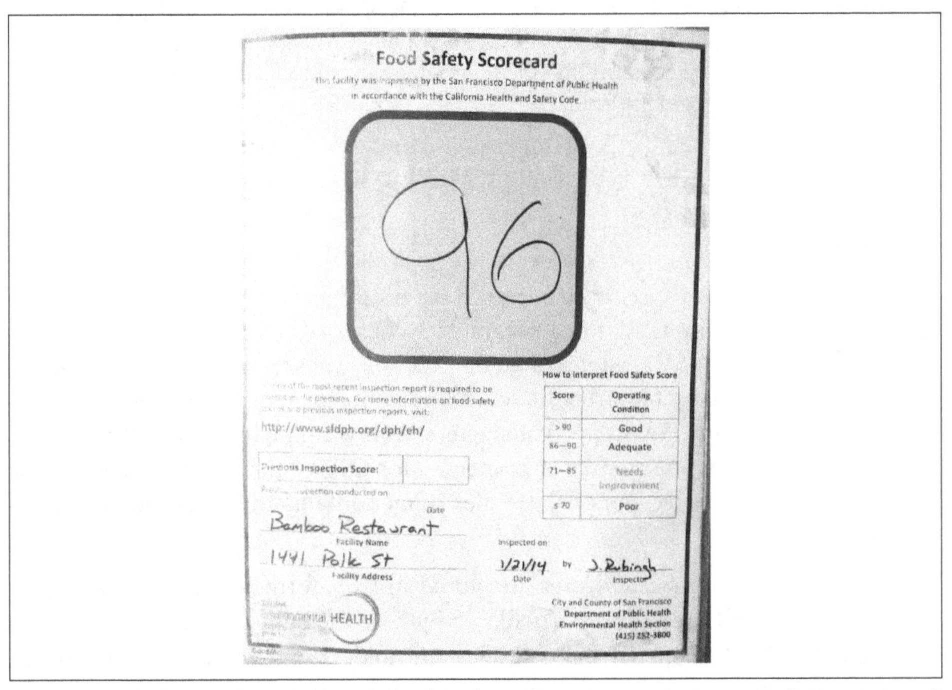

Figure 8-1. A food safety scorecard displayed in a restaurant; scores range between 0 and 100

1 In 2020, the city began giving restaurants color-coded placards indicating whether the restaurant passed (green), conditionally passed (yellow), or failed (red) the inspection. These new placards no longer display a numeric inspection score. However, a restaurant's scores and violations are still available at DataSF.

Both the DAWN survey data and the San Francisco restaurant inspection data are available online as plain-text files. However, their formats are quite different, and in the next section, we demonstrate how to figure out a file format so that we can read the data into a dataframe.

File Formats

A *file format* describes how data are stored on a computer's disk or other storage device. Understanding the file format helps us figure out how to read the data into Python in order to work with it as a data table. In this section, we introduce several popular formats used to store data tables. These are all plain-text formats, meaning they are easy for us to read with a text editor like VS Code, Sublime, Vim, or Emacs.

The file format and the *structure* of the data are two different things. We consider the data structure to be a mental representation of the data that tells us what kinds of operations we can do. For example, a table structure corresponds to data values arranged in rows and columns. But the same table can be stored in many different types of file formats.

The first format we describe is the delimited file format.

Delimited Format

Delimited formats use a specific character to separate data values. Usually, these separators are either a comma (comma-separated values, or CSV for short), a tab (tab-separated values, or TSV), whitespace, or a colon. These formats are natural for storing data that have a table structure. Each line in the file represents a record, which is delimited by newline (`\n` or `\r\n`) characters. And within a line, the record's information is delimited by the comma character (`,`) for CSV or the tab character (`\t`) for TSV, and so on. The first line of these files often contains the names of the table's columns/features.

The San Francisco restaurant scores are stored in CSV-formatted files. Let's display the first few lines of the *inspections.csv* file. In Python, the built-in `pathlib` library has a useful `Path` object to specify paths to files and folders that work across platforms. This file is within the *data* folder, so we use `Path()` to create the full pathname:

```
from pathlib import Path

# Create a Path pointing to our datafile
insp_path = Path() / 'data' / 'inspections.csv'
```

Paths are tricky when working across different operating systems (OSs). For instance, a typical path in Windows might look like *C:\files\data.csv*, while a path in Unix or macOS might look like *~/files/data.csv*. Because of this, code that works on one OS can fail to run on other OSs.

The `pathlib` Python library was created to avoid OS-specific path issues. By using it, the code shown here is more *portable*—it works across Windows, macOS, and Unix.

The `Path` object in the following code has many useful methods, such as `read_text()`, which reads in the entire contents of the file as a string:

```
text = insp_path.read_text()
# Print first five lines
print('\n'.join(text.split('\n')[:5]))
```

```
"business_id","score","date","type"
19,"94","20160513","routine"
19,"94","20171211","routine"
24,"98","20171101","routine"
24,"98","20161005","routine"
```

Notice that the field names appear in the first line of the file; these names are comma separated and in quotes. We see four fields: the business identifier, the restaurant's score, the date of the inspection, and the type of inspection. Each line in the file corresponds to one inspection, and the ID, score, date, and type values are separated by commas. In addition to identifying the file format, we also want to identify the format of the features. We see two things of note: the scores and dates both appear as strings. We will want to convert the scores to numbers so that we can calculate summary statistics and create visualizations. And we will convert the date into a date-time format so that we can make time-series plots. We show how to carry out these transformations in Chapter 9.

Displaying the first few lines of a file is something we'll do often, so we create a function as a shortcut:

```
def head(filepath, n=5, width=-1):
    '''Prints the width characters of first n lines of filepath'''
    with filepath.open() as f:
        for _ in range(n):
            (print(f.readline(), end='') if width < 0
             else print(f.readline()[:width]))
```

People often confuse CSV and TSV files with spreadsheets. This is in part because most spreadsheet software (like Microsoft Excel) will automatically display a CSV file as a table in a workbook. Behind the scenes, Excel looks at the file format and encoding just like we've done in this section. However, Excel files have a different format than CSV and TSV files, and we need to use different `pandas` functions to read these formats into Python.

All three of the restaurant source files are CSV formatted. In contrast, the DAWN source file has a fixed-width format. We describe this kind of formatting next.

Fixed-Width Format

The fixed-width format (FWF) does not use delimiters to separate data values. Instead, the values for a specific field appear in the exact same position in each line. The DAWN source file has this format. Each line in the file is very long. For display purposes, we only show the first few characters from the first five lines in the file:

```
dawn_path = Path() / 'data' / 'DAWN-Data.txt'
head(dawn_path, width=65)
```

```
1 2251082    .9426354082   3 4 1 2201141 2 865 105 1102005 1
2 2291292   5.9920106887   911 1 3201134 12077  81  82 283-8
3 7 7 251   4.7231718669   611 2 2201143 12313   1  12  -7-8
410 8 292   4.0801470012   6 2 1 3201122 1 234 358  99 215 2
5 122 942   5.1777093467  10 6 1 3201134 3 865 105 1102005 1
```

Notice how the values appear to align from one row to the next. For example, there is a decimal point in the same position (the 19th character) in each line. Notice also that some of the values seem to be squished together, and we need to know the exact position of each piece of information in a line in order to make sense of it. SAMHSA provides a 2,000-page codebook (*https://oreil.ly/a4OFo*) with all of this information, including some basic checks, so that we can confirm that we have correctly read the file. For instance, the codebook tells us that the age field appears in positions 34–35 and is coded in intervals from 1 to 11. The first two records shown in the preceding code have age categories of 4 and 11; the codebook tells us that a 4 stands for the age bracket "6 to 11" and 11 is for "65+."

Other plain-text formats that are popular include hierarchical formats and loosely formatted text (in contrast to formats that directly support table structures). These are covered in greater detail in other chapters, but for completeness, we briefly describe them here.

A widely adopted convention is to use the filename extension, such as *.csv*, *.tsv*, and *.txt*, to indicate the format of the contents of the file. Filenames that end with *.csv* are expected to contain comma-separated values, and those ending with *.tsv* are expected to contain tab-separated values; *.txt* generally denotes plain text without a designated format. However, these extension names are only suggestions. Even if a file has a *.csv* extension, the actual contents might not be formatted properly! It's a good practice to inspect the contents of the file before loading it into a dataframe. If the file is not too large, you can open and examine it with a plain-text editor. Otherwise, you can view a couple of lines using `.readline()` or shell command.

Hierarchical Formats

Hierarchical formats store data in a nested form. For instance, JavaScript Object Notation (JSON), which is commonly used for communication by web servers, includes key-value pairs and arrays that can be nested, similar to a Python dictionary. XML and HTML are other common formats for storing documents on the internet. Like JSON, these files have a hierarchical, key-value format. We cover both formats (JSON and XML) in more detail in Chapter 14.

Next, we briefly describe other plain-text files that don't fall into any of the previous categories but still have some structure to them that enables us to read and extract information.

Loosely Formatted Text

Web logs, instrument readings, and program logs typically provide data in plain text. For example, here is one line of a web log (we've split it across multiple lines for readability). It contains information such as the date, time, and type of request made to a website:

```
169.237.46.168 - -
[26/Jan/2004:10:47:58 -0800]"GET /stat141/Winter04 HTTP/1.1" 301 328
"http://anson.ucdavis.edu/courses"
"Mozilla/4.0 (compatible; MSIE 6.0; Windows NT 5.0; .NET CLR 1.1.4322)"
```

There are organizational patterns present, but not in a simple delimited format. This is what we mean by "loosely formatted." We see that the date and time appear between square brackets, and the type of request (`GET` in this case) follows the date-time information and appears in quotes. In Chapter 13, we use these observations about the web log's format and string manipulation tools to extract values of interest into a data table.

As another example, here is a single record taken from a wireless device log. The device reports the timestamp, the identifier, its location, and the signal strengths that it picks up from other devices. This information uses a combination of formats: key-value pairs, semicolon-delimited values, and comma-delimited values:

```
t=1139644637174;id=00:02:2D:21:0F:33;pos=2.0,0.0,0.0;degree=45.5;
00:14:bf:b1:97:8a=-33,2437000000,3;00:14:bf:b1:97:8a=-38,2437000000,3;
```

Like with the web logs, we can use string manipulation and the patterns in the records to extract features into a table.

We have primarily introduced formats for plain-text data that are widely used for storing and exchanging tables. The CSV format is the most common, but others, such as tab-separated and fixed-width formats, are also prevalent. And there are many types of file formats that store data!

So far, we have used the term *plain text* to broadly cover formats that can be viewed with a text editor. However, a plain-text file may have different encodings, and if we don't specify the encoding correctly, the values in the dataframe might contain gibberish. We give an overview of file encoding next.

File Encoding

Computers store data as sequences of *bits*: 0s and 1s. *Character encodings*, like ASCII, tell the computer how to translate between bits and text. For example, in ASCII, the bits `100 001` stand for the letter A and `100 010` for B. The most basic kind of plain text supports only standard ASCII characters, which includes the uppercase and lowercase English letters, numbers, punctuation symbols, and spaces.

ASCII encoding does not include a lot of special characters or characters from other languages. Other, more modern character encodings have many more characters that can be represented. Common encodings for documents and web pages are Latin-1 (ISO-8859-1) and UTF-8. UTF-8 has over a million characters and is backward compatible with ASCII, meaning that it uses the same representation for English letters, numbers, and punctuation as ASCII.

When we have a text file, we usually need to figure out its encoding. If we choose the wrong encoding to read in a file, Python either reads incorrect values or throws an error. The best way to find the encoding is by checking the data's documentation, which often explicitly says what the encoding is.

When we don't know the encoding, we have to make a guess. The `chardet` package has a function called `detect()` that infers a file's encoding. Since these guesses are imperfect, the function also returns a confidence level between 0 and 1. We use this function to look at the files from our examples:

```
import chardet

line = '{:<25} {:<10} {}'.format
```

```
# for each file, print its name, encoding & confidence in the encoding
print(line('File Name', 'Encoding', 'Confidence'))

for filepath in Path('data').glob('*'):
    result = chardet.detect(filepath.read_bytes())
    print(line(str(filepath), result['encoding'], result['confidence']))
```

```
File Name                Encoding   Confidence
data/inspections.csv     ascii      1.0
data/co2_mm_mlo.txt      ascii      1.0
data/violations.csv      ascii      1.0
data/DAWN-Data.txt       ascii      1.0
data/legend.csv          ascii      1.0
data/businesses.csv      ISO-8859-1 0.73
```

The detection function is quite certain that all but one of the files are ASCII encoded. The exception is *businesses.csv*, which appears to have an ISO-8859-1 encoding. We run into trouble if we ignore this encoding and try to read the businesses file into `pandas` without specifying the special encoding:

```
# naively reads file without considering encoding
>>> pd.read_csv('data/businesses.csv')
[...stack trace omitted...]
UnicodeDecodeError: 'utf-8' codec can't decode byte 0xd1 in
position 8: invalid continuation byte
```

To successfully read the data, we must specify the ISO-8859-1 encoding:

```
bus = pd.read_csv('data/businesses.csv', encoding='ISO-8859-1')
```

	business_id	name	address	postal_code
0	19	NRGIZE LIFESTYLE CAFE	1200 VAN NESS AVE, 3RD FLOOR	94109
1	24	OMNI S.F. HOTEL - 2ND FLOOR PANTRY	500 CALIFORNIA ST, 2ND FLOOR	94104
2	31	NORMAN'S ICE CREAM AND FREEZES	2801 LEAVENWORTH ST	94133
3	45	CHARLIE'S DELI CAFE	3202 FOLSOM ST	94110

File encoding can be a bit mysterious to figure out, and unless there is metadata that explicitly gives us the encoding, guesswork comes into play. When an encoding is not 100% confirmed, it's a good idea to seek additional documentation.

Another potentially important aspect of a source file is its size. If a file is huge, then we might not be able to read it into a dataframe. In the next section, we discuss how to figure out a source file's size.

File Size

Computers have finite resources. You have likely encountered these limits firsthand if your computer has slowed down from having too many applications open at once. We want to make sure that we do not exceed the computer's limits while working with data, and we might choose to examine a file differently depending on its size. If we know that our dataset is relatively small, then a text editor or a spreadsheet can be convenient for looking at the data. On the other hand, for large datasets, a more programmatic exploration or even distributed computing tools may be needed.

In many situations, we analyze datasets downloaded from the internet. These files reside on the computer's *disk storage*. In order to use Python to explore and manipulate the data, we need to read the data into the computer's *memory*, also known as random access memory (RAM). All Python code requires the use of RAM, no matter how short the code is. A computer's RAM is typically much smaller than its disk storage. For example, one computer model released in 2018 had 32 times more disk storage than RAM. Unfortunately, this means that datafiles can often be much bigger than what is feasible to read into memory.

Both disk storage and RAM capacity are measured in terms of *bytes* (eight 0s and 1s). Roughly speaking, each character in a text file adds one byte to a file's size. To succinctly describe the sizes of larger files, we use the prefixes described in Table 8-1; for example, a file that contains 52,428,800 characters will take up $5,242,8800/1,024^2 = 50$ mebibytes, or 50 MiB on disk.

Table 8-1. Prefixes for common file sizes

Multiple	Notation	Number of bytes
Kibibyte	KiB	1,024
Mebibyte	MiB	$1,024^2$
Gibibyte	GiB	$1,024^3$
Tebibyte	TiB	$1,024^4$
Pebibyte	PiB	$1,024^5$

Why use multiples of 1,024 instead of simple multiples of 1,000 for these prefixes? This is a historical result of the fact that most computers use a binary number scheme where powers of 2 are simpler to represent ($1,024 = 2^{10}$). You also see the typical SI prefixes used to describe size—kilobytes, megabytes, and gigabytes, for example. Unfortunately, these prefixes are used inconsistently. Sometimes a kilobyte refers to 1,000 bytes; other times, a kilobyte refers to 1,024 bytes. To avoid confusion, we stick to kibi-, mebi-, and gibibytes, which clearly represent multiples of 1,024.

It is not uncommon to have a datafile happily stored on a computer that will overflow the computer's memory if we attempt to manipulate it with a program. So we often begin our data work by making sure the files are of manageable size. To do this, we use the built-in `os` library:

```
from pathlib import Path
import os

kib = 1024
line = '{:<25} {}'.format

print(line('File', 'Size (KiB)'))
for filepath in Path('data').glob('*'):
    size = os.path.getsize(filepath)
    print(line(str(filepath), np.round(size / kib)))
```

```
File                      Size (KiB)
data/inspections.csv      455.0
data/co2_mm_mlo.txt       50.0
data/violations.csv       3639.0
data/DAWN-Data.txt        273531.0
data/legend.csv           0.0
data/businesses.csv       645.0
```

We see that the *businesses.csv* file takes up 645 KiB on disk, making it well within the memory capacities of most systems. Although the *violations.csv* file takes up 3.6 MiB of disk storage, most machines can easily read it into a `pandas` `DataFrame` too. But *DAWN-Data.txt*, which contains the DAWN survey data, is much larger.

The DAWN file takes up roughly 270 MiB of disk storage, and while some computers can work with this file in memory, it can slow down other systems. To make this data more manageable in Python, we can, for example, load in a subset of the columns rather than all of them.

Sometimes we are interested in the total size of a folder instead of the size of individual files. For example, we have three restaurant files, and we might like to see whether we can combine all the data into a single dataframe. In the following code, we calculate the size of the *data* folder, including all files in it:

```
mib = 1024**2

total = 0
for filepath in Path('data').glob('*'):
    total += os.path.getsize(filepath) / mib

print(f'The data/ folder contains {total:.2f} MiB')
```

```
The data/ folder contains 271.80 MiB
```

As a rule of thumb, reading in a file using `pandas` usually requires at least five times the available memory as the file size. For example, reading in a 1 GiB file typically requires at least 5 GiB of available memory. Memory is shared by all programs running on a computer, including the operating system, web browsers, and Jupyter notebook itself. A computer with 4 GiB total RAM might have only 1 GiB available RAM with many applications running. With 1 GiB available RAM, it is unlikely that `pandas` will be able to read in a 1 GiB file.

There are several strategies for working with data that are far larger than what is feasible to load into memory. We describe a few of them next.

The popular term *big data* generally refers to the scenario where the data are large enough that even top-of-the-line computers can't read the data directly into memory. This is a common scenario in scientific domains like astronomy, where telescopes capture images of space that can be petabytes (2^{50}) in size. While not quite as big, social media giants, health-care providers, and other companies can also struggle with large amounts of data.

Figuring out how to draw insights from these datasets is an important research problem that motivates the fields of database engineering and distributed computing. While we won't cover these fields in this book, we provide a brief overview of basic approaches:

Subset the data.
: One simple approach is to work with portions of data. Rather than loading in the entire source file, we can either select a specific part of it (e.g., one day's worth of data) or randomly sample the dataset. Because of its simplicity, we use this approach quite often in this book. The natural downside is that we lose many of the benefits of analyzing a large dataset, like being able to study rare events.

Use a database system.
: As discussed in Chapter 7, relational database management systems (RDBMSs) are specifically designed to store large datasets. SQLite is a useful system for working with datasets that are too large to fit in memory but small enough to fit on disk for a single machine. For datasets that are too large to fit on a single machine, more scalable database systems like MySQL and PostgreSQL can be used. These systems can manipulate data that are too big to fit into memory by using SQL queries. Because of their advantages, RDBMSs are commonly used for data storage in research and industry settings. One downside is that they often require a separate server for the data that needs its own configuration. Another downside is that SQL is less flexible in what it can compute than Python, which becomes especially relevant for modeling. A useful hybrid approach is to use SQL

to subset, aggregate, or sample the data into batches that are small enough to read into Python. Then we can use Python for more sophisticated analyses.

Use a distributed computing system.
: Another approach to handling complex computations on large datasets is to use a distributed computing system like MapReduce, Spark, or Ray. These systems work best on tasks that can be split into many smaller parts where they divide datasets into smaller pieces and run programs on all of the smaller datasets at once. These systems have great flexibility and can be used in a variety of scenarios. Their main downside is that they can require a lot of work to install and configure properly because they are typically installed across many computers that need to coordinate with one another.

It can be convenient to use Python to determine a file format, encoding, and size. Another powerful tool for working with files is the shell; the shell is widely used and has a more succinct syntax than Python. In the next section, we introduce a few command-line tools available in the shell for carrying out the same tasks of finding out information about a file before reading it into a dataframe.

The Shell and Command-Line Tools

Nearly all computers provide access to a *shell interpreter*, such as `sh` or `bash` or `zsh`. These interpreters typically perform operations on the files on a computer with their own language, syntax, and built-in commands.

We use the term *command-line interface (CLI) tools* to refer to the commands available in a shell interpreter. Although we only cover a few CLI tools here, there are many useful CLI tools that enable all sorts of operations on files. For instance, the following command in the `bash` shell produces a list of all the files in the *figures/* folder for this chapter along with their file sizes:

```
$ ls -l -h figures/
```

The dollar sign is the shell prompt, showing the user where to type. It's not part of the command itself.

The basic syntax for a shell command is:

```
command -options arg1 arg2
```

CLI tools often take one or more *arguments*, similar to how Python functions take arguments. In the shell, we wrap arguments with spaces, not with parentheses or commas. The arguments appear at the end of the command line, and they are usually the name of a file or some text. In the `ls` example, the argument to `ls` is `figures/`. Additionally, CLI tools support *flags* that provide additional options. These flags are specified immediately following the command name using a dash as a delimiter. In the `ls` example, we provided the flags `-l` (to provide extra information about each file) and `-h` (to provide file sizes in a more human-readable format). Many commands have default arguments and options, and the `man` tool prints a list of acceptable options, examples, and defaults for any command. For example, `man ls` describes the 30 or so flags available for `ls`.

All CLI tools we cover in this book are specific to the `sh` shell interpreter, the default interpreter for Jupyter installations on macOS and Linux systems at the time of this writing. Windows systems have a different interpreter, and the commands shown in the book may not run on Windows, although Windows gives access to an `sh` interpreter through its Linux Subsystem.

The commands in this section can be run in a terminal application, or through a terminal opened by Jupyter.

We begin with an exploration of the filesystem containing the content for this chapter, using the `ls` tool:

```
$ ls

data                               wrangling_granularity.ipynb
figures                            wrangling_intro.ipynb
wrangling_command_line.ipynb       wrangling_structure.ipynb
wrangling_datasets.ipynb           wrangling_summary.ipynb
wrangling_formats.ipynb
```

To dive deeper and list the files in the *data/* directory, we provide the directory name as an argument to `ls`:

```
$ ls -l -L -h data/

total 556664
-rw-r--r--  1 nolan  staff   267M Dec 10 14:03 DAWN-Data.txt
-rw-r--r--  1 nolan  staff   645K Dec 10 14:01 businesses.csv
-rw-r--r--  1 nolan  staff    50K Jan 22 13:09 co2_mm_mlo.txt
-rw-r--r--  1 nolan  staff   455K Dec 10 14:01 inspections.csv
-rw-r--r--  1 nolan  staff   120B Dec 10 14:01 legend.csv
-rw-r--r--  1 nolan  staff   3.6M Dec 10 14:01 violations.csv
```

We added the `-l` flag to the command to get more information about each file. The file size appears in the fifth column of the listing, and it's more readable as specified by the `-h` flag. When we have multiple simple option flags like `-l`, `-h`, and `-L`, we can combine them together as a shorthand:

```
ls -lLh data/
```

When working with datasets in this book, our code will often use an additional `-L` flag for `ls` and other CLI tools, such as `du`. We do this because we set up the datasets in the book using shortcuts (called *symlinks*). Usually, your code won't need the `-L` flag unless you're working with symlinks too.

Other CLI tools for checking file size are `wc` and `du`. The command `wc` (short for *word count*) provides helpful information about a file's size in terms of the number of lines, words, and characters in the file:

```
$ wc data/DAWN-Data.txt
```

```
229211 22695570 280095842 data/DAWN-Data.txt
```

We can see from the output that *DAWN-Data.txt* has 229,211 lines and 280,095,842 characters. (The middle value is the file's word count, which is useful for files that contain sentences and paragraphs but not very useful for files containing data, such as FWF-formatted values.)

The `ls` tool does not calculate the cumulative size of the contents of a folder. To properly calculate the total size of a folder, including the files in the folder, we use `du` (short for *disk usage*). By default, the `du` tool shows the size in units called *blocks*:

```
$ du -L data/
```

```
556664  data/
```

We commonly add the `-s` flag to `du` to show the file sizes for both files and folders and the `-h` flag to display quantities in the standard KiB, MiB, or GiB format. The asterisk in `data/*` in the following code tells `du` to show the size of every item in the_data_ folder:

```
$ du -Lsh data/*
```

```
267M    data/DAWN-Data.txt
648K    data/businesses.csv
 52K    data/co2_mm_mlo.txt
456K    data/inspections.csv
4.0K    data/legend.csv
3.6M    data/violations.csv
```

To check the formatting of a file, we can examine the first few lines with the `head` command or the last few lines with `tail`. These CLIs are very useful for peeking at a file's contents to determine whether it's formatted as CSV, TSV, and so on. As an example, let's look at the *inspections.csv* file:

```
$ head -4 data/inspections.csv

"business_id","score","date","type"
19,"94","20160513","routine"
19,"94","20171211","routine"
24,"98","20171101","routine"
```

By default, `head` displays the first 10 lines of a file. If we want to show, say, four lines, then we add the option `-n 4` to our command (or just `-4` for short).

We can print the entire contents of the file using the `cat` command. However, you should take care when using this command, as printing a large file can cause a crash. The *legend.csv* file is small, and we can use `cat` to concatenate and print its contents:

```
$ cat data/legend.csv

"Minimum_Score","Maximum_Score","Description"
0,70,"Poor"
71,85,"Needs Improvement"
86,90,"Adequate"
91,100,"Good"
```

In many cases, using `head` or `tail` alone gives us a good enough sense of the file structure to proceed with loading it into a dataframe.

Finally, the `file` command can help us determine a file's encoding:

```
$ file -I data/*

data/DAWN-Data.txt:      text/plain; charset=us-ascii
data/businesses.csv:     application/csv; charset=iso-8859-1
data/co2_mm_mlo.txt:     text/plain; charset=us-ascii
data/inspections.csv:    application/csv; charset=us-ascii
data/legend.csv:         application/csv; charset=us-ascii
data/violations.csv:     application/csv; charset=us-ascii
```

We see (again) that all of the files are ASCII, except for *businesses.csv*, which has an ISO-8859-1 encoding.

Commonly, we open a terminal program to start a shell interpreter. However, Jupyter notebooks provide a convenience: if a line of code in a Python code cell is prefixed with the `!` character, the line will go directly to the system's shell interpreter. For example, running `!ls` in a Python cell lists the files in the current directory.

Shell commands give us a programmatic way to work with files, rather than a point-and-click "manual" approach. They are useful for the following:

Documentation
: If you need to record what you did.

Error reduction
: If you want to reduce typographical errors and other simple but potentially harmful mistakes.

Reproducibility
: If you need to repeat the same process in the future or you plan to share your process with others. This gives you a record of your actions.

Volume
: If you have many repetitive operations to perform, the size of the file you are working with is large, or you need to perform things quickly. CLI tools can help in all these cases.

After the data have been loaded into a dataframe, our next task is to figure out the table's shape and granularity. We start by finding the number of rows and columns in the table (its shape). Then we need to understand what a row represents before we begin to check the quality of the data. We cover these topics in the next section.

Table Shape and Granularity

As described earlier, we refer to a dataset's *structure* as a mental representation of the data, and in particular, we represent data that have a *table* structure by arranging values in rows and columns. We use the term *granularity* to describe what each row in the table represents, and the term *shape* quantifies the table's rows and columns.

Now that we have determined the format of the restaurant-related files, we load them into dataframes and examine their shapes:

```
bus = pd.read_csv('data/businesses.csv', encoding='ISO-8859-1')
insp = pd.read_csv("data/inspections.csv")
viol = pd.read_csv("data/violations.csv")

print(" Businesses:", bus.shape, "\t Inspections:", insp.shape,
     "\t Violations:", viol.shape)
```

```
 Businesses: (6406, 9)   Inspections: (14222, 4)   Violations: (39042, 3)
```

We find that the table with the restaurant information (the business table) has 6,406 rows and 9 columns. Now let's figure out the granularity of this table. To start, we can look at the first two rows:

	business_id	name	address	city	...	postal_code	latitude	longitude	phone_number
0	19	NRGIZE LIFESTYLE CAFE	1200 VAN NESS AVE, 3RD FLOOR	San Francisco	...	94109	37.79	-122.42	+14157763262

	business_id	name	address	city	...	postal_code	latitude	longitude	phone_number
1	24	OMNI S.F. HOTEL - 2ND FLOOR PANTRY	500 CALIFORNIA ST, 2ND FLOOR	San Francisco	...	94104	37.79	-122.40	+14156779494

```
2 rows × 9 columns
```

These two rows give us the impression that each record represents a particular restaurant. But, we can't tell from just two records whether or not this is the case. The field named `business_id` implies that it is the unique identifier for the restaurant. We can confirm this by checking whether the number of records in the dataframe matches the number of unique values in the field `business_id`:

```
print("Number of records:", len(bus))
print("Number of unique business ids:", len(bus['business_id'].unique()))
```

```
Number of records: 6406
Number of unique business ids: 6406
```

The number of unique `business_ids` matches the number of rows in the table, so it seems safe to assume that each row represents a restaurant. Since `business_id` uniquely identifies each record in the dataframe, we treat `business_id` as the *primary key* for the table. We can use primary keys to join tables (see Chapter 6). Sometimes a primary key consists of two (or more) features. This is the case for the other two restaurant files. Let's continue the examination of the inspections and violations dataframes and find their granularity.

Granularity of Restaurant Inspections and Violations

We just saw that there are many more rows in the inspection table than the business table. Let's take a closer look at the first few inspections:

	business_id	score	date	type
0	19	94	20160513	routine
1	19	94	20171211	routine
2	24	98	20171101	routine
3	24	98	20161005	routine

```
(insp
 .groupby(['business_id', 'date'])
 .size()
 .sort_values(ascending=False)
 .head(5)
)
```

```
business_id  date
64859        20150924    2
```

```
87440        20160801    2
77427        20170706    2
19           20160513    1
71416        20171213    1
dtype: int64
```

The combination of restaurant ID and inspection date uniquely identifies each record in this table, with the exception of three restaurants that have two records for their ID-date combination. Let's examine the rows for restaurant 64859:

```
insp.query('business_id == 64859 and date == 20150924')
```

	business_id	score	date	type
7742	64859	96	20150924	routine
7744	64859	91	20150924	routine

This restaurant got two different inspection scores on the same day! How could this happen? It may be that the restaurant had two inspections in one day, or it might be an error. We address these sorts of questions when we consider data quality in Chapter 9. Since there are only three of these double-inspection days, we can ignore the issue until we clean the data. So the primary key would be the combination of restaurant ID and inspection date if same-day inspections are removed from the table.

Note that the `business_id` field in the inspections table acts as a reference to the primary key in the business table. So `business_id` in `insp` is a *foreign key* because it links each record in the inspections table to a record in the business table. This means that we can readily join these two tables together.

Next, we examine the granularity of the third table, the one that contains the violations:

	business_id	date	description
0	19	20171211	Inadequate food safety knowledge or lack of ce...
1	19	20171211	Unapproved or unmaintained equipment or utensils
2	19	20160513	Unapproved or unmaintained equipment or utensi...
...	...	...	...
39039	94231	20171214	High risk vermin infestation [date violation...
39040	94231	20171214	Moderate risk food holding temperature [dat...
39041	94231	20171214	Wiping cloths not clean or properly stored or ...

```
39042 rows × 3 columns
```

Looking at the first few records in this table, we see that each inspection has multiple entries. The granularity appears to be at the level of a violation found in an inspection. Reading the descriptions, we see that if corrected, a date is listed in the description within square brackets:

```
viol.loc[39039, 'description']
```

```
'High risk vermin infestation  [ date violation corrected: 12/15/2017 ]'
```

In brief, we have found that the three food safety tables have different granularities. Since we have identified primary and foreign keys for them, we can potentially join these tables. If we are interested in studying inspections, we can join the violations and inspections together using the business ID and inspection date. This would let us connect the number of violations found during an inspection to the inspection score.

We can also reduce the inspection table to one per restaurant by selecting the most recent inspection for each restaurant. This reduced data table essentially has a granularity of restaurant and may be useful for a restaurant-based analysis. In Chapter 9, we cover these kinds of actions that reshape a data table, transform columns, and create new columns.

We conclude this section with a look at the shape and granularity of the DAWN survey data.

DAWN Survey Shape and Granularity

As noted earlier in this chapter, the DAWN file has fixed-width formatting, and we need to rely on a codebook to find out where the fields are. As an example, a snippet of the codebook in Figure 8-2 tells us that age appears in positions 34 and 35 in a row, and it is categorized into 11 age groups: 1 stands for age 5 and under, 2 for ages 6 to 11, ..., and 11 for ages 65 and older. Also, –8 represents a missing value.

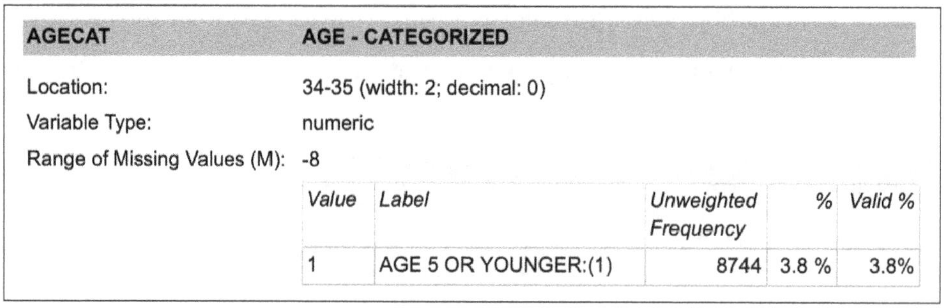

AGECAT **AGE - CATEGORIZED**

Location: 34-35 (width: 2; decimal: 0)

Variable Type: numeric

Range of Missing Values (M): -8

Value	*Label*	*Unweighted Frequency*	*%*	*Valid %*
1	AGE 5 OR YOUNGER:(1)	8744	3.8 %	3.8%

Figure 8-2. Screenshot of a portion of the DAWN coding for age

Earlier, we determined that the file contains 200,000 lines and over 280 million characters, so on average, there are about 1,200 characters per line. This might be why they used a fixed-width rather than a CSV format. Think how much larger the file would be if there was a comma between every field!

Given the tremendous amount of information on each line, let's read just a few features into a dataframe. We can use the `pandas.read_fwf` method to do this. We specify the exact positions of the fields to extract, and we provide names for these fields and other information about the header and index:

```
colspecs = [(0,6), (14,29), (33,35), (35, 37), (37, 39), (1213, 1214)]
varNames = ["id", "wt", "age", "sex", "race","type"]
dawn = pd.read_fwf('data/DAWN-Data.txt', colspecs=colspecs,
                   header=None, index_col=0, names=varNames)
```

	wt	age	sex	race	type
id					
1	0.94	4	1	2	8
2	5.99	11	1	3	4
3	4.72	11	2	2	4
4	4.08	2	1	3	4
5	5.18	6	1	3	8

We can compare the rows in the table to the number of lines in the file:

```
dawn.shape
```

```
(229211, 5)
```

The number of rows in the dataframe matches the number of lines in the file. That's good. The granularity of the dataframe is a bit complicated due to the survey design. Recall that these data are part of a large scientific study, with a complex sampling scheme. A row represents an emergency room visit, so the granularity is at the emergency room visit level. However, in order to reflect the sampling scheme and be representative of the population of all drug-related ER visits in a year, weights are provided. We must apply the weight to each record when we compute summary statistics, build histograms, and fit models. (The `wt` field contains these values.)

The weights take into account the chance of an ER visit like this one appearing in the sample. By "like this one" we mean a visit with similar features, such as the visitor age, race, visit location, and time of day. Let's examine the different values in `wt`:

```
dawn['wt'].value_counts()
```

```
wt
0.94     1719
84.26    1617
1.72     1435
         ...
1.51        1
3.31        1
3.33        1
Name: count, Length: 3500, dtype: int64
```

What Do These Weights Mean?

As a simplified example, suppose you ran a survey and 45% of your respondents were under 18 years of age, but according to the US Census Bureau, only 22% of the population is under 18. You can adjust your survey responses to reflect the US population by using a small weight (22/45) for those under 18 and a larger weight (78/55) for those 18 and older. To see how we might use these weights, suppose the respondents are asked whether they use Facebook:

Facebook	< 18	18+	Total
No	1	20	21
Yes	44	35	79
Total	45	55	100

Overall, 79% of the respondents say they are Facebook users, but the sample is skewed toward the younger generation. We can adjust the estimate with the weights so that the age groups match the population. Then the adjusted percentage of Facebook users drops to:

$$(22/45) \times 44 + (78/55) \times 35 = 71$$

The DAWN survey uses the same idea, except that it splits the groups much more finely.

It is critical to include the survey weights in your analysis to get data that represents the population at large. For example, we can compare the calculation of the proportion of females among the ER visits both with and without the weights:

```
print(f'Unweighted percent female: {np.average(dawn["sex"] == 2):.1%}')
print(f'  Weighted percent female:',
      f'{np.average(dawn["sex"] == 2, weights=dawn["wt"]):.1%}')
```

```
Unweighted percent female: 48.0%
  Weighted percent female: 52.3%
```

These figures differ by more than 4 percentage points. The weighted version is a more accurate estimate of the proportion of females among the entire population of drug-related ER visits.

Sometimes the granularity can be tricky to figure out, like we saw with the inspections data. And at other times, we need to take sampling weights into account, like for the DAWN data. These examples show it's important to take your time and review the data descriptions before proceeding with analysis.

Summary

Data wrangling is an essential part of data analysis. Without it, we risk overlooking problems in data that can have major consequences for future analysis. This chapter covered an important first step in data wrangling: reading data from a plain-text source file into a Python dataframe and identifying its granularity. We introduced different types of file formats and encodings, and we wrote code that can read data from these formats. We checked the size of source files and considered alternative tools for working with large datasets.

We also introduced command-line tools as an alternative to Python for checking the format, encoding, and size of a file. These CLI tools are especially handy for filesystem-oriented tasks because of their simple syntax. We've only touched the surface of what CLI tools can do. In practice, the shell is capable of sophisticated data processing and is well worth learning.

Understanding the shape and granularity of a table gives us insight into what a row in a data table represents. This helps us determine whether the granularity is mixed, aggregation is needed, or weights are required. After looking at the granularity of your dataset, you should have answers to the following questions:

What does a record represent?
: Clarity on this will help you correctly analyze data and state your findings.

Do all records in a table capture granularity at the same level?
: Sometimes a table contains additional summary rows that have a different granularity, and you want to use only those rows that are at the right level of detail.

If the data are aggregated, how was the aggregation performed?
: Summing and averaging are common types of aggregation. With averaged data, the variability in the measurements is typically reduced and relationships often appear stronger.

What kinds of aggregations might you perform on the data?
: Aggregations might be useful or necessary to combine one data table with another.

Knowing your table's granularity is a first step to cleaning your data, and it informs you of how to analyze the data. For example, we saw the granularity of the DAWN survey is an ER visit. That naturally leads us to think about comparisons of patient demographics to the US as a whole.

The wrangling techniques in this chapter help us bring data from a source file into a dataframe and understand its structure. Once we have a dataframe, further wrangling is needed to assess and improve quality and prepare the data for analysis. We cover these topics in the next chapter.

CHAPTER 9

Wrangling Dataframes

We often need to perform preparatory work on our data before we can begin our analysis. The amount of preparation can vary widely, but there are a few basic steps to move from raw data to data ready for analysis. Chapter 8 addressed the initial steps of creating a dataframe from a plain-text source. In this chapter, we assess quality. To do this, we perform validity checks on individual data values and entire columns. In addition to checking the quality of the data, we determine whether or not the data need to be transformed and reshaped to get ready for analysis. Quality checking (and fixing) and transformation are often cyclical: the quality checks point us toward transformations we need to make, and when we check the transformed columns to confirm that our data are ready for analysis, we may discover they need further cleaning.

Depending on the data source, we often have different expectations for quality. Some datasets require extensive wrangling to get them into an analyzable form, and others arrive clean and we can quickly launch into modeling. Here are some examples of data sources and how much wrangling we might expect to do:

- Data from a scientific experiment or study are typically clean, are well documented, and have a simple structure. These data are organized to be broadly shared so that others can build on or reproduce the findings. They are typically ready for analysis after little to no wrangling.
- Data from government surveys often come with very detailed codebooks and metadata describing how the data are collected and formatted, and these datasets are also typically ready for exploration and analysis right out of the box.
- Administrative data can be clean, but without inside knowledge of the source, we may need to extensively check their quality. Also, since we often use these data

for a purpose other than why they were collected in the first place, we may need to transform features or combine data tables.

- Informally collected data, such as data scraped from the web, can be quite messy and tends to come with little documentation. For example, texts, tweets, blogs, and Wikipedia tables usually require formatting and cleaning to transform them into information ready for analysis.

In this chapter, we break down data wrangling into the following stages: assess data quality, handle missing values, transform features, and reshape the data by modifying its structure and granularity. An important step in assessing the quality of the data is to consider its scope. Data scope was covered in Chapter 2, and we refer you there for a fuller treatment of the topic.

To clean and prepare data, we also rely on exploratory data analysis, especially visualizations. In this chapter, however, we focus on data wrangling and cover these other, related topics in more detail in Chapters 10 and 11.

We use the datasets introduced in Chapter 8: the DAWN government survey of emergency room visits related to drug abuse, and the San Francisco administrative data on food safety inspections of restaurants. But we begin by introducing the various data wrangling concepts through another example that is simple enough and clean enough that we can limit our focus in each of the wrangling steps.

Example: Wrangling CO_2 Measurements from the Mauna Loa Observatory

We saw in Chapter 2 that the National Oceanic and Atmospheric Administration (NOAA) (*https://www.noaa.gov*) monitors CO_2 concentrations in the air at the Mauna Loa Observatory (*https://oreil.ly/7HsQh*). We continue with this example and use it to introduce how to make data-quality checks, handle missing values, transform features, and reshape tables. These data are in the file *data/co2_mm_mlo.txt*. Let's begin by figuring out the formatting, encoding, and size of the source before we load it into a dataframe (see Chapter 8):

```
from pathlib import Path
import os
import chardet

co2_file_path = Path('data') / 'co2_mm_mlo.txt'

[os.path.getsize(co2_file_path),
 chardet.detect(co2_file_path.read_bytes())['encoding']]
```

```
[51131, 'ascii']
```

We have found that the file is plain text with ASCII encoding and about 50 KiB in size. Since the file is not particularly large, we should have no trouble loading it into a dataframe, but first we need to determine the file's format. Let's look at the first few lines in the file:

```
lines = co2_file_path.read_text().split('\n')
len(lines)
```

```
811
```

```
lines[:6]
```

```
['# --------------------------------------------------------------------',
 '# USE OF NOAA ESRL DATA',
 '# ',
 '# These data are made freely available to the public and the',
 '# scientific community in the belief that their wide dissemination',
 '# will lead to greater understanding and new scientific insights.']
```

We see that the file begins with information about the data source. We should read this documentation before starting our analysis, but sometimes the urge to plunge into the analysis wins over and we just start mucking about and discover properties of the data as we go. So let's quickly find where the actual data values are located:

```
lines[69:75]
```

```
['#',
 '#            decimal     average   interpolated    trend    #days',
 '#             date                              (season corr)',
 '1958   3    1958.208      315.71      315.71      314.62     -1',
 '1958   4    1958.292      317.45      317.45      315.29     -1',
 '1958   5    1958.375      317.50      317.50      314.71     -1']
```

We have found that the data begins on the 73rd line of the file. We also spot some relevant characteristics:

- The values are separated by whitespace, possibly tabs.
- The data line up in precise columns. For example, the month appears in the seventh to eighth position of each line.
- The column headings are split over two lines.

We can use `read_csv` to read the data into a `pandas` `DataFrame` and provide arguments to specify that the separators are whitespace, there is no header (we will set our own column names), and to skip the first 72 rows of the file:

```
co2 = pd.read_csv('data/co2_mm_mlo.txt',
                  header=None, skiprows=72, sep='\s+',
                  names=['Yr', 'Mo', 'DecDate', 'Avg', 'Int', 'Trend', 'days'])
co2.head(3)
```

	Yr	Mo	DecDate	Avg	Int	Trend	days
0	1958	3	1958.21	315.71	315.71	314.62	-1
1	1958	4	1958.29	317.45	317.45	315.29	-1
2	1958	5	1958.38	317.50	317.50	314.71	-1

We have successfully loaded the file contents into a dataframe, and we can see that the granularity of the data is a monthly average CO_2, from 1958 through 2019. Also, the table shape is 738 by 7.

Since scientific studies tend to have very clean data, it's tempting to jump right in and make a plot to see how CO_2 monthly averages have changed. The field `DecDate` conveniently represents the month and year as a numeric feature, so we can easily make a line plot:

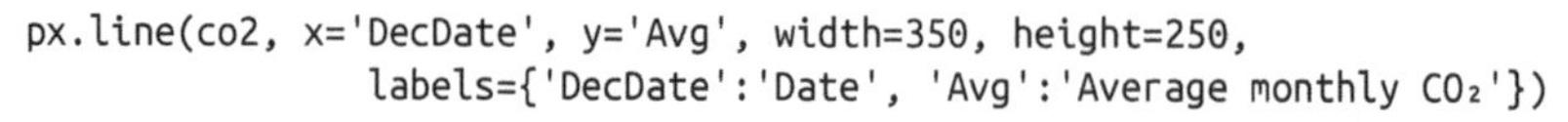

```
px.line(co2, x='DecDate', y='Avg', width=350, height=250,
        labels={'DecDate':'Date', 'Avg':'Average monthly CO₂'})
```

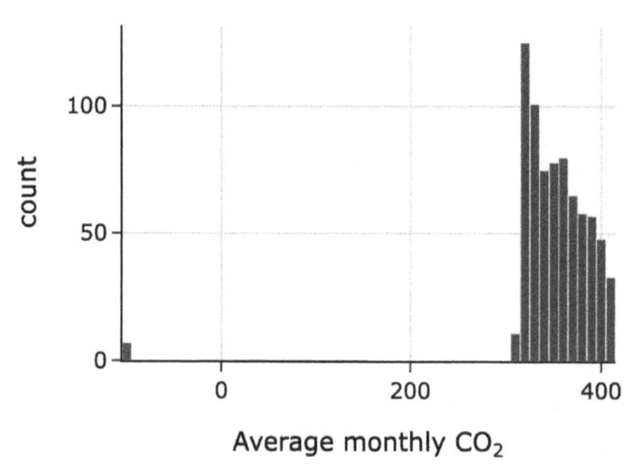

Yikes! Plotting the data has uncovered a problem. The four dips in the line plot look odd. What happened here? We can check a few percentiles of the dataframe to see if we can spot the problem:

```
co2.describe()[3:]
```

	Yr	Mo	DecDate	Avg	Int	Trend	days
min	1958.0	1.0	1958.21	-99.99	312.66	314.62	-1.0
25%	1973.0	4.0	1973.56	328.59	328.79	329.73	-1.0
50%	1988.0	6.0	1988.92	351.73	351.73	352.38	25.0
75%	2004.0	9.0	2004.27	377.00	377.00	377.18	28.0
max	2019.0	12.0	2019.62	414.66	414.66	411.84	31.0

This time, looking a bit more closely at the range of values, we see that some data have unusual values like `-1` and `-99.99`. If we read the information at the top of the file more carefully, we find that `-99.99` denotes a missing monthly average and `-1` signifies a missing value for the number of days the equipment was in operation that month. Even with relatively clean data, it's a good practice to read the documentation and make a few quality checks before jumping into the analysis stage.

Quality Checks

Let's step back for a moment and perform some quality checks. We might confirm that we have the expected number of observations, look for unusual values, and cross-check anomalies that we find against the values in other features.

First, we consider the shape of the data. How many rows should we have? From looking at the head and tail of the dataframe, the data appear to be in chronological order, beginning with March 1958 and ending with August 2019. This means we should have $12 \times (2019 - 1957) - 2 - 4 = 738$ records, which we can check against the shape of the dataframe:

```
co2.shape
```

```
(738, 7)
```

Our calculations match the number of rows in the data table.

Next, let's check the quality of the features, starting with `Mo`. We expect the values to range from 1 to 12, and each month should have 2019 – 1957 = 62 or 61 instances (since the recordings begin in March of the first year and end in August of the most recent year):

```
co2["Mo"].value_counts().reindex(range(1,13)).tolist()
```

```
[61, 61, 62, 62, 62, 62, 62, 62, 61, 61, 61, 61]
```

As expected, Jan, Feb, Sep, Oct, Nov, and Dec have 61 occurrences and the rest 62.

Now let's examine the column called `days` with a histogram:

```
px.histogram(co2, x='days', width=350, height=250,
             labels={'days':'Days operational in a month'})
```

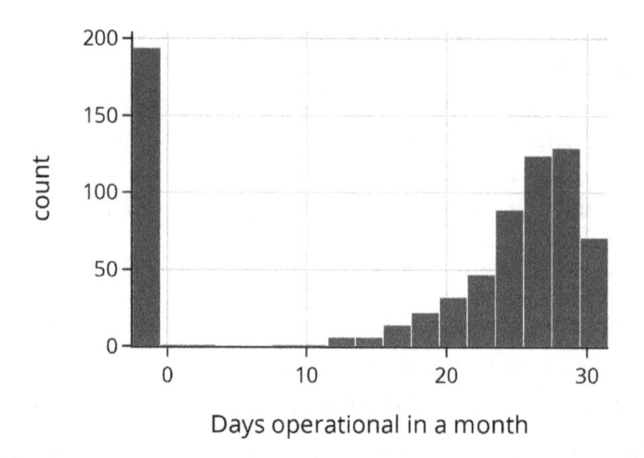

We see that a handful of months have averages based on measurements taken on fewer than half the days. In addition, there are nearly 200 missing values. A scatterplot can help us cross-check missing data against the year of the recording:

```
px.scatter(co2, x='Yr', y='days', width=350, height=250,
           labels={'Yr':'Year', 'days':'Days operational in month' })
```

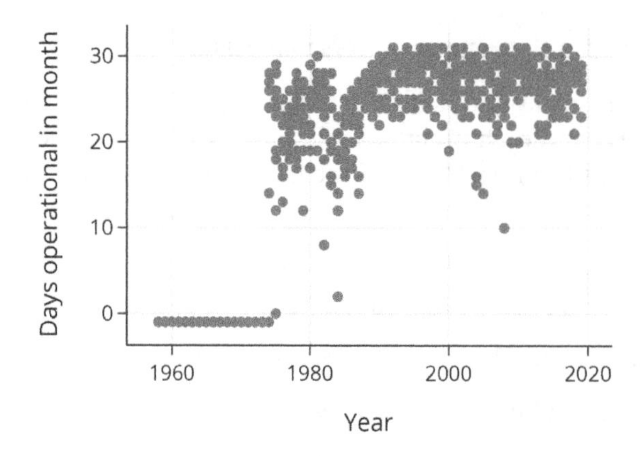

The line along the bottom left of the plot shows us that all of the missing data are in the early years of operation. The number of days of operation of the equipment may not have been collected in the early days. It also appears that there might have been problems with the equipment in the mid- to late '80s. What do we do with these conjectures? We can try to confirm them by looking through documentation about the historical readings. If we are concerned about the impact on the CO_2 averages for records with missing values for the number of days of operation, then a simple solution would be to drop the earliest recordings. However, we would want to delay such action until after we have examined the time trends and assess whether there are any potential problems with the CO_2 averages in those early days.

Next, let's return to the `-99.99` values for the average CO_2 measurement and begin our checks with a histogram:

```
px.histogram(co2, x='Avg', width=350, height=250,
             labels={'Avg':'Average monthly CO₂'})
```

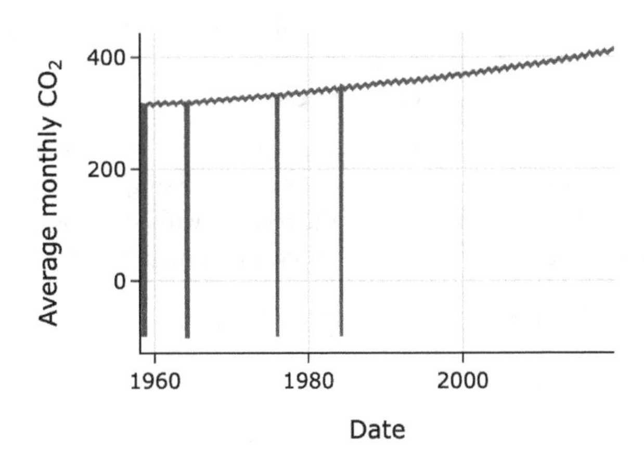

The recorded values are in the 300–400 range, which is what we expect based on our research into CO_2 levels. We also see that there are only a few missing values. Since there aren't many missing values, we can examine all of them:

```
co2[co2["Avg"] < 0]
```

	Yr	Mo	DecDate	Avg	Int	Trend	days
3	1958	6	1958.46	-99.99	317.10	314.85	-1
7	1958	10	1958.79	-99.99	312.66	315.61	-1
71	1964	2	1964.12	-99.99	320.07	319.61	-1
72	1964	3	1964.21	-99.99	320.73	319.55	-1
73	1964	4	1964.29	-99.99	321.77	319.48	-1
213	1975	12	1975.96	-99.99	330.59	331.60	0
313	1984	4	1984.29	-99.99	346.84	344.27	2

We are faced with the question of what to do with the `-99.99` values. We have seen already the problems of leaving these values as is in a line plot. There are several options, and we describe them next.

Addressing Missing Data

The `-99.99`s for average CO_2 levels indicate missing recordings. These interfere with our statistical summaries and plots. It's good to know which values are missing, but we need to do something about them. We might drop those records, replace `-99.99` with `NaN`, or substitute `99.99` with a likely value for the average CO_2. Let's examine each of these three options.

Note that the table already comes with a substitute value for the `-99.99`. The column labeled `Int` has values that exactly match those in `Avg`, except when `Avg` is `-99.99`, and then a "reasonable" estimate is used instead.

To see the effect of each option, let's zoom in on a short time period—say the measurements in 1958—where we know we have two missing values. We can create a time-series plot for the three cases: drop the records with `-99.99`s (left plot), use `NaN` for missing values (middle plot), and substitute an estimate for `-99.99` (right plot):

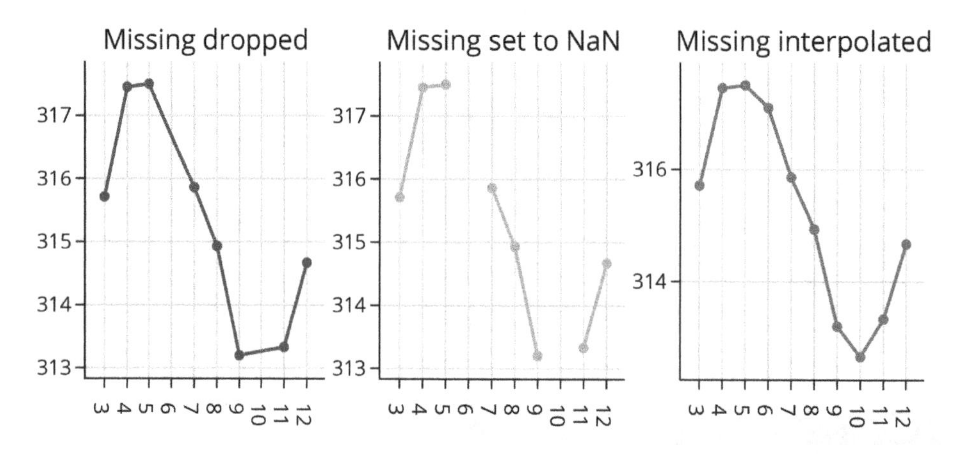

When we look closely, we can see the difference between each of these plots. The leftmost plot connects dots across a two-month time period, rather than one month. In the middle plot, the line breaks where the data are missing, and on the right, we can see that months 6 and 10 now have values. In the big picture, since there are only seven values missing from the 738 months, all of these options work. However, there is some appeal to the right plot since the seasonal trends are more cleanly discernible.

The method used to interpolate the CO_2 measurements for the missing values is an averaging process that takes into consideration the month and year. The idea is to reflect both seasonal changes and the long-term trend. This technique is described in greater detail in the documentation at the top of the datafile.

These plots have shown the granularity of the data to be monthly measurements, but other granularity options are available to us. We discuss this next.

Reshaping the Data Table

The CO_2 measurements taken at the Mauna Loa Observatory are also available both daily and hourly. The hourly data has a *finer granularity* than the daily data; reciprocally, the daily data is *coarser* than the hourly data.

Why not always just use the data with the finest granularity available? On a computational level, fine-grained data can become quite large. The Mauna Loa Observatory started recording CO_2 levels in 1958. Imagine how many rows the data table would contain if the facility provided measurements every single second! But more importantly, we want the granularity of the data to match our research question. Suppose we want to see whether CO_2 levels have risen over the past 50+ years, consistent with global warming predictions. We don't need a CO_2 measurement every second. In fact, we might well be content with yearly averages where the seasonal patterns are smoothed away. We can aggregate the monthly measurements, changing the granularity to annual averages, and make a plot to display the general trend. We can use *aggregation* to go to a coarser granularity—in `pandas`, we use `.groupby()` and `.agg()`:

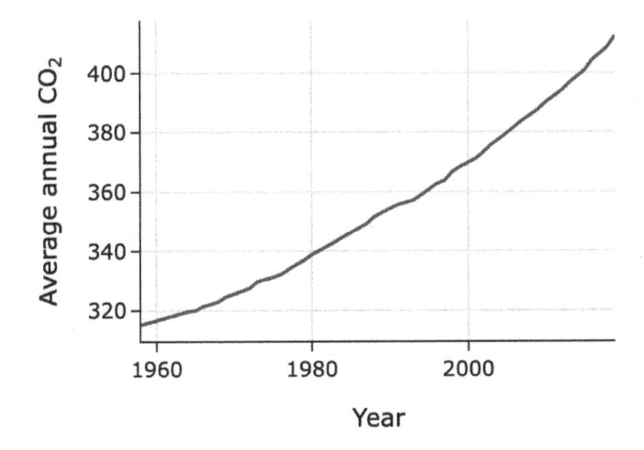

Indeed, we see a rise by nearly 100 ppm of CO_2 since Mauna Loa began recording in 1958.

To recap, after reading the whitespace-separated, plain-text file into a dataframe, we began to check its quality. We used the scope and context of the data to affirm that its shape matched the range of dates of collection. We confirmed that the values and counts for the month were as expected. We ascertained the extent of missing values in the features, and we looked for connections between missing values and other features. We considered three approaches to handling the missing data: drop records, work with `NaN` values, and impute values to have a full table. And, finally, we changed the granularity of the dataframe by rolling it up from a monthly to an annual average.

This change in granularity removed seasonal fluctuations and focused on the long-term trend in the level of CO_2 in the atmosphere. The next four sections of this chapter expand on these actions to wrangle data into a form suitable for analysis: quality checks, missing value treatments, transformations, and shape adjustments. We begin with quality checks.

Quality Checks

Once your data are in a table and you understand the scope and granularity, it's time to inspect for quality. You may have come across errors in the source as you examined and wrangled the file into a dataframe. In this section, we describe how to continue this inspection and carry out a more comprehensive assessment of the quality of the features and their values. We consider data quality from four vantage points:

Scope
: Do the data match your understanding of the population?

Measurements and values
: Are the values reasonable?

Relationships
: Are related features in agreement?

Analysis
: Which features might be useful in a future analysis?

We describe each of these points in turn, beginning with scope.

Quality Based on Scope

In Chapter 2, we addressed whether or not the data that have been collected can adequately address the problem at hand. There, we identified the target population, access frame, and sample in collecting the data. That framework helps us consider possible limitations that might impact the generalizability of our findings.

While these broader data-scope considerations are important as we deliberate our final conclusions, they are also useful for checking data quality. For example, for the San Francisco restaurant inspections data introduced in Chapter 8, a side investigation tells us that zip codes in the city should start with 941. But a quick check shows that several zip codes begin with other digits:

```
bus['postal_code'].value_counts().tail(10)
```

```
92672        1
64110        1
94120        1
            ..
94621        1
```

```
941033148    1
941          1
Name: postal_code, Length: 10, dtype: int64
```

This verification using scope helps us spot potential problems.

As another example, a bit of background reading at Climate.gov and NOAA (*https://oreil.ly/UBPDY*) on the topic of atmospheric CO_2 reveals that typical measurements are about 400 ppm worldwide. So we can check whether the monthly averages of CO_2 at Mauna Loa lie between 300 and 450 ppm.

Next, we check data values against codebooks and the like.

Quality of Measurements and Recorded Values

We can use also check the quality of measurements by considering what might be a reasonable value for a feature. For example, imagine what might be a reasonable range for the number of violations in a restaurant inspection: possibly, 0 to 5. Other checks can be based on common knowledge of ranges: a restaurant inspection score must be between 0 and 100; months run between 1 and 12. We can use documentation to tell us the expected values for a feature. For example, the type of emergency room visit in the DAWN survey, introduced in Chapter 8, has been coded as 1, 2, ..., 8 (see Figure 9-1). So we can confirm that all values for the type of visit are indeed integers between 1 and 8.

CASETYPE **TYPE OF VISIT**

Location: 1214-1214 (width: 1; decimal: 0)

Variable Type: numeric

Value	*Label*	*Unweighted Frequency*	*%*	*Valid %*
1	SUICICDE ATTEMPT:(1)	9033	3.9 %	3.9%
2	SEEKING DETOX:(2)	14841	6.5 %	6.5%
3	ALCOHOL ONLY (AGE < 21):(3)	7421	3.2 %	3.2%
4	ADVERSE REACTION:(4)	88096	38.4 %	38.4%
5	OVERMEDICATION:(5)	18146	7.9 %	7.9%
6	MALICIOUS POISONING:(6)	793	0.3 %	0.3%
7	ACCIDENTAL INGESTION:(7)	3253	1.4 %	1.4%
8	OTHER:(8)	87628	38.2 %	38.2%

Based upon 229211 valid cases out of 229211 total cases.

Figure 9-1. Screenshot of the description of the emergency room visit type (CASETYPE) variable in the DAWN survey (the typo SUICICDE appears in the actual codebook)

We also want to ensure that the data type matches our expectations. For example, we expect a price to be a number, whether or not it's stored as integer, floating point, or string. Confirming that the units of measurement match what is expected can be another useful quality check to perform (for example, weight values recorded in pounds, not kilograms). We can devise checks for all of these situations.

Other checks can be devised by comparing two related features.

Quality Across Related Features

At times, two features have built-in conditions on their values that we can cross-check for internal consistency. For example, according to the documentation for the DAWN study, alcohol consumption is only considered a valid reason for a visit to the ER for patients under age 21, so we can check that any record with "alcohol" for the type of visit has an age under 21. A cross-tabulation of the features `type` and `age` can confirm that this constraint is met:

```
display_df(pd.crosstab(dawn['age'], dawn['type']), rows=12)
```

type age	1	2	3	4	5	6	7	8
-8	2	2	0	21	5	1	1	36
1	0	6	20	6231	313	4	2101	69
2	8	2	15	1774	119	4	119	61
3	914	121	2433	2595	1183	48	76	4563
4	817	796	4953	3111	1021	95	44	6188
5	983	1650	0	4404	1399	170	48	9614
6	1068	1965	0	5697	1697	140	62	11408
7	957	1748	0	5262	1527	100	60	10296
8	1847	3411	0	10221	2845	113	115	18366
9	1616	3770	0	12404	3407	75	150	18381
10	616	1207	0	12291	2412	31	169	7109
11	205	163	0	24085	2218	12	308	1537

The cross-tabulation confirms that all of the alcohol cases (`type` is 3) have an age under 21 (these are coded as 1, 2, 3, and 4). The data values are as expected.

One last type of quality check pertains to the amount of information found in a feature.

Quality for Analysis

Even when data pass the previous quality checks, problems can arise with its usefulness. For example, if all but a handful of values for a feature are identical, then that feature adds little to the understanding of underlying patterns and relationships. Or if there are too many missing values, especially if there is a discernible pattern in the missing values, our findings may be limited. Plus, if a feature has many bad/corrupted values, then we might question the accuracy of even those values that fall in the appropriate range.

We see in the following code that the type of restaurant inspection in San Francisco can be either routine or from a complaint. Since only one of the 14,000+ inspections was from a complaint, we lose little if we drop this feature, and we might also want to drop that single inspection since it represents an anomaly:

```
pd.value_counts(insp['type'])
```

```
routine      14221
complaint        1
Name: type, dtype: int64
```

Once we find problems with our data, we need to figure out what to do.

Fixing the Data or Not

When you uncover problems with the data, essentially you have four options: leave the data as is, modify values, remove features, or drop records.

Leave it as is
: Not every unusual aspect of the data needs to be fixed. You might have discovered a characteristic of your data that will inform you about how to do your analysis and otherwise does not need correcting. Or you might find that the problem is relatively minor and most likely will not impact your analysis, so you can leave the data as is. Or, you might want to replace corrupted values with `NaN`.

Modify individual values
: If you have figured out what went wrong and can correct the value, then you can opt to change it. In this case, it's a good practice to create a new feature with the modified value and preserve the original feature, like in the CO_2 example.

Remove a column
: If many values in a feature have problems, then consider eliminating that feature entirely. Rather than excluding a feature, there may be a transformation that allows you to keep the feature while reducing the level of detail recorded.

Drop records
: In general, we do not want to drop a large number of observations from a dataset without good reason. Instead, try to scale back your investigation to a particular subgroup of the data that is clearly defined by some criteria, and does not simply correspond dropped records with corrupted values. When you discover that an unusual value is in fact correct, you still might decide to exclude the record from

your analysis because it's so different from the rest of your data and you do not want it to overly influence your analysis.

Whatever approach you take, you will want to study the possible impact of the changes that you make on your analysis. For example, try to determine whether the records with corrupted values are similar to one another, and different from the rest of the data.

Quality checks can reveal issues in the data that need to be addressed before proceeding with analysis. One particularly important type of check is to look for missing values. We suggested that there may be times when you want to replace corrupted data values with `NaN`, and hence treat them as missing. At other times, data might arrive missing. What to do with missing data is an important topic, and there is a lot of research on this problem; we cover ways to address missing data in the next section.

Missing Values and Records

In Chapter 3, we considered the potential problems when the population and the access frame are not in alignment, so we can't access everyone we want to study. We also described problems when someone refuses to participate in the study. In these cases, entire records/rows are missing, and we discussed the kinds of bias that can occur due to missing records. If nonrespondents differ in critical ways from respondents or if the nonresponse rate is not negligible, then our analysis may be seriously flawed. The example in Chapter 3 on election polls showed that increasing the sample size without addressing nonresponse does not reduce nonresponse bias. Also in that chapter, we discussed ways to prevent nonresponse. These preventive measures include using incentives to encourage response, keeping surveys short, writing clear questions, training interviewers, and investing in extensive follow-up procedures. Unfortunately, despite these efforts, some amount of nonresponse is unavoidable.

When a record is not entirely missing, but a particular field in a record is unavailable, we have nonresponse at the field level. Some datasets use a special coding to signify that the information is missing. We saw that the Mauna Loa data uses `-99.99` to indicate a missing CO_2 measurement. We found only seven of these values among 738 rows in the table. In this case, we showed that these missing values have little impact on the analysis.

The values for a feature are called *missing completely at random* when those records with the missing data are like a randomly chosen subset of records. That is, whether or not a record has a missing value does not depend on the unobserved feature, the values of other features, or the sampling design. For example, if someone accidentally breaks the laboratory equipment at Mauna Loa and CO_2 is not recorded for a day, there is no reason to think that the level of CO_2 that day had something to do with the lost measurements.

At other times, we consider values *missing at random given covariates* (covariates are other features in the dataset). For example, the type of an ER visit in the DAWN survey is missing at random given covariates if, say, the nonresponse depends only on race and sex (and not on the type of visit or anything else). In these limited cases, the observed data can be weighted to accommodate for nonresponse.

In some surveys, missing information is further categorized as to whether the respondent refused to answer, the respondent was unsure of the answer, or the interviewer didn't ask the question. Each of these types of missing values is recorded using a different value. For example, according to the codebook (*https://oreil.ly/lwBYh*), many questions in the DAWN survey use a code of `-7` for not applicable, `-8` for not documented, and `-9` for missing. Codings such as these can help us further refine our study of nonresponse.

After nonresponse has occurred, it is sometimes possible to use models to predict the missing data. We describe this process next. But remember, predicting missing observations is never as good as observing them in the first place.

At times, we substitute a reasonable value for a missing one to create a "clean" dataframe. This process is called *imputation*. Some common approaches for imputing values are *deductive*, *mean*, and *hot-deck* imputation.

In deductive imputation, we fill in a value through logical relationships with other features. For example, here is a row in the business dataframe for San Francisco restaurant inspections. The zip code is erroneously marked as "Ca" and latitude and longitude are missing:

```
bus[bus['postal_code'] == "Ca"]
```

	business_id	name	address	city	...	postal_code	latitude	longitude	phone_number
5480	88139	TACOLICIOUS	2250 CHESTNUT ST	San Francisco	...	Ca	NaN	NaN	+14156496077

```
1 row × 9 columns
```

We can look up the address on the USPS website to get the correct zip code, and we can use Google Maps to find the latitude and longitude of the restaurant to fill in these missing values.

Mean imputation uses an average value from rows in the dataset that aren't missing. As a simple example, if a dataset on test scores is missing scores for some students, mean imputation would fill in the missing value using the mean of the nonmissing scores. A key issue with mean imputation is that the variability in the imputed feature will be smaller because the feature now has values that are identical to the mean. This affects later analysis if not handled properly—for instance, confidence intervals will be smaller than they should be (these topics are covered in Chapter 17). The missing values for CO_2 in Mauna Loa used a more sophisticated averaging technique that included neighboring seasonal values.

Hot-deck imputation uses a chance process to select a value at random from rows that have values. As a simple example, hot-deck imputation could fill in missing test scores by randomly choosing another test score in the dataset. A potential problem with hot-deck imputation is that the strength of a relationship between the features might weaken because we have added randomness.

For mean and hot-deck imputation, we often impute values based on other records in the dataset that have similar values in other features. More sophisticated imputation techniques use nearest-neighbor methods to find similar subgroups of records and others use regression techniques to predict the missing value.

With all of these types of imputation, we should create a new feature that contains the altered data or a new feature to indicate whether or not the response in the original feature has been imputed so that we can track our changes.

Decisions to keep or drop a record with a missing value, to change a value, or to remove a feature may seem small, but they can be critical. One anomalous record can seriously impact your findings. Whatever you decide, be sure to check the impact of dropping or changing features and records. And be transparent and thorough in reporting any modifications you make to the data. It's best to make these changes programmatically to reduce potential errors and enable others to confirm exactly what you have done by reviewing your code.

The same transparency and reproducible precautions hold for data transformations, which we discuss next.

Transformations and Timestamps

Sometimes a feature is not in a form well-suited for analysis, and so we transform it. There are many reasons a feature might need a transformation: the value codings might not be useful for analysis, we may want to apply a mathematical function to a feature, or we might want to pull information out of a feature and create a new feature. We describe these three basic kinds of transformations: type conversions, mathematical transformations, and extractions:

Type conversion
: This kind of transformation occurs when we convert the data from one format to another to make the data more useful for analysis. We might convert information stored as a string to another format. For example, we would want to convert prices reported as strings to numbers (like changing the string `"$2.17"` to the number 2.17) so that we can compute summary statistics. Or we might want to

convert a time stored as a string, such as `"1955-10-12"`, to a `pandas Timestamp` object. Yet another example occurs when we lump categories together, such as reducing the 11 categories for age in DAWN to 5 groupings.

Mathematical transformation
: One kind of mathematical transformation is when we change the units of a measurement from, say, pounds to kilograms. We might make unit conversions so that statistics on our data can be directly compared to statistics on other datasets. Yet another reason to transform a feature is to make its distribution more symmetric (this notion is covered in more detail in Chapter 10). The most common transformation for handling asymmetry is the logarithm. Lastly, we might want to create a new feature from arithmetic operations. For example, we can combine heights and weights to create body mass indexes by calculating height/weight^2.

Extraction
: Sometimes we want to create a feature by extraction, where the new feature contains partial information taken from another feature. For example, the inspection violations consist of strings with descriptions of violations, and we may only be interested in whether the violation is related to, say, vermin. We can create a new feature that is `True` if the violation contains the word *vermin* in its text description and `False` otherwise. This conversion of information to logical values (or 0–1 values) is extremely useful in data science. The upcoming example in this chapter gives a concrete use-case for these binary features.

We cover many other examples of useful transformations in Chapter 10. For the rest of this section, we explain one more kind of transformation related to working with dates and times. Dates and times appear in many kinds of data, so it's worth learning how to work with these data types.

Transforming Timestamps

A *timestamp* is a data value that records a specific date and time. For instance, a timestamp could be recorded like `Jan 1 2020 2pm` or `2021-01-31 14:00:00` or `2017 Mar 03 05:12:41.211 PDT`. Timestamps come in many different formats! This kind of information can be useful for analysis, because it lets us answer questions like, "What times of day do we have the most website traffic?" When we work with timestamps, we often need to parse them for easier analysis.

Let's take a look at an example. The inspections dataframe for the San Francisco restaurants includes the date when restaurant inspections happened:

```
insp.head(4)
```

	business_id	score	date	type
0	19	94	20160513	routine
1	19	94	20171211	routine
2	24	98	20171101	routine
3	24	98	20161005	routine

By default, however, `pandas` reads in the `date` column as an integer:

```
insp['date'].dtype
```

```
dtype('int64')
```

This storage type makes it hard to answer some useful questions about the data. Let's say we want to know whether inspections happen more often on weekends or weekdays. To answer this question, we want to convert the `date` column to the `pandas` `Timestamp` storage type and extract the day of the week.

The date values appear to come in the format `YYYYMMDD`, where `YYYY`, `MM`, and `DD` correspond to the four-digit year, two-digit month, and two-digit day, respectively. The `pd.to_datetime()` method can parse the date strings into objects, where we can pass in the format of the dates as a date format (*https://oreil.ly/TFWcU*) string:

```
date_format = '%Y%m%d'

insp_dates = pd.to_datetime(insp['date'], format=date_format)
insp_dates[:3]
```

```
0   2016-05-13
1   2017-12-11
2   2017-11-01
Name: date, dtype: datetime64[ns]
```

We can see that `insp_dates` now has a `dtype` of `datetime64[ns]`, which means that the values were successfully converted into `pd.Timestamp` objects.[1]

`pandas` has special methods and properties for `Series` objects that hold timestamps using the `.dt` accessor. For instance, we can easily pull out the year for each timestamp:

```
insp_dates.dt.year[:3]
```

```
0    2016
1    2017
2    2017
Name: date, dtype: int32
```

1 This means that each uses 64 bits of memory for each value and that each is accurate to the nanosecond (or ns, for short).

The pandas documentation has the complete details on the .dt accessor (*https://oreil.ly/_ceNL*). By looking at the documentation, we see that the .dt.day_of_week attribute gets the day of the week for each timestamp (Monday = 0, Tuesday = 1, ..., Sunday = 6). So let's assign new columns to the dataframe that contain both the parsed timestamps and the day of the week:

```
insp = insp.assign(timestamp=insp_dates,
                   dow=insp_dates.dt.dayofweek)
insp.head(3)
```

	business_id	score	date	type	timestamp	dow
0	19	94	20160513	routine	2016-05-13	4
1	19	94	20171211	routine	2017-12-11	0
2	24	98	20171101	routine	2017-11-01	2

Now we can see whether restaurant inspectors favor a certain day of the week by grouping on the day of the week:

```
insp['dow'].value_counts().reset_index()
```

	dow	count
0	2	3281
1	1	3264
2	3	2497
3	0	2464
4	4	2101
5	6	474
6	5	141

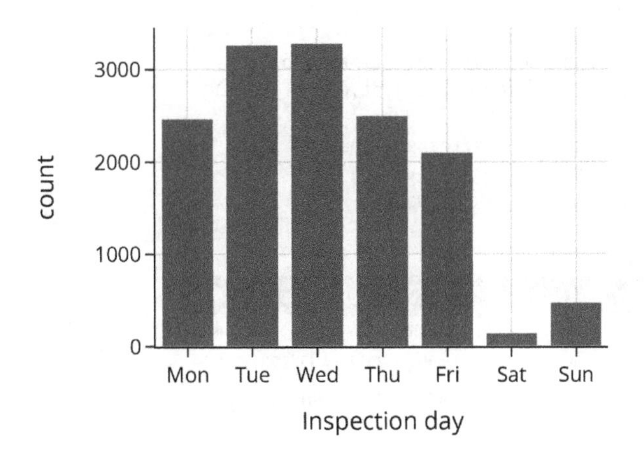

As expected, inspections rarely happen on the weekend. We also find that Tuesday and Wednesday are the most popular days for an inspection.

We have performed many wranglings on the inspections table. One approach to tracking these modifications is to pipe these actions from one to the next. We describe the idea of piping next.

Piping for Transformations

In data analyses, we typically apply many transformations to the data, and it is easy to introduce bugs when we repeatedly mutate a dataframe, in part because Jupyter notebooks let us run cells in any order we want. As a good practice, we recommend putting transformation code into functions with helpful names and using the `DataFrame.pipe()` method to chain transformations together.

Let's rewrite the earlier timestamp parsing code into a function and add the timestamps back into the dataframe as a new column, along with a second column containing the year of the timestamp:

```
date_format = '%Y%m%d'

def parse_dates_and_years(df, column='date'):
    dates = pd.to_datetime(df[column], format=date_format)
    years = dates.dt.year
    return df.assign(timestamp=dates, year=years)
```

Now we can pipe the `insp` dataframe through this function using `.pipe()`:

```
insp = (pd.read_csv("data/inspections.csv")
        .pipe(parse_dates_and_years))
```

We can chain many `.pipe()` calls together. For example, we can extract the day of the week from the timestamps:

```
def extract_day_of_week(df, col='timestamp'):
    return df.assign(dow=df[col].dt.day_of_week)

insp = (pd.read_csv("data/inspections.csv")
        .pipe(parse_dates_and_years)
        .pipe(extract_day_of_week))
insp
```

	business_id	score	date	type	timestamp	year	dow
0	19	94	20160513	routine	2016-05-13	2016	4
1	19	94	20171211	routine	2017-12-11	2017	0
2	24	98	20171101	routine	2017-11-01	2017	2
...	...	...	...	...	...	...	...
14219	94142	100	20171220	routine	2017-12-20	2017	2
14220	94189	96	20171130	routine	2017-11-30	2017	3

	business_id	score	date	type	timestamp	year	dow
14221	94231	85	20171214	routine	2017-12-14	2017	3

```
14222 rows × 7 columns
```

There are several key advantages of using `pipe()`. When there are many transformations on a single dataframe, it's easier to see what transformations happen since we can simply read the function names. Also, we can reuse transformation functions for different dataframes. For instance, the `viol` dataframe, which contains restaurant safety violations, also has a `date` column. This means we can use `.pipe()` to reuse the timestamp parsing function without needing to write extra code. Convenient!

```
viol = (pd.read_csv("data/violations.csv")
        .pipe(parse_dates_and_years))
viol.head(2)
```

	business_id	date	description	timestamp	year
0	19	20171211	Inadequate food safety knowledge or lack of ce...	2017-12-11	2017
1	19	20171211	Unapproved or unmaintained equipment or utensils	2017-12-11	2017

A different sort of transformation changes the shape of a dataframe by dropping unneeded columns, taking a subset of the rows, or rolling up the rows to a coarser granularity. We describe these structural changes next.

Modifying Structure

If a dataframe has an inconvenient structure, it can be difficult to do the analysis that we want. The wrangling process often reshapes the dataframe in some way to make the analysis easier and more natural. These changes can simply take a subset of the rows and/or columns from the table or change the table's granularity in a more fundamental way. In this section, we use the techniques from Chapter 6 to show how to modify structure in the following ways:

Simplify the structure
: If a dataframe has features that are not needed in our analysis, then we may want to drop these extraneous columns to make handling the dataframe easier. Or if we want to focus on a particular period of time or geographic area, we may want to take a subset of the rows (subsetting is covered in Chapter 6). In Chapter 8, we'll read into our dataframe a small set of features from the hundreds available in the DAWN survey because we are interested in understanding the patterns of types of ER visit by demographics of the patient. In Chapter 10, we'll restrict an analysis of home sale prices to one year and a few cities in an effort to reduce the impact of inflation and to better study the effect of location on sale price.

Adjust the granularity

In an earlier example in this chapter, CO_2 measurements were aggregated from monthly averages to yearly averages in order to better visualize annual trends. In the next section, we provide another example where we aggregate violation-level data to the inspection level so that it can be combined with the restaurant inspection scores. In both of these examples, we adjust the granularity of the dataframe to work with a coarser granularity by grouping together records and aggregating values. With the CO_2 measurements, we grouped the monthly values from the same year and then averaged them. Other common aggregations of a group are the number of records, sum, minimum, maximum, and first or last value in the group. The details of adjusting granularity of `pandas` dataframes can be found in Chapter 6, including how to group by multiple column values.

Address mixed granularity

At times, a dataset might have mixed granularity, where records are at different levels of detail. A common case is in data provided by government agencies where data at the county and state levels are included in the same file. When this happens, we usually want to split the dataframe into two, one at the county level and the other at the state level. This makes county-level and state-level analyses much easier, even feasible, to perform.

Reshape the structure

Data, especially from government sources, can be shared as pivot tables. These *wide* tables have data values as column names and are often difficult to use in analysis. We may need to reshape them into a *long* form. Figure 9-2 depicts the same data stored in both wide and long data tables. Each row of the wide data table maps to three rows in the long data table, as highlighted in the tables. Notice that in the wide data table, each row has three values, one for each month. In the long data table, each row only has a value for one month. Long data tables are generally easier to aggregate for future analysis. Because of this, long-form data is also frequently called *tidy data* (*https://doi.org/10.18637/jss.v059.i10*).

Wide-form:

	Jan	Feb	March
2001	10	20	30
2002	130	200	340

Long-form:

Year	Month	Set
2001	Jan	10
2001	Feb	20
2001	Mar	30
2002	Jan	130
2002	Feb	200
2002	Mar	340

Figure 9-2. An example of a wide data table (top) and a long data table (bottom) containing the same data

To demonstrate reshaping, we can put the CO_2 data into a wide dataframe that is like a pivot table in shape. There is a column for each month and a row for each year:

```
co2_pivot = pd.pivot_table(
    co2[10:34],
    index='Yr',    # Column to turn into new index
    columns='Mo',  # Column to turn into new columns
    values='Avg') # Column to aggregate

co2_wide = co2_pivot.reset_index()

display_df(co2_wide, cols=10)
```

Mo	Yr	1	2	3	4	...	8	9	10	11	12
0	1959	315.62	316.38	316.71	317.72	...	314.80	313.84	313.26	314.8	315.58
1	1960	316.43	316.97	317.58	319.02	...	315.91	314.16	313.83	315.0	316.19

```
2 rows × 13 columns
```

The column headings are months, and the cell values in the grid are the CO_2 monthly averages. We can turn this dataframe back into a long, aka *tall*, dataframe, where the column names become a feature, called `month`, and the values in the grid are reorganized into a second feature, called `average`:

```
co2_long = co2_wide.melt(id_vars=['Yr'],
                         var_name='month',
                         value_name='average')

display_df(co2_long, rows=4)
```

	Yr	month	average
0	1959	1	315.62
1	1960	1	316.43
...	...	...	...
22	1959	12	315.58
23	1960	12	316.19

```
24 rows × 3 columns
```

Notice that the data has been recaptured in its original shape (although the rows are not in their original order). Wide-form data is more common when we expect readers to look at the data table itself, like in an economics article or news story. But long-form data is more useful for data analysis. For instance, `co2_long` lets us write short `pandas` code to group by either year or month, while the wide-form data makes it difficult to group by year. The `.melt()` method is particularly useful for converting wide-form into long-form data.

These structural modifications have focused on a single table. However, we often want to combine information that is spread across multiple tables. In the next section, we combine the techniques introduced in this chapter to wrangle the restaurant inspection data and address joining tables.

Example: Wrangling Restaurant Safety Violations

We wrap up this chapter with an example that demonstrates many data wrangling techniques. Recall from Chapter 8 that the San Francisco restaurant inspection data are stored in three tables: `bus` (for businesses/restaurants), `insp` (for inspections), and `viol` (for safety violations). The violations dataset contains detailed descriptions of violations found during an inspection. We would like to capture some of this information and connect it to the inspection score, which is an inspection-level dataset.

Our goal is to figure out the kinds of safety violations associated with lower restaurant safety scores. This example covers several key ideas in data wrangling related to changing structure:

- Filtering to focus on a narrower segment of data
- Aggregation to modify the granularity of a table
- Joining to bring together information across tables

Additionally, an important part of this example demonstrates how we transform text data into numeric quantities for analysis.

As a first step, let's simplify the structure by reducing the data to inspections from one year. (Recall that this dataset contains four years of inspection information.) In the following code, we tally the number of records for each year in the inspections table:

```
pd.value_counts(insp['year'])
```

```
year
2016    5443
2017    5166
2015    3305
2018     308
Name: count, dtype: int64
```

Reducing the data to cover one year of inspections will simplify our analysis. Later, if we want, we can return to carry out an analysis with all four years of data.

Narrowing the Focus

We restrict our data wrangling to inspections that took place in 2016. Here, we can use the `pipe` function again in order to apply the same reshaping to both the inspections and violations dataframes:

```
def subset_2016(df):
    return df.query('year == 2016')

vio2016 = viol.pipe(subset_2016)
ins2016 = insp.pipe(subset_2016)
```

```
ins2016.head(5)
```

	business_id	score	date	type	timestamp	year
0	19	94	20160513	routine	2016-05-13	2016
3	24	98	20161005	routine	2016-10-05	2016
4	24	96	20160311	routine	2016-03-11	2016
6	45	78	20160104	routine	2016-01-04	2016
9	45	84	20160614	routine	2016-06-14	2016

In Chapter 8, we found that `business_id` and `timestamp` together uniquely identify the inspections (with a couple of exceptions). We also see here that restaurants can receive multiple inspections in a year—business #24 had two inspections in 2016, one in March and another in October.

Next, let's look at a few records from the violations table:

```
vio2016.head(5)
```

	business_id	date	description	timestamp	year
2	19	20160513	Unapproved or unmaintained equipment or utensi...	2016-05-13	2016
3	19	20160513	Unclean or degraded floors walls or ceilings ...	2016-05-13	2016
4	19	20160513	Food safety certificate or food handler card n...	2016-05-13	2016
6	24	20161005	Unclean or degraded floors walls or ceilings ...	2016-10-05	2016
7	24	20160311	Unclean or degraded floors walls or ceilings ...	2016-03-11	2016

Notice that the first few records are for the same restaurant. If we want to bring violation information into the inspections table, we need to address the different granularities of these tables. One approach is to aggregate the violations in some way. We discuss this next.

Aggregating Violations

One simple aggregation of the violations is to count them and add that count to the inspections data table. To find the number of violations at an inspection, we can group the violations by `business_id` and `timestamp` and then find the size of each group. Essentially, this grouping changes the granularity of violations to an inspection level:

```
num_vios = (vio2016
            .groupby(['business_id', 'timestamp'])
            .size()
            .reset_index()
            .rename(columns={0: 'num_vio'}));
num_vios.head(3)
```

	business_id	timestamp	num_vio
0	19	2016-05-13	3
1	24	2016-03-11	2
2	24	2016-10-05	1

Now we need to merge this new information with `ins2016`. Specifically, we want to *left-join* `ins2016` with `num_vios` because there could be inspections that do not have any violations and we don't want to lose them:

```
def left_join_vios(ins):
    return ins.merge(num_vios, on=['business_id', 'timestamp'], how='left')

ins_and_num_vios = ins2016.pipe(left_join_vios)
ins_and_num_vios
```

	business_id	score	date	type	timestamp	year	num_vio
0	19	94	20160513	routine	2016-05-13	2016	3.0
1	24	98	20161005	routine	2016-10-05	2016	1.0
2	24	96	20160311	routine	2016-03-11	2016	2.0
...	...	...	...	...	...	...	...
5440	90096	91	20161229	routine	2016-12-29	2016	2.0
5441	90268	100	20161229	routine	2016-12-29	2016	NaN
5442	90269	100	20161229	routine	2016-12-29	2016	NaN

```
5443 rows × 7 columns
```

When there are no violations at an inspection, the feature `num_vio` has a missing value (`NaN`). We can check how many values are missing:

```
ins_and_num_vios['num_vio'].isnull().sum()
```

```
833
```

About 15% of restaurant inspections in 2016 had no safety violations recorded. We can correct these missing values by setting them to 0 if the restaurant had a perfect safety score of 100. This is an example of deductive imputation since we're using domain knowledge to fill in missing values:

```
def zero_vios_for_perfect_scores(df):
    df = df.copy()
    df.loc[df['score'] == 100, 'num_vio'] = 0
    return df

ins_and_num_vios = (ins2016.pipe(left_join_vios)
                    .pipe(zero_vios_for_perfect_scores))
```

We can count the number of inspections with missing violation counts again:

```
ins_and_num_vios['num_vio'].isnull().sum()
```

```
65
```

We have corrected a large number of missing values. With further investigation, we find that some of the businesses have inspection dates that are close but don't quite match. We could do a fuzzy match where inspections with dates that are only one or two days apart are matched. But for now, we just leave them as `NaN`.

Let's examine the relationship between the number of violations and the inspection score:

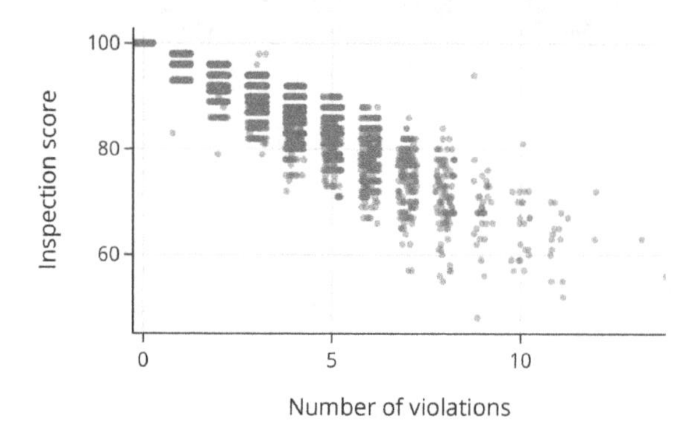

As we might expect, there is a negative relationship between the inspection score and the number of violations. We can also see variability in scores. The variability in scores grows with the number of violations. It appears that some violations are more serious than others and have a greater impact on the score. We extract information about the kinds of violations next.

Extracting Information from Violation Descriptions

We saw earlier that the feature description in the violations dataframe has a lot of text, including information in square brackets about when the violation was corrected. We can tally the descriptions and examine the most common violations:

```
display_df(vio2016['description'].value_counts().head(15).to_frame(), rows=15)
```

	description
Unclean or degraded floors walls or ceilings	161
Unapproved or unmaintained equipment or utensils	99
Moderate risk food holding temperature	95
Inadequate and inaccessible handwashing facilities	93
Inadequately cleaned or sanitized food contact surfaces	92
Improper food storage	81
Wiping cloths not clean or properly stored or inadequate sanitizer	71
Food safety certificate or food handler card not available	64
Moderate risk vermin infestation	58
Foods not protected from contamination	56
Unclean nonfood contact surfaces	54
Inadequate food safety knowledge or lack of certified food safety manager	52
Permit license or inspection report not posted	41
Improper storage of equipment utensils or linens	41
Low risk vermin infestation	34

Reading through these wordy descriptions, we see that some are related to the cleanliness of facilities, others to food storage, and still others to cleanliness of the staff.

Since there are many types of violations, we can try to group them together into larger categories. One way to do this is to create a simple boolean flag depending on whether the text contains a special term, like *vermin*, *hand*, or *high risk*.

With this approach, we create eight new features for different categories of violations. Don't worry about the particular details of the code for now—this code uses regular expressions, covered in Chapter 13. The important idea is that this code creates features containing `True` or `False` based on whether the violation description contains specific words:

```
def make_vio_categories(vio):
    def has(term):
        return vio['description'].str.contains(term)

    return vio[['business_id', 'timestamp']].assign(
        high_risk        = has(r"high risk"),
        clean            = has(r"clean|sanit"),
        food_surface     = (has(r"surface") & has(r"\Wfood")),
        vermin           = has(r"vermin"),
        storage          = has(r"thaw|cool|therm|storage"),
        permit           = has(r"certif|permit"),
        non_food_surface = has(r"wall|ceiling|floor|surface"),
        human            = has(r"hand|glove|hair|nail"),
    )

vio_ctg = vio2016.pipe(make_vio_categories)
vio_ctg
```

	business_id	timestamp	high_risk	clean	...	storage	permit	non_food_surface	human
2	19	2016-05-13	False	False	...	False	False	False	False
3	19	2016-05-13	False	True	...	False	False	True	False
4	19	2016-05-13	False	False	...	False	True	False	True
...	...	...	...	...	...	...	...	...	...
38147	89900	2016-12-06	False	False	...	False	False	False	False
38220	90096	2016-12-29	False	False	...	False	False	False	False
38221	90096	2016-12-29	False	True	...	False	False	True	False

```
15624 rows × 10 columns
```

Now that we have these new features in `vio_ctg`, we can find out whether certain violation categories are more impactful than others. For example, are restaurant scores impacted more for vermin-related violations than permit-related violations?

To do this, we want to first count up the violations per business. Then we can merge this information with the inspection information. First, let's sum the number of violations for each business:

```
vio_counts = vio_ctg.groupby(['business_id', 'timestamp']).sum().reset_index()
vio_counts
```

	business_id	timestamp	high_risk	clean	...	storage	permit	non_food_surface	human
0	19	2016-05-13	0	1	...	0	1	1	1
1	24	2016-03-11	0	2	...	0	0	2	0
2	24	2016-10-05	0	1	...	0	0	1	0
...	...	...	...	...	...	...	...	...	...
4803	89790	2016-11-29	0	0	...	0	0	0	1
4804	89900	2016-12-06	0	0	...	0	0	0	0
4805	90096	2016-12-29	0	1	...	0	0	1	0

```
4806 rows × 10 columns
```

Once again, we use a left join to merge these new features into the inspection-level dataframe. And for the special case of a score of 100, we set all of the new features to `0`:

```
feature_names = ['high_risk', 'clean', 'food_surface', 'vermin',
                 'storage', 'permit', 'non_food_surface', 'human']
def left_join_features(ins):
    return (ins[['business_id', 'timestamp', 'score']]
            .merge(vio_counts, on=['business_id', 'timestamp'], how='left'))

def zero_features_for_perfect_scores(ins):
    ins = ins.copy()
    ins.loc[ins['score'] == 100, feature_names] = 0
    return ins

ins_and_vios = (ins2016.pipe(left_join_features)
                .pipe(zero_features_for_perfect_scores))
ins_and_vios.head(3)
```

	business_id	timestamp	score	high_risk	...	storage	permit	non_food_surface	human
0	19	2016-05-13	94	0.0	...	0.0	1.0	1.0	1.0
1	24	2016-10-05	98	0.0	...	0.0	0.0	1.0	0.0
2	24	2016-03-11	96	0.0	...	0.0	0.0	2.0	0.0

```
3 rows × 11 columns
```

To see how each violation category relates to the score, we can make a collection of box plots that compare the score distributions with and without each violation. Since our focus here is on the data's patterns, not the visualization code, we hide the code (you can see it larger online (*https://oreil.ly/go29H*)):

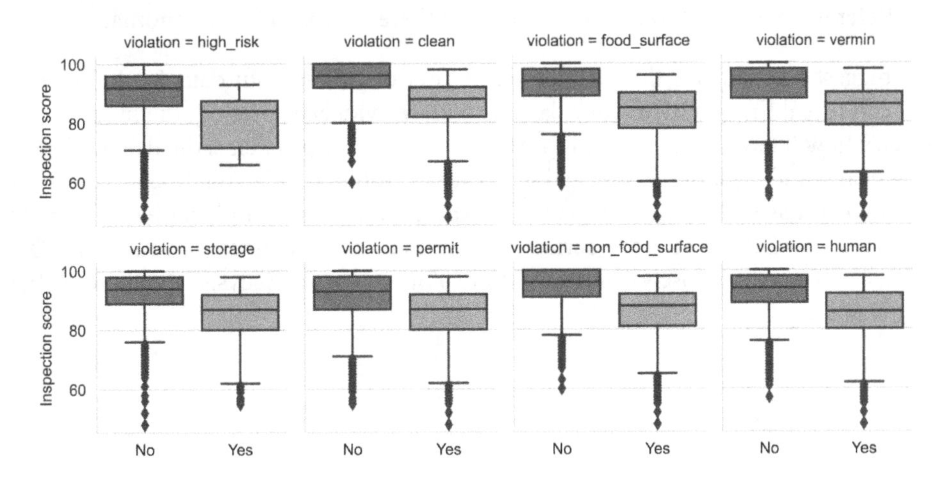

Summary

Data wrangling is an essential part of data analysis. Without it, we risk overlooking problems in data that can have major consequences for our future analysis. This chapter covered several important data wrangling steps that we use in nearly every analysis.

We described what to look for in a dataset after we've read it into a dataframe. Quality checks help us spot problems in the data. To find bad and missing values, we can take many approaches:

- Check summary statistics, distributions, and value counts. Chapter 10 provides examples and guidance on how to go about checking the quality of your data using visualizations and summary statistics. We briefly mentioned a few approaches here. A table of counts of unique values in a feature can uncover unexpected encodings and lopsided distributions, where one option is a rare occurrence. Percentiles can be helpful in revealing the proportion of values with unusually high (or low) values.
- Use logical expressions to identify records with values that are out of range or relationships that are out of whack. Simply computing the number of records that do not pass the quality check can quickly reveal the size of the problem.
- Examine the whole record for those records with problematic values in a particular feature. At times, an entire record is garbled when, for example, a comma is misplaced in a CSV-formatted file. Or the record might represent an unusual

situation (such as ranches being included in data on house sales), and you will need to decide whether it should be included in your analysis.

- Refer to an external source to figure out if there's a reason for the anomaly.

The biggest takeaway for this chapter is to be curious about your data. Look for clues that can reveal the quality of your data. The more evidence you find, the more confidence you will have in your findings. And if you uncover problems, dig deeper. Try to understand and explain any unusual phenomena. A good understanding of your data will help you assess whether an issue that you found is small and can be ignored or corrected, or whether it poses a serious limitation on the usefulness of your data. This curiosity mindset is closely connected to exploratory data analysis, the topic of the next chapter.

CHAPTER 10

Exploratory Data Analysis

More than 50 years ago, John Tukey avidly promoted an alternative type of data analysis that broke from the formal world of confidence intervals, hypothesis tests, and modeling. Today Tukey's *exploratory data analysis* (*https://oreil.ly/Himzi*) (EDA) is widely practiced. Tukey describes EDA (*https://oreil.ly/UO9F8*) as a philosophical approach to working with data:

> Exploratory data analysis is actively incisive, rather than passively descriptive, with real emphasis on the discovery of the unexpected.

As a data scientist, you will want to use EDA in every stage of the data lifecycle, from checking the quality of your data to preparing for formal modeling to confirming that your model is reasonable. Indeed, the work described in Chapter 9 to clean and transform the data relied heavily on EDA to guide our quality checks and transformations.

In EDA, we enter a process of discovery, continually asking questions and diving into uncharted territory to explore ideas. We use plots to uncover features of the data, examine distributions of values, and reveal relationships that cannot be detected from simple numerical summaries. This exploration involves transforming, visualizing, and summarizing data to build and confirm our understanding, identify and address potential issues with the data, and inform subsequent analysis.

EDA is fun! But it takes practice. One of the best ways to learn how to carry out EDA is to learn from others as they describe their thought process while they explore data, and we attempt to reveal EDA thinking in our examples and case studies in this book.

EDA can provide valuable insights, but you need to be cautious about the conclusions that you draw. It is important to recognize that EDA can bias your analysis. EDA is a winnowing process and a decision-making process that can impact the replicability of your later, model-based findings. With enough data and if you look hard, you often can dredge up something interesting that is entirely spurious.

The role of EDA in the scientific reproducibility crisis has been noted, and data scientists have cautioned against overdoing it. For example, Gelman and Loken (*https://doi.org/10.1511/2014.111.460*) note:

> Even in settings where a single analysis has been carried out on the given data, the issue of multiple comparisons [data dredging] emerges because different choices about combining variables, inclusion and exclusion of cases, transformations of variables, tests for interactions in the absence of main effects, and many other steps in the analysis could well have occurred with different data.

It's a good practice to report and provide the code from your EDA so that others are aware of the choices you made and the paths you took in learning about your data.

The topic of visualization is split across three chapters. In Chapter 9, we used plots to inform us in our data wrangling. The plots there were basic and the findings straightforward. We didn't dwell on interpretations and choices of plots. In this chapter, we spend more time learning how to choose the right plot and interpret it. We usually take the default parameter settings of the plotting functions since our goal is to make plots quickly as we carry out EDA. In Chapter 11, we'll provide guidelines for making effective and informative plots and give advice on how to make our visual argument clear and compelling.

According to Tukey (*https://oreil.ly/AIWW5*), visualization is central to EDA:

> The greatest gains from data come from surprises... The unexpected is best brought to our attention by pictures.

To make these pictures, we need to choose an appropriate type of plot, and our choice depends on the kinds of data that have been collected. This mapping between feature type and plot choice is the topic of the next section. From there, we go on to describe how to "read" a plot, what to look for, and how to interpret what you see. We first discuss what to look for in a one-feature plot, then focus on reading relationships between two features, and finally describe plots for three or more features. After we have introduced the visualization tools for EDA, we provide guidelines for carrying out an EDA and then walk through an example as we follow these guidelines.

Feature Types

Before making an exploratory plot, or any plot for that matter, it's a good idea to examine the feature (or features) and decide on its *feature type*. (Sometimes we refer to a feature as a *variable* and its type as *variable type*.) Although there are multiple ways of categorizing feature types, in this book we consider three basic ones. Ordinal and nominal data are subtypes of *categorical* data. Another name for categorical data is *qualitative*. In contrast, we also have *quantitative* features:

Nominal

A feature that represents "named" categories, where the categories do not have a natural ordering, is called nominal. Examples include political party affiliation (Democrat, Republican, Green, other); dog type (herding, hound, non-sporting, sporting, terrier, toy, working); and computer operating system (Windows, macOS, Linux).

Ordinal

Measurements that represent ordered categories are called ordinal. Examples of ordinal features are T-shirt size (small, medium, large), Likert-scale response (disagree, neutral, agree), and level of education (high school, college, graduate school). It is important to note that with an ordinal feature, the difference between, say, small and medium need not be the same as the difference between medium and large. Also, the differences between consecutive categories may not even be quantifiable. Think of the number of stars in a restaurant review and what one star means in comparison to two stars.

Quantitative

Data that represent numeric measurements or quantities are called quantitative. Examples include height measured to the nearest cm, price reported in USD, and distance measured to the nearest km. Quantitative features can be further divided into *discrete*, meaning that only a few values of the feature are possible, and *continuous*, meaning that the quantity could in principle be measured to arbitrary precision. The number of siblings in a family takes on a discrete set of values (such as 0, 1, 2, …, 8). In contrast, height can theoretically be reported to any number of decimal places, so we consider it continuous. There is no hard and fast rule to determine whether a quantity is discrete or continuous. In some cases, it can be a judgment call, and in others, we may want to purposefully consider a continuous feature to be discrete.

A feature type is not the same thing as a data storage type. Each column in a `pandas DataFrame` has its own *storage type*. These types can be integer, floating point, boolean, date-time format, category, and object (strings of varying length are stored as objects in Python with pointers to the strings). We use the term *feature type* to refer to a conceptual notion of the information and the term *storage type* to refer to the representation of the information in the computer.

A feature stored as an integer can represent nominal data, strings can be quantitative (like `"\$100.00"`), and, in practice, boolean values often represent nominal features that have only two possible values.

pandas calls the storage type `dtype`, which is short for data type. We refrain from using the term *data type* here because it can be confused with both storage type and feature type.

In order to determine a feature type, we often need to consult a dataset's *data dictionary* or *codebook*. A data dictionary is a document included with the data that describes what each column in the data table represents. In the following example, we take a look at the storage and feature types of the columns in a dataframe about various dog breeds, and we find that the storage type is often not a good indicator of the kind of information contained in a field.

Example: Dog Breeds

We use the American Kennel Club (AKC) (*https://www.akc.org*) data on registered dog breeds to introduce the various concepts related to EDA. The AKC, a nonprofit that was founded in 1884, has the stated mission to "advance the study, breeding, exhibiting, running and maintenance of purebred dogs." The AKC organizes events like the National Championship, Agility Invitational, and Obedience Classic, and mixed-breed dogs are welcome to participate in most events. The Information Is Beautiful (*https://informationisbeautiful.net*) website provides a dataset with information from the AKC on 172 breeds. Its visualization, Best in Show (*https://oreil.ly/amksD*), incorporates many features of the breeds and is fun to look at.

The AKC dataset contains several different kinds of features, and we have extracted a handful of them that show a variety of types of information. These features include the name of the breed; its longevity, weight, and height; and other information such as its suitability for children and the number of repetitions needed to learn a new trick. Each record in the dataset is a breed of dog, and the information provided is meant to be typical of that breed. Let's read the data into a dataframe:

```
dogs = pd.read_csv('data/akc.csv')
dogs
```

	breed	group	score	longevity	...	size	weight	height	repetition
0	Border Collie	herding	3.64	12.52	...	medium	NaN	51.0	<5
1	Border Terrier	terrier	3.61	14.00	...	small	6.0	NaN	15-25
2	Brittany	sporting	3.54	12.92	...	medium	16.0	48.0	5-15
...	...	...	...	...	...	...	...	...	...
169	Wire Fox Terrier	terrier	NaN	13.17	...	small	8.0	38.0	25-40
170	Wirehaired Pointing Griffon	sporting	NaN	8.80	...	medium	NaN	56.0	25-40
171	Xoloitzcuintli	non-sporting	NaN	NaN	...	medium	NaN	42.0	NaN

```
172 rows × 12 columns
```

A cursory glance at the table shows us that breed, group, and size appear to be strings, and the other columns are numbers. The summary of the dataframe, shown here, provides the index, name, count of non-null values, and `dtype` for each column:

```
dogs.info()
```

```
<class 'pandas.core.frame.DataFrame'>
RangeIndex: 172 entries, 0 to 171
Data columns (total 12 columns):
 #   Column          Non-Null Count  Dtype
---  ------          --------------  -----
 0   breed           172 non-null    object
 1   group           172 non-null    object
 2   score           87 non-null     float64
 3   longevity       135 non-null    float64
 4   ailments        148 non-null    float64
 5   purchase_price  146 non-null    float64
 6   grooming        112 non-null    float64
 7   children        112 non-null    float64
 8   size            172 non-null    object
 9   weight          86 non-null     float64
 10  height          159 non-null    float64
 11  repetition      132 non-null    object
dtypes: float64(8), object(4)
memory usage: 16.2+ KB
```

Several columns of this dataframe have a numeric computational type, as signified by `float64`, which means that the column can contain numbers other than integers. We also confirm that `pandas` encodes the string columns as the `object dtype`, rather than a `string dtype`. Notice that we guessed incorrectly that `repetition` is quantitative. Looking a bit more carefully at the data table, we see that `repetition` contains string values for ranges, such as `"<5"`, `"15-25"`, and `"25-40"`, so this feature is ordinal.

In computer architecture, a floating-point number, or "float" for short, refers to a number that can have a decimal component. We won't go in depth into computer architecture in this book, but we will point it out when it affects terminology, as in this case. The `dtype float64` says that the column contains decimal numbers that each take up 64 bits of space when stored in computer memory.

Additionally, `pandas` uses optimized storage types for numeric data, like `float64` or `int64`. However, it doesn't have optimizations for Python objects like strings, dictionaries, or sets, so these are all stored as the `object dtype`. This means that the storage type is ambiguous, but in most settings we know whether `object` columns contain strings or some other Python type.

Looking at the column storage types, we might guess `ailments` and `children` are quantitative features because they are stored as `float64` dtypes. But let's tally their unique values:

```
display_df(dogs['ailments'].value_counts(), rows=8)
```

```
ailments
0.0    61
1.0    42
2.0    24
4.0    10
3.0     6
5.0     3
9.0     1
8.0     1
Name: count, dtype: int64
```

```
dogs['children'].value_counts()
```

```
children
1.0    67
2.0    35
3.0    10
Name: count, dtype: int64
```

Both `ailments` and `children` only take on a few integer values. What does a value of `3.0` for `children` or `9.0` for `ailments` mean? We need more information to figure this out. The name of the column and how the information is stored in the dataframe is not enough. Instead, we consult the data dictionary shown in Table 10-1.

Table 10-1. AKC dog breed codebook

Feature	Description
`breed`	Dog breed, e.g., Border Collie, Dalmatian, Vizsla
`group`	American Kennel Club grouping (herding, hound, non-sporting, sporting, terrier, toy, working)
`score`	AKC score
`longevity`	Typical lifetime (years)
`ailments`	Number of serious genetic ailments
`purchase_price`	Average purchase price from puppyfind.com
`grooming`	Grooming required once every: 1 = day, 2 = week, 3 = few weeks
`children`	Suitability for children: 1 = high, 2 = medium, 3 = low
`size`	Size: small, medium, large
`weight`	Typical weight (kg)
`height`	Typical height from the shoulder (cm)
`repetition`	Number of repetitions to understand a new command: <5, 5–15, 15–25, 25–40, 40–80, >80

Although the data dictionary does not explicitly specify the feature types, the description is enough for us to figure out that the feature `children` represents the suitability of the breed for children, and a value of `1.0` corresponds to "high" suitability. We also find that the feature `ailments` is a count of the number of serious genetic ailments that dogs of this breed tend to have. Based on the codebook, we treat `children` as a categorical feature, even though it is stored as a floating-point number, and since low < medium < high, the feature is ordinal. Since `ailments` is a count, we treat it as a quantitative (numeric) type, and for some analyses we further define it as discrete because there are only a few possible values that `ailments` can take on.

The codebook also confirms that the features `score`, `longevity`, `purchase_price`, `weight`, and `height` are quantitative. The idea here is that numeric features have values that can be compared through differences. It makes sense to say that chihuahuas typically live about four years longer than dachshunds (16.5 versus 12.6 years). Another check is whether it makes sense to compare ratios of values: a dachshund is usually about five times heavier than a chihuahua (11 kg versus 2 kg). All of these quantitative features are continuous; only `ailments` is discrete.

The data dictionary descriptions for `breed`, `group`, `size`, and `repetition` suggest that these features are qualitative. Each variable has different, and yet commonly found, characteristics that are worth exploring a bit more. We do this by examining the counts of each unique value for the various features. We begin with `breed`:

```
dogs['breed'].value_counts()
```

```
breed
Border Collie       1
Great Pyrenees      1
English Foxhound    1
                   ..
Saluki              1
Giant Schnauzer     1
Xoloitzcuintli      1
Name: count, Length: 172, dtype: int64
```

The `breed` feature has 172 unique values—that's the same as the number of records in the dataframe—so we can think of `breed` as the *primary key* for the data table. By design, each dog breed has one record, and this `breed` feature determines the dataset's granularity. Although `breed` is also considered a nominal feature, it doesn't really make sense to analyze it. We do want to confirm that all values are unique and clean, but otherwise we would only use it to, say, label unusual values in a plot.

Next, we examine the feature `group`:

```
dogs['group'].value_counts()
```

```
group
terrier         28
```

```
sporting        28
working         27
hound           26
herding         25
toy             19
non-sporting    19
Name: count, dtype: int64
```

This feature has seven unique values. Since a dog breed labeled as "sporting" and another considered to be "toy" differ from each other in several ways, the categories cannot be easily reduced to an ordering. So we consider `group` a nominal feature. Nominal features do not provide meaning in even the direction of the differences.

Next, we examine the unique values and their counts for `size`:

```
dogs['size'].value_counts()
```

```
size
medium    60
small     58
large     54
Name: count, dtype: int64
```

The `size` feature has a natural ordering: small < medium < large, so it is ordinal. We don't know how the category "small" is determined, but we do know that a small breed is in some sense smaller than a medium-sized breed, which is smaller than a large one. We have an ordering, but differences and ratios don't make sense conceptually for this feature.

The `repetition` feature is an example of a quantitative variable that has been collapsed into categories to become ordinal. The codebook tells us that `repetition` is the number of times a new command needs to be repeated before the dog understands it:

```
dogs['repetition'].value_counts()
```

```
repetition
25-40     39
15-25     29
40-80     22
5-15      21
80-100    11
<5        10
Name: count, dtype: int64
```

The numeric values have been lumped together as `<5`, `5-15`, `15-25`, `25-40`, `40-80`, `80-100`, and notice that these categories have different widths. The first has 5 repetitions, while others are 10, 15, and 40 repetitions wide. The ordering is clear, but the gaps from one category to the next are not of the same magnitude.

Now that we have double-checked the values in the variables against the descriptions in the codebook, we can augment the data dictionary to include this additional information about the feature types. Our revised dictionary appears in Table 10-2.

Table 10-2. Revised AKC dog breed codebook

Feature	Description	Feature type	Storage type
`breed`	Dog breed, e.g., Border Collie, Dalmatian, Vizsla	primary key	string
`group`	AKC group (herding, hound, non-sporting, sporting, terrier, toy, working)	qualitative - nominal	string
score	AKC `score`	quantitative	floating point
`longevity`	Typical lifetime (years)	quantitative	floating point
`ailments`	Number of serious genetic ailments (0, 1, ..., 9)	quantitative - discrete	floating point
`purchase_price`	Average purchase price from puppyfind.com	quantitative	floating point
`grooming`	Groom once every: 1 = day, 2 = week, 3 = few weeks	qualitative - ordinal	floating point
`children`	Suitability for children: 1 = high, 2 = medium, 3 = low	qualitative - ordinal	floating point
`size`	Size: small, medium, large	qualitative - ordinal	string
`weight`	Typical weight (kg)	quantitative	floating point
`height`	Typical height from the shoulder (cm)	quantitative	floating point
`repetition`	Number of repetitions to understand a new command: <5, 5–15, 15–25, 25–40, 40–80, 80–100	qualitative - ordinal	string

This sharper understanding of the feature types of the AKC data helps us make quality checks and transformations. We discussed transformations in Chapter 9, but there are a few additional transformations that were not covered. These pertain to categories of qualitative features, and we describe them next.

Transforming Qualitative Features

Whether a feature is nominal or ordinal, we may find it useful to relabel categories so that they are more informative, collapse categories to simplify a visualization, and even convert a numeric feature to ordinal to focus on particular transition points. We explain when we may want to make each of these transformations and give examples.

Relabel categories

Summary statistics, like the mean and the median, make sense for quantitative data, but typically not for qualitative data. For example, the average price for toy breeds makes sense to calculate ($687), but the "average" breed suitability for children doesn't. However, `pandas` will happily compute the mean of the values in the `children` column if we ask it to:

```
# Don't use this value in actual data analysis!
dogs["children"].mean()
```

```
1.4910714285714286
```

Instead, we want to consider the distribution of 1s, 2s, and 3s of `children`.

The key difference between storage types and feature types is that storage types say what operations we can write code to *compute*, while feature types say what operations *make sense for the data*.

We can transform `children` by replacing the numbers with their string descriptions. Changing 1, 2, and 3 into high, medium, and low makes it easier to recognize that `children` is categorical. With strings, we would not be tempted to compute a mean, the categories would be connected to their meaning, and labels for plots would have reasonable values by default. For example, let's focus on just the toy breeds and make a bar plot of suitability for children. First, we create a new column with the categories of suitability as strings:

```
kids = {1:"high", 2:"medium", 3:"low"}
dogs = dogs.assign(kids=dogs['children'].replace(kids))
```

```
dogs
```

	breed	group	score	longevity	...	weight	height	repetition	kids
0	Border Collie	herding	3.64	12.52	...	NaN	51.0	<5	low
1	Border Terrier	terrier	3.61	14.00	...	6.0	NaN	15-25	high
2	Brittany	sporting	3.54	12.92	...	16.0	48.0	5-15	medium
...	...	...	...	...	...	...	...	...	...
169	Wire Fox Terrier	terrier	NaN	13.17	...	8.0	38.0	25-40	NaN
170	Wirehaired Pointing Griffon	sporting	NaN	8.80	...	NaN	56.0	25-40	NaN
171	Xoloitzcuintli	non-sporting	NaN	NaN	...	NaN	42.0	NaN	NaN

```
172 rows × 13 columns
```

Then we can make the bar plot of counts of each category of suitability among the toy breeds:

```
toy_dogs = dogs.query('group == "toy"').groupby('kids').count().reset_index()
px.bar(toy_dogs, x='kids', y='breed', width=350, height=250,
       category_orders={"kids": ["low", "medium", "high"]},
       labels={"kids": "Suitability for children", "breed": "count"})
```

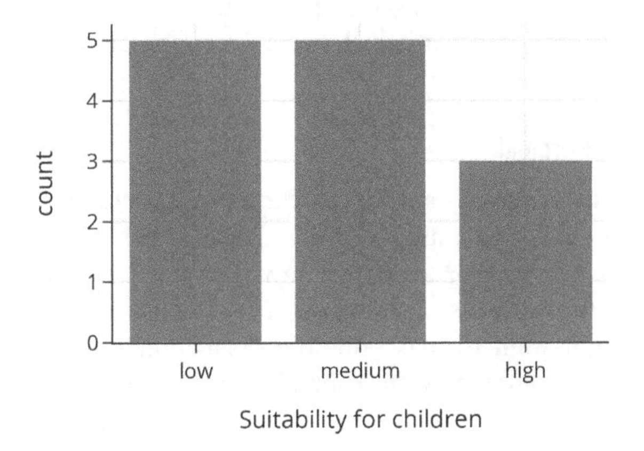

We do not always want to have categorical data represented by strings. Strings generally take up more space to store, which can greatly increase the size of a dataset if it contains many categorical features.

At times, a qualitative feature has many categories and we prefer a higher-level view of the data, so we collapse categories.

Collapse categories

Let's create a new column, called `play`, to represent the groups of dogs whose "purpose" is to play (or not). (This is a fictitious distinction used for demonstration purposes.) This category consists of the toy and non-sporting breeds. The new feature, `play`, is a transformation of the feature `group` that collapses categories: toy and non-sporting are combined into one category, and the remaining categories are placed in a second, non-play category. The boolean (`bool`) storage type is useful to indicate the presence or absence of this characteristic:

```
with_play = dogs.assign(play=(dogs["group"] == "toy") |
                              (dogs["group"] == "non-sporting"))
```

Representing a two-category qualitative feature as a boolean has a few advantages. For example, the mean of `play` makes sense because it returns the fraction of `True` values. When booleans are used for numeric calculations, `True` becomes 1 and `False` becomes 0:

```
with_play['play'].mean()
```

```
0.22093023255813954
```

This storage type gives us a shortcut to compute counts and averages of boolean values. In Chapter 15, we'll see that it's also a handy encoding for modeling.

There are also times, like when a discrete quantitative feature has a long tail, that we want to truncate the higher values, which turns the quantitative feature into an ordinal. We describe this next.

Convert quantitative to ordinal

Finally, another transformation that we sometimes find useful is to convert numeric values into categories. For example, we might collapse the values in `ailments` into categories: 0, 1, 2, 3, 4+. In other words, we turn `ailments` from a quantitative feature into an ordinal feature with the mapping 0→0, 1→1, 2→2, 3→3, and any value 4 or larger→4+. We might want to make this transformation because few breeds have more than three genetic ailments. This simplification can be clearer and adequate for an investigation.

As of this writing (late 2022), `pandas` also implements a `category dtype` that is designed to work with qualitative data. However, this storage type is not yet widely adopted by the visualization and modeling libraries, which limits its usefulness. For that reason, we do not transform our qualitative variables into the `category dtype`. We expect that future readers may want to use the `category dtype` as more libraries support it.

When we convert a quantitative feature to ordinal, we lose information. We can't go back. That is, if we know the number of ailments for a breed is four or more, we can't re-create the actual numeric value. The same thing happens when we collapse categories. For this reason, it's a good practice to keep the original feature. If we need to check our work or change categories, we can document and re-create our steps.

In general, the feature type helps us figure out what kind of plot is most appropriate. We discuss the mapping between feature type and plots next.

The Importance of Feature Types

Feature types guide us in our data analysis. They help specify the operations, visualizations, and models we can meaningfully apply to the data. Table 10-3 matches the feature type(s) to the various kinds of plots that are typically good options. Whether the variable(s) are quantitative or qualitative generally determines the set of viable plots to make, although there are exceptions. Other factors that enter into the decision are the number of observations and whether the feature takes on only a few distinct values. For example, we might make a bar chart, rather than a histogram, for a discrete quantitative variable.

Table 10-3. Mapping feature types to plots

Feature type	Dimension	Plot
Quantitative	One feature	Rug plot, histogram, density curve, box plot, violin plot
Qualitative	One feature	Bar plot, dot plot, line plot, pie chart
Quantitative	Two features	Scatterplot, smooth curve, contour plot, heat map, quantile-quantile plot
Qualitative	Two features	Side-by-side bar plots, mosaic plot, overlaid lines
Mixed	Two features	Overlaid density curves, side-by-side box plots, overlaid smooth curves, quantile-quantile plot

The feature type also helps us decide the kind of summary statistics to calculate. With qualitative data, we usually don't compute means or standard deviations, and instead compute the count, fraction, or percentage of records in each category. With a quantitative feature, we compute the mean or median as a measure of center, and, respectively, the standard deviation or inner quartile range (75th percentile to 25th percentile) as a measure of spread. In addition to the quartiles, we may find other percentiles informative.

The *n*th percentile is that value *q* such that *n% of the data values fall at or below it.* The value *q* might not be unique, and there are several approaches to select a unique value from the possibilities. With enough data, there should be little difference between these definitions.

To compute percentiles in Python, we prefer using:

```
np.percentile(data, method='lower')
```

When exploring data, we need to know how to interpret the shapes that our plots reveal. The next three sections give guidance with this interpretation. We also introduce many of the types of plots listed in Table 10-3 through the examples. Others are introduced in Chapter 11.

What to Look For in a Distribution

Visual displays of a feature can help us see patterns in observations; they are often much better than direct examination of the numbers or strings themselves. The simple rug plot locates each observation as a "yarn" in a "rug" along an axis. The rug plot can be useful when we have a handful of observations, but it soon gets difficult to distinguish high-density (most-populated) regions with, say, even 100 values. The following figure shows a rug plot with about 150 longevity values for dog breeds along the top of a histogram:

```
px.histogram(dogs, x="longevity", marginal="rug", nbins=20,
             histnorm='percent', width=350, height=250,
             labels={'longevity':'Typical lifespan (yr)'})
```

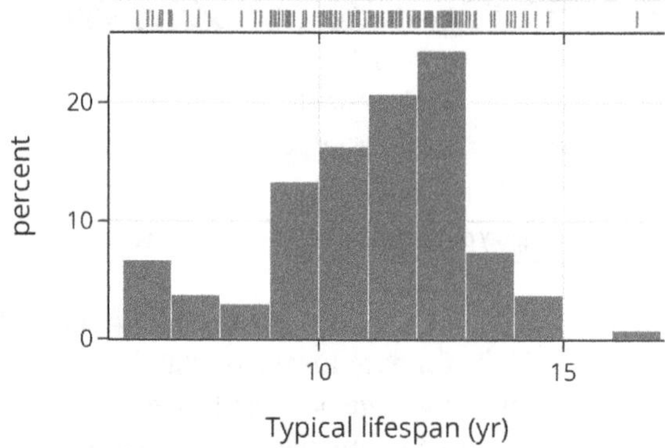

Although we can see an unusually large value that's greater than 16 in the rug plot, it's hard to compare the density of yarns in the other regions. Instead, the histogram gives a much better sense of the density of observations for various longevity values. Similarly, the *density curve* shown in the following figure gives a picture of the regions of high and low density:

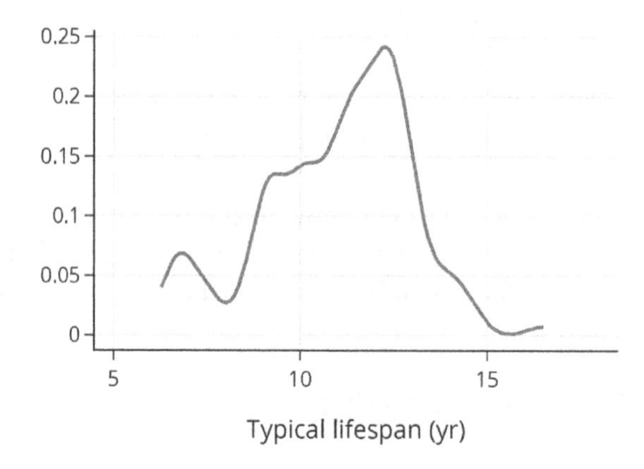

In both the histogram and density curve, we can see that the distribution of longevity is asymmetric. There is one main mode around 12 years and a shoulder in the 9-to-11-year range, meaning that while 12 is the most common longevity, many breeds have a longevity one to three years shorter than 12. We also see a small secondary mode around 7, and a few breeds with longevity as long as 14 to 16 years.

When interpreting a histogram or density curve, we examine the symmetry and skewness of the distribution; the number, location, and size of high-frequency regions (modes); the length of tails (often in comparison to a bell-shaped curve); gaps where no values are observed; and unusually large or anomalous values. Figure 10-1 provides a characterization of a distribution with several of these features. When we read a distribution, we connect the features that we see in the plot to the quantity measured.

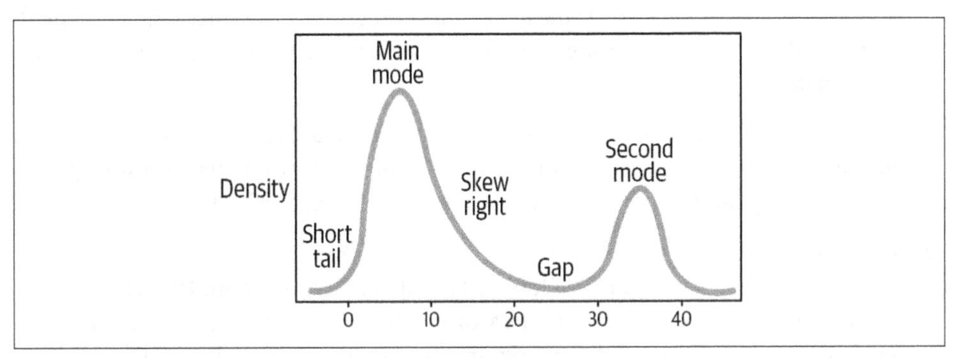

Figure 10-1. Example density plot identifying qualities of a distribution based on its shape

As another example, the distribution of the number of ailments in dog breeds appears in the following histogram:

```
bins = [-0.5, 0.5, 1.5, 2.5, 3.5, 9.5]
g = sns.histplot(data=dogs, x="ailments", bins=bins, stat="density")
g.set(xlabel='Number of ailments', ylabel='density');
```

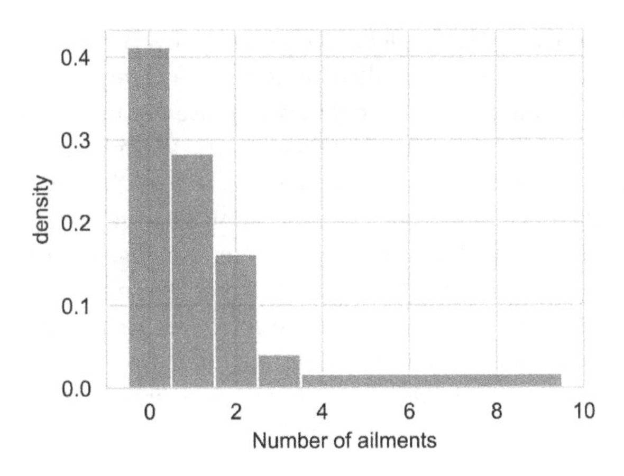

A value of 0 means this breed has no genetic ailments, 1 corresponds to one genetic ailment, and so on. From the histogram, we see that the distribution of ailments is unimodal with a peak at 0. We also see that the distribution is heavily skewed to the right, with a long right tail indicating that few breeds have between four and nine genetic ailments. Although quantitative, ailments is discrete because only a few integer values are possible. For this reason, we centered the bins on the integers so that the bin from 1.5 to 2.5 contains only those breeds with two ailments. We also made the rightmost bin wider. We lumped into one bin all of the breeds with four to nine ailments. When bin counts are small, we use wider bins to further smooth the distribution because we do not want to read too much into the fluctuations of small numbers. In this case, none of the breeds have six or seven ailments, but some have four, five, eight, or nine.

Next, we point out three key aspects of histograms and density curves: the y-axis should be on a density scale, smoothing hides unimportant details, and histograms are fundamentally different from bar plots. We describe each in turn:

Density in the y-axis
: The y-axes in the histograms of longevity and ailments are both labeled "density." This label implies that the total area of the bars in the histogram equals 1. To explain, we can think of the histogram as a skyline with tall buildings having denser populations, and we find the fraction of observations in any bin from the area of the rectangle. For example, the rectangle that runs from 3.5 to 9.5 in the ailments histogram contains about 10% of the breeds: 6 (width) × 0.017 (height) is roughly 0.10. If all of the bins are the same width, then the "skyline" will look the same whether the y-axis represents counts or density. But changing the y-axis to counts in this histogram would give a misleading picture of a very large rectangle in the right tail.

Smoothing
: With a histogram we hide the details of individual yarns in a rug plot in order to view the general features of the distribution. Smoothing refers to this process of replacing sets of points with rectangles; we choose not to show every single point in the dataset in order to reveal broader trends. We might want to smooth out these points because this is a sample and we believe that other values near the ones we observed are reasonable, and/or we want to focus on general structure rather than individual observations. Without the rug, we can't tell where the points are in a bin. Smooth density curves, like the one we showed earlier for longevity, also have the property that the total area under the curve sums to 1. The density curve uses a smooth *kernel* function to spread out the individual yarns and is sometimes referred to as a *kernel density estimate* (KDE).

Bar plot ≠ histogram
: With qualitative data, the bar plot serves a similar role to the histogram. The bar plot gives a visual presentation of the "popularity" or frequency of different groups. However, we cannot interpret the shape of the bar plot in the same way as a histogram. Tails and symmetry do not make sense in this setting. Also, the frequency of a category is represented by the height of the bar, and the width carries no information. The two bar charts that follow display identical information

about the number of breeds in a category; the only difference is in the width of the bars. In the extreme, the rightmost plot eliminates the bars entirely and represents each count by a single dot. (Without the connecting lines, this figure is called a *dot plot.*) Reading this line plot, we see that only a few breeds are unsuitable for children:

```
kid_counts = dogs.groupby(['kids']).count()
kid_counts = kid_counts.reindex(["high", "medium", "low"])
```

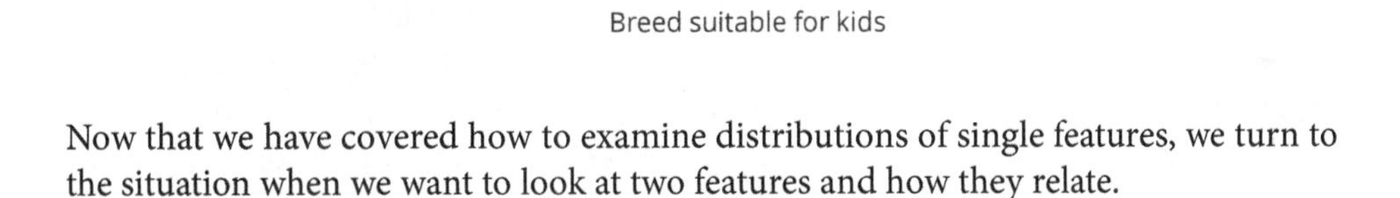

Now that we have covered how to examine distributions of single features, we turn to the situation when we want to look at two features and how they relate.

What to Look For in a Relationship

When we investigate multiple variables, we examine the relationships between them, in addition to their distributions. In this section, we consider pairs of features and describe what to look for. Table 10-3 provides guidelines for the type of plot to make based on the feature types. For two features, the combination of types (both quantitative, both qualitative, or a mix) matters. We consider each combination in turn.

Two Quantitative Features

If both features are quantitative, then we often examine their relationship with a scatterplot. Each point in a scatterplot marks the position of a pair of values for an observation. So we can think of a scatterplot as a two-dimensional rug plot.

With scatter plots, we look for linear and simple nonlinear relationships, and we examine the strength of the relationships. We also look to see if a transformation of one or the other or both features leads to a linear relationship.

The following scatterplot displays the weight and height of dog breeds (both are quantitative):

```
px.scatter(dogs, x='height', y='weight',
           marginal_x="rug", marginal_y="rug",
           labels={'height':'Height (cm)', 'weight':'Weight (kg)'},
           width=350, height=250)
```

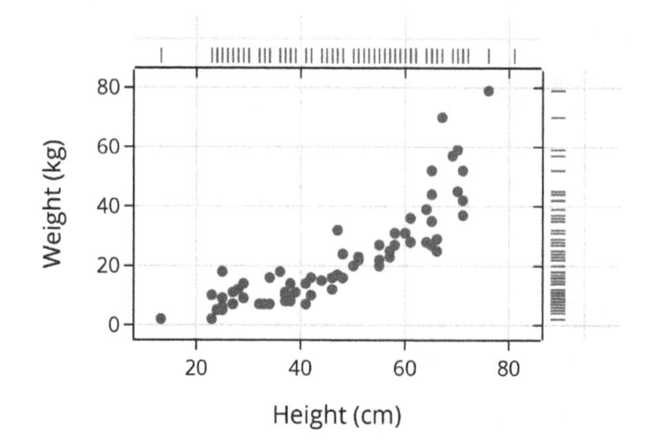

We observe that dogs that are above average in height tend to be above average in weight. This relationship appears nonlinear: the change in weight for taller dogs grows faster than for shorter dogs. Indeed, that makes sense if we think of a dog as basically shaped like a box: for similarly proportioned boxes, the weight of the contents of the box has a cubic relationship to its length.

It's important to note that two univariate plots are missing information found in a bivariate plot—information about how the two features vary together. Practically, histograms for two quantitative features do not contain enough information to create a scatterplot of the features. We must exercise caution and not read too much into a pair of univariate plots. Instead, we need to use one of the plots listed in the appropriate row of Table 10-3 (scatterplot, smooth curve, contour plot, heat map, quantile–quantile plot) to get a sense of the relationship between two quantitative features.

When one feature is numeric and the other qualitative, Table 10-3 makes different recommendations. We describe them next.

One Qualitative and One Quantitative Variable

To examine the relationship between a quantitative and a qualitative feature, we often use the qualitative feature to divide the data into groups and compare the distribution of the quantitative feature across these groups. For example, we can compare the distribution of height for small, medium, and large dog breeds with three overlaid density curves:

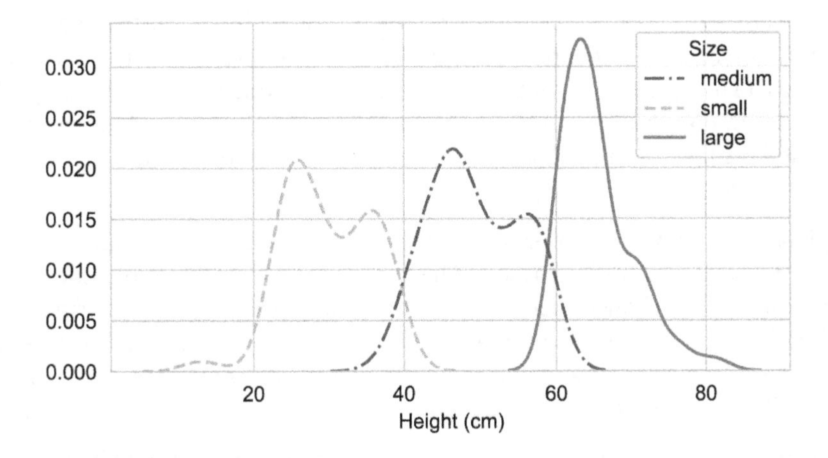

We see that the distribution of height for the small and medium breeds both appear bimodal, with the left mode the larger in each group. Also, the small and medium groups have a larger spread in height than the large group of breeds.

Side-by-side box plots offer a similar comparison of distributions across groups. The box plot offers a simpler approach that can give a crude understanding of a distribution. Likewise, violin plots sketch density curves along an axis for each group. The curve is flipped to create a symmetric "violin" shape. The violin plot aims to bridge the gap between the density curve and box plot. We create box plots (left) and violin plots (right) for the height of breeds given the size labeling:

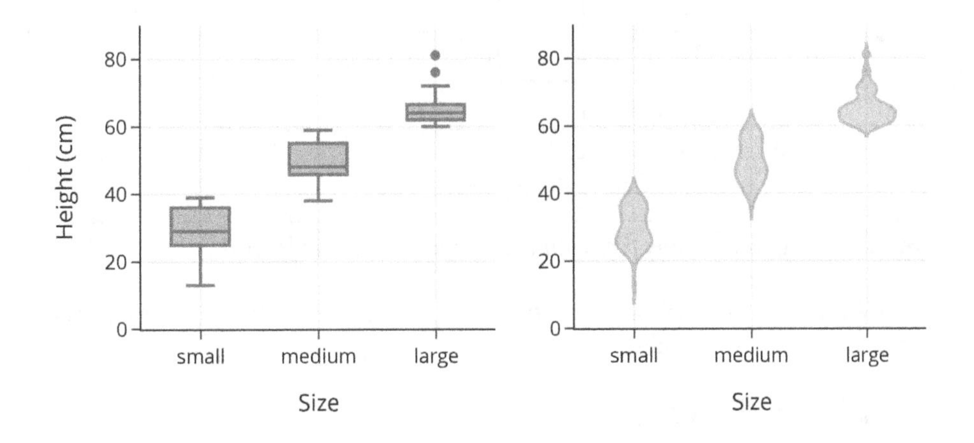

The three box plots of height, one for each size of dog, make it clear that the size categorization is based on height because there is almost no overlap in height ranges for the groups. (This was not evident in the density curves due to the smoothing.) What we don't see in these box plots is the bimodality in the small and medium groups, but we can still see that the large dogs have a narrower spread compared to the other two groups.

Box plots (also known as box-and-whisker plots) give a visual summary of a few important statistics of a distribution. The box denotes the 25th percentile, median, and 75th percentile, the whiskers show the tails, and unusually large or small values are also plotted. Box plots cannot reveal as much shape as a histogram or density curve. They primarily show symmetry and skew, long/short tails, and unusually large/small values (also known as *outliers*).

Figure 10-2 is a visual explanation of the parts of a box plot. Asymmetry is evident from the median not being in the middle of the box, the sizes of the tails are shown by the length of the whiskers, and outliers are shown by the points that appear beyond the whiskers. The maximum is considered an outlier because it appears beyond the whisker on the right.

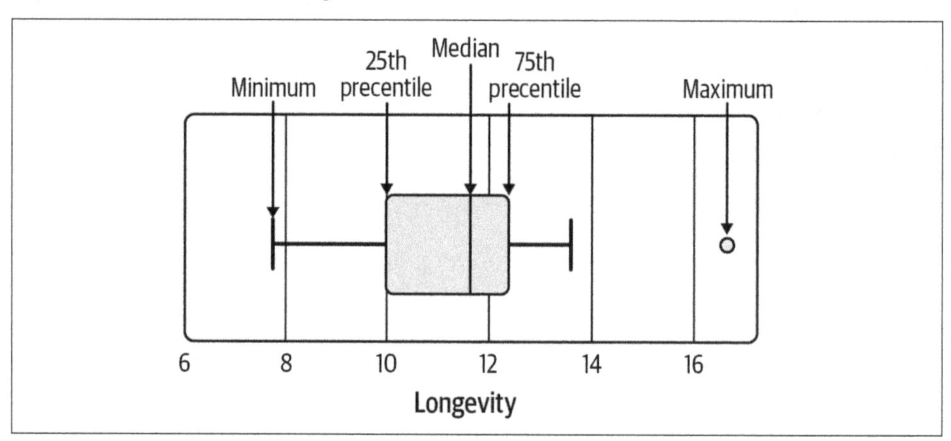

Figure 10-2. Diagram of a box plot with the summary statistics labeled

When we examine the relationship between two qualitative features, our focus is on proportions, as we explain next.

Two Qualitative Features

With two qualitative features, we often compare the distribution of one feature across subgroups defined by the other feature. In effect, we hold one feature constant and plot the distribution of the other one. To do this, we can use some of the same plots we used to display the distribution of one qualitative feature, such as a line plot or bar plot. As an example, let's examine the relationship between the suitability of a breed for children and the size of the breed.

To examine the relationship between these two qualitative features, we calculate three sets of proportions (one each for low, medium, and high suitability). Within each suitability category, we find the proportion of small, medium, and large dogs. These proportions are displayed in the following table. Notice that each column sums to 1 (equivalent to 100%):

```
prop_table_t
```

kids size	high	medium	low
large	0.37	0.29	0.1
medium	0.36	0.34	0.2
small	0.27	0.37	0.7

The line plot that follows provides a visualization of these proportions. There is one "line" (set of connected dots) for each suitability level. The connected dots give the breakdown of size within a suitability category. We see that breeds with low suitability for kids are primarily small:

```
fig = px.line(prop_table_t, y=prop_table_t.columns,
        x=prop_table_t.index, line_dash='kids',
        markers=True, width=500, height=250)

fig.update_layout(
    yaxis_title="proportion", xaxis_title="Size",
    legend_title="Suitability <br>for children"
)
```

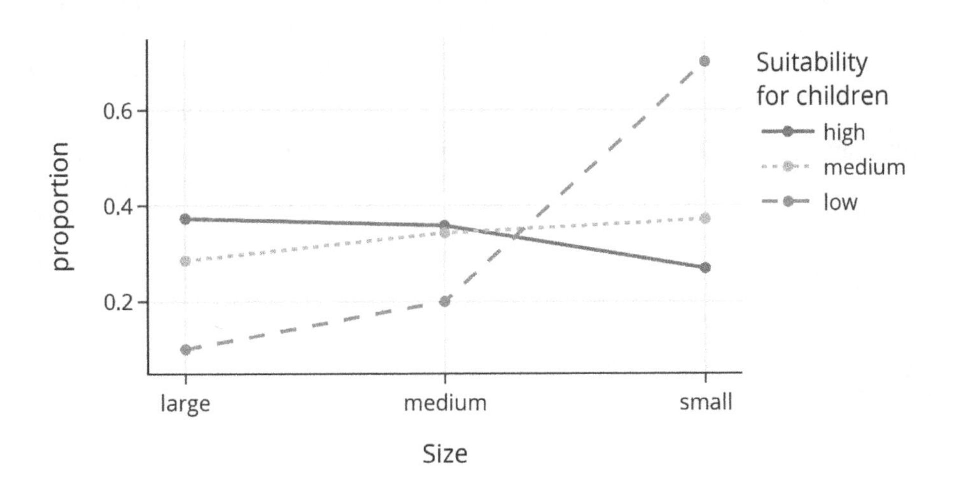

We can also present these proportions as a collection of side-by-side bar plots, as shown here:

```
fig = px.bar(prop_table_t, y=prop_table_t.columns, x=prop_table_t.index,
             barmode='group', width=500, height=250)

fig.update_layout(
    yaxis_title="proportion", xaxis_title="Size",
    legend_title="Suitability <br>for children"
)
```

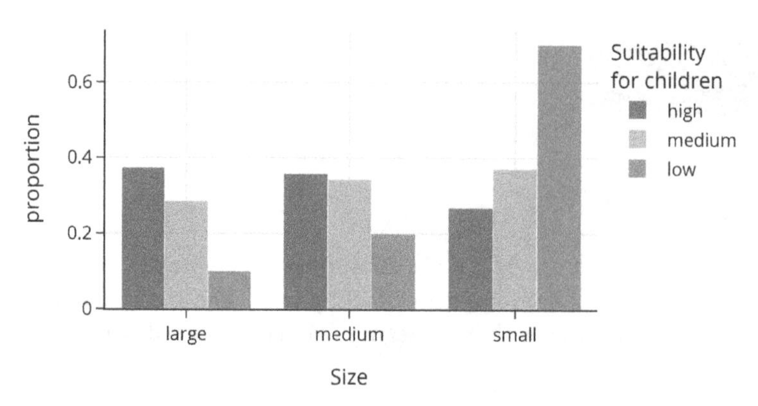

So far, we've covered visualizations that incorporate one or two features. In the next section, we discuss visualizations that incorporate more than two features.

Comparisons in Multivariate Settings

When we examine a distribution or relationship, we often want to compare it across subgroups of the data. This process of conditioning on additional factors often leads to visualizations that involve three or more variables. In this section, we explain how to read plots that are commonly used to visualize multiple variables.

As an example, let's compare the relationship between height and longevity across repetition categories. First, we collapse repetition (the typical number of times it takes for a dog to learn a new command) from six categories into four: <15, 15–25, 25–40, and 40+:

```
rep_replacements = {
    '80-100': '40+', '40-80': '40+',
    '<5': '<15', '5-15': '<15',
}
dogs = dogs.assign(
    repetition=dogs['repetition'].replace(rep_replacements))
```

Now each group has about 30 breeds in it, and having fewer categories makes it easier to decipher relationships. These categories are conveyed by differently shaped symbols in a scatterplot:

```
px.scatter(dogs.dropna(subset=['repetition']), x='height', y='longevity',
           symbol='repetition', width=450, height=250,
           labels={'height':'Height (cm)',
                   'longevity':'Typical lifespan (yr)',
                  'repetition':'Repetition'},
          )
```

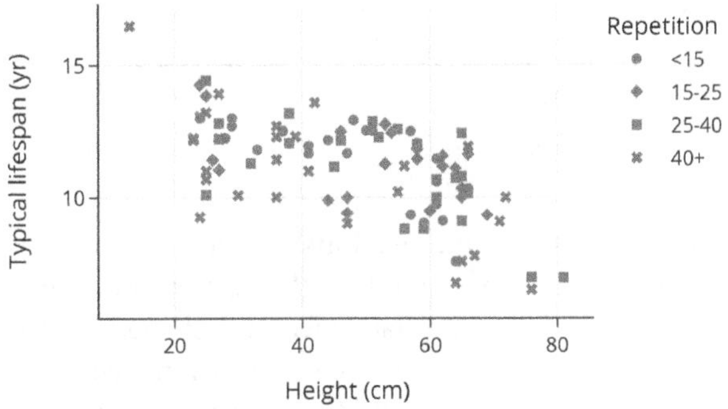

This plot would be challenging to interpret if there were more levels within the `repetition` feature.

Facet plots offer an alternative approach to display these three features:

```
px.scatter(dogs.dropna(subset=['repetition']),
           x='height', y='longevity', trendline='ols',
           facet_col='repetition', facet_col_wrap=2,
           labels={'height':'Height (cm)',
                   'longevity':'Typical lifespan (yr)'})
```

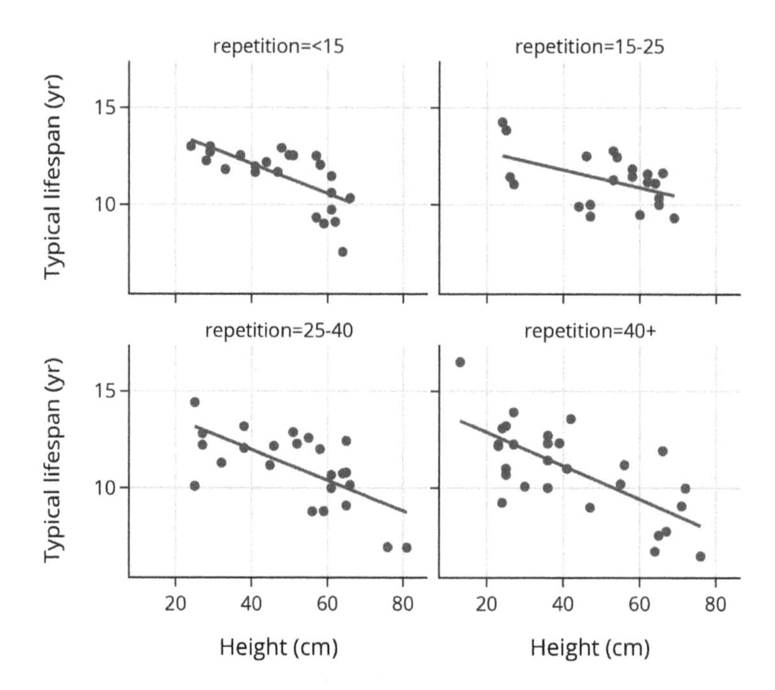

Each of the four scatterplots shows the relationship between longevity and height for a different range of repetitions. By separating the scatterplots, we can better assess how the relationship between two quantitative features changes across the subgroups. And we can more easily see the range of height and longevity for each repetition range. We can see that the larger breeds tend to have shorter lifespans. Another interesting feature is that the lines are similar in slope, but the line for the 40+ repetitions sits about 1.5 years below the others. Those breeds tend to live about 1.5 years less on average than the other repetition categories, no matter the height.

Here we summarize the various plotting techniques for making comparisons when we have three (or more) features:

Two quantitative and one qualitative
: We demonstrated this case already with a scatterplot that varies the markers according to the qualitative feature's categories, or by the panels of scatterplots, with one for each category.

Two qualitative and one quantitative feature
: We have seen in the collections of box plots of height according to breed size that we can compare the basic shape of a distribution across subgroups with side-by-side box plots. When we have two or more qualitative features, we can organize the box plots into groups according to one of the qualitative features.

Three quantitative features

We can use a similar technique when we plot two quantitative features and one qualitative. This time, we convert one of the quantitative features into an ordinal feature, where each category typically has roughly the same number of records. Then we make faceted scatterplots of the other two features. We again look for similarities in relationships across the facets.

Three qualitative features

When we examine relationships between qualitative features, we examine proportions of one feature within subgroups defined by another. In the previous section, the three line plots in one figure and the side-by-side bar plots both display such comparisons. With three (or more) qualitative features, we can continue to subdivide the data according to the combinations of levels of the features and compare these proportions using line plots, dot plots, side-by-side bar charts, and so forth. But these plots tend to get increasingly difficult to understand with further subdivisions.

It's a good practice to break down a visualization to see whether a relationship changes for subgroups of the data determined by a qualitative feature. This technique is called *controlling for* a feature. You might get a surprise when, for example, a linear relationship in a scatterplot that has an upward trend reverses to downward trends in some or all facets of the scatterplot. This phenomenon is known as *Simpson's paradox*. The paradox can happen with qualitative features as well. A famous case (*https://oreil.ly/h9tMw*) occurred at Berkeley when the admissions to graduate school for men were higher than for women, but when examined within each program the rates favored women. The issue was that women were applying in greater numbers to programs that had lower admission rates.

Comparisons that involve more than one categorical variable can quickly become cumbersome as the number of possible combinations of categories grows. For example, there are 3 × 4 = 12 size-repetition combinations (if we had kept the original categories for repetitions, we would have 18 combinations). Examining a distribution across 12 subgroups can be difficult. Further, we come up against the problem of having too few observations in subgroups. Although there are nearly 200 rows in the dogs dataframe, half of the size-repetition combinations have 10 or fewer observations. (This is compounded by losing an observation when one feature has a missing value.) This *curse of dimensionality* also arises when we compare relationships with quantitative data. With just three quantitative variables, some of the scatterplots in a facet plot can easily have too few observations to confirm the shape of the relationship between two variables for the subgroups.

Now that we've seen practical examples of visualizations that are commonly used in exploratory data analysis, we proceed to discuss some high-level guidelines for EDA.

Guidelines for Exploration

So far in this chapter, we have introduced the notion of feature types, seen how the feature type can help to figure out what plot to make, and described how to read distributions and relationships in a visualization. EDA relies on building these skills and flexibly developing your understanding of the data.

You saw EDA in action in Chapter 9 when we developed checks for data quality and feature transformations to improve their usefulness in data analysis. Following are questions to guide you when making plots to explore the data:

- How are the values of Feature X distributed?
- How do Feature X and Feature Y relate to each other?
- Is the distribution of Feature X the same across subgroups defined by Feature Z?
- Are there any unusual observations in X? In the combination of (X,Y)? In X for a subgroup of Z?

As you answer each of these questions, it is important to tie your answer back to the features measured and the context. It is also important to adopt an active, inquisitive approach to the investigation. To guide your explorations, ask yourself "what next" and "so what" questions, such as the following:

- Do you have reason to expect that one group/observation might be different?
- Why might your finding about shape matter?
- What additional comparison might bring added value to the investigation?
- Are there any potentially important features to create comparisons with/against?

In this process, it's important to step away from the computer at times to mull over your work. You may want to read additional literature on the subject or go to an expert in the field to discuss your findings. For example, there could be good reasons for an unusual observation, and someone in the field can help clear up and provide more background.

We put these guidelines into practice with a concrete example of EDA next.

Example: Sale Prices for Houses

In this final section, we carry out an exploratory analysis using the questions in the previous section to direct our investigations. Although EDA typically begins in the data wrangling stage, for demonstration purposes the data we work with here have already been partially cleaned so that we can focus on exploring the features of interest. Note also that we do not discuss refining the visualizations in much detail; that topic is covered in Chapter 11.

Our data were scraped from the *San Francisco Chronicle* (*https://oreil.ly/tP9Xp*) (SFChron) website. The data comprise a complete list of homes sold in the area from April 2003 to December 2008. Since we have no plans to generalize our findings beyond the time period and the location and we are working with a census, the population matches the access frame and the sample consists of the entire population.

As for granularity, each record represents a sale of a home in the SF Bay Area during the specified time period. This means that if a home was sold twice during this time, then there are two records in the table. And if a home in the Bay Area was not up for sale during this time, then it does not appear in the dataset.

The data are in the dataframe `sfh_df`:

```
sfh_df
```

	city	zip	street	price	br	lsqft	bsqft	timestamp
0	Alameda	94501.0	1001 Post Street	689000.0	4.0	4484.0	1982.0	2004-08-29
1	Alameda	94501.0	1001 Santa Clara Avenue	880000.0	7.0	5914.0	3866.0	2005-11-06
2	Alameda	94501.0	1001 Shoreline Drive \#102	393000.0	2.0	39353.0	1360.0	2003-09-21
...	...	...	...	...	...	...	...	...
521488	Windsor	95492.0	9998 Blasi Drive	392500.0	NaN	3111.0	NaN	2008-02-17
521489	Windsor	95492.0	9999 Blasi Drive	414000.0	NaN	2915.0	NaN	2008-02-17
521490	Windsor	95492.0	999 Gemini Drive	325000.0	3.0	7841.0	1092.0	2003-09-21

```
521491 rows × 8 columns
```

The dataset does not have an accompanying codebook, but we can determine the features and their storage types by inspection:

```
sfh_df.info()
```

```
<class 'pandas.core.frame.DataFrame'>
RangeIndex: 521491 entries, 0 to 521490
Data columns (total 8 columns):
 #   Column     Non-Null Count   Dtype
---  ------     --------------   -----
 0   city       521491 non-null  object
 1   zip        521462 non-null  float64
 2   street     521479 non-null  object
```

```
 3   price      521491 non-null  float64
 4   br         421343 non-null  float64
 5   lsqft      435207 non-null  float64
 6   bsqft      444465 non-null  float64
 7   timestamp  521491 non-null  datetime64[ns]
dtypes: datetime64[ns](1), float64(5), object(2)
memory usage: 31.8+ MB
```

Based on the names of the fields, we expect the primary key to consist of some combination of city, zip code, street address, and date.

Sale price is our focus, so let's begin by exploring its distribution. To develop your intuition about distributions, make a guess about the shape of the distribution before you start reading the next section. Don't worry about the range of prices, just sketch the general shape.

Understanding Price

It seems that a good guess for the shape of the distribution of sale price might be highly skewed to the right with a few very expensive houses. The following summary statistics confirm this skewness:

```
percs = [0, 25, 50, 75, 100]
prices = np.percentile(sfh_df['price'], percs, method='lower')
pd.DataFrame({'price': prices}, index=percs)
```

	price
0	22000.00
25	410000.00
50	555000.00
75	744000.00
100	20000000.00

The median is closer to the lower quartile than the upper quartile. Also, the maximum is 40 times the median! We might wonder whether that $20M sale price is simply an anomalous value or whether there are many houses that sold at such a high price. To find out, we can zoom in on the right tail of the distribution and compute a few high percentiles:

```
percs = [95, 97, 98, 99, 99.5, 99.9]
prices = np.percentile(sfh_df['price'], percs, method='lower')
pd.DataFrame({'price': prices}, index=percs)
```

	price
95.00	1295000.00
97.00	1508000.00
98.00	1707000.00
99.00	2110000.00
99.50	2600000.00
99.90	3950000.00

We see that 99.9% of the houses sold for under $4M, so the $20M sale is indeed a rarity. Let's examine the histogram of sale prices below $4M:

```
under_4m = sfh_df[sfh_df['price'] < 4_000_000].copy()

px.histogram(under_4m, x='price', nbins=50, width=350, height=250,
             labels={'price':'Sale price (USD)'})
```

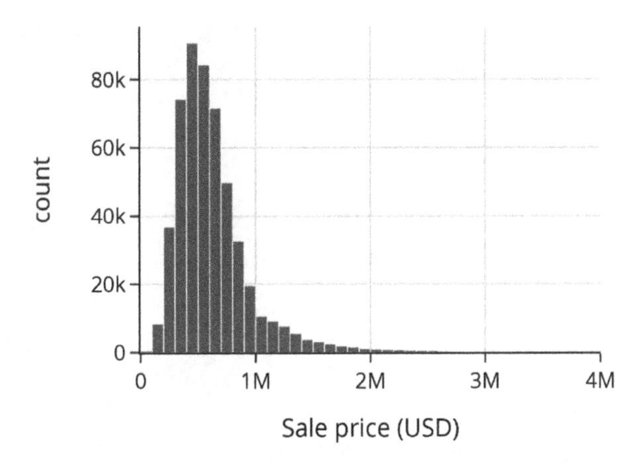

Even without the top 0.1%, the distribution remains highly skewed to the right, with a single mode around $500,000. Let's plot the histogram of the log-transformed sale price. The logarithm transformation often does a good job at converting a right-skewed distribution into one that is more symmetric:

```
under_4m['log_price'] = np.log10(under_4m['price'])

px.histogram(under_4m, x='log_price', nbins=50, width=350, height=250,
             labels={'log_price':'Sale price (log10 USD)'})
```

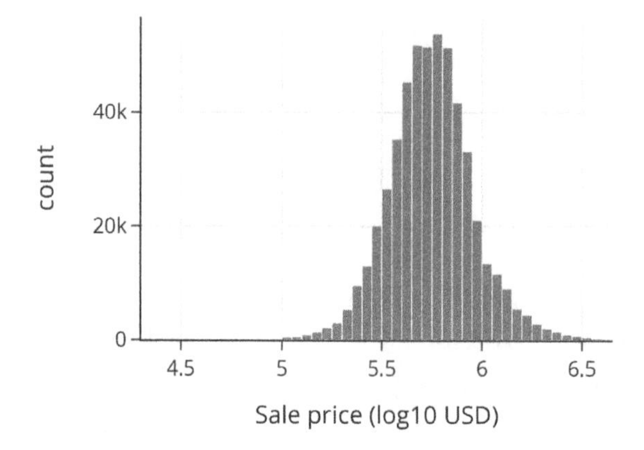

We see that the distribution of log-transformed sale price is roughly symmetric. Now that we have an understanding of the distribution of sale price, let's consider the so-what questions posed in the previous section on EDA guidelines.

What Next?

We have a description of the shape of the sale price, but we need to consider why the shape matters and look for comparison groups where distributions might differ.

Shape matters because models and statistics based on symmetric distributions tend to have more robust and stable properties than highly skewed distributions. (We address this issue more when we cover linear models in Chapter 15.) For this reason, we primarily work with the log-transformed sale price. And we might also choose to limit our analysis to sale prices under $4M since the super-expensive houses may behave quite differently.

As for possible comparisons to make, we look to the context. The housing market rose rapidly during this time and then the bottom fell out of the market. So the distribution of sale price in, say, 2004 might be quite different than in 2008, right before the crash. To explore this notion further, we can examine the behavior of prices over time. Alternatively, we can fix time, and examine the relationships between price and the other features of interest. Both approaches are potentially worthwhile.

We narrow our focus to one year (in Chapter 11 we look at the time dimension). We reduce the data to sales made in 2004, so rising prices should have a limited impact on the distributions and relationships that we examine. To limit the influence of the very expensive and large houses, we also restrict the dataset to sales below $4M and houses smaller than 12,000 ft^2. This subset still contains large and expensive houses, but not outrageously so. Later, we further narrow our exploration to a few cities of interest:

```
def subset(df):
    return df.loc[(df['price'] < 4_000_000) &
                  (df['bsqft'] < 12_000) &
                  (df['timestamp'].dt.year == 2004)]


sfh = sfh_df.pipe(subset)
sfh
```

	city	zip	street	price	br	lsqft	bsqft	timestamp
0	Alameda	94501.00	1001 Post Street	689000.00	4.00	4484.00	1982.00	2004-08-29
3	Alameda	94501.00	1001 Shoreline Drive \#108	485000.00	2.00	39353.00	1360.00	2004-09-05
10	Alameda	94501.00	1001 Shoreline Drive \#306	390000.00	2.00	39353.00	1360.00	2004-01-25
...	...	...	...	...	...	...	...	...
521467	Windsor	95492.00	9960 Herb Road	439000.00	3.00	9583.00	1626.00	2004-04-04
521471	Windsor	95492.00	9964 Troon Court	1200000.00	3.00	20038.00	4281.00	2004-10-31
521478	Windsor	95492.00	9980 Brooks Road	650000.00	3.00	45738.00	1200.00	2004-10-24

```
105996 rows × 8 columns
```

For these data, the shape of the distribution of sale price remains the same—price is still highly skewed to the right. We continue to work with this subset to address the question of whether there are any potentially important features to study along with price.

Examining Other Features

In addition to the sale price, which is our main focus, a few other features that might be important to our investigation are the size of the house, lot (or property) size, and number of bedrooms. We explore the distributions of these features and their relationship to sale price and to each other.

Since the size of the house and the property are likely related to its price, it seems reasonable to guess that these features are also skewed to the right, so we apply a log transformation to the building size:

```
sfh = sfh.assign(log_bsqft=np.log10(sfh['bsqft']))
```

We compare the distribution of building size on the regular and logged scales:

```
fig = make_subplots(1,2)
fig.add_trace(go.Histogram(x=sfh['bsqft'], histnorm='percent',
                           nbinsx=60), row=1, col=1)
fig.add_trace(go.Histogram(x=sfh['log_bsqft'], histnorm='percent',
                           nbinsx=60), row=1, col=2)

fig.update_xaxes(title='Building size (ft²)', row=1, col=1)
fig.update_xaxes(title='Building size (ft², log10)', row=1, col=2)
fig.update_yaxes(title="percent", row=1, col=1)
```

```
fig.update_yaxes(range=[0, 18])
fig.update_layout(width=450, height=250, showlegend=False)
fig
```

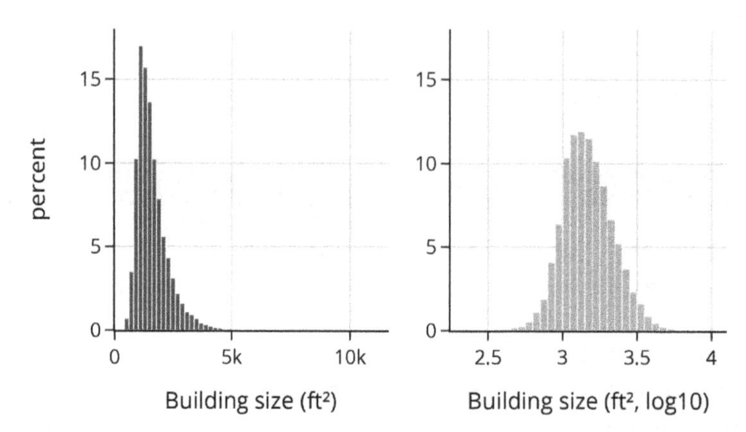

The distribution is unimodal with a peak at about 1,500 ft², and many houses are over 2,500 ft² in size. We have confirmed our intuition: the log-transformed building size is nearly symmetric, although it maintains a slight skew. The same is the case for the distribution of lot size.

Given that both house and lot size have skewed distributions, a scatterplot of the two should most likely be on log scale too:

```
sfh = sfh.assign(log_lsqft=np.log10(sfh['lsqft']))
```

We compare the plot with and without the log transformation:

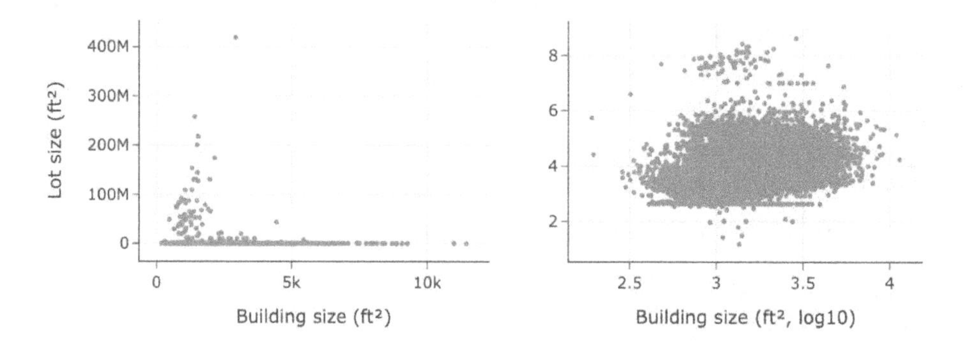

The scatterplot on the left is in the original units, which makes it difficult to discern the relationship because most of the points are crowded into the bottom of the plotting region. In contrast, the scatterplot on the right reveals a few interesting features: there is a horizontal line along the bottom of the scatterplot, where it appears that many houses have the same lot size, no matter the building size; and there appears to be a slight positive log–log linear association between lot and building size.

Let's look at some lower quantiles of lot size to try to figure out this unusual value:

```
percs = [0.5, 1, 1.5, 2, 2.5, 3]
lots = np.percentile(sfh['lsqft'].dropna(), percs, method='lower')
pd.DataFrame({'lot_size': lots}, index=percs)
```

	lot_size
0.50	436.00
1.00	436.00
1.50	436.00
2.00	436.00
2.50	436.00
3.00	782.00

We found something interesting: about 2.5% of the houses have a lot size of 436 ft^2. This is tiny and makes little sense, so we make a note of the anomaly for further investigation.

Another measure of house size is the number of bedrooms. Since this is a discrete quantitative variable, we can treat it as a qualitative feature and make a bar plot.

Houses in the Bay Area tend to be on the smaller side, so we venture to guess that the distribution will have a peak at three and skew to the right, with a few houses having five or six bedrooms. Let's check:

```
br_cat = sfh['br'].value_counts().reset_index()
px.bar(br_cat, x="br", y="count", width=350, height=250,
       labels={'br':'Number of bedrooms'})
```

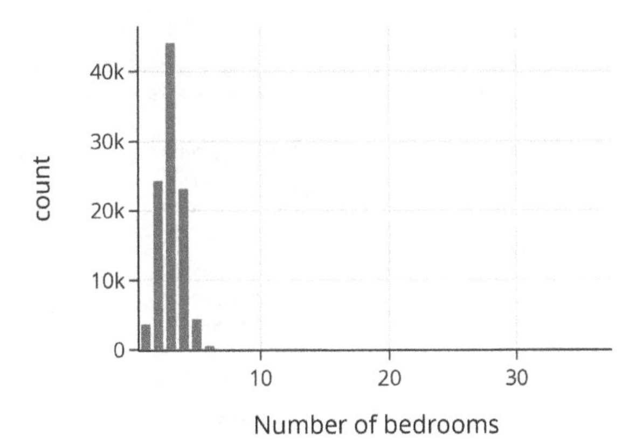

The bar plot confirms that we generally had the right idea. However, we find that there are some houses with over 30 bedrooms! That's a bit hard to believe and points to another possible data quality problem. Since the records include the addresses of the houses, we can double-check theses values on a real estate app.

In the meantime, let's just transform the number of bedrooms into an ordinal feature by reassigning all values larger than 8 to 8+, and re-create the bar plot with the transformed data:

```
eight_up = sfh.loc[sfh['br'] >= 8, 'br'].unique()
sfh['new_br'] = sfh['br'].replace(eight_up, 8)

br_cat = sfh['new_br'].value_counts().reset_index()
px.bar(br_cat, x="new_br", y="count", width=350, height=250,
       labels={'new_br':'Number of bedrooms'})
```

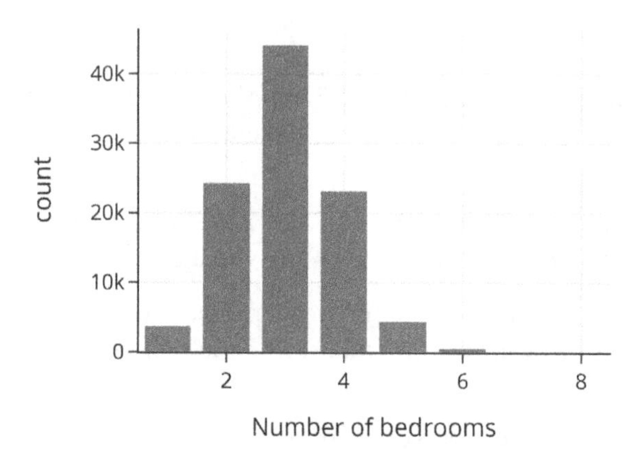

We can see that even if we lump all of the houses with 8+ bedrooms together, they do not amount to many. The distribution is nearly symmetric with a peak at 3, nearly the same proportion of houses have two or four bedrooms, and nearly the same have one or five. There is asymmetry present, with a few houses having six or more bedrooms.

Now we examine the relationship between the number of bedrooms and sale price. Before we proceed, we save the transformations done thus far:

```
def log_vals(df):
    return df.assign(log_price=np.log10(df['price']),
                     log_bsqft=np.log10(df['bsqft']),
                     log_lsqft=np.log10(df['lsqft']))

def clip_br(df):
    eight_up = df.loc[df['br'] >= 8, 'br'].unique()
    new_br = df['br'].replace(eight_up, 8)
    return df.assign(new_br=new_br)
```

```
sfh = (sfh_df
  .pipe(subset)
  .pipe(log_vals)
  .pipe(clip_br)
)
```

Now we're ready to consider relationships between the number of bedrooms and other variables.

Delving Deeper into Relationships

Let's begin by examining how the distribution of price changes for houses with different numbers of bedrooms. We can do this with box plots:

```
px.box(sfh, x='new_br', y='price', log_y=True, width=450, height=250,
       labels={'new_br':'Number of bedrooms','price':'Sale price (USD)'})
```

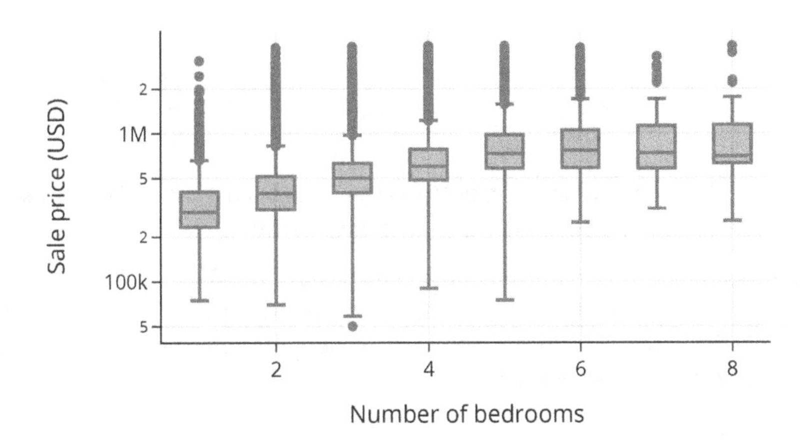

The median sale price increases with the number of bedrooms from one to five, but for the largest houses (those with more than six bedrooms), the distribution of log-transformed sale price appears nearly the same.

We would expect houses with one bedroom to be smaller than houses with, say, four bedrooms. We might also guess that houses with six or more bedrooms are similar in size and price. To dive deeper, we consider a kind of transformation that divides price by building size to give us the price per square foot. We want to check if this feature is constant for all houses; in other words, if price is primarily determined by size. To do this we look at the relationship between the two pairs of size and price, and price per square foot and size:

```
sfh = sfh.assign(
    ppsf=sfh['price'] / sfh['bsqft'],
    log_ppsf=lambda df: np.log10(df['ppsf']))
```

We create two scatterplots. The one on the left shows price against building size (both log-transformed), and the plot on the right shows price per square foot (log-transformed) against building size. In addition, each plot has an added smooth curve that reflects the local average price or price per square foot for buildings of roughly the same size:

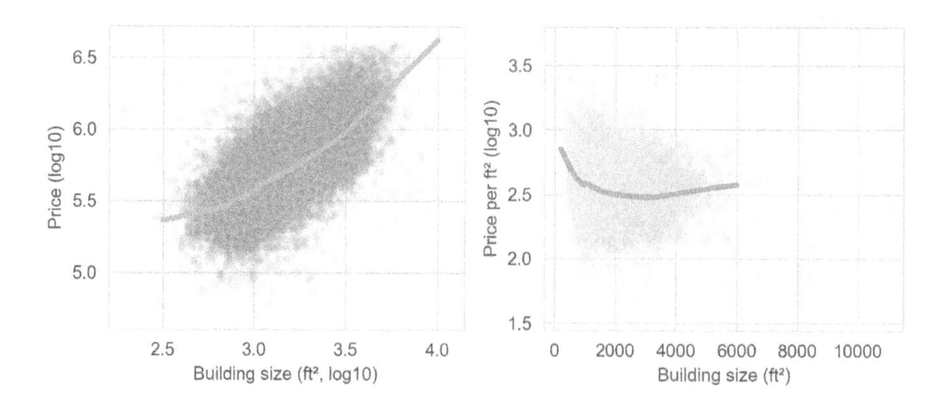

The lefthand plot shows what we expect—larger houses cost more. We also see that there is roughly a log–log association between these features.

The righthand plot in this figure is interestingly nonlinear. We see that smaller houses cost more per square foot than larger ones, and the price per square foot for larger houses is relatively flat. This feature appears to be quite interesting, so we save the price per square foot transforms into `sfh`:

```
def compute_ppsf(df):
    return df.assign(
        ppsf=df['price'] / df['bsqft'],
        log_ppsf=lambda df: np.log10(df['ppsf']))
```

So far we haven't considered the relationship between prices and location. There are house sales from over 150 different cities in this dataset. Some cities have a handful of sales and others have thousands. We continue our narrowing down of the data and examine relationships for a few cities next.

Fixing Location

You may have heard the expression: there are three things that matter in real estate—*location, location, location*. Comparing prices across cities might bring additional insights to our investigation.

We examine data for some cities in the San Francisco East Bay: Richmond, El Cerrito, Albany, Berkeley, Walnut Creek, Lamorinda (which is a combination of Lafayette, Moraga, and Orinda, three neighboring bedroom communities), and Piedmont.

Let's begin by comparing the distribution of sale price for these cities:

```
cities = ['Richmond', 'El Cerrito', 'Albany', 'Berkeley',
          'Walnut Creek', 'Lamorinda', 'Piedmont']

px.box(sfh.query('city in @cities'), x='city', y='price',
       log_y=True, width=450, height=250,
       labels={'city':'', 'price':'Sale price (USD)'})
```

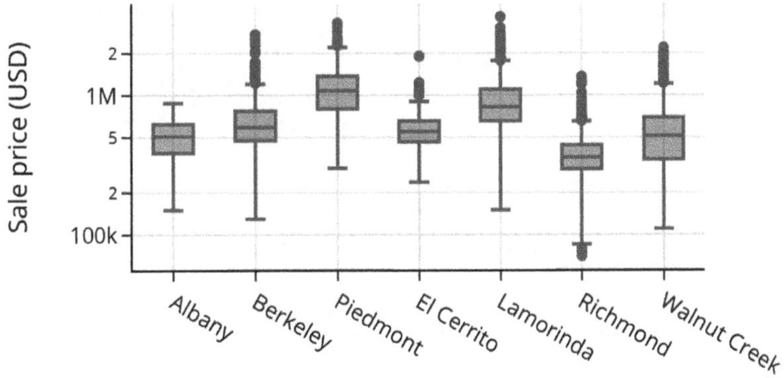

The box plots show that Lamorinda and Piedmont tend to have more expensive homes and Richmond has the least expensive, but there is overlap in sale price for many cities.

Next, we examine the relationship between price per square foot and house size more closely with faceted scatterplots, one for each of four cities:

```
four_cities = ["Berkeley", "Lamorinda", "Piedmont", "Richmond"]
fig = px.scatter(sfh.query("city in @four_cities"),
    x="bsqft", y="log_ppsf", facet_col="city", facet_col_wrap=2,
    labels={'bsqft':'Building size (ft^2)',
            'log_ppsf': "Price per square foot"},
    trendline="ols", trendline_color_override="black",
)

fig.update_layout(xaxis_range=[0, 5500], yaxis_range=[1.5, 3.5],
                  width=450, height=400)
fig.show()
```

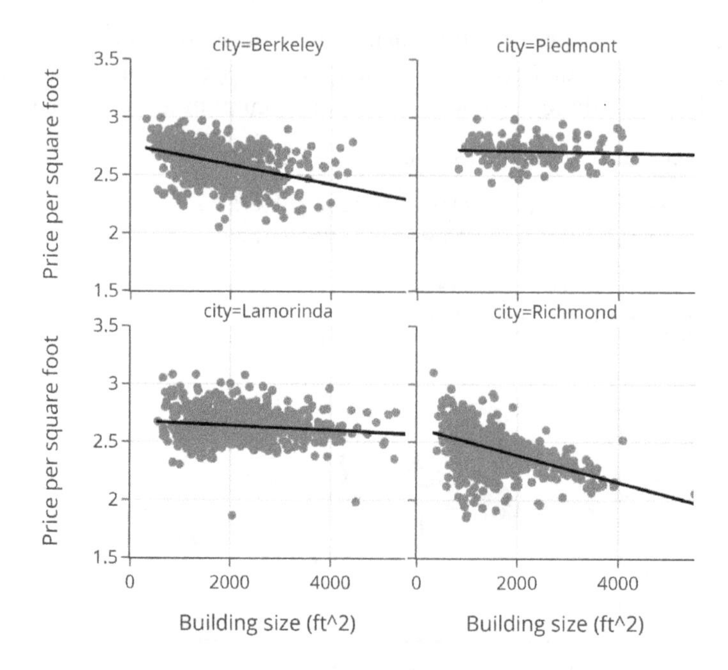

The relationship between price per square foot and building size is roughly log-linear, with a negative association for each of the four locations. While not parallel, it does appear that there is a location boost for houses, regardless of size, where, say, a house in Berkeley costs about $250 more per square foot than a house in Richmond. We also see that Piedmont and Lamorinda are more expensive cities, and in both cities there is not the same reduction in price per square foot for larger houses in comparison to smaller ones. These plots support the "location, location, location" adage.

In EDA, we often revisit earlier plots to check whether new findings add insights to previous visualizations. It is important to continually take stock of our findings and use them to guide us in further explorations. Let's summarize our findings so far.

EDA Discoveries

Our EDA has uncovered several interesting phenomena. Briefly, some of the most notable are:

- Sale price and building size are highly skewed to the right with one mode.
- Price per square foot decreases nonlinearly with building size, with smaller houses costing more per square foot than larger houses and price per square foot being roughly constant for large houses.

- More desirable locations add a bump in sale price that is roughly the same amount for houses of different sizes.

There are many additional explorations we can (and should) perform, and there are several checks that we should make. These include investigating the 436 value for lot size and cross-checking unusual houses, like the 30-bedroom house and the $20M house, with online real estate apps.

We narrowed our investigation down to one year and later to a few cities. This narrowing helped us control for features that might interfere with finding simple relationships. For example, since the data were collected over several years, the date of sale may confound the relationship between sale price and number of bedrooms. At other times, we want to consider the effect of time on prices. To examine price changes over time, we often make line plots, and we adjust for inflation. We revisit these data in Chapter 11 when we consider data scope and look more closely at trends in time.

Despite being brief, this section conveys the basic approach of EDA in action. For an extended case study on a different dataset, see Chapter 12.

Summary

In this chapter, we introduced the nominal, ordinal, and numerical feature types and their importance for data analysis. When presented with a dataset, we demonstrated how to consult the data dictionary and the data itself to determine the feature types for each column. We also explained how the storage type is not to be confused with feature type. Since much of EDA is carried out with statistical graphs, we described how to recognize and interpret the shapes and patterns that emerge and how to connect these to the data being plotted. Finally, we provided guidelines for how you might conduct an EDA, and provided an example.

One approach that you may find helpful in developing your intuition about distributions and relationships of features is to make a guess about what you will see before you make the plot. Try to sketch or describe what you think the shape of distribution will be, and then make the plot. For example, variables that have a natural lower/upper bound on their values tend to have a long tail on the opposite of the bound. The distribution of income (bounded below by 0) tends to have a long right tail, and exam scores (bounded above by 100) tend to have a long left tail. You can make similar guesses for the shape of a relationship. We saw that price and house size had nearly a log–log linear relationship. As you gain intuition about shapes, it becomes easier to carry out an EDA; you can more easily identify when a plot shows a surprising shape.

Our focus in this chapter was on "reading" visualizations. In Chapter 11, we provide style guidelines for how to create informative, effective, and beautiful graphs. Many of the ideas in that chapter were followed here, but we have not called attention to them.

CHAPTER 11

Data Visualization

As data scientists, we create data visualizations in order to understand our data and explain our analyses to other people. A plot should have a message, and it's our job to communicate this message as clearly as possible.

In Chapter 10, we connected the choice of a statistical graph to the kind of data being plotted; we also introduced many standard plots and showed how to read them. In this chapter, we discuss the principles of effective data visualization that make it easier for our audience to grasp the message in our plot. We talk about how to choose scales for axes, handle large amounts of data with smoothing and aggregation, facilitate meaningful comparisons, incorporate study design, and add contextual information. We also show how to create plots using `plotly`, a popular package for plotting in Python.

One tricky part about writing a chapter on data visualization is that software packages for visualization change all the time, so any code we display can quickly get out of date. Because of this, some books avoid code entirely. We instead strike a balance, where we cover high-level data visualization principles that are broadly useful. Then we separately include practical plotting code to implement these principles. When new software becomes available, readers can still use our principles to guide the creation of their visualizations.

Choosing Scale to Reveal Structure

In Chapter 10, we explored prices for houses sold in the San Francisco Bay Area between 2003 and 2009. Let's revisit that example and take a look at a histogram of sale prices:

```
px.histogram(sfh, x='price', nbins=100,
             labels={'price':"Sale price (USD)"}, width=350, height=250)
```

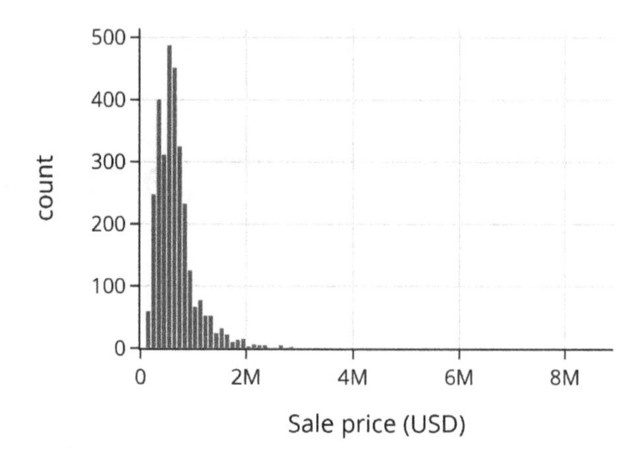

While this plot accurately displays the data, most of the visible bins are crammed into the left side of the plot. This makes it hard to understand the distribution of prices.

Through data visualization, we want to reveal important features of the data, like the shape of a distribution and the relationship between two or more features. As this example shows, after we produce an initial plot, there are still other aspects we need to consider. In this section, we cover *principles of scale* that help us decide how to adjust the axis limits, place tick marks, and apply transformations. We begin by examining when and how we might adjust a plot to reduce empty space; in other words, we try to fill the data region of our plot with data.

Filling the Data Region

As we can see from the histogram of sale prices, it's hard to read a distribution when most of the data appear in a small portion of the plotting region. When this happens, important features of the data, like multiple modes and skewness, can be obscured. A similar issue happens for scatterplots. When all the points are bunched together in the corner of a scatterplot, it's hard to see the shape of the distribution and therefore glean any insights the shape would impart.

This issue can crop up when there are a few unusually large observations. In order to get a better view of the main portion of the data, we can drop those observations from the plot by adjusting the x- or y-axis limits, or we can remove the outlier values from the data before plotting. In either case, we mention this exclusion in the caption or on the plot itself.

Let's use this idea to improve the histogram of sale prices. In the side-by-side plots that follow, we clip the data by changing the limits of the x-axis. On the left, we've excluded houses that cost over $2 million. The shape of the distribution for the bulk of the houses is much clearer in this plot. For instance, we can more easily observe the skewness and a smaller secondary mode. On the right, we separately show detail in the long right tail of the distribution (see it larger online (*https://oreil.ly/lVDrE*)):

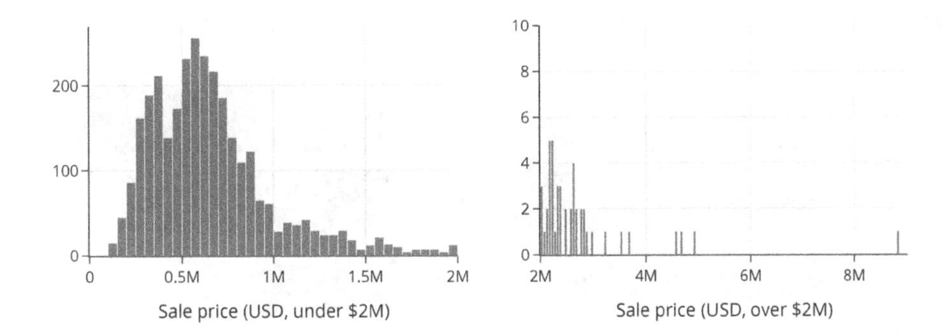

Notice that the x-axis in the left plot includes 0, but the x-axis in the right plot begins at $2M. We consider when to include or exclude 0 on an axis next.

Including Zero

We often don't need to include 0 on an axis, especially if including it makes it difficult to fill the data region. For example, let's make a scatterplot of average longevity plotted against average height for dog breeds. (This dataset was first introduced in Chapter 10; it includes several features for 172 breeds.)

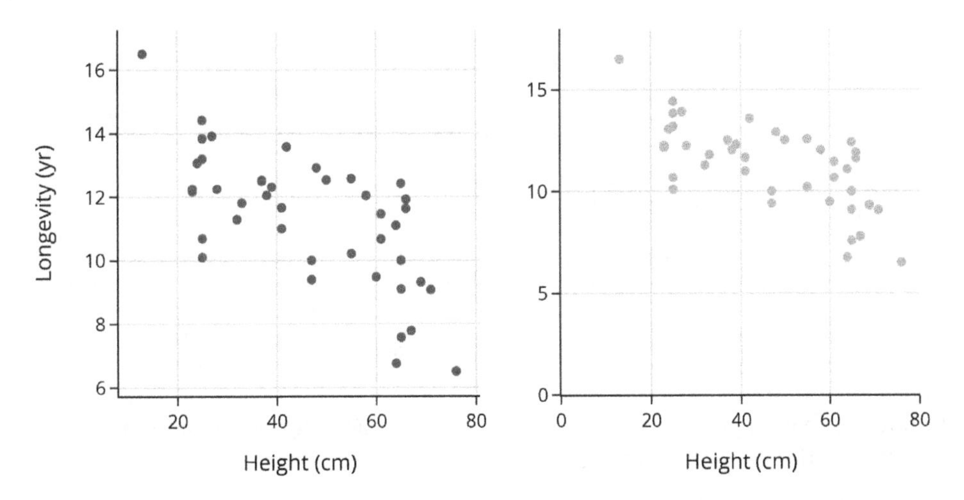

The x-axis of the plot on the left starts at 10 cm since all dogs are at least that tall, and similarly, the y-axis begins at 6 years. The scatterplot on the right includes 0 on both axes. This pushes the data up to the top of the data region and leaves empty space that doesn't help us see the linear relationship.

There are some cases where we usually want to include 0. For bar charts, including 0 is important so that the heights of the bars directly relate to the data values. As an example, we've created two bar charts that compare the longevity of dog breeds. The left plot includes 0, but the right plot doesn't:

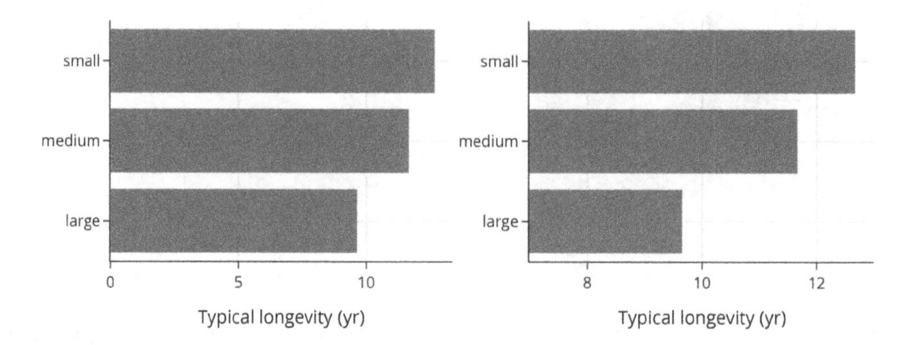

It's easy to incorrectly conclude from the right plot that small breeds live twice as long as large breeds.

We also typically want to include 0 when working with proportions, since proportions range from 0 to 1. The following plot shows the proportion of breeds in each type:

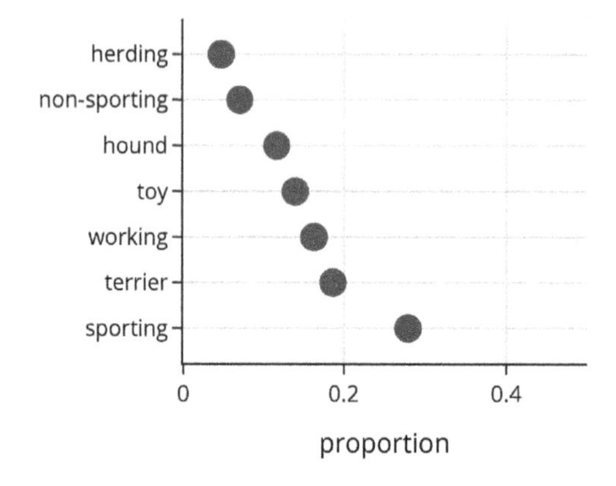

In both the bar and dot plots, including 0 makes it easier for you to accurately compare the relative sizes of the categories.

Earlier, when we adjusted axes, we essentially dropped data from our plotting region. While this is a useful strategy when a handful of observations are unusually large (or small), it is less effective with skewed distributions. In this situation, we often need to transform the data to gain a better view of its shape.

Revealing Shape Through Transformations

Another common way to adjust scale is to transform the data or the plot's axes. We use transformations for skewed data so that it is easier to inspect the distribution. And when the transformation produces a symmetric distribution, the symmetry carries with it useful properties for modeling (see Chapter 15).

There are multiple ways to transform data, but the log transformation tends to be especially useful. For instance, in the following charts we reproduced two histograms of San Francisco house sale prices. The top histogram is the original data. For the histogram below, we took the log (base 10) of the prices before plotting:

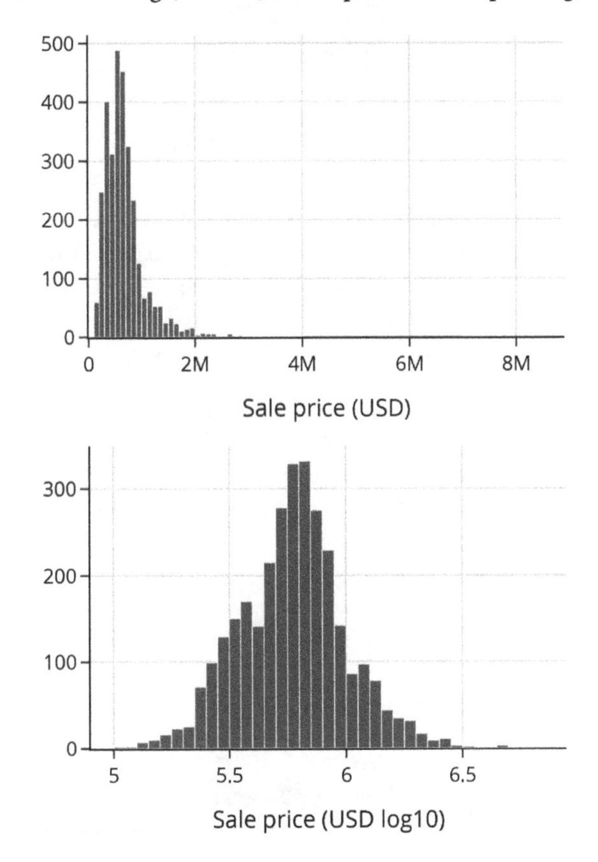

The log transformation makes the distribution of prices more symmetric. Now we can more easily see important features of the distribution, like the mode at around $10^{5.85}$, which is about 700,000, and the secondary mode near $10^{5.55}$, or 350,000.

The downside of using the log transform is that the actual values aren't as intuitive—in this example, we needed to convert the values back to dollars to understand the sale price. Therefore, we often favor transforming the axis to a log scale, rather than the data. This way, we can see the original values on the axis:

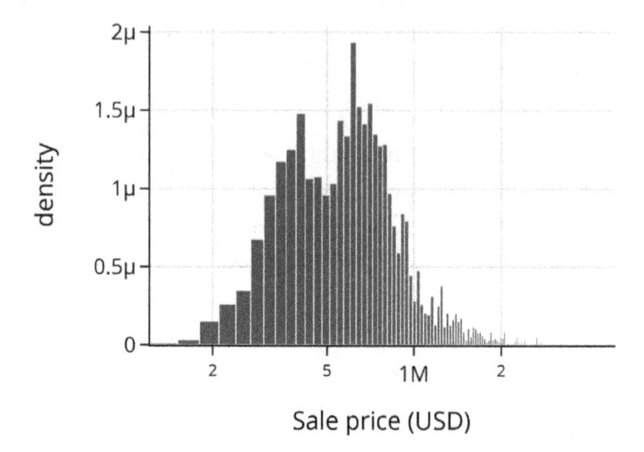

This histogram with its log-scaled x-axis essentially shows the same shape as the histogram of the transformed data. But since the axis is displayed in the original units, we can directly read off the location of the modes in dollars. Note that the bins get narrower to the right because the bin widths are equal on the USD scale but plotted on the log USD scale. Also note that μ on the y-axis is 10^{-6}.

The log transform can also reveal shape in scatterplots. Here, we've plotted building size on the x-axis and lot size on the y-axis. It's hard to see the shape in this plot since many of the points are crammed along the bottom of the data region:

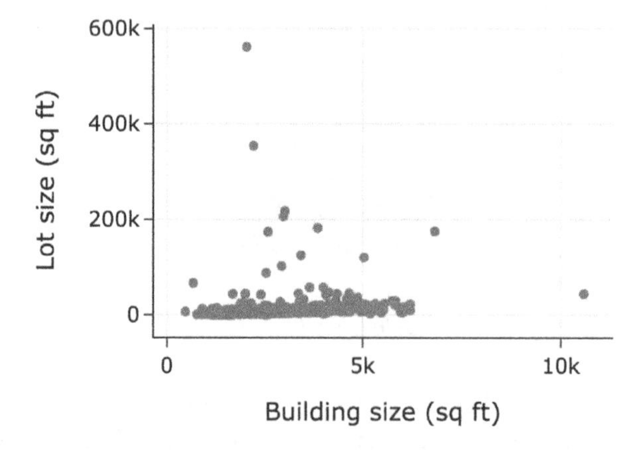

However, when we use a log scale for both the x- and y-axes, the shape of the relationship is much easier to see:

```
px.scatter(sfh, x='bsqft', y='lsqft',
           log_x=True, log_y=True,
           labels={"bsqft": "Building size (sq ft)",
                   "lsqft": "Lot size (sq ft)"},
           width=350, height=250)
```

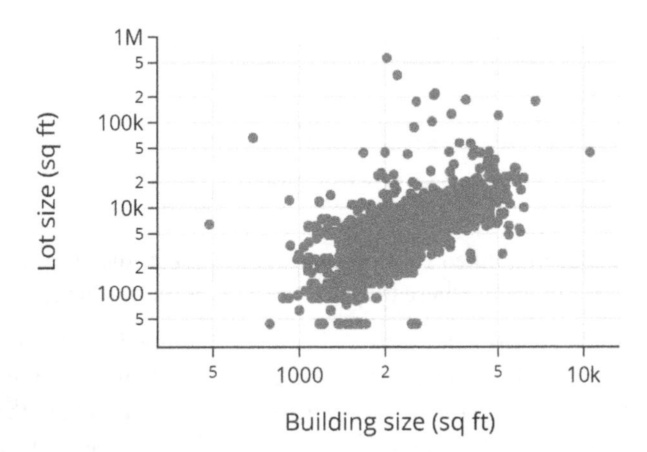

With the transformed axes, we can see that the lot size increases roughly linearly with building size (on the log scale). The log transformation pulls large values—values that are orders of magnitude larger than others—in toward the center. This transformation can help fill the data region and uncover hidden structure, as we saw for both the distribution of house price and the relationship between house size and lot size.

In addition to setting the limits of an axis and transforming an axis, we also want to consider the aspect ratio of the plot—the length compared to the width. Adjusting the aspect ratio is called *banking*, and in the next section, we show how banking can help reveal relationships between features.

Banking to Decipher Relationships

With scatterplots, we try to choose scales so that the relationship between the two features roughly follows a 45-degree line. This scaling is called *banking to 45 degrees*. It makes it easier to see shape and trends because our eyes can more easily pick up deviations from a line this way. As an example of this, we've reproduced the plot that shows longevity of dog breeds against height:

```
px.scatter(dogs, x='height', y='longevity', width=300, height=250,
           labels={"height": "Height (cm)",
                   "longevity": "Typical lifespan (yr)"})
```

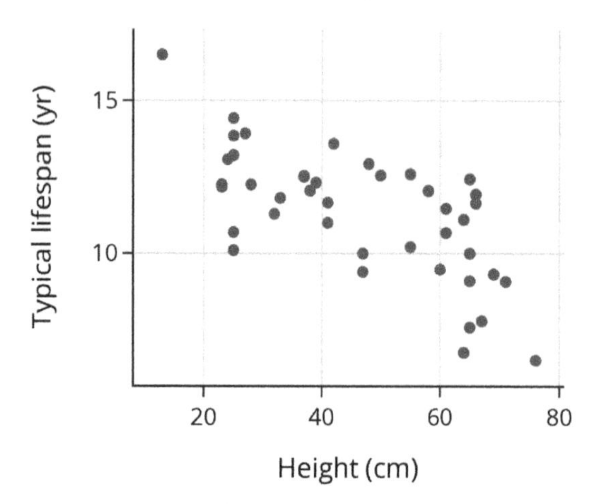

The scatterplot has been banked to 45 degrees, and we can more easily see how the data roughly follow a line and where they deviate a bit at the extremes.

While banking to 45 degrees helps us see whether or not the data follow a linear relationship, when there is clear curvature it can be hard to figure out the form of the relationship. When this happens, we try transformations that will get the data to fall along a straight line (see, for example, Figure 11-1). The log transformation can be useful in uncovering the general form of curvilinear relationships.

Revealing Relationships Through Straightening

We often use scatter plots to look at the relationship between two features. For instance, here we've plotted height against weight for the dog breeds:

```
px.scatter(dogs, x='height', y='weight', width=350, height=250,
           labels={"height": "Height (cm)", "weight": "Weight (lb)"})
```

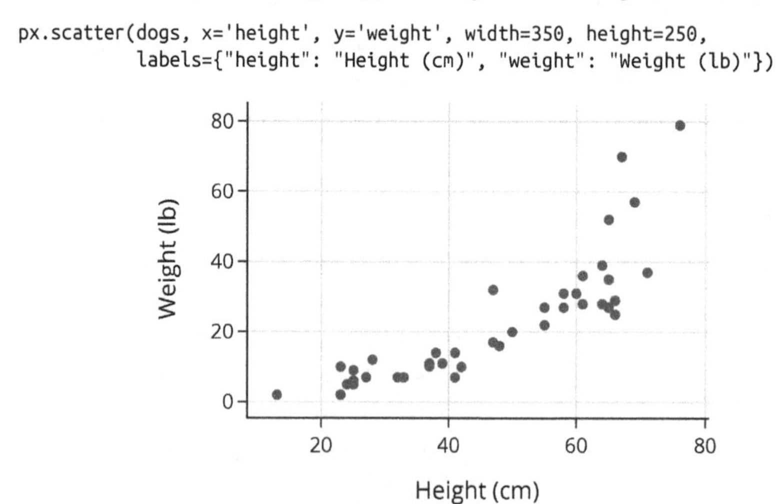

We see that taller dogs weigh more, but this relationship isn't linear.

When it looks like two variables have a nonlinear relationship, it's useful to try applying a log scale to the x-axis, y-axis, or both. Let's look for a linear relationship in the scatterplot with transformed axes. Here we re-created the plot of weight by height for dog breeds, but this time we applied a log scale to the y-axis:

```
px.scatter(dogs, x='height', y='weight', log_y=True,
           labels={"height": "Height (cm)", "weight": "Weight (lb)"},
           width=300, height=300)
```

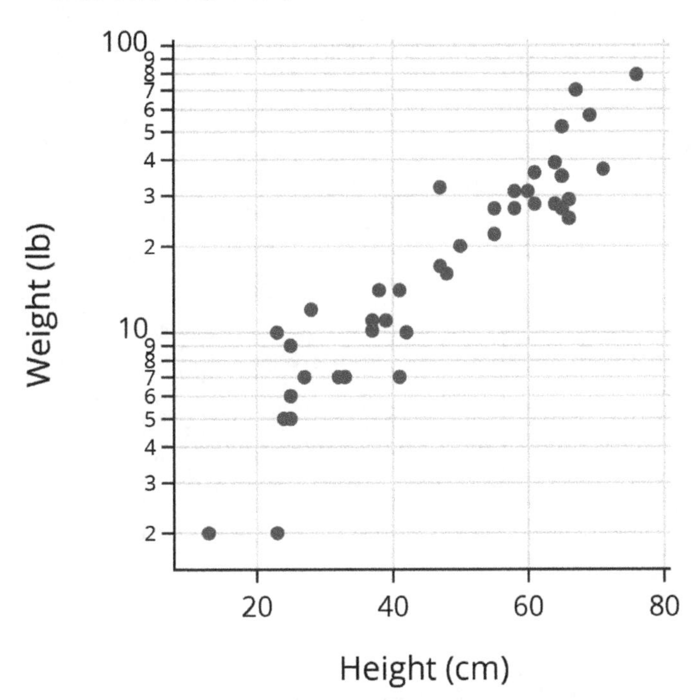

This plot shows a roughly linear relationship, and in this case, we say that there's a log–linear relationship between dog weight and height.

In general, when we see a linear relationship after transforming one or both axes, we can use Table 11-1 to reveal what relationship the original variables have (in the table, a and b are constants). We make these transformations because it is easier for us to see if points fall along a line than to see if they follow a power law compared to an exponential.

Table 11-1. Relationships between two variables when transformations are applied

x-axis	y-axis	Relationship	Also known as
No transform	No transform	Linear: $y = ax + b$	Linear
Log–scale	No transform	Log: $y = a\log x + b$	Linear–log
No transform	Log–scale	Exponential: $y = ba^x$	Log–linear
Log–scale	Log–scale	Power: $y = bx^a$	Log–log

As Table 11-1 shows, the log transform can reveal several common types of relationships. Because of this, the log transform is considered the jackknife of transformations. As another, albeit artificial, example, the leftmost plot in Figure 11-1 reveals a curvilinear relationship between *x* and *y*. The middle plot shows a different curvilinear relationship between *log(y)* and *x*; this plot also appears nonlinear. A further log transformation, at the far right, displays a plot of *log(y)* against *log(x)*. This plot confirms that the data have a log–log (or power) relationship because the transformed points fall along a line.

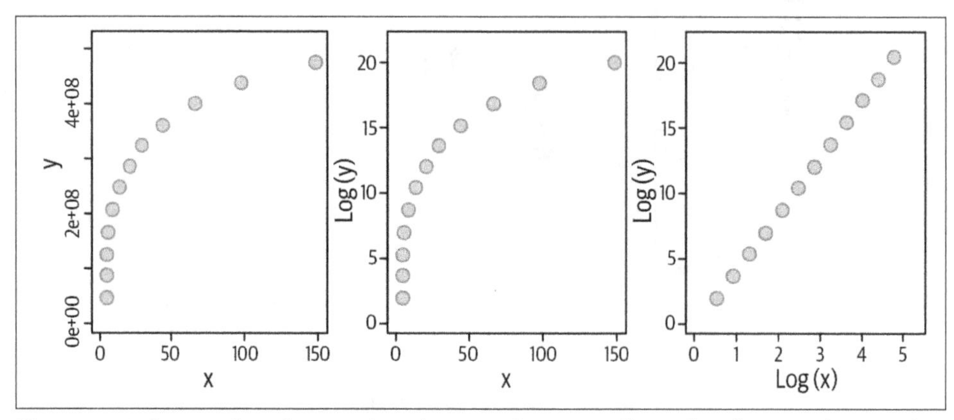

Figure 11-1. Scatterplots showing how log transforms can "straighten" a curvilinear relationship between two variables

Adjusting scale is an important practice in data visualization. While the log transform is versatile, it doesn't handle all situations where skew or curvature occurs. For example, at times the values are all roughly the same order of magnitude and the log transformation has little impact. Another transformation to consider is the square root transformation, which is often useful for count data.

In the next section, we look at principles of smoothing, which we use when we need to visualize lots of data.

Smoothing and Aggregating Data

When we have lots of data, we often don't want to plot all of the individual data points. The following scatter plot shows data from Cherry Blossom (*https://www.cherryblossom.org*), an annual 10-mile race that takes place in Washington, DC, in April, when the cherry trees are in bloom. These data were scraped from the race's website and include official times and other information for all registered male runners from 1999 to 2012. We've plotted the runners' ages on the x-axis and race times on the y-axis:

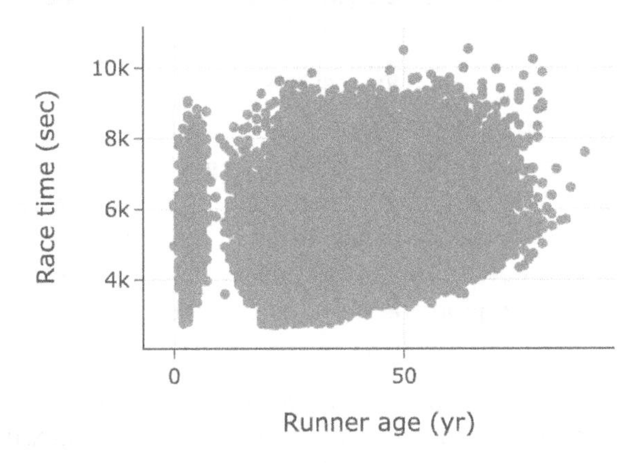

This scatterplot contains over 70,000 points. With so many points, many of them overlap. This is a common problem called *overplotting*. In this case, overplotting prevents us from seeing how time and age are related. About the only thing that we can see in this plot is a group of very young runners, which points to possible issues with data quality. To address overplotting, we use smoothing techniques that aggregate data before plotting.

Smoothing Techniques to Uncover Shape

The histogram is a familiar type of plot that uses smoothing. A histogram aggregates data values by putting points into bins and plotting one bar for each bin. Smoothing here means that we cannot differentiate the location of individual points in a bin: the points are smoothly allocated across their bins. With histograms, the area of a bin corresponds to the percentage (or count or proportion) of points in the bin. (Often the bins are equal in width, and we take a shortcut to label the height of a bin as the proportion.)

The following histogram plots the distribution of lifespans for dog breeds:

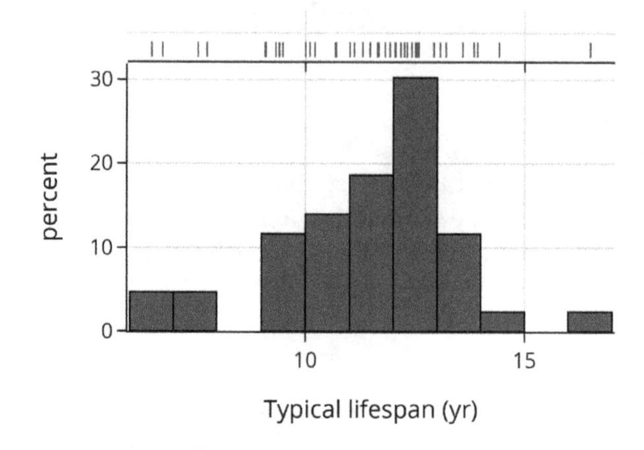

Above this histogram is a rug plot that draws a single line for every data value. We can see in the tallest bin that even a small amount of data can cause overplotting in the rug plot. By smoothing out the points in the rug plot, the histogram reveals the general shape of the distribution. In this case, we see that many breeds have a longevity of about 12 years. For more on how to read and interpret histograms, see Chapter 10.

Another common smoothing technique is *kernel density estimation* (KDE). A KDE plot shows the distribution using a smooth curve rather than bars. In the following plot, we show the same histogram of dog longevity with a KDE curve overlaid. The KDE curve has a similar shape as the histogram:

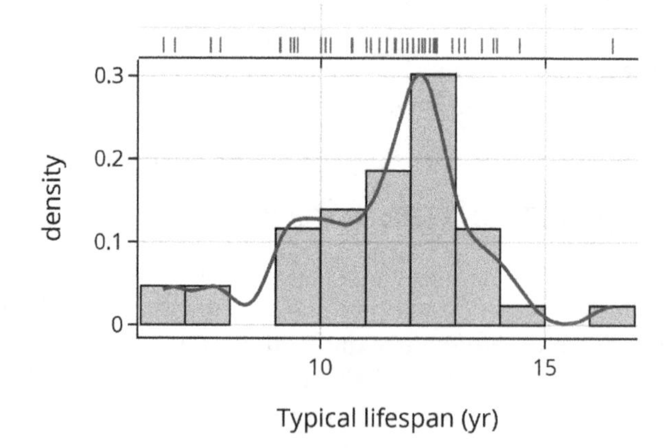

It might come as a surprise to think of a histogram as a smoothing method. Both the KDE and histogram aim to help us see important features in the distribution of values. Similar smoothing techniques can be used with scatterplots. This is the topic of the next section.

Smoothing Techniques to Uncover Relationships and Trends

We can find high-density regions of a scatterplot by binning data, like in a histogram. The following plot remakes the earlier scatterplot of the Cherry Blossom race times against age. This plot uses hexagonal bins to aggregate points together, and then shades the hexagons based on how many points fall in them:

```
runners_over_17 = runners[runners["age"] > 17]

plt.figure(figsize=(4, 4))
plt.hexbin(data=runners_over_17, x='age', y='time', gridsize=35, cmap='Blues')

sns.despine()
plt.grid(False)
plt.xlabel("Runner age (yr)")
plt.ylabel("Race time (sec)");
```

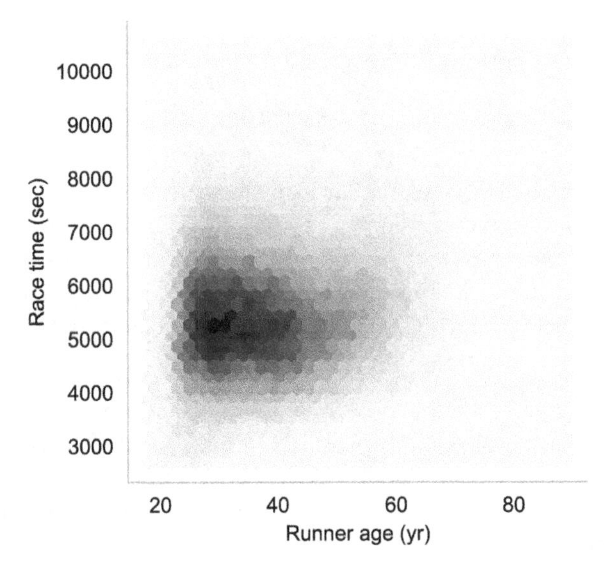

Notice the high-density region in the 25-to-40 age group, signified by the dark region in the plot. The plot shows us that many of the runners in this age range complete the race in around 5,000 seconds (about 80 minutes). (Note that we drop the dubious young runners from this plot.) We can also see upward curvature in the region corresponding to the 40-to-60 age group, which indicates that these runners are generally slower than those in the 25-to-40 age group. This plot is similar to a heat map, where the higher-density regions are conveyed through hotter or brighter colors.

Kernel density estimation also works in two dimensions. When we use KDE in two dimensions, we typically plot the contours of the resulting three-dimensional shape, and we read the plot like a topographical map:

```
plt.figure(figsize=(5, 3))
fig = sns.kdeplot(data=runners_over_17, x='age', y='time')
plt.xlabel("Runner age (yr)")
plt.ylabel("Race time (sec)");
```

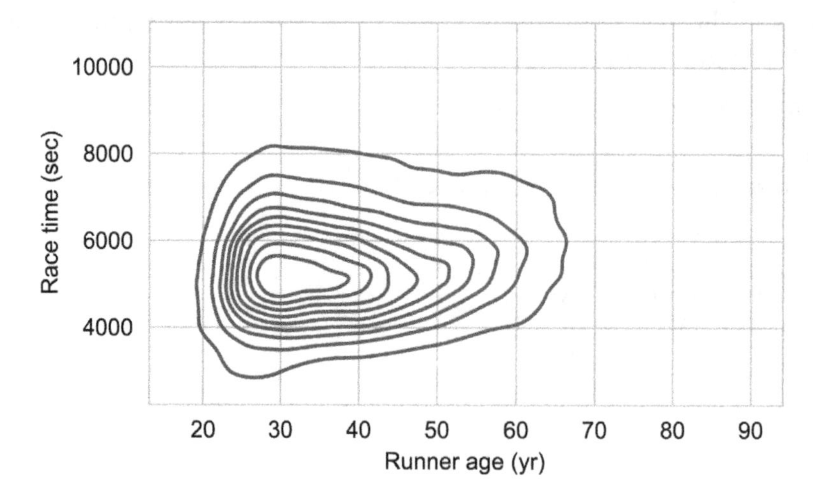

This two-dimensional KDE gives similar insights as the shaded squares of the previous plot. We see a high concentration of runners in the 25-to-40 age group, and these runners have times that appear to be roughly 5,000 seconds. Smoothing lets us get a better picture when there's lots of data because it can reveal the locations of highly concentrated data values and the shape of these high-concentration regions. These regions may be impossible to see otherwise.

Another smoothing approach that is often more informative smooths the y-values for points with a similar x-value. To explain, let's group together runners with similar ages, using five-year increments: 20–25, 25–30, 30–35, and so on. Then, for each five-year bin of runners, we average their race times, plot the average time for each group, and connect the points to form a "curve":

```
times = (
    runners_over_17.assign(age_5yr=runners_over_17['age'] // 5 * 5)
    .groupby('age_5yr')['time'].mean().reset_index()
)

px.line(times, x='age_5yr', y='time',
        labels={'time':"Average race time (sec)", 'age_5yr':"Runner age (5-yr)"},
        markers=True,
        width=350, height=250)
```

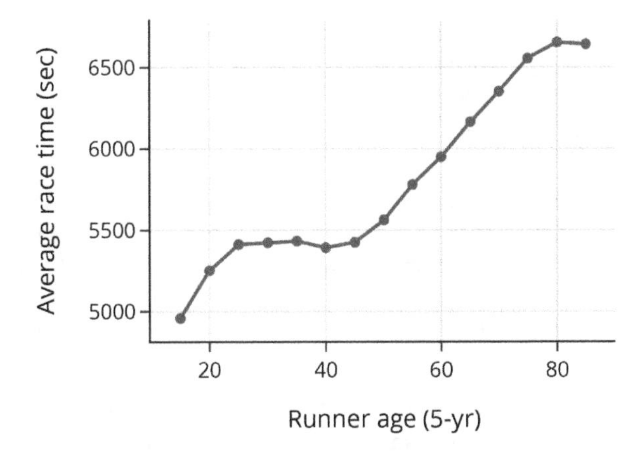

This plot shows once again that runners in the 25-to-40 age range have typical run times of about 5,400 seconds. It also shows that older runners took longer to complete the race on average (not really a surprise, but it wasn't nearly as evident in the earlier plots). The dip in times for runners under age 20 and the flattening of the curve at age 80 may simply be the result of fewer and fitter runners in these groups. Another smoothing technique uses kernel smoothing similar to the KDE. We don't go into the details here.

The binning and kernel smoothing techniques rely on a tuning parameter that specifies the width of the bin or the spread of the kernel, and we often need to specify this parameter when making a histogram, KDE, or smooth curve. This is the topic of the next section.

Smoothing Techniques Need Tuning

Now that we've seen how smoothing is useful for plotting, we turn to the issue of tuning. For histograms, the width of the bins or, for equal-width bins, the number of bins affects the look of the histogram. The left histogram of longevity shown here has a few wide bins, and the right histogram has many narrow bins (see it larger online (*https://oreil.ly/SmLYq*)):

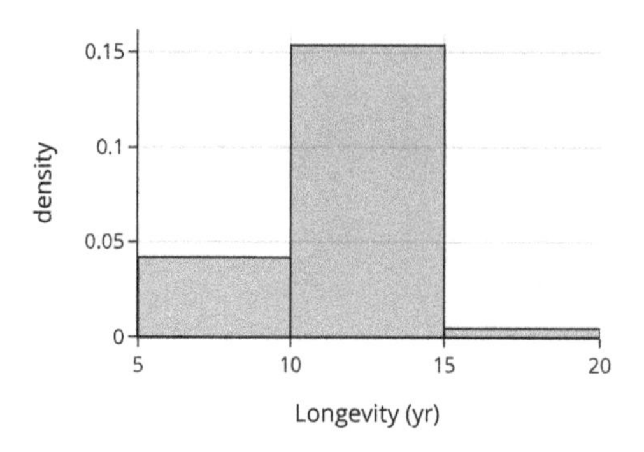

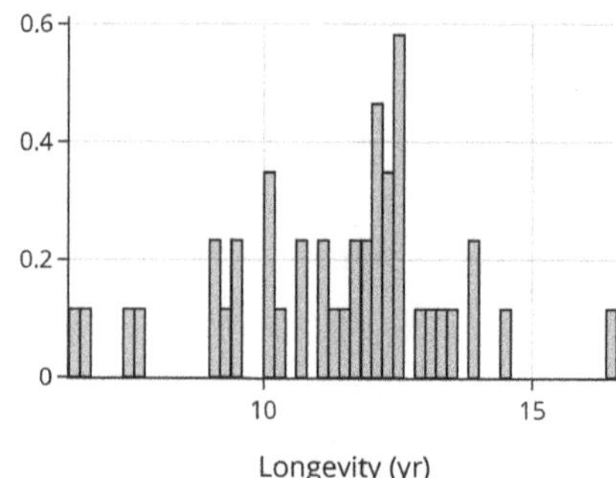

In both histograms, it's hard to see the shape of the distribution. With a few wide bins (the plot on the left), we have oversmoothed the distribution, which makes it impossible to discern modes and tails. On the other hand, too many bins (the plot on the right) gives a plot that's little better than a rug plot. KDE plots have a parameter called the *bandwidth* that works similarly to the bin width of a histogram.

Most histogram and KDE software automatically chooses the bin width for the histogram and the bandwidth for the kernel. However, these parameters often need a bit of fiddling to create the most useful plot. When you create visualizations that rely on tuning parameters, it's important to try a few different values before settling on one.

A different approach to data reduction is to examine quantiles. This is the topic of the next section.

Reducing Distributions to Quantiles

We found in Chapter 10 that while box plots aren't as informative as histograms, they can be useful when comparing the distributions of many groups at once. A box plot reduces the data to a few essential features based on the data quartiles. More generally, quantiles (the lower quartile, median, and upper quartile are the 25th, 50th, and 75th quantiles) can provide a useful reduction in the data when comparing distributions.

When two distributions are roughly similar in shape, it can be hard to compare them with histograms. For instance, the histograms that follow show the price distributions for two- and four-bedroom houses in the San Francisco housing data. The distributions look roughly similar in shape. But a plot of their quantiles handily compares the distributions' center, spread, and tails (see it larger online (*https://oreil.ly/zVagW*)):

```
px.histogram(sfh.query('br in [2, 4]'),
             x='price', log_x=True, facet_col='br',
             labels={'price':"Sale price (USD)"},
             width=700, height=250)
```

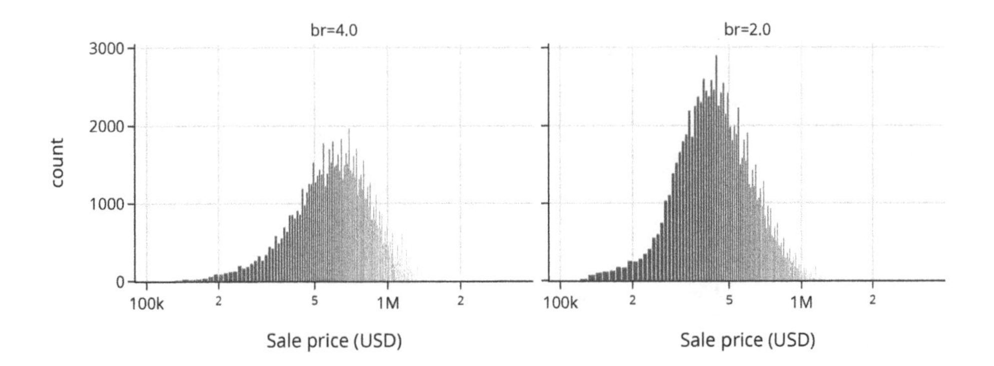

We can compare quantiles with a *quantile–quantile* plot, called *q–q plot* for short. To make this plot, we first compute percentiles (also called quantiles) for both the two- and four-bedroom distributions of price:

```
br2 = sfh.query('br == 2')
br4 = sfh.query('br == 4')
percs = np.arange(1, 100, 1)
perc2 = np.percentile(br2['price'], percs, method='lower')
perc4 = np.percentile(br4['price'], percs, method='lower')
perc_sfh = pd.DataFrame({'percentile': percs, 'br2': perc2, 'br4': perc4})
perc_sfh
```

	percentile	br2	br4
0	1	1.50e+05	2.05e+05
1	2	1.82e+05	2.50e+05
2	3	2.03e+05	2.75e+05
...	...	...	...
96	97	1.04e+06	1.75e+06
97	98	1.20e+06	1.95e+06
98	99	1.44e+06	2.34e+06

```
99 rows × 3 columns
```

Then we plot the matching percentiles on a scatterplot. We usually also show the reference line y = x to help with the comparison:

```
fig = px.scatter(perc_sfh, x='br2', y='br4', log_x=True, log_y=True,
                 labels={'br2': 'Price of 2-bedroom house',
                         'br4': 'Price of 4-bedroom house'},
                 width=350, height=250)

fig.add_trace(go.Scatter(x=[1e5, 2e6], y=[1e5, 2e6],
              mode='lines', line=dict(dash='dash')))

fig.update_layout(showlegend=False)
fig
```

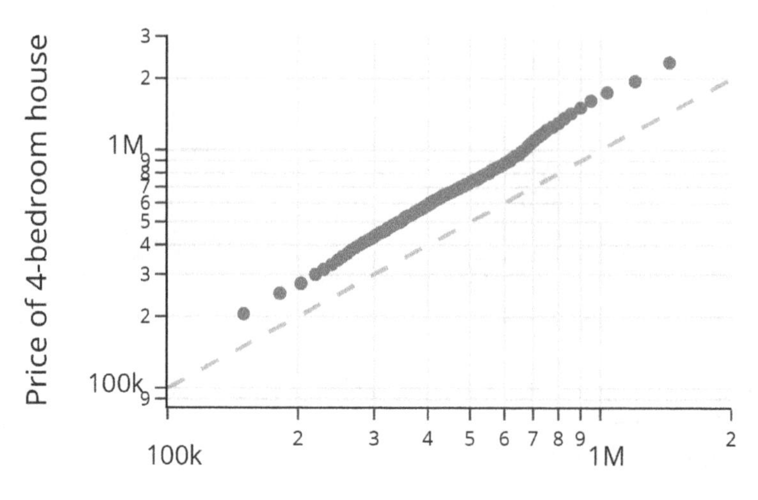

When the quantile points fall along a line, the variables have similarly shaped distributions. Lines parallel to the reference indicate a difference in center, lines with slopes other than 1 indicate a difference in spread, and curvature indicates a difference in shape. From the preceding q–q plot, we see that the distribution of price for four-bedroom houses is similar in shape to the two-bedroom distribution, except for a shift of about $100K and a slightly longer right tail (indicated by the upward bend for large values). Reading a q–q plot takes practice. Once you get the hang of it, though, it can be a handy way to compare distributions. Notice that the housing data have over 100,000 observations, and the q–q plot has reduced the data to 99 percentiles. This data reduction is quite useful. However, we don't always want to use smoothers. This is the topic of the next section.

When Not to Smooth

Smoothing and aggregating can help us see important features and relationships, but when we have only a handful of observations, smoothing techniques can be misleading. With just a few observations, we prefer rug plots over histograms, box plots, and density curves, and we use scatterplots rather than smooth curves and density contours. This may seem obvious, but when we have a large amount of data, the amount of data in a subgroup can quickly dwindle. This phenomenon is an example of the *curse of dimensionality*.

One of the most common misuses of smoothing happens with box plots. As an example, here is a collection of seven box plots of longevity, one for each of seven types of dog breed:

```
px.box(dogs, x='group', y='longevity',
       labels={'group':"", 'longevity':"Longevity (yr)"},
       width=500, height=250)
```

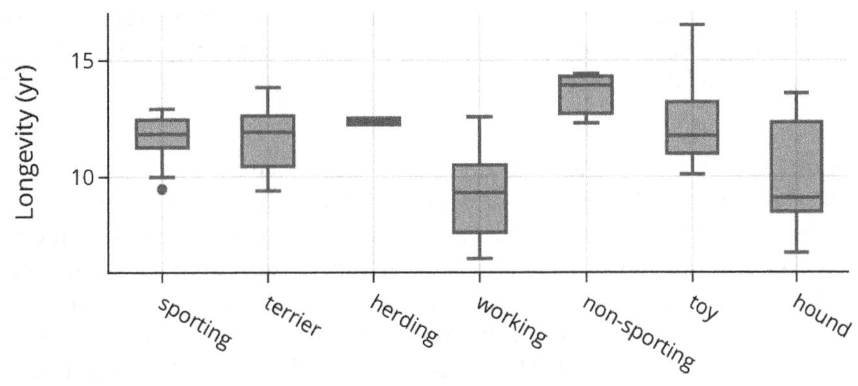

Some of these box plots have as few as two or three observations. The strip plot that follows is a preferable visualization:

```
px.strip(dogs, x="group", y="longevity",
         labels={'group':"", 'longevity':"Longevity (yr)"},
         width=400, height=250)
```

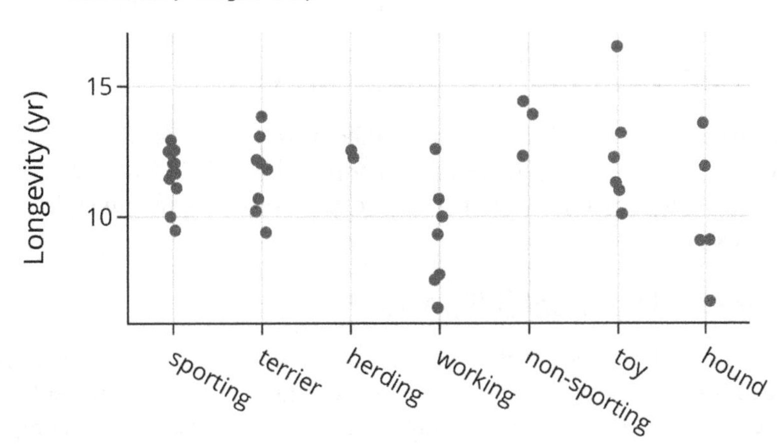

In this plot, we can still compare the groups, but we also see the exact values in each group. Now we can tell that there are only three breeds in the non-sporting group; the impression of a skewed distribution, based on the box plot, reads too much into the shape of the box.

This section introduced the problem of overplotting, where we have overlapping points because of a large dataset. To address this issue, we introduced smoothing techniques that aggregate data. We saw two common examples of smoothing—

binning and kernel smoothing—and applied them in the one- and two-dimensional settings. In one dimension, these are histograms and kernel density curves, respectively, and they both help us see the shape of a distribution. In two dimensions, we found it useful to smooth y-values while keeping x-values fixed in order to visualize trends. We addressed the need to tune the smoothing amount to get more informative histograms and density curves, and we cautioned against smoothing with too few data.

There are many other ways to reduce overplotting in scatter plots. For instance, we can make the dots partially transparent so that overlapping points appear darker. If many observations have the same values (like when longevity is rounded to the nearest year), then we can add a small amount of random noise to the values to reduce the amount of overplotting. This procedure is called *jittering*, and it is used in the strip plot of longevity. Transparency and jittering are convenient for medium-sized data. However, they don't work well for large datasets since plotting all the points still overwhelms the visualization.

The quantile–quantile plot we introduced offers one way to compare distributions with far fewer points; another is to use side-by-side box plots and yet another is to overlay KDE curves in the same plot. We often aim to compare distributions and relationships across subsets (or groups) of data, and we next discuss several design principles that facilitate meaningful comparisons for a variety of plot types.

Facilitating Meaningful Comparisons

The same data can be visualized in many different ways, and deciding which plot to make can be daunting. Generally speaking, our plot should help a reader make meaningful comparisons. In this section, we go over several useful principles that can improve the clarity of our plots.

Emphasize the Important Difference

Whenever we make a plot that compares groups, we consider whether the plot emphasizes the important difference. As a rule of thumb, it's easier for readers to see differences when plotted objects are aligned in ways that make these comparisons easier to read. Let's look at an example.

The US Bureau of Labor Statistics (*https://oreil.ly/b0YMJ*) publishes data on income. We took the 2020 median full-time-equivalent weekly earnings for people over age 25 and plotted them. We split people into groups by education level and sex:[1]

1 US government surveys still collect data based on a binary definition of gender, but progress is being made. For example, starting in 2022, US citizens are allowed to select an "X" as their gender marker on their passport application.

```
labels = {"educ": "Education",
          "income": "Weekly earnings (USD)",
          "gender": "Sex"}
fig = px.bar(earn, x="educ", y="income",
             facet_col="gender", labels=labels,
             width=450, height=250)
fig.update_layout(margin=dict(t=30))
```

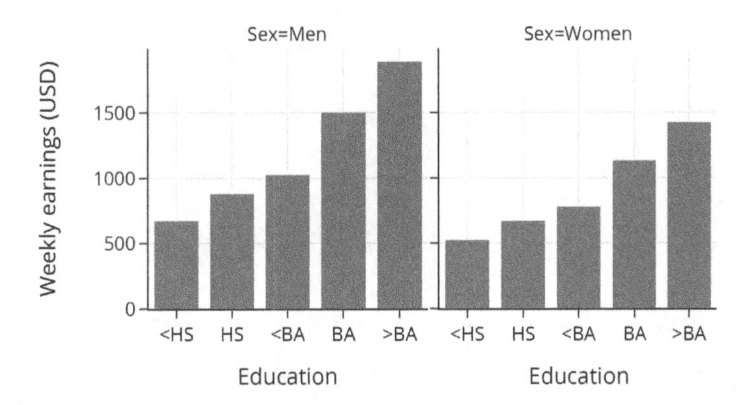

These bar plots show that earnings increase with more education. But arguably, a more interesting comparison is between men and women of the same education level. We can group the bars differently to focus instead on this comparison:

```
px.bar(earn, x='educ', y='income', color='gender',
       barmode='group', labels=labels,
       width=450, height=250)
```

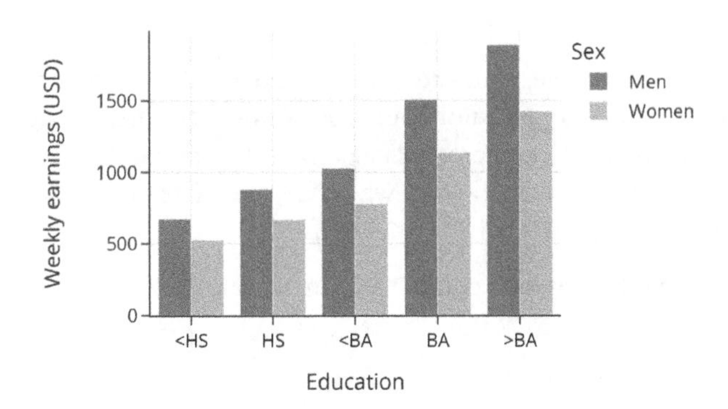

This plot is much better; we can more easily compare the earnings of men and women for each level of education. However, we can make this difference even clearer using vertical alignment. Instead of bars, we use dots for groups of men and women that align vertically at each education level:

```
fig = px.line(earn, x='educ', y='income', symbol='gender',
              color='gender', labels=labels, width=450, height=250)
fig.update_traces(marker_size=10)
```

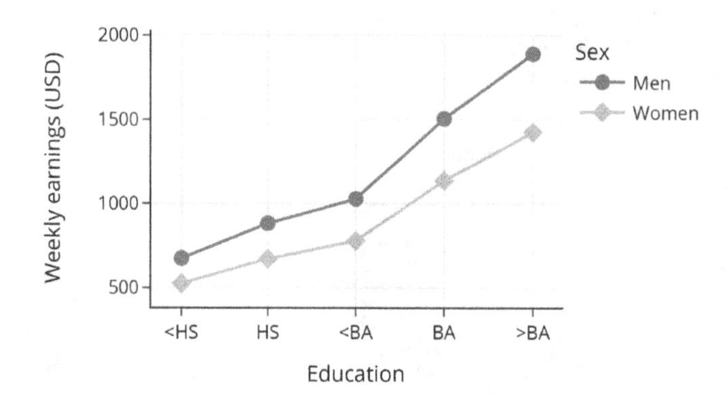

This plot more clearly reveals an important difference: the earnings gap between men and women grows with education. We considered three plots that visualize the same data, but they differ in how readily we can see the message in the plot. We prefer the last one because it aligns the income differences vertically, making them easier to compare.

Notice that in making all three plots, we ordered the education categories from the least to greatest number of years of education. This ordering makes sense because education level is ordinal. When we compare nominal categories, we use other approaches to ordering.

Ordering Groups

With ordinal features, we keep the categories in their natural order when we make plots, but the same principle does not apply for nominal features. Instead, we choose an ordering that helps us make comparisons. With bar plots, it's a good practice to order the bars according to their height, while for box plots and strip plots, we typically order the boxes/strips according to medians.

The two bar plots that follow each compare the mean lifespan for types of dog breeds:

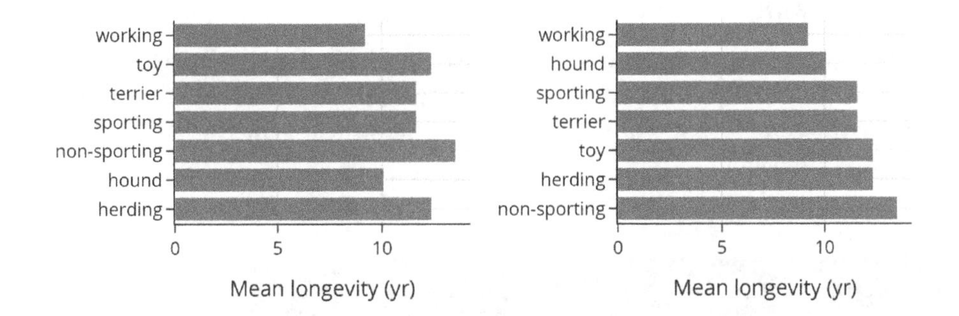

The plot on the left orders the bars alphabetically. We prefer the plot on the right because it orders the bars by longevity, which makes it easier to compare longevity across the categories. We don't have to bounce back and forth or squint to guess whether herding breeds have a shorter lifespan than toy breeds.

As another example, the following two sets of box plots each compare the distribution of sale price for houses in different cities in the San Francisco East Bay area:

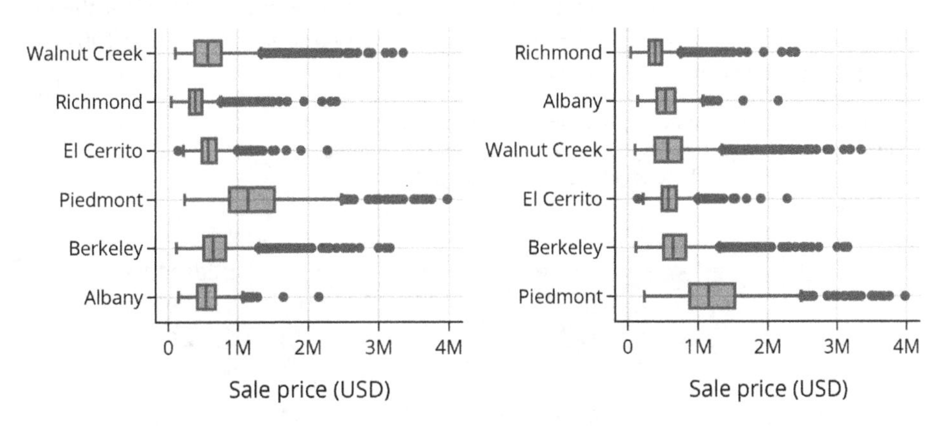

We prefer the plot on the right since it has boxes ordered according to the median price for each city. Again, this ordering makes it easier to compare distributions across groups, in this case cities. We see that the lower quartile and median price in Albany and Walnut Creek are roughly the same, but the prices in Walnut Creek have a greater right skew.

When possible, ordering bars in a bar plot by height and boxes in a box plot by median makes it easier for us to make comparisons across groups. Another technique used for presenting grouped data is stacking. We describe stacking in the next section and provide examples to convince you to steer away from this sort of plot.

Avoid Stacking

The figure that follows shows a stacked bar plot in which there is one bar for each city and these bars are divided according to the proportion of houses sold that have from one to eight or more bedrooms. This is called a *stacked bar plot*. The bar plot is based on a cross-tabulation:

```
br_crosstab
```

br	1.0	2.0	3.0	4.0	5.0	6.0	7.0	8.0
city								
Albany	1.21e-01	0.56	0.25	0.05	9.12e-03	1.01e-03	2.03e-03	4.05e-03
Berkeley	6.91e-02	0.38	0.31	0.16	4.44e-02	1.42e-02	6.48e-03	7.23e-03
El Cerrito	1.81e-02	0.34	0.47	0.14	2.20e-02	6.48e-03	0.00e+00	6.48e-04
Piedmont	8.63e-03	0.22	0.40	0.26	9.50e-02	1.29e-02	7.19e-03	1.44e-03
Richmond	3.60e-02	0.36	0.42	0.15	2.52e-02	7.21e-03	7.72e-04	7.72e-04
Walnut Creek	1.16e-01	0.35	0.30	0.18	4.37e-02	5.08e-03	4.12e-04	2.75e-04

Each bar in the plot has the same height of 1 because the segments represent the proportion of houses with one or more bedrooms in a city and so add to 1 or 100% (see it in color online (*https://oreil.ly/cIba_*)):

```
fig = px.bar(br_crosstab, width=450, height=300)
fig.update_layout(yaxis_title=None, xaxis_title=None,
                  legend_title="# Bedrooms")
fig.show()
```

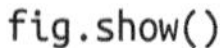

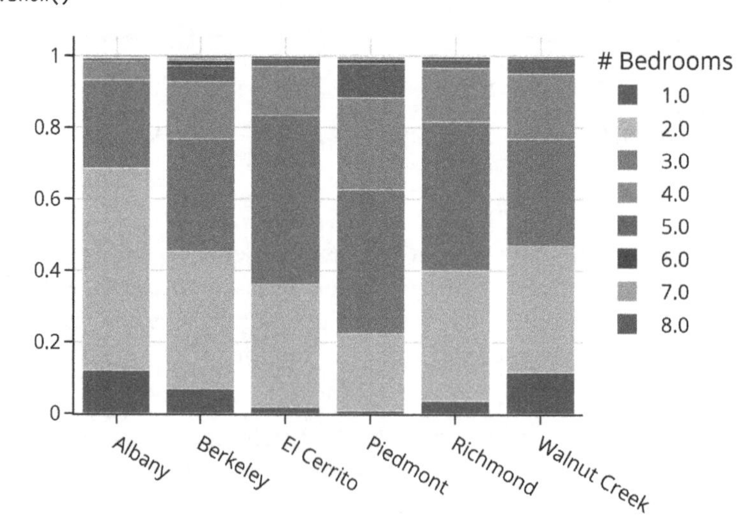

It's easy to compare the proportion of one-bedroom houses in each of the cities by simply scanning across the top of the first segment in each column. But the comparison of four-bedroom houses is more difficult. The bottoms of the segments are not aligned horizontally, so our eyes must judge the lengths of segments that move up and down across the plot. This up-and-down movement is called *jiggling the baseline*. (We recognize that with so many colors, this plot does not render well in grayscale, but our goal is to steer you away from plots like this one and the next, so we have kept all of the colors for those of you reading the online version.)

Stacked line plots are even more difficult to read because we have to judge the gap between curves as they jiggle up and down. The following plot shows carbon dioxide (CO_2) emissions (*https://oreil.ly/kjk9N*) from 1950 to 2012 for the 10 highest emitters (see it in color online (*https://oreil.ly/VUyuz*)):

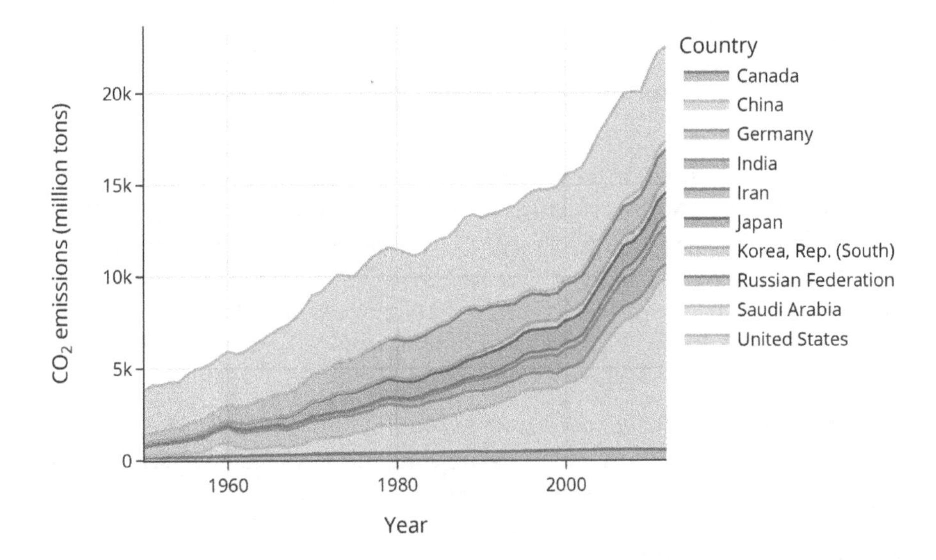

Since the lines are stacked on top of each other, it's very hard to see how the emissions for a particular country have changed and it's hard to compare countries. Instead, we can plot each country's line without stacking, as the next plot illustrates (see it in color online (*https://oreil.ly/erWuU*)):

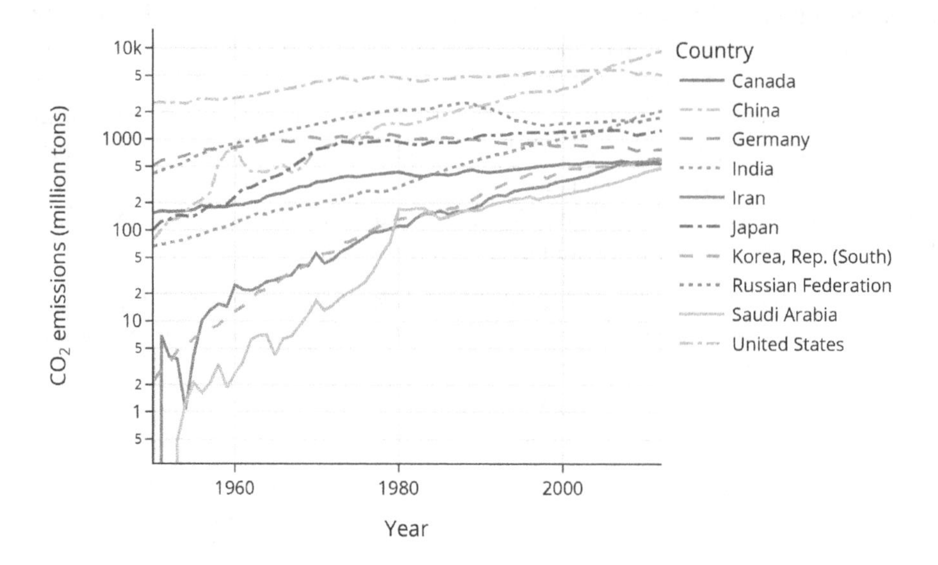

Now it's much easier to see changes for individual countries and to compare countries because we need judge only y-axis positions rather than short vertical segments with different baselines. We also used a log scale on the y-axis. We can now see that some countries have had flat rates of growth in CO_2 emissions, such as the United States and Japan, while others have increased much more quickly, like China and India, and Germany has slowed its CO_2 emissions. These aspects were nearly impossible to detect when each country's baseline jiggled across the plot.

In both of these plots, to make it easier to tell one country from the next, we have used different line types and colors. Choosing colors to facilitate comparisons relies on many considerations. This is the topic of the next section.

Selecting a Color Palette

Choosing colors also plays an important role in data visualization. We want to avoid overly bright or dark colors so that we don't strain the reader's eyes. We should also avoid color palettes that might be difficult for colorblind people—7% to 10% of people (mostly males) are red-green colorblind.

For categorical data, we want to use a color palette that can clearly distinguish between categories. One example is shown at the top in Figure 11-2. From top to bottom, these palettes are qualitative for categorical data, diverging for numeric data where you want to draw attention to both large and small values, and sequential for numeric data where you want to emphasize either large or small values.

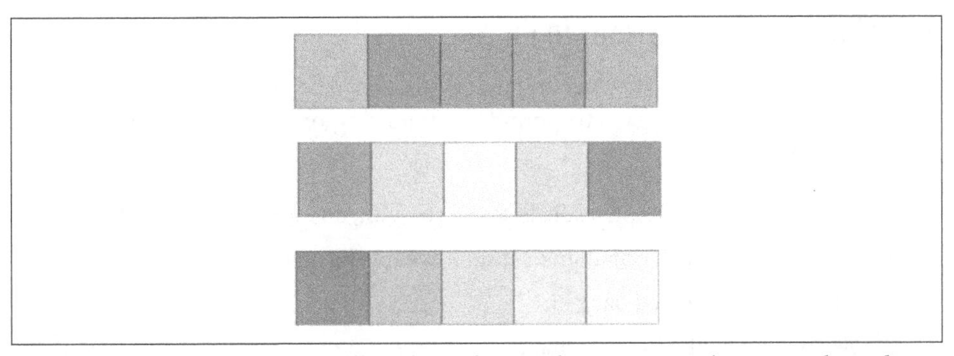

Figure 11-2. Three printer-friendly palettes from ColorBrewer 2.0 (see it in color online (https://oreil.ly/0ub2H))

For numeric data, we want to use a sequential color palette that emphasizes one side of the spectrum more than the other or a diverging color palette that equally emphasizes both ends of the spectrum and deemphasizes the middle. A sequential palette is shown at the bottom and a diverging palette is shown in the middle of Figure 11-2.

We choose a sequential palette when we want to emphasize either low or high values, like cancer rates. We choose a diverging palette when we want to emphasize both extremes, like for two-party election results.

It's important to choose a perceptually uniform color palette. By this we mean that when a data value is doubled, the color in the visualization looks twice as colorful to the human eye. We also want to avoid colors that create an afterimage when we look from one part of the graph to another, colors of different intensities that make one attribute appear more important than another, and colors that colorblind people have trouble distinguishing between. We strongly recommend using a palette or a palette generator made specifically for data visualizations.

Plots are meant to be examined for long periods of time, so we should choose colors that don't impede the reader's ability to carefully study a plot. Even more so, the use of color should not be gratuitous—colors should represent information. On a related note, people typically have trouble distinguishing between more than about seven colors, so we limit the number of colors in a plot. Finally, colors can appear quite different when printed on paper in grayscale than when viewed on a computer screen. When we choose colors, we keep in mind how our plots will be displayed.

Making accurate comparisons in a visualization is such an important goal that researchers have studied how well people perceive differences in colors and other plotting features such as angles and lengths. This is the topic of the next section.

Guidelines for Comparisons in Plots

Researchers have studied how accurately people can read information displayed in different types of plots. They have found the following ordering, from most to least accurately judged:

- Positions along a common scale, like in a rug plot, strip plot, or dot plot
- Positions on identical, nonaligned scales, like in a bar plot
- Length, like in a stacked bar plot
- Angle and slope, like in a pie chart
- Area, like in a stacked line plot or bubble chart
- Volume and density, like in a three-dimensional bar plot
- Color saturation and hue, like when overplotting with semitransparent points

As an example, here is a pie chart that shows the proportion of houses sold in San Francisco that have from one to eight or more bedrooms, and a bar chart with the same proportions:

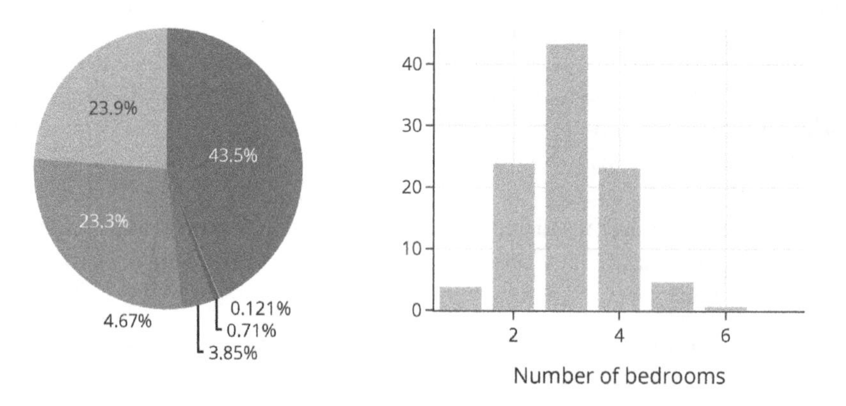

It's hard to judge the angles in the pie chart, and the annotations with the actual percentages are needed. We also lose the natural ordering of the number of bedrooms. The bar chart doesn't suffer from these issues.

However, there are exceptions to any rule. Multiple pie charts with only two or three slices in each pie can provide effective visualizations. For example, a set of pie charts of the proportion of two-bedroom houses sold in each of six cities in the San Francisco East Bay Area, ordered according to the proportion, can be an impactful visualization. Yet, sticking with a bar chart will generally always be at least as clear as any pie chart.

Given these guidelines, we recommend sticking to position and length for making comparisons. Readers can more accurately judge comparisons based on position or length, rather than angle, area, volume, or color. But, if we want to add additional information to a plot, we often use color, symbols, and line styles, in addition to position and length. We have shown several examples in this chapter.

We next turn to the topic of data design and how to reflect the aspects of when, where, and how the data were collected in a visualization. This is a subtle but important topic. If we ignore the data scope, we can get very misleading plots.

Incorporating the Data Design

When we create a visualization, it's important to consider the data scope, especially the data design (see Chapter 2). Considering the question of how the data were collected can impact our plot choice and the comparisons we portray. These considerations include the time and place where the data were collected and the design used to select a sample. We look at a few examples of how the data scope can inform the visualizations we make.

Data Collected Over Time

When data are collected over time, we typically make a line plot that puts timestamps on the x-axis and a feature of interest on the y-axis to look for trends in time. As an example, let's revisit the data on San Francisco housing prices. These data were collected from 2003 through 2008 and show the crash in 2008/2009 of the US housing bubble (*https://oreil.ly/PUPiQ*). Since time is a key aspect of the scope of these data, let's visualize sale price as a time series. Our earlier explorations showed that sale price is highly skewed, so let's work with percentiles rather than averages. We plot the median price (this is a form of smoothing we saw earlier in this chapter):

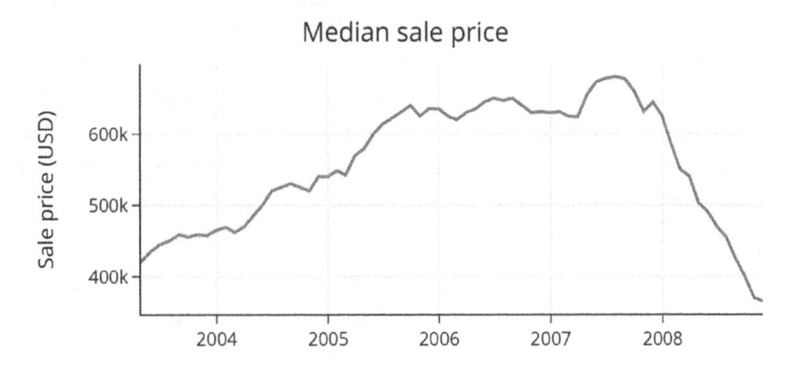

This plot shows the rise in prices from 2003 to 2007 and the fall in 2008. But we can show more information by plotting a few additional percentiles instead of just the median. Let's draw separate lines for the 10th, 30th, 50th (median), 70th, and 90th percentile sale prices. When we examine prices over time, we typically need to adjust for inflation so that the comparisons are on the same footing. In addition to adjusting for inflation, let's plot the prices relative to the starting price in 2003 for each of the percentiles. This means that all the lines start at y = 1 in 2003. (A value of 1.5 for the 90th percentile in 2006 indicates that the sale price is 1.5 times the 90th percentile in 2003.) This normalization lets us see how the housing crash affected home owners in the different parts of the market:

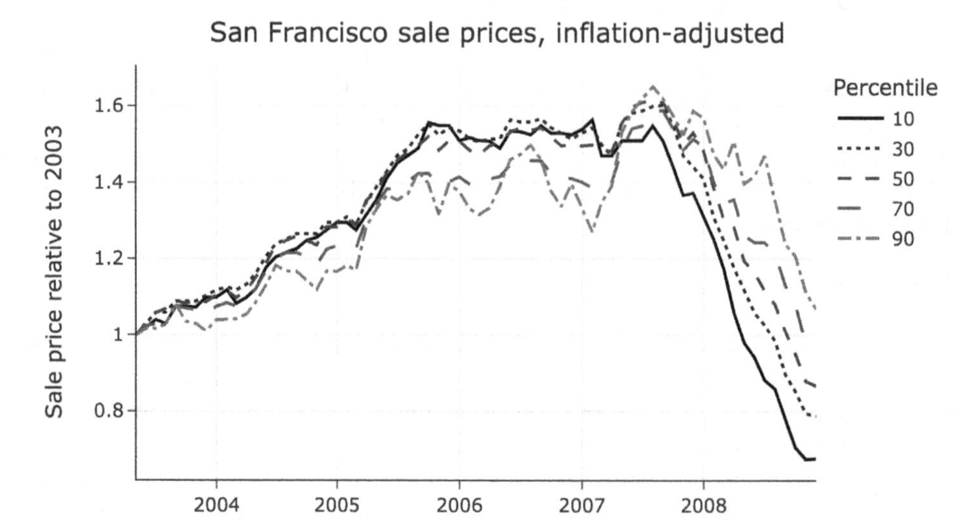

When we follow the 10th percentile line plot over time, we see that it increases quickly in 2005, stays high relative to its 2003 value for a few years, and then drops earlier and more quickly than the other percentiles. This tells us that the less expensive houses, such as starter homes, suffered greater volatility and lost much more value in the housing market crash. In contrast, higher-end homes were affected less by the crash; at the end of 2008, the 90th percentile home prices were still higher than the 2003 prices. Applying this bit of domain knowledge helps reveal trends in the data that we might otherwise miss, and shows how we can use the data design to improve a visualization.

The housing data are an example of observational data that form a complete census in a geographic region over a specific period of time. Next we consider another observational study where self-selection and the time period impact the visualization.

Observational Studies

We need to be particularly cautious with data that do not form a census or scientific sample. We should also take care with cross-sectional studies, whether from a census or scientific sample. For this example, we revisit the data from the Cherry Blossom 10-mile run. Earlier in this chapter, we made a smoothed curve to examine the relationship between race time and age. We reproduce this plot here to highlight a potential pitfall in interpretation:

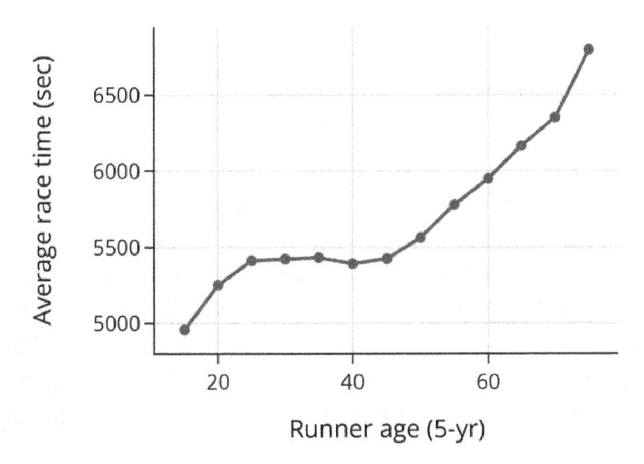

It's tempting to look at this plot and conclude that, for instance, a runner at age 60 can typically expect to take an additional 600 seconds to finish the run than when they were 40. However, this is a *cross-sectional* study, not a *longitudinal* study. The study does not follow people over time; instead, it gets a snapshot of a cross-section of people. The 60-year-old runners represented in the plot are different people than the 40-year-old runners. These two groups could be different in ways that affect the relationship between race time and age. As a group, the 60-year-olds in the race are likely to be fitter for their age than the 40-year-olds. In other words, the data design doesn't let us make conclusions about individual runners. The visualization isn't wrong, but we need to be careful about the conclusions we draw from it.

The design is even more complex because we have race results from many years. Each year forms a cohort, a group of racers, and from one year to the next, the cohort changes. We create a visualization that makes this message clear by comparing runners in different race years. Here, we've separately plotted lines for the runners in 1999, 2005, and 2010:

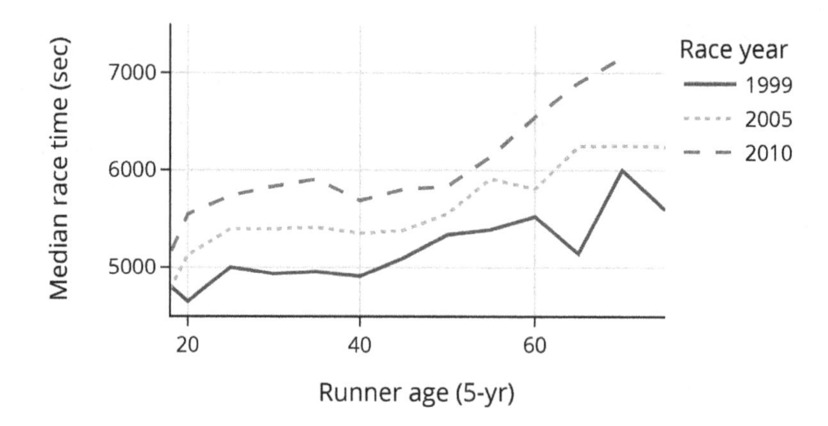

We see that the median race times in 2010 are higher at every age group than the times for the runners in 2005, and in turn, the times are higher for the runners in 2005 than for the runners in 1999. It's interesting that race times have slowed over the years. This is quite likely due to the increased popularity of the race, where there is higher participation from novice runners in more recent years. This example has shown how we need to be aware of the data scope when interpreting patterns. We also need to keep data scope in mind with scientific studies. This is the topic of the next section.

Unequal Sampling

In a scientific study, we must consider the sample design because it can impact our plots. Some samples draw individuals at unequal rates, and this needs to be accounted for in our visualizations. We saw an example of a scientific study in Chapter 8 and Chapter 9: the Drug Abuse Warning Network (DAWN) survey. These data are from a complex randomized study of drug-related emergency room visits, and each record comes with a weight that we must use in order to accurately represent the emergency room visits in the population. The two bar plots that follow show the distribution of the type of ER visit. (See them larger online (*https://oreil.ly/LYAol*).) The one on the left doesn't use the survey weights and the one on the right does:

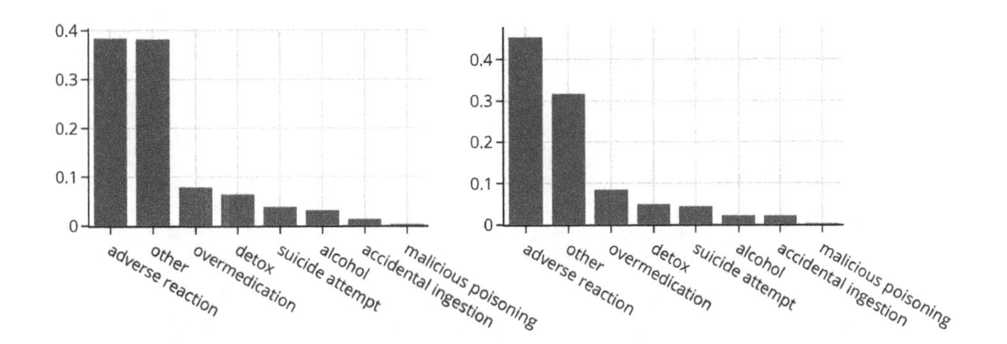

In the unweighted bar plot, the "Other" category is as frequent as the "Adverse reaction" category. However, when weighted, "Other" drops to about two-thirds of "Adverse reaction." Ignoring sampling weights can give a misleading presentation of a distribution. Whether for a histogram, bar plot, box plot, two-dimensional contour, or smooth curve, we need to use the weights to get a representative plot. Another aspect of the data scope that can impact our choice of plots is where the data are collected, which is the topic of the next section.

Geographic Data

When our data contains geographic information like latitude and longitude, we should consider making a map, in addition to the typical plots. For example, the following map shows the locations for US air quality sensors, which is the focus of the case study in Chapter 12:

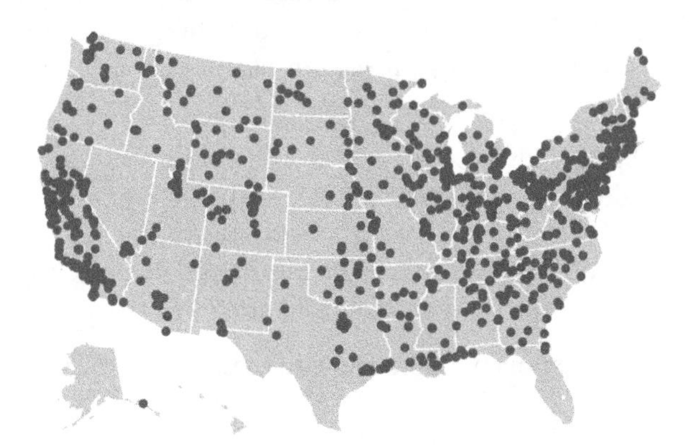

Notice that there are many more points in California and the Eastern Seaboard. A simple histogram of air quality with data from all of these sensors would misrepresent the distribution of air quality in the US. To incorporate the spatial aspect into the distribution, we can add air quality measurements to the map with different color markers, and we can facet the histograms of air quality by location.

In addition to plotting features like bars, color, and line styles, we also have the option to add text with contextual information to make our plot more informative. This is the topic of the next section.

Adding Context

We have used text in our graphs throughout this chapter to provide meaningful axis labels that include units of measurement, tick-mark labels for categories, and titles. This is a good practice when sharing a visualization more broadly. A good goal is to include enough context in a plot that it can stand alone—a reader should be able to get the gist of the plot without needing to search for explanation elsewhere. That said, every element of a statistical graph should have a purpose. Superfluous text or plot features, often referred to as *chartjunk*, should be eliminated. In this section, we provide a brief overview of ways we can add helpful context to our plots and an example where we create a publication-ready plot by adding context.

Text context includes *labels* and *captions*. It is a good practice to consistently use informative labels on tick marks and axes. For example, axis labels often benefit from including units of measurement. Graphs should contain titles and legends when needed. Informative labels are especially important for plots that other people will see and interpret. However, even when we're doing exploratory data analysis just for us, we often want to include enough context that when we later return to an analysis we can easily figure out what we plotted.

Captions serve several purposes. They describe what has been plotted and orient the reader. Captions also point out important features of the plot and comment on their implications. It's OK for the caption to repeat information found in the text. Readers often skim a publication and focus on section headings and visualizations, so plot captions should be self-contained.

Reference markers bring additional context to the plotting region. Reference points and lines that provide benchmarks, historical values, and other external information help form comparisons and interpretations. For example, we often add a reference line with slope 1 to a quantile–quantile plot. We might also add a vertical line on a time-series plot to mark a special event, like a natural disaster.

The following example demonstrates how to add these context elements to a plot.

Example: 100m Sprint Times

The following figure shows the race times in the men's 100-meter sprint since 1968. These data include only races that were electronically timed and held outdoors in normal wind conditions, and the times included are only for those runners who came in under 10 seconds. The plot is a basic scatterplot showing race time against year. Beginning with this plot, we augment it to create a plot featured in a FiveThirtyEight article (*https://oreil.ly/pxHr4*) about the 100-meter sprint:

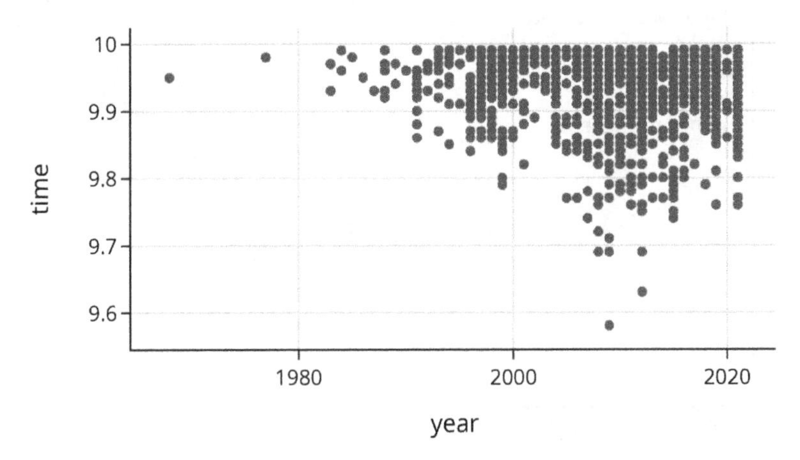

When we want to prepare a plot for others to read, we consider the takeaway message. In this case, our main message is twofold: the best runners have been getting faster over the past 50 years, and Usain Bolt's remarkable record time of 9.58 seconds set in 2009 remains untouched. (In fact, the second-best race time also belongs to Bolt.) We provide context to this plot by adding a title that directly states the main takeaway, units of measurement in the y-axis label, and annotations to key points in the scatterplot, including the two best race times that belong to Usain Bolt. In addition, we add a horizontal reference line at 10 seconds to clarify that only times below 10 seconds are plotted, and we use a special symbol for the world record time to draw the reader's attention to this crucial point:

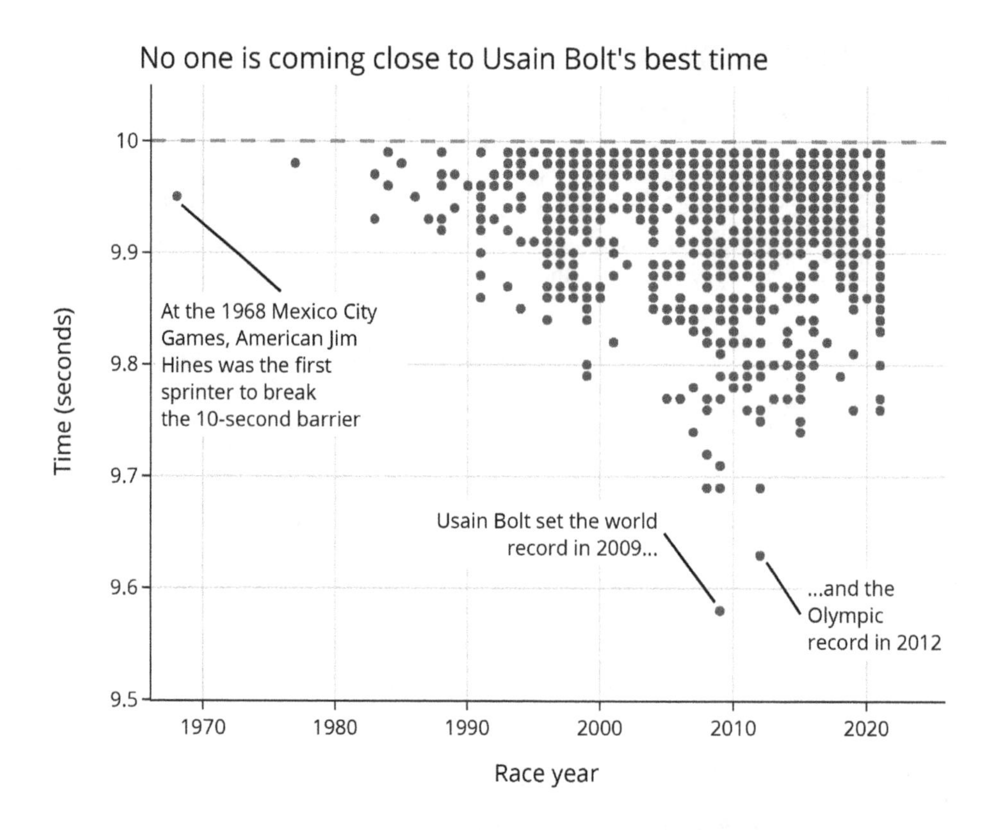

These bits of context describe what we have plotted, help readers see the main takeaway, and point out several interesting features in the data. The plot can now be a useful part of a slideshow, technical report, or social media post. In our experience, people who look at our data analyses remember our plots, not paragraphs of text or equations. It's important to go the extra mile and add context to the plots we prepare for others.

In the next section, we move on to specifics on how to create plots using the `plotly` Python package.

Creating Plots Using plotly

In this section, we cover the basics of the `plotly` Python package, the main tool we use in this book to create plots.

The plotly package has several advantages over other plotting libraries. It creates interactive plots rather than static images. When you create a plot in plotly, you can pan and zoom to see parts of the plot that are too small to see normally. You can also hover over plot elements, like the symbols in a scatterplot, to see the raw data values. Also, plotly can save plots using the SVG file format, which means that images appear sharp even when zoomed in. If you're reading this chapter in a PDF or paper copy of the book, we used this feature to render plot images. Finally, it has a simple "express" API for creating basic plots, which helps when you're doing exploratory analysis and want to quickly create many plots.

We go over the fundamentals of plotly in this section. We recommend using the official plotly documentation (*https://plotly.com/python*) if you encounter something that isn't covered here.

Figure and Trace Objects

Every plot in plotly is wrapped in a Figure object. Figure objects keep track of what to draw. For instance, a single Figure can draw a scatterplot on the left and a line plot on the right. Figure objects also keep track of the plot layout, which includes the plot's size, title, legend, and annotations.

The plotly.express module provides a concise API for making plots:

```
import plotly.express as px
```

We use plotly.express in the following code to make a scatterplot of weight against height for the data on dog breeds. Notice that the return value from .scatter() is a Figure object:

```
fig = px.scatter(
    dogs, x="height", y="weight",
    labels=dict(height="Height (cm)", weight="Weight (kg)"),
    width=350, height=250,
)

fig.__class__
```

```
plotly.graph_objs._figure.Figure
```

Displaying a Figure object renders it to the screen:

```
fig
```

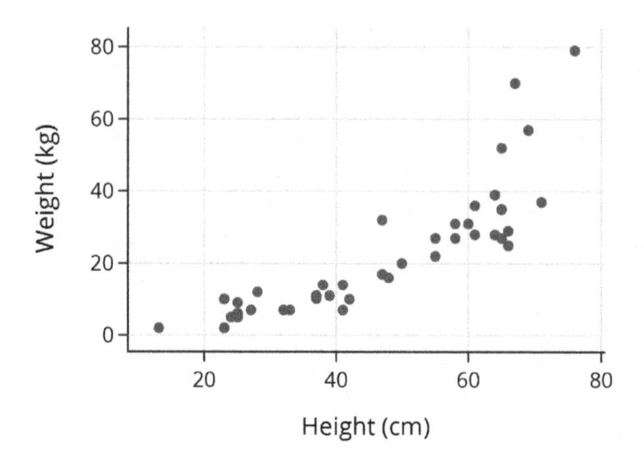

This particular `Figure` holds one plot, but `Figure` objects can hold any number of plots. Here, we create a facet of three scatterplots:

```
# The plot titles are cut off; we'll fix them in the next snippet
px.scatter(dogs, x='height', y='weight',
           facet_col='size',
           labels=dict(height="Height (cm)", weight="Weight (kg)"),
           width=550, height=250)
```

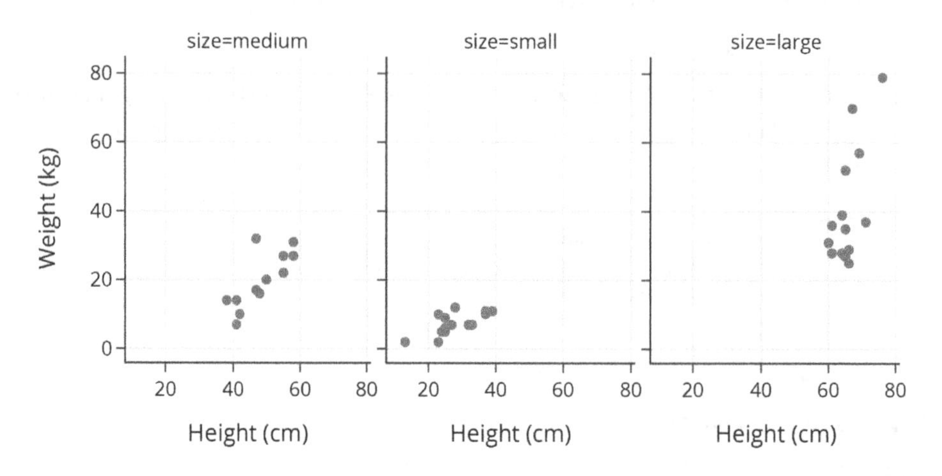

These three plots are stored in `Trace` objects. However, we try to avoid manipulating `Trace` objects manually. Instead, `plotly` provides functions that automatically create faceted subplots, like the `px.scatter` function we used here. Now that we have seen how to make a simple plot, we next show how to modify plots.

Modifying Layout

We often need to change a figure's layout. For instance, we might want to adjust the figure's margins or the axis range. We can use the `Figure.update_layout()` method to do this. In the facet scatterplot that we made, the title is cut off because the plot doesn't have large enough margins. We can correct this with `Figure.update_layout()`:

```
fig = px.scatter(dogs, x='height', y='weight',
                 facet_col='size',
                 labels=dict(height="Height (cm)", weight="Weight (kg)"),
                 width=550, height=250)

fig.update_layout(margin=dict(t=40))
fig
```

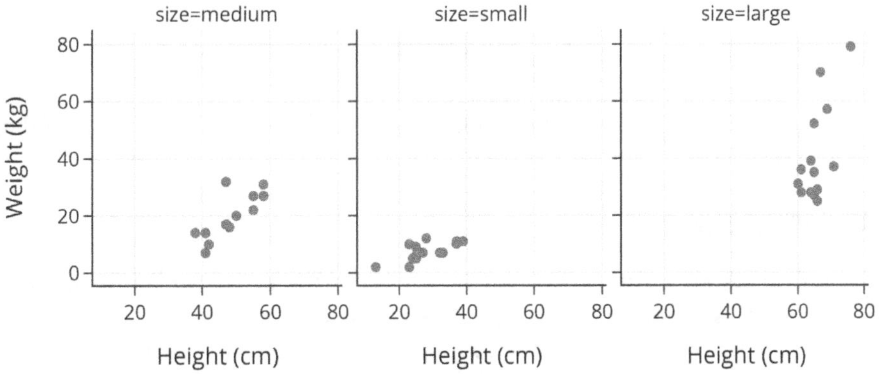

The `.update_layout()` method lets us modify any property of a layout. This includes the plot title (`title`), margins (`margins` dictionary), and whether to display a legend (`showlegend`). The `plotly` documentation has the full list of layout properties (*https://oreil.ly/aBLxx*).

`Figure` objects also have `.update_xaxes()` and `.update_yaxes()` functions, which are similar to `.update_layout()`. These two functions let us modify properties of the axes, like the axis limits (`range`), number of ticks (`nticks`), and axis label (`title`). Here, we adjust the range of the y-axis and change the title on the x-axis. We also add a title to the plot and update the layout so that the title is not cut off:

```
fig = px.scatter(
    dogs, x="weight", y="longevity",
    title="Smaller dogs live longer",
    width=350, height=250,
)

fig.update_yaxes(range=[5, 18], title="Typical lifespan (yr)")
```

```
fig.update_xaxes(title="Average weight (kg)")
fig.update_layout(margin=dict(t=30))
fig
```

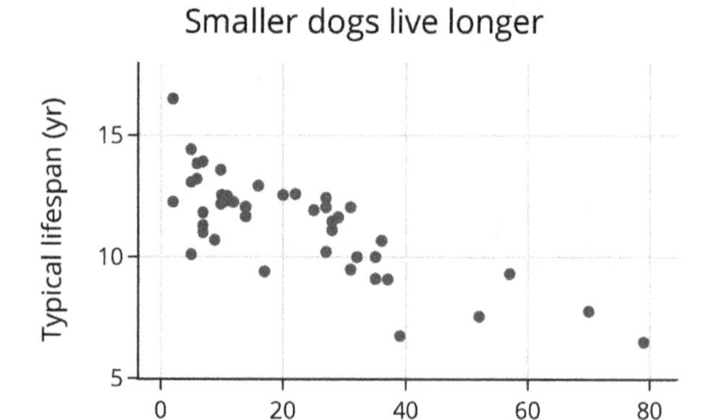

The `plotly` package comes with many plotting methods; we describe several of them in the next section.

Plotting Functions

The `plotly` methods includes line plots, scatterplots, bar plots, box plots, and histograms. The API is similar for each type of plot. The dataframe is the first argument. Then we can specify a column of the dataframe to place on the x-axis and a column to place on the y-axis using the x and y keyword arguments.

We begin with a line plot of median time each year for the runners in the Cherry Blossom race:

```
px.line(medians, x='year', y='time', width=350, height=250)
```

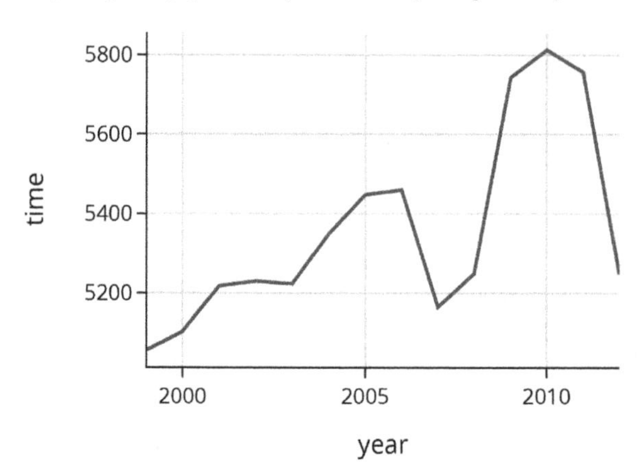

Next, we make a bar plot of average longevity for different size dog breeds:

```
lifespans = dogs.groupby('size')['longevity'].mean().reset_index()

px.bar(lifespans, x='size', y='longevity',
       width=350, height=250)
```

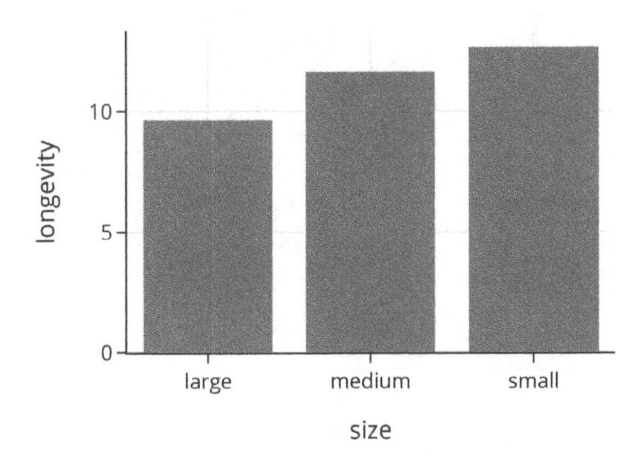

Plotting methods in `plotly` also contain arguments for making facet plots. We can facet using color on the same plot, plotting symbol, or line style. Or we can facet into multiple subplots. Following are examples of each. We first make a scatterplot of height and weight of dog breeds and use different plotting symbols and colors to facet within the plot by size:

```
fig = px.scatter(dogs, x='height', y='weight',
                 color='size', symbol='size',
                 labels=dict(height="Height (cm)",
                             weight="Weight (kg)", size="Size"),
                 width=350, height=250)
fig
```

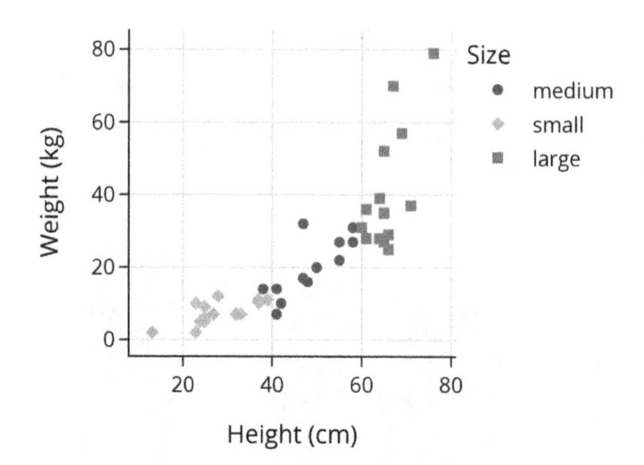

The next plot shows side-by-side histograms of longevity for each breed size. Here we facet by columns:

```
fig = px.histogram(dogs, x='longevity', facet_col='size',
                   width=550, height=250)
fig.update_layout(margin=dict(t=30))
```

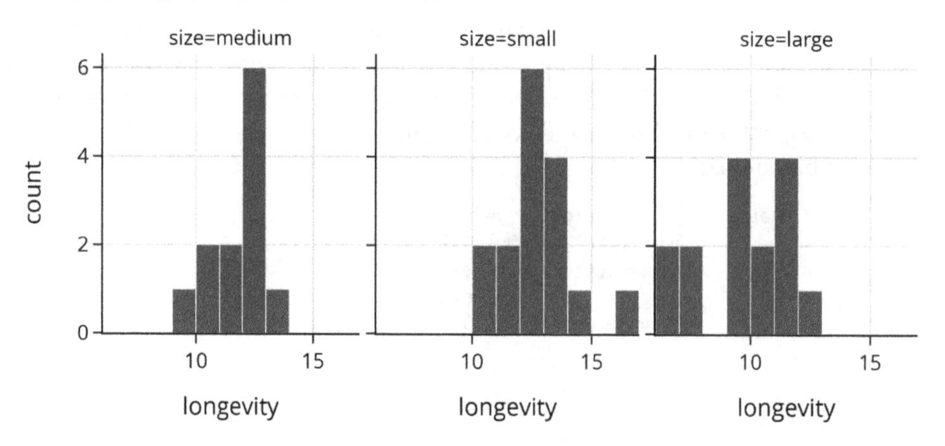

For the complete list of plotting functions, see the main documentation for `plotly` (*https://oreil.ly/GxvpT*) or `plotly.express` (*https://oreil.ly/DhU9j*), the submodule of `plotly` that we primarily use in the book.

To add context to a plot, we use the `plotly` annotation methods; these are described next.

Annotations

The `Figure.add_annotation()` method places annotations on a `plotly` figure. These annotations are line segments with text and an optional arrow. The location of the arrow is set using the `x` and `y` parameters, and we can shift the text from its default position using the `ax` and `ay` parameters. Here, we annotate the scatter diagram with information about one of the points:

```
fig = px.scatter(dogs, x='weight', y='longevity',
                 labels=dict(weight="Weight (kg)",
                 longevity="Typical lifespan (yr)"),
                 width=350, height=250)

fig.add_annotation(text='Chihuahuas live 16.5 years on average!',
                   x=2, y=16.5,
                   ax=30, ay=5,
                   xshift=3,
                   xanchor='left')
fig
```

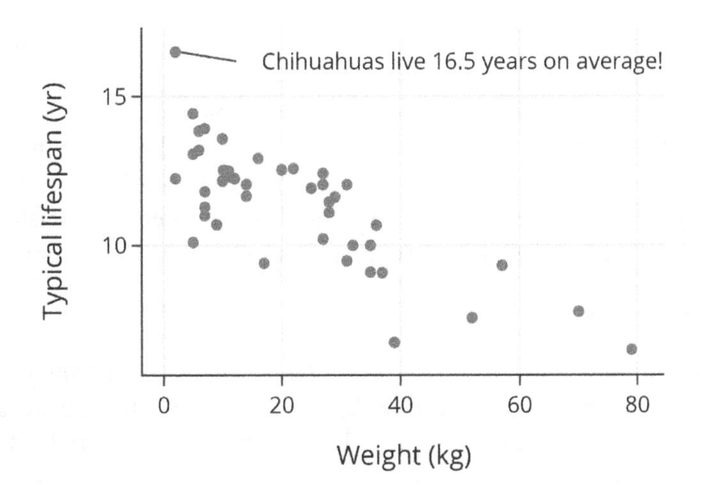

This section covered the basics of creating plots using the `plotly` Python package. We introduced the `Figure` object, which is the object `plotly` uses to store plots and their layouts. We covered the basic plot types that `plotly` makes available, and a few ways to customize plots by adjusting the layout and axes and by adding annotations. In the next section, we briefly compare `plotly` to other common tools for creating visualizations in Python.

Other Tools for Visualization

There are many software packages and tools for creating data visualizations. In this book, we primarily use `plotly`. But it's worth knowing about a few other commonly used tools. In this section, we compare `plotly` to `matplotlib` and to the grammar of graphics tools.

matplotlib

The library `matplotlib` (*https://matplotlib.org/*) is one of the first data visualization tools created for Python. Because of this, it is widely used and has a large ecosystem of packages. Notably, the built-in plotting methods for `pandas` `DataFrames` make plots using `matplotlib`. One popular package that builds on top of `matplotlib` is called `seaborn` (*https://seaborn.pydata.org*). Compared to `matplotlib` alone, `seaborn` provides a much simpler API to create statistical plots, like dot plots with confidence intervals. In fact, `seaborn`'s API was used as an inspiration for `plotly`'s API. If you look at `plotly` code and `seaborn` code side by side, you'll find that the methods to create basic plots use similar code.

One advantage of using `matplotlib` is its popularity. It's relatively easy to find help creating or fine-tuning plots online because many existing projects use it. For this book, the main advantage of using `plotly` is that the plots we create are interactive. Plots in `matplotlib` are usually static images, which don't allow for panning, zooming, or hovering over marks. Still, we expect that `matplotlib` will continue to be used for data analyses. In fact, several of the plots in this book were made using `seaborn` and `matplotlib` because `plotly` doesn't yet support all the plots we want to make.

Grammar of Graphics

The grammar of graphics (*https://dl.acm.org/doi/book/10.5555/1088896*) is a theory developed by Lee Wilkinson for creating data visualizations. The basic idea is to use common building blocks for making plots. For instance, a bar plot and a dot plot are nearly identical, except that a bar plot draws rectangles and a dot plot draws points. This idea is captured in the grammar of graphics, which would say that a bar plot and a dot plot differ only in their "geometry" component. The grammar of graphics is an elegant system that we can use to describe nearly every kind of plot we wish to make.

This system is implemented in the popular plotting libraries `ggplot2` (*https://ggplot2.tidyverse.org*) for the R programming language and `Vega` (*https://vega.github.io/vega*) for JavaScript. `Vega-Altair` (*https://altair-viz.github.io*), a Python package, provides a way to create `Vega` plots using Python, and we encourage interested readers to look over its documentation.

Using a grammar of graphics tool like `Vega-Altair` enables flexibility in visualizations. And like `plotly`, `altair` also creates interactive visualizations. However, the Python API for these tools can be less straightforward than `plotly`'s API. In this book, we don't typically need plots outside of what `plotly` is capable of creating, so we have opted for `plotly`'s simpler API.

There are many more plotting tools for Python that we've left out for brevity. But for the purposes of this book, relying on `plotly` provides a useful balance of interactivity and flexibility.

Summary

When we analyze a dataset, we use visualizations to uncover patterns in the data that are difficult to detect otherwise. Data visualization is an iterative process. We create a plot, then decide whether to make adjustments or choose an entirely new type of plot. This chapter covered principles that we use to make these decisions.

We started by covering principles of scale, and saw how adjusting the scale by changing or transforming plot axes can reveal hidden structure in the data. We then discussed smoothing and aggregating techniques that help us work with large datasets that would otherwise result in overplotting. To facilitate meaningful comparisons, we applied principles of perception, like aligning baselines to make lines, bars, and points easier to compare. We showed how to take the data design into account to improve visualizations. And we saw how adding context to a plot helps a reader understand our message.

After this chapter, you should be able to create a plot and understand what kinds of adjustments would make the plot more effective. As you learn how to make informative visualizations, be patient and iterate. None of us make the perfect plot on the first try, and as we make discoveries in an analysis, we continue to refine our plots. Then, when it comes time to present our findings to others, we sift through our work to get the few plots that best convince our future reader of the correctness and importance of our analysis. This can even lead to the creation of a new plot that better conveys our findings, which we iteratively develop.

In the next chapter, we walk through an extended case study that combines everything we've learned in the book so far. We hope that you find yourself surprised by how much you can already do.

CHAPTER 12
Case Study: How Accurate Are Air Quality Measurements?

California is prone to wildfires, so much so that its residents (like the authors of this book) sometimes say that California is "always on fire." In 2020, 40 separate fires covered the state in smoke, forced thousands of people to evacuate, and caused more than $12 billion in damages (Figure 12-1).

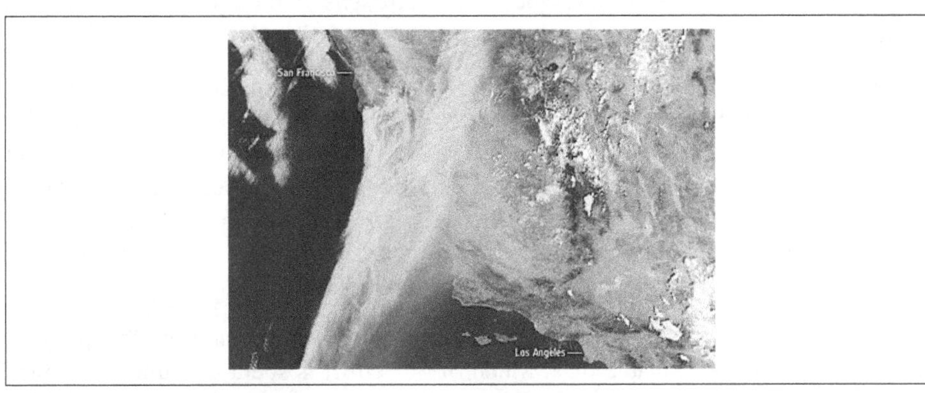

Figure 12-1. Satellite image from August 2020 showing smoke covering California (image from Wikipedia (https://oreil.ly/CrDld) licensed under CC BY-SA 3.0 IGO)

In places like California, people use air quality measurements to learn what kinds of protective measures they need to take. Depending on conditions, people may wish to wear a mask, use air filters, or avoid going outside altogether.

In the US, one important source of air quality information is the Air Quality System (*https://www.epa.gov/aqs*) (AQS), run by the US government. AQS places high-quality sensors at locations across the US and makes their data available to the public.

These sensors are carefully calibrated to strict standards—in fact, the AQS sensors are generally seen as the gold standard for accuracy. However, they have a few downsides. The sensors are expensive: typically between $15,000 and $40,000 each. This means that there are fewer sensors, and they are farther apart. Someone living far away from a sensor might not be able to access AQS data for their personal use. Also, AQS sensors do not provide real-time data. Since the data undergo extensive calibration, they are only released hourly and have a time lag of one to two hours. In essence, the AQS sensors are accurate but not timely.

In contrast, PurpleAir (*https://www2.purpleair.com*) sensors, which we introduced in Chapter 3, sell for about $250 and can be easily installed at home. With the lower price point, thousands of people across the US have purchased these sensors for personal use. The sensors can connect to a home WiFi network so that air quality can be easily monitored, and they can report data back to PurpleAir. In 2020, thousands of owners of PurpleAir sensors made their sensors' measurements publicly available. Compared to the AQS sensors, PurpleAir sensors are timelier. They report measurements every two minutes rather than every hour. Since there are more deployed PurpleAir sensors, more people live close enough to a sensor to make use of the data. However, PurpleAir sensors are less accurate. To make the sensors affordable, PurpleAir uses a simpler method to count particles in the air. This means that PurpleAir measurements can report that air quality is worse than it really is (see Josh Hug's blog post (*https://oreil.ly/ZH5aj*)). In essence, PurpleAir sensors tend to be timely but less accurate.

In this chapter, we plan to use the AQS sensor measurements to improve the PurpleAir measurements. It's a big task, and we follow the analysis first developed by Karoline Barkjohn, Brett Gantt, and Andrea Clements (*https://oreil.ly/XPxZu*) from the US Environmental Protection Agency. Barkjohn and her colleagues' work was so successful that, as of this writing, the official US government maps, like the AirNow Fire and Smoke (*https://fire.airnow.gov*) map, include both AQS and PurpleAir sensors and apply Barkjohn's correction to the PurpleAir data.

Our work follows the data science lifecycle, beginning with considering the question and the scope of the available data. Much of our effort is spent cleaning and wrangling the data into shape for analysis, but we also carry out an exploratory data analysis and build a model for generalization. We begin by considering the question and the design and scope of the data.

Question, Design, and Scope

Ideally, measures of air quality should be both *accurate* and *timely*. Inaccurate or biased measurements can mean people do not take air quality as seriously as they should. Delayed alerts can expose people to harmful air. The context provided in the introduction about the popularity of inexpensive air quality sensors got us wondering about their quality and usefulness.

Two different kinds of instruments measure a natural phenomenon—the amount of particulate matter in the air. The AQS sensor has the advantage of small measurement error and negligible bias (see Chapter 2). On the other hand, the PurpleAir instrument is less accurate; the measurements have greater variability and are also biased. Our initial question is: can we use the AQS measurements to make the PurpleAir measurements better?

We are in the situation where we have a lot of data available to us. We have access to a small number of high-quality measurements from AQS, and we can get data from thousands of PurpleAir sensors. To narrow the focus of our question, we consider how we might use these two sources of data to improve PurpleAir measurements.

The data from these two sources includes the locations of the sensors. So we can try to pair them up, finding a PurpleAir sensor close to each AQS sensor. If they're close, then these sensors are essentially measuring the same air. We can treat the AQS sensors as the ground truth (because they are so accurate) and study the variation in the PurpleAir measurements given the true air quality.

Even though there are relatively few pairs of collocated AQS and PurpleAir sensors, it seems reasonable to generalize any relationship we find to other PurpleAir sensors. If there's a simple relationship between AQS and PurpleAir measurements, then we can use this relationship to adjust measurements from any PurpleAir sensor so that they are more accurate.

We have narrowed down our question quite a bit: can we model the relationship between PurpleAir sensor readings and neighboring AQS sensor readings? If yes, then hopefully we can use the model to improve PurpleAir readings. Spoiler alert: indeed we can!

This case study nicely integrates the concepts introduced in this part of the book. It gives us an opportunity to see how data scientists wrangle, explore, and visualize data in a real-world setting. In particular, we see how a large, less-accurate dataset can amplify the usefulness of a small, accurate dataset. Combining large and small datasets like this is particularly exciting to data scientists and applies broadly to other domains ranging from social science to medicine.

In the next section, we begin our wrangling by finding the pairs of AQS and PurpleAir sensors that are near each other. We focus specifically on readings for PM2.5 particles, which are particles that are smaller than 2.5 micrometers in diameter. These particles are small enough to be inhaled into the lungs, pose the greatest risk to health, and are especially common in wood smoke.

Finding Collocated Sensors

Our analysis begins by finding collocated pairs of AQS and PurpleAir sensors—sensors that are placed essentially next to each other. This step is important because it lets us reduce the effects of other variables that might cause differences in sensor readings. Consider what would happen if we compared an AQS sensor placed in a park with a PurpleAir sensor placed along a busy freeway. The two sensors would have different readings, in part because the sensors are exposed to different environments. Ensuring that sensors are truly collocated lets us claim the differences in sensor readings are due to how the sensors are built and to small, localized air fluctuations, rather than other potential confounding variables.

Barkjohn's analysis conducted by the EPA group found pairs of AQS and PurpleAir sensors that are installed within 50 meters of each other. The group contacted each AQS site to see whether the PurpleAir sensor was also maintained there. This extra effort gave them confidence that their sensor pairs were truly collocated.

In this section, we explore and clean location data from AQS and PurpleAir. Then we perform a join of sorts to construct a list of potentially collocated sensors. We won't contact AQS sites ourselves; instead, we proceed in later sections with Barkjohn's list of confirmed collocated sensors.

We downloaded a list of AQS and PurpleAir sensors and saved the data in the files *data/list_of_aqs_sites.csv* and *data/list_of_purpleair_sensors.json*. Let's begin by reading these files into `pandas DataFrames`. First, we check file sizes to see whether they are reasonable to load into memory:

```
!ls -lLh data/list_of*
```

```
-rw-r--r--  1 sam  staff   4.8M Oct 27 16:54 data/list_of_aqs_sites.csv
-rw-r--r--  1 sam  staff   3.8M Oct 22 16:10 data/list_of_purpleair_sensors.json
```

Both files are relatively small. Let's start with the list of AQS sites.

Wrangling the List of AQS Sites

We have filtered the AQS map of sites (*https://oreil.ly/EkZcB*) to show only the AQS sites that measure PM2.5, and then downloaded the list of sites as a CSV file using the map's web app. Now we can load it into a `pandas DataFrame`:

```
aqs_sites_full = pd.read_csv('data/list_of_aqs_sites.csv')
aqs_sites_full.shape
```

```
(1333, 28)
```

There are 28 columns in the table. Let's check the column names:

```
aqs_sites_full.columns
```

```
Index(['AQS_Site_ID', 'POC', 'State', 'City', 'CBSA', 'Local_Site_Name',
       'Address', 'Datum', 'Latitude', 'Longitude', 'LatLon_Accuracy_meters',
       'Elevation_meters_MSL', 'Monitor_Start_Date', 'Last_Sample_Date',
       'Active', 'Measurement_Scale', 'Measurement_Scale_Definition',
       'Sample_Duration', 'Sample_Collection_Frequency',
       'Sample_Collection_Method', 'Sample_Analysis_Method',
       'Method_Reference_ID', 'FRMFEM', 'Monitor_Type', 'Reporting_Agency',
       'Parameter_Name', 'Annual_URLs', 'Daily_URLs'],
      dtype='object')
```

To find out which columns are most useful for us, we reference the data dictionary (*https://oreil.ly/GvMPI*) that the AQS provides on its website. There we confirm that the data table contains information about the AQS sites. So we might expect the granularity corresponds to an AQS site, meaning each row represents a single site and the column labeled `AQS_Site_ID` is the primary key. We can confirm this with a count of records for each ID:

```
aqs_sites_full['AQS_Site_ID'].value_counts()
```

```
06-071-0306    4
19-163-0015    4
39-061-0014    4
              ..
46-103-0020    1
19-177-0006    1
51-680-0015    1
Name: AQS_Site_ID, Length: 921, dtype: int64
```

It looks like some sites appear multiple times in this dataframe. Unfortunately, this means that the granularity is finer than the individual site level. To figure out why sites are duplicated, let's take a closer look at the rows for one duplicated site:

```
dup_site = aqs_sites_full.query("AQS_Site_ID == '19-163-0015'")
```

We select a few columns to examine based on their names—those that sound like they might shed some light on the reason for duplicates:

```
some_cols = ['POC', 'Monitor_Start_Date',
             'Last_Sample_Date', 'Sample_Collection_Method']
dup_site[some_cols]
```

	POC	Monitor_Start_Date	Last_Sample_Date	Sample_Collection_Method
458	1	1/27/1999	8/31/2021	R & P Model 2025 PM-2.5 Sequential Air Sampler...
459	2	2/9/2013	8/26/2021	R & P Model 2025 PM-2.5 Sequential Air Sampler...
460	3	1/1/2019	9/30/2021	Teledyne T640 at 5.0 LPM
461	4	1/1/2019	9/30/2021	Teledyne T640 at 5.0 LPM

The POC column looks to be useful for distinguishing between rows in the table. The data dictionary states this about the column:

> This is the "Parameter Occurrence Code" used to distinguish different instruments that measure the same parameter at the same site.

So, the site 19-163-0015 has four instruments that all measure PM2.5. The granularity of the dataframe is at the level of a single instrument.

Since our aim is to match AQS and PurpleAir sensors, we can adjust the granularity by selecting one instrument from each AQS site. To do this, we group rows according to site ID, then take the first row in each group:

```
def rollup_dup_sites(df):
    return (
        df.groupby('AQS_Site_ID')
        .first()
        .reset_index()
    )

aqs_sites = (aqs_sites_full
             .pipe(rollup_dup_sites))
aqs_sites.shape
```

```
(921, 28)
```

Now the number of rows matches the number of unique IDs.

To match AQS sites with PurpleAir sensors, we only need the site ID, latitude, and longitude. So we further adjust the structure and keep only those columns:

```
def cols_aqs(df):
    subset = df[['AQS_Site_ID', 'Latitude', 'Longitude']]
    subset.columns = ['site_id', 'lat', 'lon']
    return subset

aqs_sites = (aqs_sites_full
             .pipe(rollup_dup_sites)
             .pipe(cols_aqs))
```

Now the aqs_sites dataframe is ready, and we move to the PurpleAir sites.

Wrangling the List of PurpleAir Sites

Unlike the AQS sites, the file containing PurpleAir sensor data comes in a JSON format. We address this format in more detail in Chapter 14. For now, we use shell tools (see Chapter 8) to peek at the file contents:

```
!head data/list_of_purpleair_sensors.json | cut -c 1-60
```

```
{"version":"7.0.30",
"fields":
["ID","pm","pm_cf_1","pm_atm","age","pm_0","pm_1","pm_2","pm
"data":[
```

```
[20,0.0,0.0,0.0,0,0.0,0.0,0.0,0.0,0.0,0.0,0.0,97,0.0,0.0,0.0
[47,null,null,null,4951,null,null,null,null,null,null,null,9
[53,0.0,0.0,0.0,0,0.0,0.0,0.0,0.0,1.2,5.2,6.0,97,0.0,0.5,702
[74,0.0,0.0,0.0,0,0.0,0.0,0.0,0.0,0.0,0.0,0.0,97,0.0,0.0,0.0
[77,9.8,9.8,9.8,1,9.8,10.7,11.0,11.2,13.8,15.1,15.5,97,9.7,9
[81,6.5,6.5,6.5,0,6.5,6.1,6.1,6.6,8.1,8.3,9.7,97,5.9,6.8,405
```

From the first few lines of the file, we can guess that the data are stored in the "data" key and the column labels in the "fields" key. We can use Python's json library to read in the file as a Python dict:

```
import json

with open('data/list_of_purpleair_sensors.json') as f:
    pa_json = json.load(f)

list(pa_json.keys())
```

```
['version', 'fields', 'data', 'count']
```

We can create a dataframe from the values in data and label the columns with the content of fields:

```
pa_sites_full = pd.DataFrame(pa_json['data'], columns=pa_json['fields'])
pa_sites_full.head()
```

	ID	pm	pm_cf_1	pm_atm	...	Voc	Ozone1	Adc	CH
0	20	0.0	0.0	0.0	...	NaN	NaN	0.01	1
1	47	NaN	NaN	NaN	...	NaN	0.72	0.72	0
2	53	0.0	0.0	0.0	...	NaN	NaN	0.00	1
3	74	0.0	0.0	0.0	...	NaN	NaN	0.05	1
4	77	9.8	9.8	9.8	...	NaN	NaN	0.01	1

```
5 rows × 36 columns
```

Like the AQS data, there are many more columns in this dataframe than we need:

```
pa_sites_full.columns
```

```
Index(['ID', 'pm', 'pm_cf_1', 'pm_atm', 'age', 'pm_0', 'pm_1', 'pm_2', 'pm_3',
       'pm_4', 'pm_5', 'pm_6', 'conf', 'pm1', 'pm_10', 'p1', 'p2', 'p3', 'p4',
       'p5', 'p6', 'Humidity', 'Temperature', 'Pressure', 'Elevation', 'Type',
       'Label', 'Lat', 'Lon', 'Icon', 'isOwner', 'Flags', 'Voc', 'Ozone1',
       'Adc', 'CH'],
      dtype='object')
```

In this case, we can guess that the columns we're most interested in are the sensor IDs (ID), sensor labels (Label), latitude (Lat), and longitude (Lon). But we did consult the data dictionary on the PurpleAir website to double-check.

Now let's check the ID column for duplicates, as we did for the AQS data:

```
pa_sites_full['ID'].value_counts()[:3]
```

```
85829     1
117575    1
118195    1
Name: ID, dtype: int64
```

Since the `value_counts()` method lists the counts in descending order, we can see that every ID was included only once. So we have verified the granularity is at the individual sensor level. Next, we keep only the columns needed to match sensor locations from the two sources:

```
def cols_pa(df):
    subset = df[['ID', 'Label', 'Lat', 'Lon']]
    subset.columns = ['id', 'label', 'lat', 'lon']
    return subset

pa_sites = (pa_sites_full
            .pipe(cols_pa))
pa_sites.shape
```

```
(23138, 4)
```

Notice there are tens of thousands more PurpleAir sensors than AQS sensors. Our next task is to find the PurpleAir sensor close to each AQS sensor.

Matching AQS and PurpleAir Sensors

Our goal is to match sensors in the two dataframes by finding a PurpleAir sensor near each AQS instrument. We consider near to mean within 50 meters. This kind of matching is a bit more challenging than the joins we've seen thus far. For instance, the naive approach to use the `merge` method of `pandas` fails us:

```
aqs_sites.merge(pa_sites, left_on=['lat', 'lon'], right_on=['lat', 'lon'])
```

	site_id	lat	lon	id	label
0	06-111-1004	34.45	-119.23	48393	VCAPCD OJ

We cannot simply match instruments with the exact same latitude and longitude; we need to find the PurpleAir sites that are close enough to the AQS instrument.

To figure out how far apart two locations are, we use a basic approximation: `111,111` meters in the north-south direction roughly equals one degree of latitude, and `111,111 * cos(latitude)` in the east-west direction corresponds to one degree of longitude.[1] So we can find the latitude and longitude ranges that correspond to 25 meters in each direction (to make a 50-meter-by-50-meter rectangle around each point):

```
magic_meters_per_lat = 111_111
offset_in_m = 25
offset_in_lat = offset_in_m / magic_meters_per_lat
offset_in_lat
```

```
0.000225000225000225
```

To simplify even more, we use the median latitude for the AQS sites:

```
median_latitude = aqs_sites['lat'].median()
magic_meters_per_lon = 111_111 * np.cos(np.radians(median_latitude))
offset_in_lon = offset_in_m / magic_meters_per_lon
offset_in_lon
```

```
0.000291515219937587
```

Now we can match coordinates to within the `offset_in_lat` and `offset_in_lon`. Doing this in SQL is much easier than in `pandas`, so we push the tables into a temporary SQLite database, then run a query to read the tables back into a dataframe:

```
import sqlalchemy

db = sqlalchemy.create_engine('sqlite://')

aqs_sites.to_sql(name='aqs', con=db, index=False)
pa_sites.to_sql(name='pa', con=db, index=False)

query = f'''
SELECT
  aqs.site_id AS aqs_id,
  pa.id AS pa_id,
  pa.label AS pa_label,
  aqs.lat AS aqs_lat,
  aqs.lon AS aqs_lon,
  pa.lat AS pa_lat,
  pa.lon AS pa_lon
FROM aqs JOIN pa
  ON  pa.lat - {offset_in_lat} <= aqs.lat
  AND                             aqs.lat <= pa.lat + {offset_in_lat}
  AND pa.lon - {offset_in_lon} <= aqs.lon
  AND                             aqs.lon <= pa.lon + {offset_in_lon}
'''
```

1 This estimation works by assuming that the Earth is perfectly spherical. Then, one degree of latitude is the radius of the Earth in meters. Plugging in the average radius of the Earth gives 111,111 meters per degree of latitude. Longitude is the same, but the radius of each "ring" around the Earth decreases as we get closer to the poles, so we adjust by a factor of cos (lat). It turns out that the Earth isn't perfectly spherical, so these estimations can't be used for precise calculations, like landing a rocket. But for our purposes, they do just fine.

```
matched = pd.read_sql(query, db)
matched
```

	aqs_id	pa_id	pa_label	aqs_lat	aqs_lon	pa_lat	pa_lon
0	06-019-0011	6568	IMPROVE_FRES2	36.79	-119.77	36.79	-119.77
1	06-019-0011	13485	AMTS_Fresno	36.79	-119.77	36.79	-119.77
2	06-019-0011	44427	Fresno CARB CCAC	36.79	-119.77	36.79	-119.77
...	...	...	...	...	...	...	...
146	53-061-1007	3659	Marysville 7th	48.05	-122.17	48.05	-122.17
147	53-063-0021	54603	Augusta 1 SRCAA	47.67	-117.36	47.67	-117.36
148	56-021-0100	50045	WDEQ-AQD Cheyenne NCore	41.18	-104.78	41.18	-104.78

```
149 rows × 7 columns
```

We've achieved our goal—we matched 149 AQS sites with PurpleAir sensors. Our wrangling of the locations is complete, and we turn to the task of wrangling and cleaning the sensor measurements. We start with the measurements taken from an AQS site.

Wrangling and Cleaning AQS Sensor Data

Now that we have located sensors that are near each other, we are ready to wrangle and clean the files that contain the measurement data for these sites. We demonstrate the tasks involved with one AQS instrument and its matching PurpleAir sensor. We picked a pair located in Sacramento, California. The AQS sensor ID is `06-067-0010`, and the PurpleAir sensor name is `AMTS_TESTINGA`.

The AQS provides a website and API (*https://oreil.ly/tl_nc*) to download sensor data. We downloaded the daily measurements from May 20, 2018, to December 29, 2019, into the *data/aqs_06-067-0010.csv* file. Let's begin by loading this file into a dataframe:

```
aqs_full = pd.read_csv('data/aqs_06-067-0010.csv')
aqs_full.shape
```

```
(2268, 31)
```

From the data dictionary (*https://oreil.ly/e1PjI*), we find out that the column called `arithmetic_mean` corresponds to the actual PM2.5 measurements. Some AQS sensors take a measurement every hour. For our analysis, we downloaded the 24-hour averages (the arithmetic mean) of the hourly sensor measurements.

Let's carry out some quality checks and clean the data where necessary. We focus on checks related to scope and quality of values:

1. Check and correct the granularity of the data.
2. Remove unneeded columns.

3. Check values in the `date_local` column.
4. Check values in the `arithmetic_mean` column.

For the sake of brevity, we've chosen a few important quality checks that specifically reinforce ideas we've covered in data wrangling, EDA, and visualization.

Checking Granularity

We would like each row of our data to correspond to a single date with an average PM2.5 reading for that date. As we saw earlier, a simple way to check is to see whether there are repeat values in the `date_local` column:

```
aqs_full['date_local'].value_counts()
```

```
date_local
2019-01-03    12
2018-12-31    12
2018-12-28    12
              ..
2018-11-28    12
2018-11-25    12
2018-11-22    12
Name: count, Length: 189, dtype: int64
```

Indeed, there are 12 rows for each date, so the granularity is *not* at the individual date level.

From the data dictionary, we learn that there are multiple standards for computing the final measurements from the raw sensor data. The `pollutant_standard` column contains the name of each standard. The `event_type` column marks whether data measured during "exceptional events" are included in the measurement. Let's check how different these average values are by calculating the range of 12 measurements:

```
(aqs_full
 .groupby('date_local')
 ['arithmetic_mean']
 .agg(np.ptp) # np.ptp computes max() - min()
 .value_counts()
)
```

```
arithmetic_mean
0.0    189
Name: count, dtype: int64
```

For all 189 dates, the max PM2.5–min PM2.5 is 0. This means that we can simply take the first PM2.5 measurement for each date:

```
def rollup_dates(df):
    return (
        df.groupby('date_local')
        .first()
```

```
        .reset_index()
    )

aqs = (aqs_full
       .pipe(rollup_dates))
aqs.shape
```

```
(189, 31)
```

This data-cleaning step gives us the desired granularity: each row represents a single date, with an average PM2.5 measurement for that date. Next, we further modify the structure of the dataframe and drop unneeded columns.

Removing Unneeded Columns

We plan to match the PM2.5 measurements in the AQS dataframe with the PurpleAir PM2.5 measurements for each date. To simplify the structure, we can drop all but the date and PM2.5 columns. We also rename the PM2.5 column so that it's easier to understand:

```
def drop_cols(df):
    subset = df[['date_local', 'arithmetic_mean']]
    return subset.rename(columns={'arithmetic_mean': 'pm25'})

aqs = (aqs_full
       .pipe(rollup_dates)
       .pipe(drop_cols))
aqs.head()
```

	date_local	pm25
0	2018-05-20	6.5
1	2018-05-23	2.3
2	2018-05-29	11.8
3	2018-06-01	6.0
4	2018-06-04	8.0

Now that we have the desired shape for our data table, we turn to checking the data values.

Checking the Validity of Dates

Let's take a closer look at the dates. We have already seen that there are gaps when there are no PM2.5 readings, so we expect there are missing dates. Let's parse the dates as timestamps objects to make it easier to figure out which dates are missing. As we did in Chapter 9, we check the format:

```
aqs['date_local'].iloc[:3]
```

```
0    2018-05-20
1    2018-05-23
```

```
2    2018-05-29
Name: date_local, dtype: object
```

The dates are represented as YYYY-MM-DD, so we describe the format in the Python representation `'%Y-%m-%d'`. To parse the dates, we use the `pd.to_datetime()` function, and we reassign the `date_local` column as `pd.TimeStamps`:

```
def parse_dates(df):
    date_format = '%Y-%m-%d'
    timestamps = pd.to_datetime(df['date_local'], format=date_format)
    return df.assign(date_local=timestamps)

aqs = (aqs_full
       .pipe(rollup_dates)
       .pipe(drop_cols)
       .pipe(parse_dates))
```

The method runs without erroring, indicating that all the strings matched the format.

Just because the dates can be parsed doesn't mean that the dates are immediately ready to use for further analysis. For instance, the string `9999-01-31` can be parsed into a `pd.TimeStamp`, but the date isn't valid.

Now that the dates have been converted to timestamps, we can calculate how many dates are missing. We find the number of days between the earliest and latest dates—this corresponds to the maximum number of measurements we could have recorded:

```
date_range = aqs['date_local'].max() - aqs['date_local'].min()
date_range.days
```

```
588
```

Subtracting timestamps gives `Timedelta` objects, which as we see have a few useful properties. There are many dates missing from the data. However, when we combine these data for this sensor with other sensors, we expect to have enough data to fit a model.

Our final wrangling step is to check the quality of the PM2.5 measurements.

Checking the Quality of PM2.5 Measurements

Particulate matter is measured in micrograms per cubic meter of air (μg/m^3). (There are 1 million micrograms in 1 gram, and 1 pound is equal to about 450 grams.) The EPA has set a standard (*https://oreil.ly/XqVqG*) of 35 μg/m^3 for a daily average of PM2.5 and 12 μg/m^3 for an annual average. We can use this information to make a few basic checks on the PM2.5 measurements. First, PM2.5 can't go below 0. Second, we can look for abnormally high PM2.5 values and see whether they correspond to major events like a wildfire.

One visual way to perform these checks is to plot the PM2.5 measurement against the date:

```
px.scatter(aqs, x='date_local', y='pm25',
           labels={'date_local':'Date', 'pm25':'AQS daily avg PM2.5'},
           width=500, height=250)
```

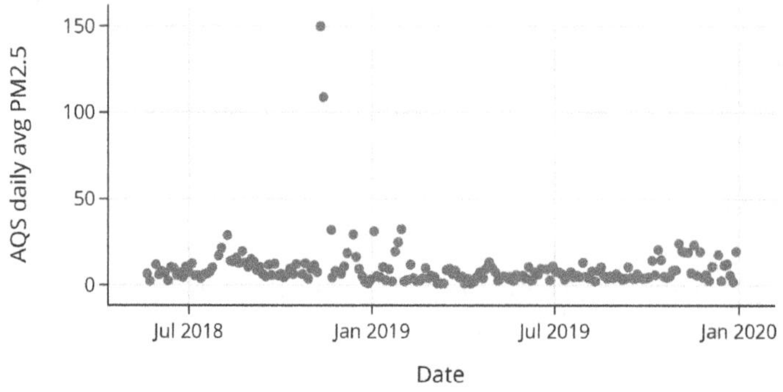

We see that the PM2.5 measurements don't go below 0 and are typically lower than the EPA level. We also found a large spike in PM2.5 around mid-November of 2018. This sensor is located in Sacramento, so we can check if there was a fire around that area.

Indeed, November 8, 2018, marks the start of the Camp Fire, the "deadliest and most destructive wildfire in California history" (see the Camp Fire page (*https://oreil.ly/tqxtH*) managed by the US Census Bureau). The fire started just 80 miles north of Sacramento, so this AQS sensor captured the dramatic spike in PM2.5.

We've cleaned and explored the data for one AQS sensor. In the next section, we do the same for its collocated PurpleAir sensor.

Wrangling PurpleAir Sensor Data

In the previous section, we analyzed data from AQS site `06-067-0010`. The matching PurpleAir sensor is named `AMTS_TESTINGA`, and we've used the PurpleAir website to download the data for this sensor into the *data/purpleair_AMTS* folder:

```
!ls -alh data/purpleair_AMTS/* | cut -c 1-72
```

```
-rw-r--r--  1 nolan  staff    50M Jan 25 16:35 data/purpleair_AMTS/AMTS_
-rw-r--r--  1 nolan  staff    50M Jan 25 16:35 data/purpleair_AMTS/AMTS_
-rw-r--r--  1 nolan  staff    48M Jan 25 16:35 data/purpleair_AMTS/AMTS_
-rw-r--r--  1 nolan  staff    50M Jan 25 16:35 data/purpleair_AMTS/AMTS_
```

There are four CSV files. Their names are quite long, and the beginning of each is identical. The data dictionary for the PurpleAir data says that each sensor has two separate instruments, A and B, that each record data. Note that the PurpleAir site we used to collect these data and the accompanying data dictionary has been downgraded. The data are now available through a REST API. The site that documents the API (*https://oreil.ly/WSciR*) also contains information about the fields. (The topic of REST is covered in Chapter 14.) Let's examine the later portions of the filenames:

```
!ls -alh data/purpleair_AMTS/* | cut -c 73-140
```

```
TESTING (outside) (38.568404 -121.493163) Primary Real Time 05_20_20
TESTING (outside) (38.568404 -121.493163) Secondary Real Time 05_20_
TESTING B (undefined) (38.568404 -121.493163) Primary Real Time 05_2
TESTING B (undefined) (38.568404 -121.493163) Secondary Real Time 05
```

We can see that the first two CSV files correspond to instrument A and the last two to B. Having two instruments is useful for data cleaning; if A and B disagree about a measurement, we might question the integrity of the measurement and decide to remove it.

The data dictionary also mentions that each instrument records Primary and Secondary data. The Primary data contains the fields we're interested in: PM2.5, temperature, and humidity. The Secondary data contains data for other particle sizes, like PM1.0 and PM10. So we work only with the Primary files.

Our tasks are similar to those of the previous section, with the addition of addressing readings from two instruments.

We begin by loading in the data. When CSV files have long names, we can assign the filenames into a Python variable to more easily load the files:

```
from pathlib import Path

data_folder = Path('data/purpleair_AMTS')
pa_csvs = sorted(data_folder.glob('*.csv'))
pa_csvs[0]
```

```
PosixPath('data/purpleair_AMTS/AMTS_TESTING (outside) (38.568404 -121.493163)
Primary Real Time 05_20_2018 12_29_2019.csv')
```

```
pa_full = pd.read_csv(pa_csvs[0])
pa_full.shape
```

```
(672755, 11)
```

Let's look at the columns to see which ones we need:

```
pa_full.columns
```

```
Index(['created_at', 'entry_id', 'PM1.0_CF1_ug/m3', 'PM2.5_CF1_ug/m3',
       'PM10.0_CF1_ug/m3', 'UptimeMinutes', 'RSSI_dbm', 'Temperature_F',
       'Humidity_%', 'PM2.5_ATM_ug/m3', 'Unnamed: 10'],
      dtype='object')
```

Although we're interested in PM2.5, it appears there are two columns that contain PM2.5 data: `PM2.5_CF1_ug/m3` and `PM2.5_ATM_ug/m3`. We investigate the difference between these two columns to find that PurpleAir sensors use two different methods to convert a raw laser recording into a PM2.5 number. These two calculations correspond to the CF1 and ATM columns. Barkjohn found that using CF1 produced better results than ATM, so we keep that column, along with the date, temperature, and relative humidity:

```
def drop_and_rename_cols(df):
    df = df[['created_at', 'PM2.5_CF1_ug/m3', 'Temperature_F', 'Humidity_%']]
    df.columns = ['timestamp', 'PM25cf1', 'TempF', 'RH']
    return df

pa = (pa_full
      .pipe(drop_and_rename_cols))
pa.head()
```

	timestamp	PM25cf1	TempF	RH
0	2018-05-20 00:00:35 UTC	1.23	83.0	32.0
1	2018-05-20 00:01:55 UTC	1.94	83.0	32.0
2	2018-05-20 00:03:15 UTC	1.80	83.0	32.0
3	2018-05-20 00:04:35 UTC	1.64	83.0	32.0
4	2018-05-20 00:05:55 UTC	1.33	83.0	32.0

Next we check granularity.

Checking the Granularity

In order for the granularity of these measurements to match the AQS data, we want one average PM2.5 for each date (a 24-hour period). PurpleAir states that sensors take measurements every two minutes. Let's double-check the granularity of the raw measurements before we aggregate them to 24-hour periods.

To do this we convert the column containing the date information from strings to `pd.TimeStamp` objects. The format of the date is different than the AQS format, which we describe as `'%Y-%m-%d %X %Z'`. As we soon see, `pandas` has special support for dataframes with an index of timestamps:

```
def parse_timestamps(df):
    date_format = '%Y-%m-%d %X %Z'
    times = pd.to_datetime(df['timestamp'], format=date_format)
    return (df.assign(timestamp=times)
            .set_index('timestamp'))

pa = (pa_full
      .pipe(drop_and_rename_cols)
```

```
        .pipe(parse_timestamps))
pa.head(2)
```

timestamp	PM25cf1	TempF	RH
2018-05-20 00:00:35+00:00	1.23	83.0	32.0
2018-05-20 00:01:55+00:00	1.94	83.0	32.0

Timestamps are tricky—notice that the original timestamps were given in the UTC time zone. However, the AQS data were averaged according to the *local time in California*, which is either seven or eight hours behind UTC time, depending on whether daylight saving time is in effect. This means we need to change the time zone of the PurpleAir timestamps to match the local time zone. The `df.tz_convert()` method operates on the index of the dataframe, which is one reason why we set the index of `pa` to the timestamps:

```
def convert_tz(pa):
    return pa.tz_convert('US/Pacific')

pa = (pa_full
      .pipe(drop_and_rename_cols)
      .pipe(parse_timestamps)
      .pipe(convert_tz))
pa.head(2)
```

timestamp	PM25cf1	TempF	RH
2018-05-19 17:00:35-07:00	1.23	83.0	32.0
2018-05-19 17:01:55-07:00	1.94	83.0	32.0

If we compare the first two rows of this version of the dataframe to the previous one, we see that the time has changed to indicate the seven-hour difference from UTC.

Visualizing timestamps can help us check the granularity of the data.

Visualizing timestamps

One way to visualize timestamps is to count how many appear in each 24-hour period, then plot those counts over time. To group time-series data in `pandas`, we can use the `df.resample()` method. This method works on dataframes that have an index of timestamps. It behaves like `df.groupby()`, except that we can specify how we want the timestamps to be grouped—we can group into dates, weeks, months, and many more options (the `D` argument tells `resample` to aggregate timestamps into individual dates):

```
per_day = (pa.resample('D')
           .size()
           .rename('records_per_day')
           .to_frame()
)

percs = [10, 25, 50, 75, 100]
np.percentile(per_day['records_per_day'], percs, method='lower')
```

```
array([ 293,  720, 1075, 1440, 2250])
```

We see that the number of measurements in a day varies widely. A line plot of these counts gives us a better sense of these variations:

```
px.line(per_day, x=per_day.index, y='records_per_day',
        labels={'timestamp':'Date', 'records_per_day':'Records per day'},
        width=550, height=250,)
```

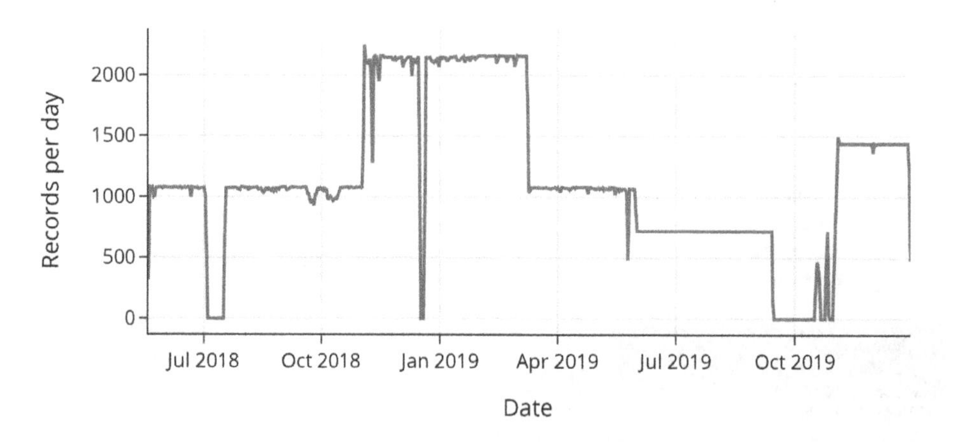

This is a fascinating plot. We see clear gaps in the data where there are no measurements. It appears that significant portions of data in July 2018 and September 2019 are missing. Even when the sensor appears to be working, the number of measurements per day is slightly different. For instance, the plot is "bumpy" between August and October 2018, where dates have a varying number of measurements. We need to decide what we want to do with missing data. But perhaps more urgently: there are strange "steps" in the plot. Some dates have around 1,000 readings, some around 2,000, some around 700, and some around 1,400. If a sensor takes measurements every two minutes, there should be a maximum of 720 measurements per day. For a perfect sensor, the plot would display a flat line at 720 measurements. This is clearly not the case. Let's investigate.

Checking the sampling rate

Deeper digging reveals that although PurpleAir sensors currently record data every 120 seconds, this was not always the case. Before May 30, 2019, sensors recorded data every 80 seconds, or 1,080 points a day. The change in sampling rate does explain the drop on May 30, 2019. Let's next look at the time periods where there were many more points than expected. This could mean that some measurements were duplicated in the data. We can check this by looking at the measurements for one day, say, January 1, 2019. We pass a string into `.loc` to filter timestamps for that date:

```
len(pa.loc['2019-01-01'])
```

```
2154
```

There are almost double the 1,080 expected readings. Let's check to see if readings are duplicated:

```
pa.loc['2019-01-01'].index.value_counts()
```

```
2019-01-01 13:52:30-08:00    2
2019-01-01 12:02:21-08:00    2
2019-01-01 11:49:01-08:00    2
                            ..
2019-01-01 21:34:10-08:00    2
2019-01-01 11:03:41-08:00    2
2019-01-01 04:05:38-08:00    2
Name: timestamp, Length: 1077, dtype: int64
```

Each timestamp appears exactly twice, and we can verify that all duplicated dates contain the same PM2.5 reading. Since this is also true for both temperature and humidity, we drop the duplicate rows from the dataframe:

```
def drop_duplicate_rows(df):
    return df[~df.index.duplicated()]

pa = (pa_full
      .pipe(drop_and_rename_cols)
      .pipe(parse_timestamps)
      .pipe(convert_tz)
      .pipe(drop_duplicate_rows))
pa.shape
```

```
(502628, 3)
```

To check, we remake the line plot of the number of records for a day, and this time we shade the regions where the counts are supposed to be contained:

```
per_day = (pa.resample('D')
 .size().rename('records_per_day')
 .to_frame()
)

fig = px.line(per_day, x=per_day.index, y='records_per_day',
              labels={'timestamp':'Date', 'records_per_day':'Records per day'},
```

```
                width=550, height=250)

fig.add_annotation(x='2019-07-24', y=720,
            text="720", showarrow=False, yshift=10)
fig.add_annotation(x='2019-07-24', y=1080,
            text="1080", showarrow=False, yshift=10)

fig.add_hline(y=1080, line_width=3, line_dash="dot", opacity=0.6)
fig.add_hline(y=720, line_width=3, line_dash="dot", opacity=0.6)
fig.add_vline(x="2019-05-30", line_width=3, line_dash="dash", opacity=0.6)

fig
```

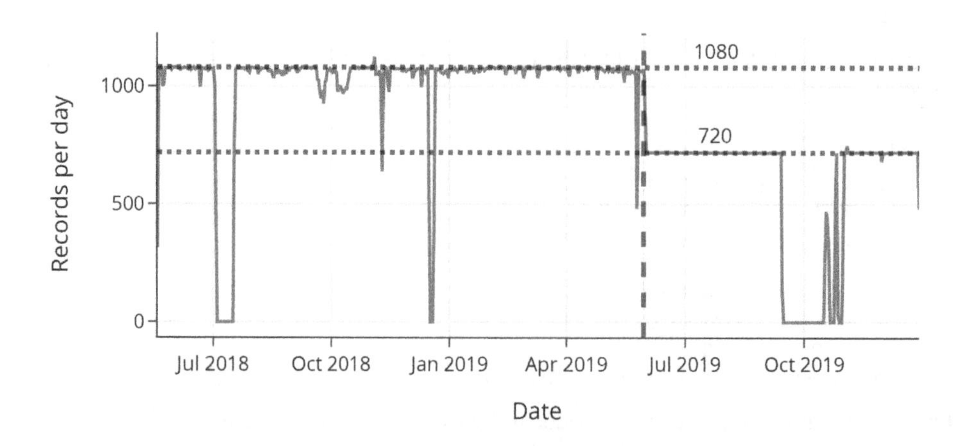

After dropping duplicate dates, the plot of measurements per day looks much more consistent with the counts we expect. Careful readers will see two spikes above the maximum measurements around November of each year when daylight saving time is no longer in effect. When clocks are rolled back one hour, that day has 25 hours instead of the usual 24 hours. Timestamps are tricky!

But there are still missing measurements, and we need to decide what to do about them.

Handling Missing Values

The plan is to create 24-hour averages of the measurements, but we don't want to use days when there are not enough measurements. We follow Barkjohn's analysis and only keep a 24-hour average if there are at least 90% of the possible points for that day. Remember that before May 30, 2019, there are 1,080 possible points in a day, and after that there are 720 possible points. We calculate the minimum number of measurements needed to keep per day:

```
needed_measurements_80s = 0.9 * 1080
needed_measurements_120s = 0.9 * 720
```

Now we can determine which of the days have enough measurements to keep:

```
cutoff_date = pd.Timestamp('2019-05-30', tz='US/Pacific')

def has_enough_readings(one_day):
    [n] = one_day
    date = one_day.name
    return (n >= needed_measurements_80s
            if date <= cutoff_date
            else n >= needed_measurements_120s)

should_keep = per_day.apply(has_enough_readings, axis='columns')
should_keep.head()
```

```
timestamp
2018-05-19 00:00:00-07:00    False
2018-05-20 00:00:00-07:00     True
2018-05-21 00:00:00-07:00     True
2018-05-22 00:00:00-07:00     True
2018-05-23 00:00:00-07:00     True
Freq: D, dtype: bool
```

We're ready to average together the readings for each day and then remove the days without enough readings:

```
def compute_daily_avgs(pa):
    should_keep = (pa.resample('D')
                   ['PM25cf1']
                   .size()
                   .to_frame()
                   .apply(has_enough_readings, axis='columns'))
    return (pa.resample('D')
            .mean()
            .loc[should_keep])

pa = (pa_full
      .pipe(drop_and_rename_cols)
      .pipe(parse_timestamps)
      .pipe(convert_tz)
      .pipe(drop_duplicate_rows)
      .pipe(compute_daily_avgs))
pa.head(2)
```

	PM25cf1	TempF	RH
timestamp			
2018-05-20 00:00:00-07:00	2.48	83.35	28.72
2018-05-21 00:00:00-07:00	3.00	83.25	29.91

Now we have the average daily PM2.5 readings for instrument A, and we need to repeat on instrument B the data wrangling we just performed on instrument A. Fortunately, we can reuse the same pipeline. For brevity, we don't include that wrangling here. But we need to decide what to do if the PM2.5 averages differ. Barkjohn dropped rows if the PM2.5 values for A and B differed by more than 61%, or by more than 5 μg m^{-3}. For this pair of sensors, that leads to dropping 12 of the 500+ rows.

As you can see, it takes a lot of work to prepare and clean these data: we handled missing data, aggregated the readings for each instrument, averaged the readings together from the two instruments, and removed rows where they disagreed. This work has given us a set of PM2.5 readings that we are more confident in. We know that each PM2.5 value in the final dataframe is the daily average from two separate instruments that generated consistent and complete readings.

To fully replicate Barkjohn's analysis, we would need to repeat this process over all the PurpleAir sensors. Then we would repeat the AQS cleaning procedure on all the AQS sensors. Finally, we would merge the PurpleAir and AQS data together. This procedure produces daily average readings for each collocated sensor pair. For brevity, we omit this code. Instead, we proceed with the final steps of the analysis using the group's dataset. We begin with an EDA with an eye toward modeling.

Exploring PurpleAir and AQS Measurements

Let's explore the cleaned dataset of matched AQS and PurpleAir PM2.5 readings and look for insights that might help us in modeling. Our main interest is in the relationship between the two sources of air quality measurements. But we want to keep in mind the scope of the data, like how these data are situated in time and place. We learned from our data cleaning that we are working with daily averages of PM2.5 for a couple of years and that we have data from dozens of locations across the US.

First we review the entire cleaned dataframe:

```
full_df
```

	date	id	region	pm25aqs	pm25pa	temp	rh	dew
0	2019-05-17	AK1	Alaska	6.7	8.62	18.03	38.56	3.63
1	2019-05-18	AK1	Alaska	3.8	3.49	16.12	49.40	5.44
2	2019-05-21	AK1	Alaska	4.0	3.80	19.90	29.97	1.73
...	...	...	...	...	...	...	...	...
12427	2019-02-20	WI6	North	15.6	25.30	1.71	65.78	-4.08
12428	2019-03-04	WI6	North	14.0	8.21	-14.38	48.21	-23.02
12429	2019-03-22	WI6	North	5.8	9.44	5.08	52.20	-4.02

```
12246 rows × 8 columns
```

We include an explanation for each of the columns in our dataframe in the following table:

Column	Description
date	Date of the observation
id	A unique label for a site, formatted as the US state abbreviation with a number (we performed data cleaning for site ID CA1)
region	The name of the region, which corresponds to a group of sites (the CA1 site is located in the West region)
pm25aqs	The PM2.5 measurement from the AQS sensor
pm25pa	The PM2.5 measurement from the PurpleAir sensor
temp	Temperature, in Celsius
rh	Relative humidity, ranging from 0% to 100%
dew	The dew point (a higher dew point means more moisture is in the air)

Let's start with making a few simple visualizations to gain insight. Since the scope involves measurements over time at particular locations, we can choose one location with many measurements and make a line plot of the weekly average air quality. To choose, let's find the sites with many records:

```
full_df['id'].value_counts()[:3]
```

```
id
IA3    830
NC4    699
CA2    659
Name: count, dtype: int64
```

The location labeled NC4 has nearly 700 observations. To smooth the line plot a bit, let's plot weekly averages:

```
nc4 = full_df.loc[full_df['id'] =='NC4']
```

```
ts_nc4 = (nc4.set_index('date')
 .resample('W')
 ['pm25aqs', 'pm25pa']
 .mean()
 .reset_index()
)
```

```
fig = px.line(ts_nc4, x='date', y='pm25aqs',
              labels={'date':'', 'pm25aqs':'PM2.5 weekly avg'},
              width=500, height=250)

fig.add_trace(go.Scatter(x=ts_nc4['date'], y=ts_nc4['pm25pa'],
                         line=dict(color='black', dash='dot')))

fig.update_yaxes(range=[0,30])
fig.update_layout(showlegend=False)
fig.show()
```

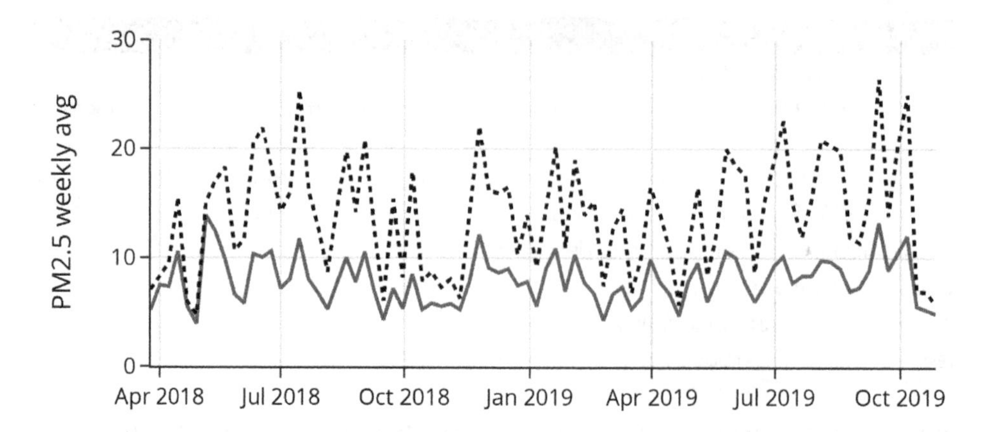

We see that most PM2.5 values for the AQS sensor (solid line) range between 5.0 and 15.0 µg m^{-3}. The PurpleAir sensor follows the up-and-down pattern of the AQS sensor, which is reassuring. But the measurements are consistently higher than AQS and, in some cases, quite a bit higher, which tells us that a correction might be helpful.

Next, let's consider the distributions of the PM2.5 readings for the two sensors:

```
left = px.histogram(nc4, x='pm25aqs', histnorm='percent')
right = px.histogram(nc4, x='pm25pa', histnorm='percent')

fig = left_right(left, right, width=600, height=250)
fig.update_xaxes(title='AQS readings', col=1, row=1)
fig.update_xaxes(title='PurpleAir readings', col=2, row=1)
fig.update_yaxes(title='percent', col=1, row=1)
fig.show()
```

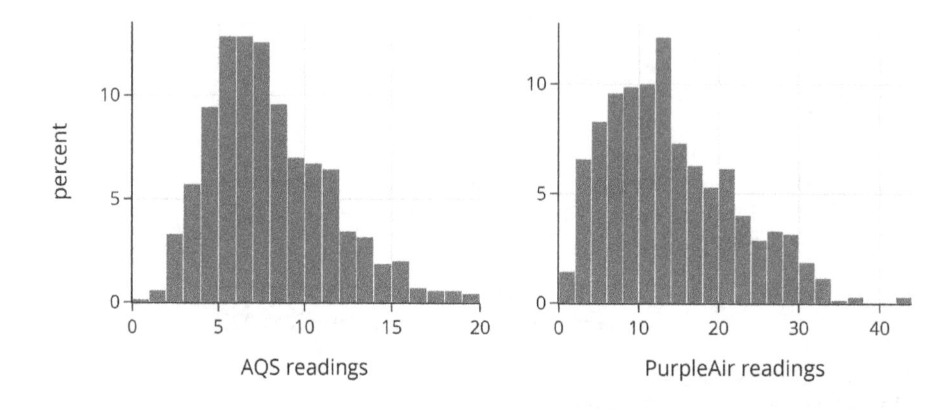

Both distributions are skewed right, which often happens when there's a lower bound on values (in this case, 0). A better way to compare these two distributions is with a quantile–quantile plot (see Chapter 10). With a q–q plot it can be easier to compare means, spreads, and tails:

```
percs = np.arange(1, 100, 1)
aqs_qs = np.percentile(nc4['pm25aqs'], percs, interpolation='lower')
pa_qs = np.percentile(nc4['pm25pa'], percs, interpolation='lower')
perc_df = pd.DataFrame({'percentile': percs, 'aqs_qs':aqs_qs, 'pa_qs':pa_qs})

fig = px.scatter(perc_df, x='aqs_qs', y='pa_qs',
                 labels={'aqs_qs': 'AQS quantiles',
                         'pa_qs': 'PurpleAir quantiles'},
                 width=350, height=250)


fig.add_trace(go.Scatter(x=[2, 13], y=[1, 25],
              mode='lines', line=dict(dash='dash', width=4)))
fig.update_layout(showlegend=False)
fig
```

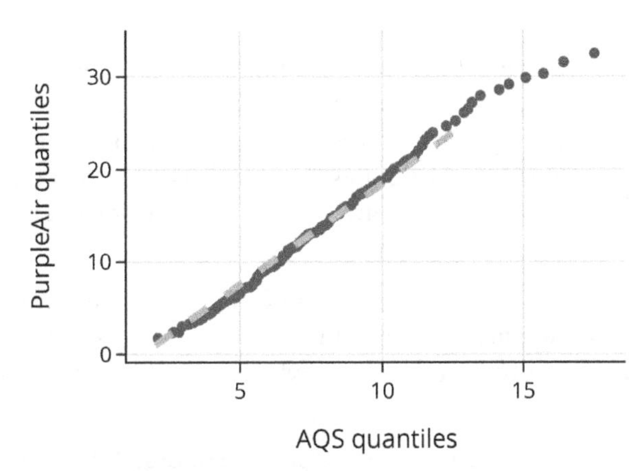

The quantile–quantile plot is roughly linear. We overlaid a dashed line with a slope of 2.2; it lines up the quantiles well, which indicates the spread of the PurpleAir measurements is about twice that of AQS.

What we can't see in the q–q plot or the side-by-side histograms is how the sensor readings vary together. Let's look at this next. First, we take a look at the distribution of difference between the two readings:

```
diffs = (nc4['pm25pa'] - nc4['pm25aqs'])

fig = px.histogram(diffs, histnorm='percent',
                   width=350, height=250)


fig.update_xaxes(range=[-10,30], title="Difference: PA&#8211;AQS reading")
fig.update_traces(xbins=dict(
        start=-10.0, end=30.0, size=2))
```

```
fig.update_layout(showlegend=False)
fig.show()
```

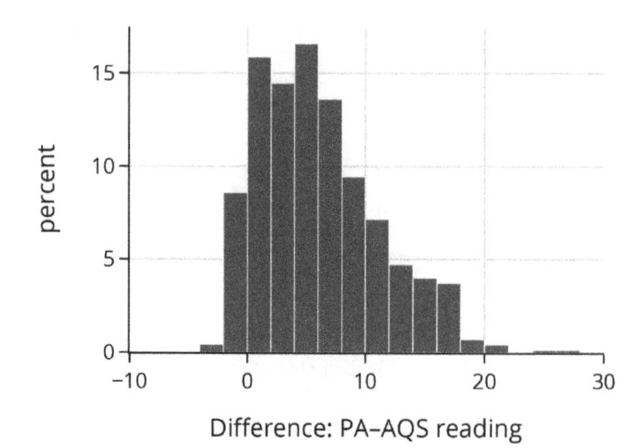

If the instruments are in perfect agreement, we will see a spike at 0. If the instruments are in agreement and there is a measurement error with no bias, we expect to see a distribution centered at 0. Instead, we see that 90% of the time, the PurpleAir measurement is larger than the AQS 24-hour average, and about 25% of the time it is more than 10 μg/m³ higher, which is a lot given the AQS averages tend to be between 5 μg/m³ and 10 μg/m³.

A scatterplot can give us additional insight into the relationship between the measurements from these two instruments. Since we are interested in finding a general relationship, regardless of time and location, we include all of our average readings in the plot:

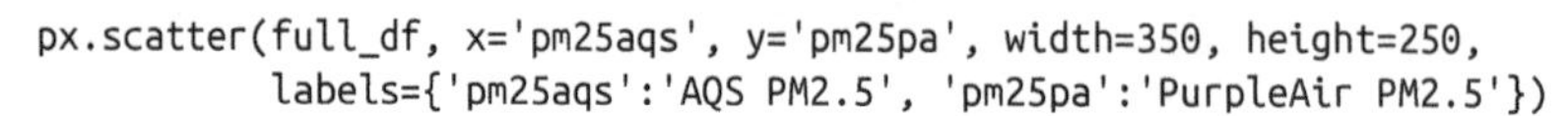

```
px.scatter(full_df, x='pm25aqs', y='pm25pa', width=350, height=250,
           labels={'pm25aqs':'AQS PM2.5', 'pm25pa':'PurpleAir PM2.5'})
```

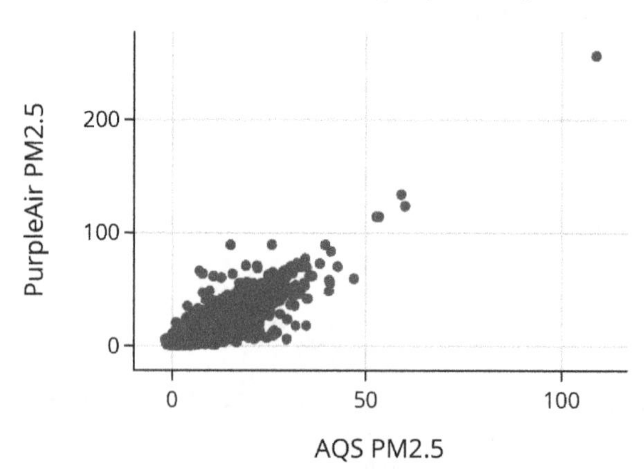

While the relationship looks linear, all but a handful of readings are in the bottom-left corner of the plot. Let's remake the scatterplot and zoom in on the bulk of the data to get a better look. We also add a smooth curve to the plot to help us see the relationship better:

```
full_df = full_df.loc[(full_df['pm25aqs'] < 50)]
```

```
px.scatter(full_df, x='pm25aqs', y='pm25pa',
           trendline='lowess', trendline_color_override="orange",
           labels={'pm25aqs':'AQS PM2.5', 'pm25pa':'PurpleAir PM2.5'},
           width=350, height=250)
```

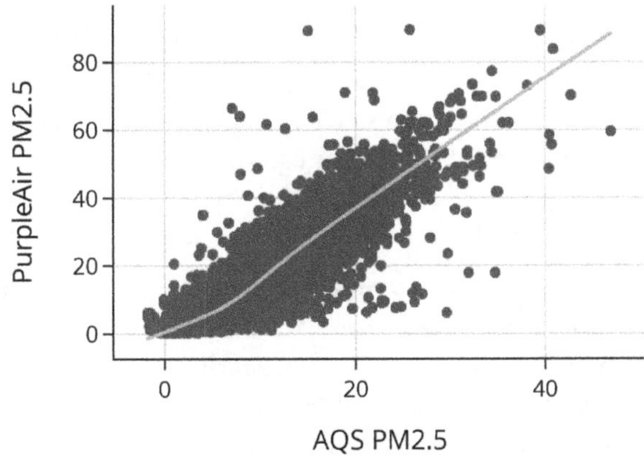

The relationship looks roughly linear, but there is a slight bend in the curve for small values of AQS. When the air is very clean, the PurpleAir sensor doesn't pick up as much particulate matter and so is more accurate. Also, we can see that the curve should go through the point (0, 0). Despite the slight bend in the relationship, the linear association (correlation) between these two measurements is high:

```
np.corrcoef(full_df['pm25aqs'], full_df['pm25pa'])
```

```
array([[1.  , 0.88],
       [0.88, 1.  ]])
```

Before starting this analysis, we expected that PurpleAir measurements would generally overestimate the PM2.5. And indeed, this is reflected in the scatterplot, but we also see that there appears to be a strong linear relationship between the measurements from these two instruments that will be helpful in calibrating the PurpleAir sensor.

Creating a Model to Correct PurpleAir Measurements

Now that we've explored the relationship between PM2.5 readings from AQS and PurpleAir sensors, we're ready for the final step of the analysis: creating a model that corrects PurpleAir measurements. Barkjohn's original analysis fits many models to the data in order to find the most appropriate one. In this section, we fit a simple linear model using the techniques from Chapter 4. We also briefly describe the final model Barkjohn chose for real-world use. Since these models use methods that we introduce later in the book, we won't explain the technical details very deeply here. Instead, we encourage you to revisit this section after reading Chapter 15.

First, let's go over our modeling goals. We want to create a model that predicts PM2.5 as accurately as possible. To do this, we build a model that adjusts PurpleAir measurements based on AQS measurements. We treat the AQS measurements as the true PM2.5 values because they are taken from carefully calibrated instruments and are actively used by the US government for decision making. So we have reason to trust the AQS PM2.5 values as being precise and close to the truth.

After we build the model that adjusts the PurpleAir measurements using AQS, we then flip the model around and use it to predict the true air quality in the future from PurpleAir measurements when we don't have a nearby AQS instrument. This is a *calibration* scenario. Since the AQS measurements are close to the truth, we fit the more variable PurpleAir measurements to them; this is the calibration procedure. Then we use the calibration curve to correct future PurpleAir measurements. This two-step process is encapsulated in the upcoming simple linear model and its flipped form.

First, we fit a line to predict a PurpleAir (PA) measurement from the ground truth, as recorded by an AQS instrument:

$$\text{PA} \approx b + m\text{AQS}$$

Next, we flip the line around to use a PA measurement to predict the air quality:

$$\text{True air quality} \approx -b/m + 1/m\text{PA}$$

The scatterplot and histograms that we made during our exploratory data analysis suggest that the PurpleAir measurements are more variable, which supports the calibration approach. And we saw that the PurpleAir measurements are about twice as high as the AQS measurements, which suggests that m may be close to 2 and $1/m$ close to $1/2$.

Why Two Steps?

This calibration procedure might seem a bit roundabout. Why not fit a linear model that directly uses PurpleAir to predict AQS? That seems a lot easier, and we wouldn't need to invert anything.

Since AQS measurements are "true" (or close to it), they have no error. Intuitively, a linear model works conditionally on the value of the variable on the x-axis, and minimizes the error in the y direction: $y - (b + mx)$. So we place the accurate measurements on the x-axis to fit the line, called the *calibration curve*. Then, in the future, we invert the line to predict the truth from our instrument's measurement. That's why this process is also called *inverse regression*. Calibration is a reasonable thing to do only when the pairs of measurements are highly correlated.

As a simpler example, let's say we want to calibrate a scale. We could do this by placing known weights—say, 1 kg, 5 kg, and 10 kg—onto the scale and seeing what the scale reports. We typically repeat this a few times, each time getting slightly different measurements from the scale. If we discover that our scale is often 10% too high ($y = 1.1x$), then when we weigh something in the future, we adjust our scale's reading downward by 90% ($1/1.1 = 0.9$). Analogously, the AQS measurements are the known quantities, and our model checks what the PurpleAir sensor reports.

Now let's fit the model. Following the notion from Chapter 4, we choose a loss function and minimize the average error. Recall that a *loss function* measures how far away our model is from the actual data. We use squared loss, which in this case is $[PA - (b + mAQS)]^2$. And to fit the model to our data, we minimize the average squared loss over our data:

$$\frac{1}{n}\sum_{i=1}^{n} [PA_i - (b + mAQS_i]^2$$

We use the linear modeling functionality provided by `scikit-learn` to do this (again, don't worry about these details for now):

```
from sklearn.linear_model import LinearRegression

AQS, PA = full_df[['pm25aqs']], full_df['pm25pa']

model = LinearRegression().fit(AQS, PA)
m, b = model.coef_[0], model.intercept_
```

By inverting the line, we get the estimate:

```
print(f"True air quality estimate = {-b/m:.2} + {1/m:.2}PA")

True air quality estimate = 1.4 + 0.53PA
```

This is close to what we expected. The adjustment to PurpleAir measurements is about 1/2.

The model that Barkjohn settled on incorporated the relative humidity:

$$\text{PA} \approx b + m_1\text{AQS} + m_2\text{RH}$$

This is an example of a multivariable linear regression model—it uses more than one variable to make predictions. We can fit it by minimizing the average squared error over the data:

$$\frac{1}{n}\sum_{i=1}^{n} [PA_i - (b + m_1 AQS_i + m_2 RH_i]^2$$

Then we invert the calibration to find the prediction model using the following equation:

$$\text{True air quality} \approx -\frac{b}{m_1} + \frac{1}{m_1}\text{PA} - \frac{m_2}{m_1}\text{RH}$$

We fit this model and check the coefficients:

```
AQS_RH, PA = full_df[['pm25aqs', 'rh']], full_df['pm25pa']
model_h = LinearRegression().fit(AQS_RH, PA)
[m1, m2], b = model_h.coef_, model_h.intercept_

print(f"True Air Quality Estimate = {-b/m:1.2} + {1/m1:.2}PA + {-m2/m1:.2}RH")

True Air Quality Estimate = 5.7 + 0.53PA + -0.088RH
```

In Chapters 15 and 16, we will learn how to compare these two models by examining things like the size of and patterns in prediction errors. For now, we note that the model that incorporates relative humidity performs the best.

Summary

In this chapter, we replicated Barkjohn's analysis. We created a model that corrects PurpleAir measurements so that they closely match AQS measurements. The accuracy of this model enables the PurpleAir sensors to be included on official US government maps, like the AirNow Fire and Smoke map. Importantly, this model gives people timely *and* accurate measurements of air quality.

We saw how crowdsourced, open data can be improved with data from precise, rigorously maintained, government-monitored equipment. In the process, we focused on cleaning and merging data from multiple sources, but we also fit models to adjust and improve air quality measurements.

For this case study, we applied many concepts covered in this part of the book. As you saw, wrangling files and data tables into a form we can analyze is a large and important part of data science. We used file wrangling and the notions of granularity from Chapter 8 to prepare two sources for merging. We got them into structures where we could match neighboring air quality sensors. This "grungy" part of data science was essential to widening the reach of data from rigorously maintained, precise government-monitored equipment by augmenting it with crowdsourced, open data.

This preparation process involved intensive, careful examination, cleaning, and improvement of the data to ensure their compatibility across the two sources and their trustworthiness in our analysis. Concepts from Chapter 9 helped us work with time data effectively and find and correct numerous issues like missing data points and even duplicated data values.

File and data wrangling, exploratory data analysis, and visualization are major parts of many analyses. While fitting models may seem to be the most exciting part of data science, getting to know and trust the data is crucial and often leads to important insights in the modeling phase. Topics related to modeling make up most of the rest of this book. However, before we begin, we cover two more topics related to data wrangling. In the next chapter, we show how to create analyzable data from text, and in the following chapter we examine other formats for source files that we mentioned in Chapter 8.

Before you head to the next chapter, take stock of what you've learned so far. Pat yourself on the back—you've already come a long way! The principles and techniques we've covered here are useful for nearly every type of data analysis, and you can readily start applying them toward analyses of your own.

PART IV

Other Data Sources

CHAPTER 13
Working with Text

Data can reside not just as numbers but also in words: names of dog breeds, restaurant violation descriptions, street addresses, speeches, blog posts, internet reviews, and much more. To organize and analyze information contained in text, we often need to do some of the following tasks:

Convert text into a standard format
: This is also referred to as *canonicalizing text*. For example, we might need to convert characters to lowercase, use common spellings and abbreviations, or remove punctuation and blank spaces.

Extract a piece of text to create a feature
: As an example, a string might contain a date embedded in it, and we want to pull it out from the string to create a date feature.

Transform text into features
: We might want to encode particular words or phrases as 0-1 features to indicate their presence in a string.

Analyze text
: In order to compare entire documents at once, we can transform a document into a vector of word counts.

This chapter introduces common techniques for working with text data. We show how simple string manipulation tools are often all we need to put text in a standard form or extract portions of strings. We also introduce regular expressions for more general and robust pattern matching. To demonstrate these text operations we use several examples. We first introduce these examples and describe the work we want to do to prepare the text for analysis.

Examples of Text and Tasks

For each type of task just introduced, we provide a motivating example. These examples are based on real tasks that we have carried out, but to focus on the concept, we've reduced the data to snippets.

Convert Text into a Standard Format

Let's say we want to study connections between population demographics and election results. To do this, we've taken election data from Wikipedia and population data from the US Census Bureau. The granularity of the data is at the county level, and we need to use the county names to join the tables. Unfortunately, the county names in these two tables don't always match:

	County	State	Voted
0	De Witt County	IL	97.8
1	Lac qui Parle County	MN	98.8
2	Lewis and Clark County	MT	95.2
3	St John the Baptist Parish	LA	52.6

	County	State	Population
0	DeWitt	IL	16,798
1	Lac Qui Parle	MN	8,067
2	Lewis & Clark	MT	55,716
3	St. John the Baptist	LA	43,044

We can't join the tables until we clean the strings to have a common format for county names. We need to change the case of characters, use common spellings and abbreviations, and address punctuation.

Extract a Piece of Text to Create a Feature

Text data sometimes has a lot of structure, especially when it was generated by a computer. As an example, the following is a web server's log entry. Notice how the entry has multiple pieces of data, but the pieces don't have a consistent delimiter—for instance, the date appears in square brackets, but other parts of the data appear in quotes and parentheses:

```
169.237.46.168 - -
[26/Jan/2004:10:47:58 -0800]"GET /stat141/Winter04 HTTP/1.1" 301 328
"http://anson.ucdavis.edu/courses"
"Mozilla/4.0 (compatible; MSIE 6.0; Windows NT 5.0; .NET CLR 1.1.4322)"
```

Even though the file format doesn't align with one of the simple formats we saw in Chapter 8, we can use text processing techniques to extract pieces of text to create features.

Transform Text into Features

In Chapter 9, we created a categorical feature based on the content of the strings. There, we examined the descriptions of restaurant violations and we created nominal variables for the presence of particular words. We've displayed a few example violations here:

```
unclean or degraded floors walls or ceilings
inadequate and inaccessible handwashing facilities
inadequately cleaned or sanitized food contact surfaces
wiping cloths not clean or properly stored or inadequate sanitizer
foods not protected from contamination
unclean nonfood contact surfaces
unclean or unsanitary food contact surfaces
unclean hands or improper use of gloves
inadequate washing facilities or equipment
```

These new features can be used in an analysis of food safety scores. Previously, we made simple features that marked whether a description contained a word like *glove* or *hair*. In this chapter, we more formally introduce the regular expression tools that we used to create these features.

Text Analysis

Sometimes we want to compare entire documents. For example, the US president gives a State of the Union speech every year. Here are the first few lines of the very first speech:

```
***

State of the Union Address
George Washington
January 8, 1790

Fellow-Citizens of the Senate and House of Representatives:
I embrace with great satisfaction the opportunity which now presents itself
of congratulating you on the present favorable prospects of our public ...
```

We might wonder: How have the State of the Union speeches changed over time? Do different political parties focus on different topics or use different language in their speeches? To answer these questions, we can transform the speeches into a numeric form that lets us use statistics to compare them.

These examples serve to illustrate the ideas of string manipulation, regular expressions, and text analysis. We start with describing simple string manipulation.

String Manipulation

There are a handful of basic string manipulation tools that we use a lot when we work with text:

- Transform uppercase characters to lowercase (or vice versa).
- Replace a substring with another or delete the substring.
- Split a string into pieces at a particular character.
- Slice a string at specified locations.

We show how we can combine these basic operations to clean up the county names data. Remember that we have two tables that we want to join, but the county names are written inconsistently.

Let's start by converting the county names to a standard format.

Converting Text to a Standard Format with Python String Methods

We need to address the following inconsistencies between the county names in the two tables:

- Capitalization: `qui` versus `Qui`.
- Omission of words: `County` and `Parish` are absent from the `census` table.
- Different abbreviation conventions: `&` versus `and`.
- Different punctuation conventions: `St.` versus `St`.
- Use of whitespace: `DeWitt` versus `De Witt`.

When we clean text, it's often easiest to first convert all of the characters to lowercase. It's easier to work entirely with lowercase characters than to try to track combinations of uppercase and lowercase. Next, we want to fix inconsistent words by replacing `&` with `and` and removing `County` and `Parish`. Finally, we need to fix up punctuation and whitespace inconsistencies.

With just two Python string methods, `lower` and `replace`, we can take all of these actions and clean the county names. These are combined into a method called `clean_county`:

```
def clean_county(county):
    return (county
            .lower()
```

```
               .replace('county', '')
               .replace('parish', '')
               .replace('&', 'and')
               .replace('.', '')
               .replace(' ', ''))
```

Although simple, these methods are the primitives that we can piece together to form more complex string operations. These methods are conveniently defined on all Python strings and do not require importing other modules. It is worth familiarizing yourself with the complete list of string methods (*https://oreil.ly/YWl9d*), but we describe a few of the most commonly used methods in Table 13-1.

Table 13-1. String methods

Method	Description
`str.lower()`	Returns a copy of a string with all letters converted to lowercase
`str.replace(a, b)`	Replaces all instances of the substring `a` in `str` with substring `b`
`str.strip()`	Removes leading and trailing whitespace from `str`
`str.split(a)`	Returns substrings of `str` split at a substring `a`
`str[x:y]`	Slices `str`, returning indices x (inclusive) to y (not inclusive)

We next verify that the `clean_county` method produces matching county names:

```
([clean_county(county) for county in election['County']],
 [clean_county(county) for county in census['County']])
```

```
(['dewitt', 'lacquiparle', 'lewisandclark', 'stjohnthebaptist'],
 ['dewitt', 'lacquiparle', 'lewisandclark', 'stjohnthebaptist'])
```

Since the county names now have consistent representations, we can successfully join the two tables.

String Methods in pandas

In the preceding code, we used a loop to transform each county name. The `pandas` `Series` objects provide a convenient way to apply string methods to each item in the series.

The `.str` property on `pandas` `Series` exposes the same Python string methods. Calling a method on the `.str` property calls the method on each item in the series. This allows us to transform each string in the series without using a loop. We save the transformed counties back into their originating tables. First we transform the county names in the election table:

```
election['County'] = (election['County']
 .str.lower()
 .str.replace('parish', '')
 .str.replace('county', '')
```

```
    .str.replace('&', 'and')
    .str.replace('.', '', regex=False)
    .str.replace(' ', ''))
```

We also transform the names in the census table so that the two tables contain the same representations of the county names. We can join these tables:

```
election.merge(census, on=['County','State'])
```

	County	State	Voted	Population
0	dewitt	IL	97.8	16,798
1	lacquiparle	MN	98.8	8,067
2	lewisandclark	MT	95.2	55,716
3	stjohnthebaptist	LA	52.6	43,044

Note that we merged on two columns: the county name and the state. We did this because some states have counties with the same name. For example, California and New York both have a county called King.

To see the complete list of string methods, we recommend looking at the Python documentation on `str` methods (*https://oreil.ly/Fb34C*) and the `pandas` documentation for the `.str` accessor (*https://oreil.ly/njVi3*). We did the canonicalization task using only `str.lower()` and multiple calls to `str.replace()`. Next, we extract text with another string method, `str.split()`.

Splitting Strings to Extract Pieces of Text

Let's say we want to extract the date from the web server's log entry:

```
log_entry
```

```
169.237.46.168 - - [26/Jan/2004:10:47:58 -0800]"GET /stat141/Winter04 HTTP/1.1"
301 328 "http://anson.ucdavis.edu/courses""Mozilla/4.0 (compatible; MSIE 6.0;
Windows NT 5.0; .NET CLR 1.1.4322)"
```

String splitting can help us home in on the pieces of information that form the date. For example, when we split the string on the left bracket, we get two strings:

```
log_entry.split('[')
```

```
['169.237.46.168 - - ',
 '26/Jan/2004:10:47:58 -0800]"GET /stat141/Winter04 HTTP/1.1" 301 328 "http://
anson.ucdavis.edu/courses""Mozilla/4.0 (compatible; MSIE 6.0; Windows NT
5.0; .NET CLR 1.1.4322)"']
```

The second string has the date information, and to get the day, month, and year, we can split that string on a colon:

```
log_entry.split('[')[1].split(':')[0]
```

```
'26/Jan/2004'
```

To separate out the day, month, and year, we can split on the forward slash. Altogether we split the original string three times, each time keeping only the pieces we are interested in:

```
(log_entry.split('[')[1]
 .split(':')[0]
 .split('/'))
```

```
['26', 'Jan', '2004']
```

By repeatedly using `split()`, we can extract many of the parts of the log entry. But this approach is complicated—if we wanted to also get the hour, minute, second, and time zone of the activity, we would need to use `split()` six times in total. There's a simpler way to extract these parts:

```
import re

pattern = r'[ \[/:\]]'
re.split(pattern, log_entry)[4:11]
```

```
['26', 'Jan', '2004', '10', '47', '58', '-0800']
```

This alternative approach uses a powerful tool called a regular expression, which we cover in the next section.

Regular Expressions

Regular expressions (or *regex* for short) are special patterns that we use to match parts of strings. Think about the format of a Social Security number (SSN) like `134-42-2012`. To describe this format, we might say that SSNs consist of three digits, then a dash, two digits, another dash, then four digits. Regexes let us capture this pattern in code. Regexes give us a compact and powerful way to describe this pattern of digits and dashes. The syntax of regular expressions is fortunately quite simple to learn; we introduce nearly all of the syntax in this section alone.

As we introduce the concepts, we tackle some of the examples described in an earlier section and show how to carry out the tasks with regular expressions. Almost all programming languages have a library to match patterns using regular expressions, making regular expressions useful in any programming language. We use some of the common methods available in the Python built-in `re` module to accomplish the tasks from the examples. These methods are summarized in Table 13-7 at the end of this section, where the basic usage and return value are briefly described. Since we only cover a few of the most commonly used methods, you may find it useful to consult the official documentation on the `re` module (*https://oreil.ly/IXWol*) as well.

Regular expressions are based on searching a string one character (aka *literal*) at a time for a pattern. We call this notion *concatenation of literals*.

Concatenation of Literals

Concatenation is best explained with a basic example. Suppose we are looking for the pattern `cat` in the string `cards scatter!`. Figure 13-1 contains a diagram that shows how the search proceeds through the string one character at a time. Notice that a "c" is found in the first position, followed by "a," but not "t," so the search backs up to the second character in the string and begins searching for a "c" again. The pattern "cat" is found within the string `cards scatter!` in positions 8–10. Once you get the hang of this process, you can move on to the richer set of patterns; they all follow this basic paradigm.

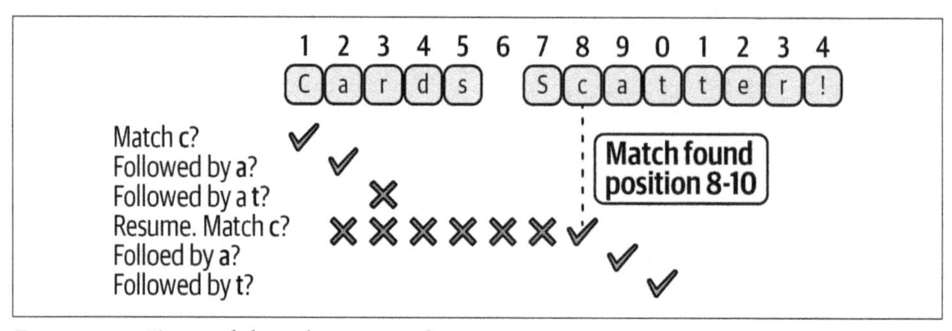

Figure 13-1. To match literal patterns, the regex engine moves along the string and checks one literal at a time for a match of the entire pattern. Notice that the pattern is found within the word `scatters` and that a partial match is found in `cards`.

In the preceding example, we observe that regular expressions can match patterns that appear anywhere in the input string. In Python, this behavior differs depending on the method used to match the regex—some methods only return a match if the regex appears at the start of the string; other methods return a match anywhere in the string.

These richer patterns are made of character classes and metacharacters like wildcards. We describe them in the subsections that follow.

Character classes

We can make patterns more flexible by using a *character class* (also known as a *character set*), which lets us specify a collection of equivalent characters to match. This allows us to create more relaxed matches. To create a character class, wrap the set of desired characters in brackets `[ ]`. For example, the pattern `[0123456789]` means "match any literal within the brackets"—in this case, any single digit. Then, the following regular expression matches three digits:

```
[0123456789][0123456789][0123456789]
```

This is such a commonly used character class that there is a shorthand notation for the range of digits, `[0-9]`. Character classes allow us to create a regex for SSNs:

```
[0-9][0-9][0-9]-[0-9][0-9]-[0-9][0-9][0-9][0-9]
```

Two other ranges that are commonly used in character classes are `[a-z]` for lowercase and `[A-Z]` for uppercase letters. We can combine ranges with other equivalent characters and use partial ranges. For example, `[a-cX-Z27]` is equivalent to the character class `[abcXYZ27]`.

Let's return to our original pattern `cat` and modify it to include two character classes:

```
c[oa][td]
```

This pattern matches `cat`, but it also matches `cot`, `cad`, and `cod`:

```
   Regex: c[oa][td]
    Text: The cat eats cod, cads, and cots, but not coats.
 Matches:     ***      ***  ***       ***
```

The idea of moving through the string one character at a time still remains the core notion, but now there's a bit more flexibility in which literal is considered a match.

Wildcard character

When we really don't care what the literal is, we can specify this with `.`, the period character. This matches any character except a newline.

Negated character classes

A *negated character class* matches any character *except* those between the square brackets. To create a negated character class, place the caret symbol as the first character after the left square bracket. For example, `[^0-9]` matches any character except a digit.

Shorthands for character classes

Some character sets are so common that there are shorthands for them. For example, `\d` is short for `[0-9]`. We can use this shorthand to simplify our search for SSN:

```
\d\d\d-\d\d-\d\d\d\d
```

Our regular expression for SSNs isn't quite bulletproof. If the string has extra digits at the beginning or end of the pattern we're looking for, then we still get a match. Note that we add the `r` character before the quotes to create a raw string, which makes regexes easier to write:

```
  Regex: \d\d\d-\d\d-\d\d\d\d
   Text: My other number is 6382-13-38420.
Matches:                     ***********
```

We can remedy the situation with a different sort of metacharacter: one that matches a word boundary.

Anchors and boundaries

At times we want to match a position before, after, or between characters. One example is to locate the beginning or end of a string; these are called *anchors*. Another is to locate the beginning or end of a word, which we call a *boundary*. The metacharacter `\b` denotes the boundary of a word. It has 0 length, and it matches whitespace or punctuation on the boundary of the pattern. We can use it to fix our regular expression for SSNs:

```
  Regex: \d\d\d-\d\d-\d\d\d\
   Text: My other number is 6382-13-38420.
Matches:

  Regex: \b\d\d\d-\d\d-\d\d\d\d\b
   Text: My reeeal number is 382-13-3842.
Matches:                     ***********
```

Escaping metacharacters

We have now seen several special characters, called *metacharacters*: `[` and `]` denote a character class, `^` switches to a negated character class, `.` represents any character, and `-` denotes a range. But sometimes we might want to create a pattern that matches one of these literals. When this happens, we must escape it with a backslash. For example, we can match the literal left bracket character using the regex `\[`:

```
  Regex: \[
   Text: Today is [2022/01/01]
Matches:          *
```

Next, we show how quantifiers can help create a more compact and clear regular expression for SSNs.

Quantifiers

To create a regex to match SSNs, we wrote:

```
\b[0-9][0-9][0-9]-[0-9][0-9]-[0-9][0-9][0-9][0-9]\b
```

This matches a "word" consisting of three digits, a dash, two more digits, a dash, and four more digits.

Quantifiers allow us to match multiple consecutive appearances of a literal. We specify the number of repetitions by placing the number in curly braces `{ }`.

Let's use Python's built-in `re` module for matching this pattern:

```
import re

ssn_re = r'\b[0-9]{3}-[0-9]{2}-[0-9]{4}\b'
re.findall(ssn_re, 'My SSN is 382-34-3840.')
```

```
['382-34-3840']
```

Our pattern shouldn't match phone numbers. Let's try it:

```
re.findall(ssn_re, 'My phone is 382-123-3842.')
```

```
[]
```

A quantifier always modifies the character or character class to its immediate left. Table 13-2 shows the complete syntax for quantifiers.

Table 13-2. Quantifier examples

Quantifier	Meaning
{m, n}	Match the preceding character m to n times.
{m}	Match the preceding character exactly m times.
{m,}	Match the preceding character at least m times.
{,n}	Match the preceding character at most n times.

Some commonly used quantifiers have a shorthand, as shown in Table 13-3.

Table 13-3. Shorthand quantifiers

Symbol	Quantifier	Meaning
*	{0,}	Match the preceding character 0 or more times.
+	{1,}	Match the preceding character 1 or more times.
?	{0,1}	Match the preceding character 0 or 1 time.

Quantifiers are greedy and will return the longest match possible. This sometimes results in surprising behavior. Since an SSN starts and ends with a digit, we might think the following shorter regex will be a simpler approach for finding SSNs. Can you figure out what went wrong in the matching?

```
ssn_re_dot = r'[0-9].+[0-9]'
re.findall(ssn_re_dot, 'My SSN is 382-34-3842 and hers is 382-34-3333.')
```

```
['382-34-3842 and hers is 382-34-3333']
```

Notice that we use the metacharacter `.` to match any character. In many cases, using a more specific character class prevents these false "overmatches." Our earlier pattern that includes word boundaries does this:

```
re.findall(ssn_re, 'My SSN is 382-34-3842 and hers is 382-34-3333.')
```

```
['382-34-3842', '382-34-3333']
```

Some platforms allow you to turn off greedy matching and use *lazy* matching, which returns the shortest string.

Literal concatenation and quantifiers are two of the core concepts in regular expressions. Next, we introduce two more core concepts: alternation and grouping.

Alternation and Grouping to Create Features

Character classes let us match multiple options for a single literal. We can use alternation to match multiple options for a group of literals. For instance, in the food safety example in Chapter 9, we marked violations related to body parts by seeing if the violation had the substring `hand`, `nail`, `hair`, or `glove`. We can use the `|` character in a regex to specify this alteration:

```
body_re = r"hand|nail|hair|glove"
re.findall(body_re, "unclean hands or improper use of gloves")
```

```
['hand', 'glove']
```

```
re.findall(body_re, "Unsanitary employee garments hair or nails")
```

```
['hair', 'nail']
```

With parentheses we can locate parts of a pattern, which are called *regex groups*. For example, we can use regex groups to extract the day, month, year, and time from the web server log entry:

```
# This pattern matches the entire timestamp
time_re = r"\[[0-9]{2}/[a-zA-z]{3}/[0-9]{4}:[0-9:\- ]*\]"
re.findall(time_re, log_entry)
```

```
['[26/Jan/2004:10:47:58 -0800]']
```

```
# Same regex, but we use parens to make regex groups...
time_re = r"\[([0-9]{2})/([a-zA-z]{3})/([0-9]{4}):([0-9:\- ]*)\]"

# ...which tells findall() to split up the match into its groups
re.findall(time_re, log_entry)
```

```
[('26', 'Jan', '2004', '10:47:58 -0800')]
```

As we can see, `re.findall` returns a list of tuples containing the individual components of the date and time of the web log.

We have introduced a lot of terminology, so in the next section we bring it all together into a set of tables for easy reference.

Reference Tables

We conclude this section with a few tables that summarize order of operation, metacharacters, and shorthands for character classes. We also provide tables summarizing the handful of methods in the `re` Python library that we have used in this section.

The four basic operations for regular expressions—concatenation, quantifying, alternation, and grouping—have an order of precedence, which we make explicit in Table 13-4.

Table 13-4. Order of operations

<table>
<tr><th>Operation</th><th>Order</th><th>Example</th><th>Matches</th></tr>
<tr><td>Concatenation</td><td>3</td><td><code>cat</code></td><td><code>cat</code></td></tr>
<tr><td>Alternation</td><td>4</td><td><code>cat|mouse</code></td><td><code>cat</code> and <code>mouse</code></td></tr>
<tr><td>Quantifying</td><td>2</td><td><code>cat?</code></td><td><code>ca</code> and <code>cat</code></td></tr>
<tr><td>Grouping</td><td>1</td><td>c(at)?</td><td><code>c</code> and <code>cat</code></td></tr>
</table>

Table 13-5 provides a list of the metacharacters introduced in this section, plus a few more. The column labeled "Doesn't match" gives examples of strings that the example regexes don't match.

Table 13-5. Metacharacters

<table>
<tr><th>Char</th><th>Description</th><th>Example</th><th>Matches</th><th>Doesn't match</th></tr>
<tr><td>.</td><td>Any character except \n</td><td><code>...</code></td><td>abc</td><td>ab</td></tr>
<tr><td>[ ]</td><td>Any character inside brackets</td><td><code>[cb.]ar</code></td><td>car
.ar</td><td>jar</td></tr>
<tr><td>[^]</td><td>Any character not inside brackets</td><td><code>[^b]ar</code></td><td>car
par</td><td>bar
ar</td></tr>
<tr><td>*</td><td>≥ 0 or more of previous symbol, shorthand for {0,}</td><td><code>[pb]*ark</code></td><td>bbark
ark</td><td>dark</td></tr>
<tr><td>+</td><td>≥ 1 or more of previous symbol, shorthand for {1,}</td><td><code>[pb]+ark</code></td><td>bbpark
bark</td><td>dark
ark</td></tr>
<tr><td>?</td><td>0 or 1 of previous symbol, shorthand for {0,1}</td><td><code>s?he</code></td><td>she
he</td><td>the</td></tr>
<tr><td>{n}</td><td>Exactly n of previous symbol</td><td><code>hello{3}</code></td><td>hellooo</td><td>hello</td></tr>
<tr><td>|</td><td>Pattern before or after bar</td><td><code>we|[ui]s</code></td><td>we
us
is</td><td>es
e
s</td></tr>
<tr><td>\</td><td>Escape next character</td><td><code>\[hi\]</code></td><td>[hi]</td><td>hi</td></tr>
<tr><td>^</td><td>Beginning of line</td><td><code>^ark</code></td><td>ark two</td><td>dark</td></tr>
<tr><td>$</td><td>End of line</td><td><code>ark$</code></td><td>noahs ark</td><td>noahs arks</td></tr>
<tr><td>\b</td><td>Word boundary</td><td><code>ark\b</code></td><td>ark of noah</td><td>noahs arks</td></tr>
</table>

Additionally, in Table 13-6, we provide shorthands for some commonly used character sets. These shorthands don't need [ ].

Table 13-6. Character class shorthands

Description	Bracket form	Shorthand
Alphanumeric character	`[a-zA-Z0-9_]`	`\w`
Not an alphanumeric character	`[^a-zA-Z0-9_]`	`\W`
Digit	`[0-9]`	`\d`
Not a digit	`[^0-9]`	`\D`
Whitespace	`[\t\n\f\r\p{Z}]`	`\s`
Not whitespace	`[^\t\n\f\r\p{z}]`	`\S`

We used the following methods in `re` in this chapter. The names of the methods are indicative of the functionality they perform: *search* or *match* a pattern in a string; *find all* cases of a pattern in a string; *sub*stitute all occurrences of a pattern with a substring; and *split* a string into pieces at the pattern. Each requires a pattern and string to be specified, and some have extra arguments. Table 13-7 provides the format of the method usage and a description of the return value.

Table 13-7. Regular expression methods

Method	Return value
`re.search(pattern, string)`	Match object if the pattern is found anywhere in the string, otherwise `None`
`re.match(pattern, string)`	Match object if the pattern is found at the beginning of the string, otherwise `None`
`re.findall(pattern, string)`	List of all matches of `pattern` in `string`
`re.sub(pattern, replacement, string)`	String where all occurrences of `pattern` are replaced by `replacement` in the `string`
`re.split(pattern, string)`	List of the pieces of `string` around the occurrences of `pattern`

As we saw in the previous section, `pandas` `Series` objects have a `.str` property that supports string manipulation using Python string methods. Conveniently, the `.str` property also supports some functions from the `re` module. Table 13-8 shows the analogous functionality from Table 13-7 of the `re` methods. Each requires a pattern. See the `pandas` docs (*https://oreil.ly/aHJRz*) for a complete list of string methods.

Table 13-8. Regular expressions in `pandas`

Method	Return value
`str.contains(pattern, regex=True)`	Series of booleans indicating whether the `pattern` is found
`str.findall(pattern, regex=True)`	List of all matches of `pattern`

Method	Return value
`str.replace(pattern, replacement, regex=True)`	Series with all matching occurrences of `pattern` replaced by `replacement`
`str.split(pattern, regex=True)`	Series of lists of strings around given `pattern`

Regular expressions are a powerful tool, but they're somewhat notorious for being difficult to read and debug. We close with some advice for using regexes:

- Develop your regular expression on simple test strings to see what the pattern matches.
- If a pattern matches nothing, try weakening it by dropping part of the pattern. Then tighten it incrementally to see how the matching evolves. (Online regex-checking tools can be very helpful here.)
- Make the pattern only as specific as it needs to be for the data at hand.
- Use raw strings whenever possible for cleaner patterns, especially when a pattern includes a backslash.
- When you have lots of long strings, consider using compiled patterns because they can be faster to match (see `compile()` in the `re` library).

In the next section, we carry out an example text analysis. We clean the data using regular expressions and string manipulation, convert the text into quantitative data, and analyze the text via these derived quantities.

Text Analysis

So far, we've used Python methods and regular expressions to clean short text fields and strings. In this section, we analyze entire documents using a technique called *text mining*, which transforms free-form text into a quantitative representation to uncover meaningful patterns and insights.

Text mining is a deep topic. Instead of a comprehensive treatment, we introduce a few key ideas through an example, where we analyze the State of the Union speeches from 1790 to 2022. Every year, the US president gives a State of the Union speech to Congress. These speeches talk about current events in the country and make recommendations for Congress to consider. The American Presidency Project (*https://oreil.ly/JbpO4*) makes these speeches available online.

Let's begin by opening the file that has all of the speeches:

```
from pathlib import Path

text = Path('data/stateoftheunion1790-2022.txt').read_text()
```

At the beginning of this chapter, we saw that each speech in the data begins with a line with three asterisks: ***. We can use a regular expression to count the number of times the string *** appears:

```
import re
num_speeches = len(re.findall(r"\*\*\*", text))
print(f'There are {num_speeches} speeches total')
```

```
There are 232 speeches total
```

In text analysis, a *document* refers to a single piece of text that we want to analyze. Here, each speech is a document. We split apart the `text` variable into its individual documents:

```
records = text.split("***")
```

Then we can put the speeches into a dataframe:

```
def extract_parts(speech):
    speech = speech.strip().split('\n')[1:]
    [name, date, *lines] = speech
    body = '\n'.join(lines).strip()
    return [name, date, body]

def read_speeches():
    return pd.DataFrame([extract_parts(l) for l in records[1:]],
                        columns = ["name", "date", "text"])

df = read_speeches()
df
```

	name	date	text
0	George Washington	January 8, 1790	Fellow-Citizens of the Senate and House of Rep...
1	George Washington	December 8, 1790	Fellow-Citizens of the Senate and House of Rep...
2	George Washington	October 25, 1791	Fellow-Citizens of the Senate and House of Rep...
...	...	...	...
229	Donald J. Trump	February 4, 2020	Thank you very much. Thank you. Thank you very...
230	Joseph R. Biden, Jr.	April 28, 2021	Thank you. Thank you. Thank you. Good to be ba...
231	Joseph R. Biden, Jr.	March 1, 2022	Madam Speaker, Madam Vice President, our First...

```
232 rows × 3 columns
```

Now that we have the speeches loaded into a dataframe, we want to transform the speeches to see how they have changed over time. Our basic idea is to look at the words in the speeches—if two speeches contain very different words, our analysis should tell us that. With some kind of measure of document similarity, we can see how the speeches differ from one another.

There are a few problems in the documents that we need to take care of first:

- Capitalization shouldn't matter: `Citizens` and `citizens` should be considered the same word. We can address this by lowercasing the text.
- There are unspoken remarks in the text: `[laughter]` points out where the audience laughed, but these shouldn't count as part of the speech. We can address this by using a regex to remove text within brackets: `\[[^\]]+\]`. Remember that `\[` and `\]` match the literal left and right brackets, and `[^\]]` matches any character that isn't a right bracket.
- We should take out characters that aren't letters or whitespace: some speeches talk about finances, but a dollar amount shouldn't count as a word. We can use the regex `[^a-z\s]` to remove these characters. This regex matches any character that isn't a lowercase letter (`a-z`) or a whitespace character (`\s`):

```
def clean_text(df):
    bracket_re = re.compile(r'\[[^\]]+\]')
    not_a_word_re = re.compile(r'[^a-z\s]')
    cleaned = (df['text'].str.lower()
               .str.replace(bracket_re, '', regex=True)
               .str.replace(not_a_word_re, ' ', regex=True))
    return df.assign(text=cleaned)

df = (read_speeches()
      .pipe(clean_text))
df
```

	name	date	text
0	George Washington	January 8, 1790	fellow citizens of the senate and house of rep...
1	George Washington	December 8, 1790	fellow citizens of the senate and house of rep...
2	George Washington	October 25, 1791	fellow citizens of the senate and house of rep...
...	...	...	...
229	Donald J. Trump	February 4, 2020	thank you very much thank you thank you very...
230	Joseph R. Biden, Jr.	April 28, 2021	thank you thank you thank you good to be ba...
231	Joseph R. Biden, Jr.	March 1, 2022	madam speaker madam vice president our first...

```
232 rows × 3 columns
```

Next, we look at some more complex issues:

- *Stop words* like `is`, `and`, `the`, and `but` appear so often that we would like to just remove them.
- `argue` and `arguing` should count as the same word, even though they appear differently in the text. To address this, we'll use *word stemming*, which transforms both words to `argu`.

To handle these issues, we can use built-in methods from the `nltk` library (*https://www.nltk.org*).

Finally, we transform the speeches into *word vectors*. A word vector represents a document using a vector of numbers. For example, one basic type of word vector counts up how many times each word appears in the text, as depicted in Figure 13-2.

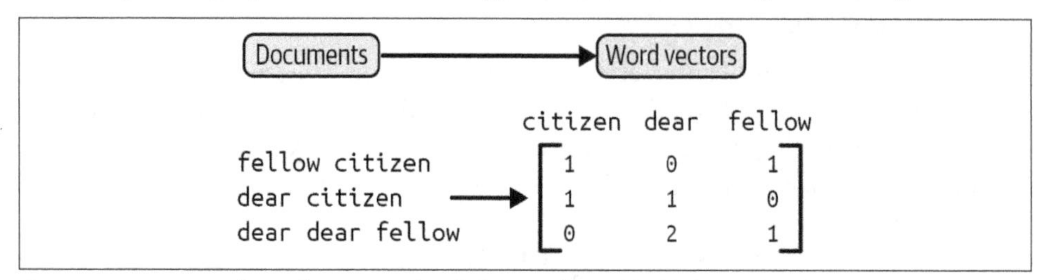

Figure 13-2. Bag-of-words vectors for three small example documents

This simple transform is called *bag-of-words*, and we apply it on all of our speeches. Then we calculate the *term frequency-inverse document frequency* (*tf-idf* for short) to normalize the counts and measure the rareness of a word. The tf-idf puts more weight on words that only appear in a few documents. The idea is that if just a few documents mention the word *sanction*, say, then this word is extra useful for distinguishing documents from each other. The `scikit-learn` library (*https://oreil.ly/3A6a5*) has a complete description of the transform and an implementation, which we use.

After applying these transforms, we have a two-dimensional array, `speech_vectors`. Each row of this array is one speech transformed into a vector:

```
import nltk
nltk.download('stopwords')
nltk.download('punkt')

from nltk.stem.porter import PorterStemmer
from sklearn.feature_extraction.text import TfidfVectorizer

stop_words = set(nltk.corpus.stopwords.words('english'))
porter_stemmer = PorterStemmer()
```

```
def stemming_tokenizer(document):
    return [porter_stemmer.stem(word)
            for word in nltk.word_tokenize(document)
            if word not in stop_words]

tfidf = TfidfVectorizer(tokenizer=stemming_tokenizer)
speech_vectors = tfidf.fit_transform(df['text'])
```

```
speech_vectors.shape
```

```
(232, 13211)
```

We have 232 speeches, and each speech was transformed into a length-13,211 vector. To visualize these speeches, we use a technique called *principal component analysis* to represent the data table of 13,211 features by a new set of features that are orthogonal to one another. The first vector accounts for the maximum variation in the original features, the second for the maximum variance that is orthogonal to the first, and so on. Often the first two components, which we can plot as pairs of points, reveal clusters and outliers.

Next, we plot the first two principal components. Each point is one speech, and we've colored the points according to the year of the speech. Points that are close together represent similar speeches, and points that are far away from one another represent dissimilar speeches:

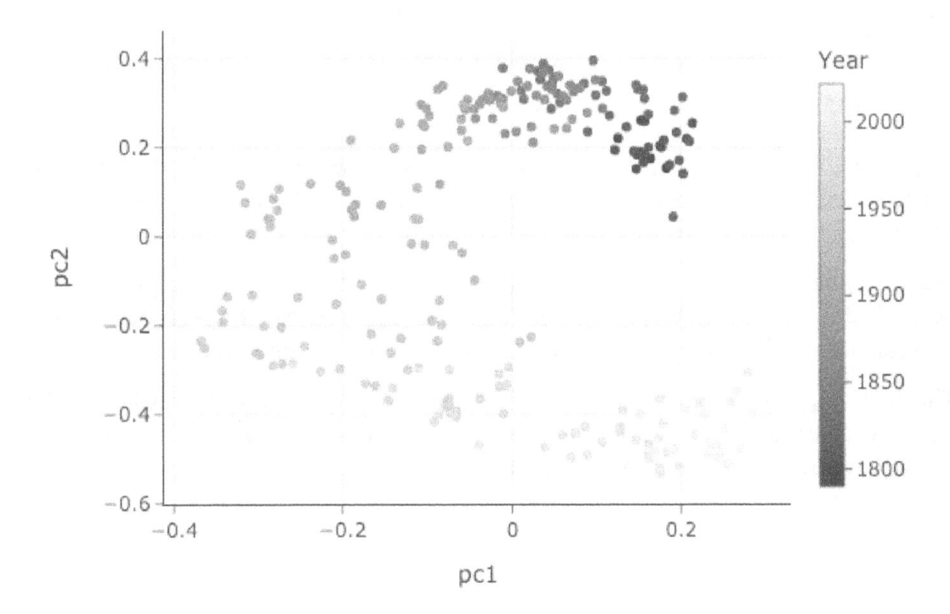

We see a clear difference in speeches over time—speeches given in the 1800s used very different words than speeches given after 2000. It's also interesting to see that the speeches cluster tightly in the same time period. This suggests that speeches within the same period sound relatively similar, even though the speakers were from different political parties.

This section gave a whirlwind introduction to text analysis. We used text manipulation tools from previous sections to clean up the presidential speeches. Then we used more advanced techniques like stemming, the tf-idf transform, and principal component analysis to compare speeches. Although we don't have enough space in this book to cover all of these techniques in detail, we hope that this section piqued your interest in the exciting world of text analysis.

Summary

This chapter introduced techniques for working with text to clean and analyze data, including string manipulation, regular expressions, and document analysis. Text data has rich information about how people live, work, and think. But this data is also hard for computers to use—think about all the creative ways people manage to spell the same word. The techniques in this chapter let us correct typos, extract features from logs, and compare documents.

We don't recommend you use regular expressions to:

- Parse hierarchical structures such as JSON or HTML; use a parser instead
- Search for complex properties, like palindromes and balanced parentheses
- Validate a complex feature, such as a valid email address

Regular expressions, while powerful, are terrible at these types of tasks. However, in our experience, even the basics of text manipulation can enable all sorts of interesting analyses—a little bit goes a long way.

We have one final caution about regular expressions: they can be computationally expensive. You will want to consider the trade-offs between these concise, clear expressions and the overhead they create if they're being put into production code.

The next chapter considers other sorts of data, such as data in binary formats, and the highly structured text of JSON and HTML. Our focus will be on loading these data into dataframes and other Python data structures.

CHAPTER 14

Data Exchange

Data can be stored and exchanged in many different formats. Thus far, we've focused on plain-text delimited and fixed-width formats (Chapter 8). In this chapter, we expand our horizons a bit and introduce a few other popular formats. While CSV, TSV, and FWF files are useful for organizing data into a dataframe, other file formats can save space or represent more complex data structures. *Binary* files (*binary* is a term for formats that aren't plaintext) can be more economical than plain-text data sources. For example, in this chapter we introduce NetCDF, a popular binary format for exchanging large amounts of scientific data. Other plain-text formats like JSON and XML can organize data in ways that are more general and useful for complex data structures. Even HTML web pages, a close cousin to XML, often contain useful information that we can scrape and wrangle into shape for analysis.

In this chapter, we introduce these popular formats, describe a mental model for their organization, and provide examples. In addition to introducing these formats, we cover programmatic ways to acquire data online. Before the internet, data scientists had to physically move disk drives to share data with one another. Now we can freely retrieve datasets from computers across the world. We introduce HTTP, the primary communication protocol for the web, and REST, an architecture to transfer data. By learning a bit about these web technologies, we can take better advantage of the web as a data source.

Throughout this book, we have set an example of reproducible code for wrangling, exploring, and modeling with data. In this chapter, we address how to acquire data that are available online in a reproducible fashion.

We begin with a description of NetCDF, followed by JSON. Then, after an overview of web protocols for data exchange, we wrap up the chapter with an introduction to XML, HTML, and XPath, a tool for extracting content from these types of files.

NetCDF Data

The Network Common Data Form (NetCDF) (*https://oreil.ly/_qZGj*) is a convenient and efficient format for storing array-oriented scientific data. A mental model for this format represents a variable by a multidimensional grid of values. The diagram in Figure 14-1 shows the concept. A variable such as rainfall is recorded daily at places around the globe. We can imagine these rainfall values arranged in a cube with longitude running along one side of the cube, latitude along another, and date in the third dimension. Each cell in the cube holds the rainfall recorded for one day at a particular location. A NetCDF file also contains information, which we call *metadata*, about the dimensions of the cube. The same information would be organized quite differently in a dataframe, where we would need three features for latitude, longitude, and date for each rainfall measurement. This would mean repeating lots of data. With a NetCDF file, we don't need to repeat the latitude and longitude values for each day, nor the dates for each location.

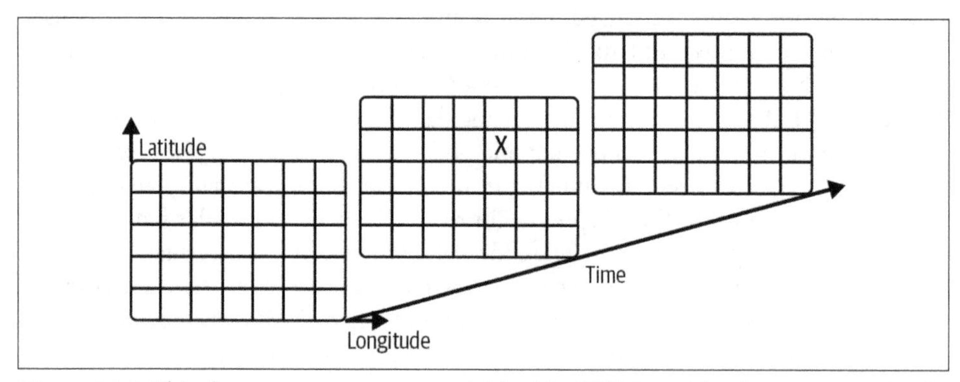

Figure 14-1. This diagram represents a model for NetCDF data. The data are organized into a three-dimensional array that contains recordings of rainfall at locations in time (latitude, longitude, and time). The "X" marks one rainfall measurement for a specific location on a particular date.

NetCDF has several other advantages, in addition to being more compact:

Scalable
: It provides efficient access to subsets of the data.

Appendable
: You can easily add new data without redefining the structure.

Sharable
: It's a common format that's independent of the coding language and operating system.

Self-describing
: The source file contains both a description of the data's organization and the data itself.

Community
: The tools are made available by a community of users.

The NetCDF format is an example of *binary* data—data that can't directly be read into a text editor like `vim` or Visual Studio Code, unlike text formats like CSV. There are a multitude of other binary data formats, including SQLite databases (from Chapter 7), Feather, and Apache Arrow. Binary data formats provide flexibility in how datasets are stored, but they also typically need special tools to open and read them in.

NetCDF variables are not limited to three dimensions. For example, elevation could be added to our earth science application so that we have recordings of, say, temperature, in time, latitude, longitude, and elevation. And dimensions need not correspond to physical dimensions. Climate scientists often run several models and store the model number in a dimension along with the model output. While NetCDF was originally developed for atmospheric scientists at the University Corporation for Atmospheric Research (UCAR), the format has gained popularity and is now used at thousands of educational, research, and government sites around the world. And the applications have expanded to other areas, such as astronomy and physics with the Smithsonian/NASA Astrophysics Data System (ADS) (*https://oreil.ly/kg9kV*) and medical imaging with Medical Image NetCDF (MINC) (*https://oreil.ly/6t3gJ*).

NetCDF files have three basic components: dimensions, variables, and various sorts of metadata. The *variable* contains what we think of as the data, such as the rainfall recordings. Each variable has a name, storage type, and shape, meaning the number of dimensions. The *dimensions* component gives each dimension's name and number of grid points. Additional information is provided by the *coordinates*—in particular, the points at which the measurements are made, such as for longitude, where these might be 0.0, 0.25, 0.50, ..., 359.75. Other metadata include *attributes*. Attributes for a variable can hold ancillary information about the variables, and other attributes contain global information about the file, such as who published the dataset, their contact information, and permissions for using the data. This global information is critical to ensure reproducible results.

The following example examines the components of a particular NetCDF file and demonstrates how to extract portions of data from variables.

The Climate Data Store (*https://oreil.ly/NAhRW*) provides a collection of datasets from various climate sectors and services. We visited their site and requested measurements of temperature and total precipitation for a two-week period in December 2022. Let's walk through a brief examination of these data to get a sense of the organization of the components in the file, how to extract subsets, and how to make visualizations.

The data are in the NetCDF file *CDS_ERA5_22-12.nc*. Let's first figure out how large the file is:

```
from pathlib import Path
import os

file_path = Path() / 'data' / 'CDS_ERA5_22-12.nc'

kib = 1024
size = os.path.getsize(file_path)
np.round(size / kib**3)
```

```
2.0
```

Despite having only three variables (total precipitation, rain rate, temperature) for two weeks, the file is two GiB in size! These climate sources often tend to be quite large.

The `xarray` package is useful for working with array-like data and, in particular, NetCDF. We use its functionality to explore the components of our climate file. First we open the file:

```
import xarray as xr

ds = xr.open_dataset(file_path)
```

Now let's check the dimensions component of the file:

```
ds.dims
```

```
Frozen(SortedKeysDict({'longitude': 1440, 'latitude': 721, 'time': 408}))
```

As in Figure 14-1, our file has three dimensions: longitude, latitude, and time. The size of each dimension tells us that there are over 400,000 cells of data values (1440 × 721 × 408). If these data were in a dataframe, then it would have 400,000 rows with latitude, longitude, and time columns in great repetition! Instead, we only need their values once, and the coordinates component gives them to us:

```
ds.coords
```

```
Coordinates:
  * longitude  (longitude) float32 0.0 0.25 0.5 0.75 ... 359.0 359.2 359.5 359.8
  * latitude   (latitude) float32 90.0 89.75 89.5 89.25 ... -89.5 -89.75 -90.0
  * time       (time) datetime64[ns] 2022-12-15 ... 2022-12-31T23:00:00
```

Each variable in our file is three-dimensional. Actually, a variable doesn't have to have all three dimensions, but in our example they do:

```
ds.data_vars
```

```
Data variables:
    t2m      (time, latitude, longitude) float32 ...
    lsrr     (time, latitude, longitude) float32 ...
    tp       (time, latitude, longitude) float32 ...
```

Metadata for a variable provides the units and a longer description, while metadata for the source gives us information such as when we retrieved the data:

```
ds.tp.attrs
```

```
{'units': 'm', 'long_name': 'Total precipitation'}
```

```
ds.attrs
```

```
{'Conventions': 'CF-1.6',
 'history': '2023-01-19 19:54:37 GMT by grib_to_netcdf-2.25.1: /opt/ecmwf/mars-
client/bin/grib_to_netcdf.bin -S param -o /cache/data6/
adaptor.mars.internal-1674158060.3800251-17201-13-c46a8ac2-f1b6-4b57-
a14e-801c001f7b2b.nc /cache/tmp/c46a8ac2-f1b6-4b57-a14e-801c001f7b2b-
adaptor.mars.internal-1674158033.856014-17201-20-tmp.grib'}
```

By keeping all of these pieces of information in the source file itself, we don't risk losing it or having the description get out of sync with the data.

Like with pandas, xarray provides many different ways to select portions of the data to work with. We show two examples. First we focus on one specific location and examine the total precipitation in time with a line plot:

```
plt.figure()
(ds.sel(latitude=37.75, longitude=237.5).tp * 100).plot(figsize=(8,3))
plt.xlabel('')
plt.ylabel('Total precipitation (cm)')
plt.show();
```

```
<Figure size 288x216 with 0 Axes>
```

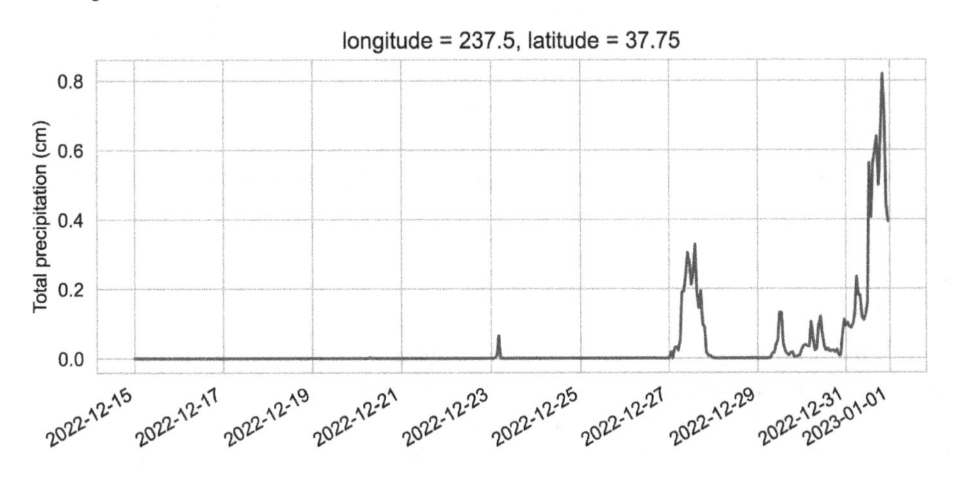

Next we choose one date, December 31, 2022, at 1 p.m., and narrow down the latitude and longitude to the continental US to make a map of temperature:

```
import datetime
one_day = datetime.datetime(2022, 12, 31, 13, 0, 0)

min_lon, min_lat, max_lon, max_lat = 232, 21, 300, 50

mask_lon = (ds.longitude > min_lon) & (ds.longitude < max_lon)
mask_lat = (ds.latitude > min_lat) & (ds.latitude < max_lat)

ds_oneday_us = ds.sel(time=one_day).t2m.where(mask_lon & mask_lat, drop=True)
```

Like `loc` for dataframes, `sel` returns a new `DataArray` whose data is determined by the index labels along the specified dimension, which for this example is the date. And like `np.where`, `xr.where` returns elements depending on the logical condition provided. We use `drop=True` to reduce the size of the dataset.

Let's make a choropleth map of temperature, where color represents the temperature:

```
ds_oneday_us.plot(figsize=(8,4))
```

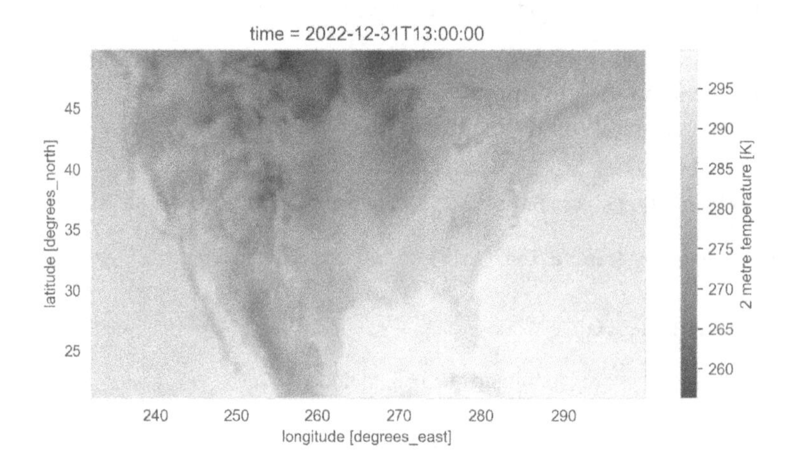

We can make out the shape of the US, the warm Caribbean, and the colder mountain ranges from this map.

We wrap up by closing the file:

```
ds.close()
```

This brief introduction to NetCDF is meant to touch on the basic concepts. Our main goal is to show that other kinds of data formats exist and can have advantages over plain-text read into a dataframe. For interested readers, NetCDF has a rich ecosystem of packages and functionality. For example, in addition to the `xarray` module,

NetCDF files can be read with other Python modules like `netCDF4` (*https://oreil.ly/UlX_k*) and `gdal` (*https://oreil.ly/fKeQh*). The NetCDF community has also provided command-line tools for interacting with NetCDF data. And to make visualizations and maps, options include `matplotlib`, `iris` (*https://oreil.ly/ozNrI*), which is built on top of `netCDF4`, and `cartopy` (*https://oreil.ly/9N7y7*).

Next we consider the JSON format, which offers more flexibility to represent hierarchical data than the CSV and FWF formats.

JSON Data

JavaScript Object Notation (JSON) is a popular format for exchanging data on the web. This plain-text format has a simple and flexible syntax that aligns well with Python dictionaries, and it is easy for machines to parse and people to read.

Briefly, JSON has two main structures, the object and the array:

Object
: Like a Python `dict`, a JSON object is an unordered collection of name-value pairs. These pairs are contained in curly braces; each is formatted as `"name":value`, and separated by commas.

Array
: Like a Python `list`, a JSON array is an ordered collection of values contained in square brackets, where the values are unnamed and separated by commas.

The values in an object and array can be of different types and can be nested. That is, an array can contain objects and vice versa. The primitive types are limited to string in double quotes, number in text representation, logical as true or false, and null.

The following short JSON file demonstrates all of these syntactical features:

```
{"lender_id":"matt",
 "loan_count":23,
 "status":[2, 1, 3],
 "sponsored": false,
 "sponsor_name": null,
 "lender_dem":{"sex":"m","age":77 }
}
```

Here we have an object that contains six name-value pairs. The values are heterogeneous; four are primitive values: string, number, logical, and null. The `status` value consists of an array of three (ordered) numbers, and `lender_dem` is an object with demographic information.

The built-in `json` package can be used to work with JSON files in Python. For example, we can load this small file into a Python dictionary:

```
import json
from pathlib import Path

file_path = Path() / 'data' / 'js_ex' / 'ex.json'

ex_dict = json.load(open(file_path))
ex_dict
```

```
{'lender_id': 'matt',
 'loan_count': 23,
 'status': [2, 1, 3],
 'sponsored': False,
 'sponsor_name': None,
 'lender_dem': {'sex': 'm', 'age': 77}}
```

The dictionary matches the format of the Kiva file. This format doesn't naturally translate to a dataframe. The `json_normalize` method can organize this semistructured JSON data into a flat table:

```
ex_df = pd.json_normalize(ex_dict)
ex_df
```

	lender_id	loan_count	status	sponsored	sponsor_name	lender_dem.sex	lender_dem.age
0	matt	23	[2, 1, 3]	False	None	m	77

Notice how the third element in this one-row dataframe is a list, whereas the nested object was converted into two columns.

There's a tremendous amount of flexibility in how data can be structured in JSON, which means that if we want to create a dataframe from JSON content, we need to understand how the data are organized in the JSON file. We provide three structures that translate easily into a dataframe in the next example.

The list of PurpleAir sites used in the case study in Chapter 12 was JSON-formatted. In that chapter, we didn't call attention to the format and simply read the file contents into a dictionary with the `json` library's `load` method and then into a dataframe. Here, we have simplified that file while maintaining the general structure so that it's easier to examine.

We begin with an examination of the original file, and then reorganize it into two other JSON structures that might also be used to represent a dataframe. With these examples we aim to show the flexibility of JSON. The diagrams in Figure 14-2 give representations of the three possibilities.

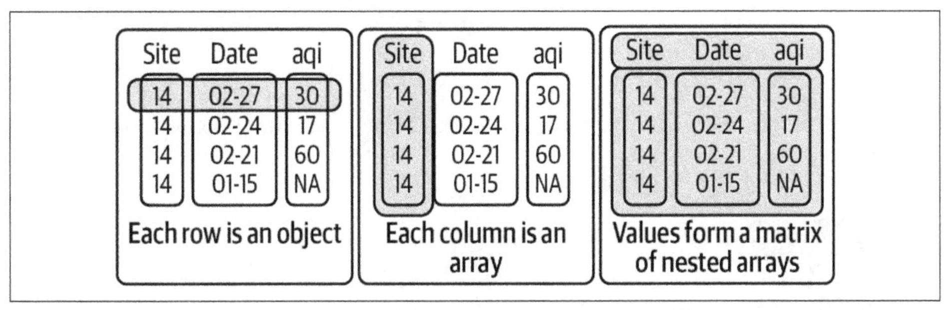

Figure 14-2. Three different approaches for a JSON-formatted file to store a dataframe.

The leftmost dataframe in the diagram shows an organization by rows. Each row is an object of named values where the name corresponds to the column name of the dataframe. Rows would then be collected in an array. This structure coincides with that of the original file. In the following code, we display the file contents:

```
{"Header": [
    {"status": "Success",
     "request_time": "2022-12-29T01:48:30-05:00",
     "url": "https://aqs.epa.gov/data/api/dailyData/...",
     "rows": 4
    }
  ],
  "Data": [
    {"site": "0014", "date": "02-27", "aqi": 30},
    {"site": "0014", "date": "02-24", "aqi": 17},
    {"site": "0014", "date": "02-21", "aqi": 60},
    {"site": "0014", "date": "01-15", "aqi": null}
  ]
}
```

We see that the file consists of one object with two elements, named `Header` and `Data`. The `Data` element is an array with an element for each row in the dataframe, and as described earlier each element is an object. Let's load the file into a dictionary and check its contents (see Chapter 8 for more on finding a pathname to a file and printing its contents):

```
from pathlib import Path
import os

epa_file_path = Path('data/js_ex/epa_row.json')

data_row = json.loads(epa_file_path.read_text())
data_row
```

```
{'Header': [{'status': 'Success',
   'request_time': '2022-12-29T01:48:30-05:00',
   'url': 'https://aqs.epa.gov/data/api/dailyData/...',
   'rows': 4}],
 'Data': [{'site': '0014', 'date': '02-27', 'aqi': 30},
```

```
  {'site': '0014', 'date': '02-24', 'aqi': 17},
  {'site': '0014', 'date': '02-21', 'aqi': 60},
  {'site': '0014', 'date': '01-15', 'aqi': None}]}
```

We can quickly convert the array of objects into a dataframe with the following call:

```
pd.DataFrame(data_row["Data"])
```

	site	date	aqi
0	0014	02-27	30.0
1	0014	02-24	17.0
2	0014	02-21	60.0
3	0014	01-15	NaN

The middle diagram in Figure 14-2 takes a column approach to organizing the data. Here the columns are provided as arrays and collected into an object with names that match the column names. The following file demonstrates the concept:

```
epa_col_path = Path('data/js_ex/epa_col.json')
print(epa_col_path.read_text())
```

```
{"site":[ "0014", "0014", "0014", "0014"],
"date":["02-27", "02-24", "02-21", "01-15"],
"aqi":[30,17,60,null]}
```

Since `pd.read_json()` expects this format, we can read the file into a dataframe directly without needing to first load it into a dictionary:

```
pd.read_json(epa_col_path)
```

	site	date	aqi
0	14	02-27	30.0
1	14	02-24	17.0
2	14	02-21	60.0
3	14	01-15	NaN

Lastly, we organize the data into a structure that resembles a matrix (the diagram on the right in the figure) and separately provide the column names for the features. The data matrix is organized as an array of arrays:

```
{'vars': ['site', 'date', 'aqi'],
 'data': [['0014', '02-27', 30],
  ['0014', '02-24', 17],
  ['0014', '02-21', 60],
  ['0014', '01-15', None]]}
```

We can provide `vars` and `data` to create the dataframe:

```
pd.DataFrame(data_mat["data"], columns=data_mat["vars"])
```

	site	date	aqi
0	0014	02-27	30.0
1	0014	02-24	17.0
2	0014	02-21	60.0
3	0014	01-15	NaN

We've included these examples to show the versatility of JSON. The main takeaway is that JSON files can arrange data in different ways, so we typically need to examine the file before we can read the data into a dataframe successfully. JSON files are very common for data stored on the web: the examples in this section were files downloaded from the PurpleAir and Kiva websites. Although we downloaded the data manually in this section, we often want to download many datafiles at a time, or we want a reliable and reproducible record of the download. In the next section, we introduce HTTP, a protocol that will let us write programs to download data from the web automatically.

HTTP

HTTP (HyperText Transfer Protocol) is an all-purpose infrastructure to access resources on the web. There are a tremendous number of datasets available to us on the internet, and with HTTP we can acquire these datasets.

The internet allows computers to communicate with each other, and HTTP places a structure on the communication. HTTP is a simple *request-response* protocol, where a client submits a *request* to a server in a specially formatted text message, and the server sends a specially formatted text *response* back. The client might be a web browser or our Python session.

An HTTP request has two parts: a header and an optional body. The header must follow a specific syntax. An example request to obtain the Wikipedia page shown in Figure 14-3 looks like the following:

```
GET /wiki/1500_metres_world_record_progression HTTP/1.1
Host: en.wikipedia.org
User-Agent: curl/7.65.2
Accept: */*
{blank_line}
```

The first line contains three pieces of information: it starts with the method of the request, which is `GET`; this is followed by the URL of the web page we want; and last is the protocol and version. Each of the three lines that follow give auxiliary information for the server. This information has the format `name: value`. Finally, a blank line marks the end of the header. Note that we've marked the blank line with `{blank_line}` in the preceding snippet; in the actual message, this is a blank line.

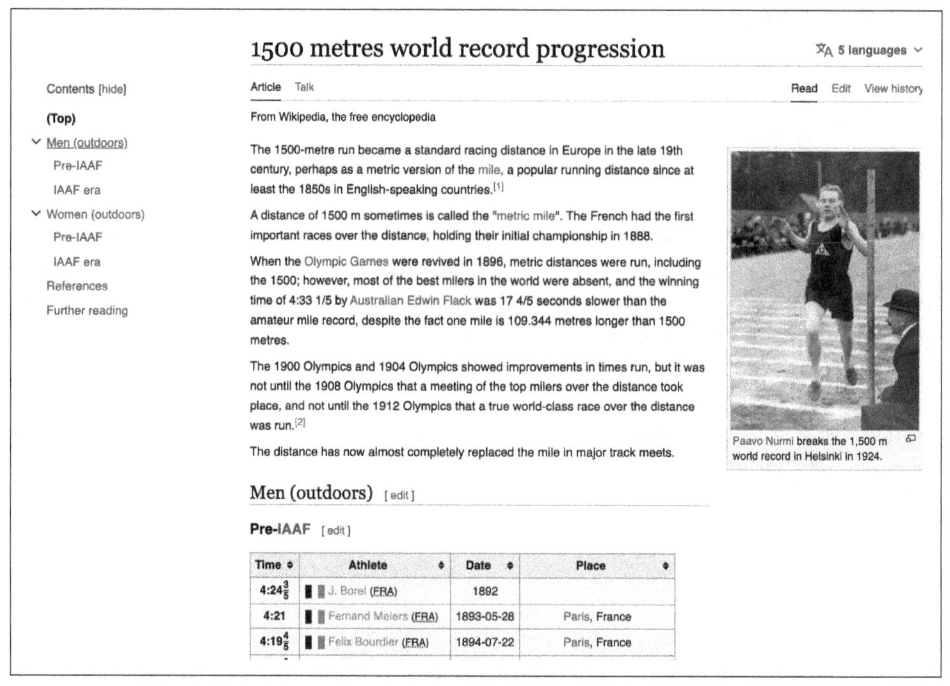

1500 metres world record progression

5 languages

Contents [hide]

(Top)

Men (outdoors)

Pre-IAAF

IAAF era

Women (outdoors)

Pre-IAAF

IAAF era

References

Further reading

Article Talk Read Edit View history

From Wikipedia, the free encyclopedia

The 1500-metre run became a standard racing distance in Europe in the late 19th century, perhaps as a metric version of the mile, a popular running distance since at least the 1850s in English-speaking countries.[1]

A distance of 1500 m sometimes is called the "metric mile". The French had the first important races over the distance, holding their initial championship in 1888.

When the Olympic Games were revived in 1896, metric distances were run, including the 1500; however, most of the best milers in the world were absent, and the winning time of 4:33 1/5 by Australian Edwin Flack was 17 4/5 seconds slower than the amateur mile record, despite the fact one mile is 109.344 metres longer than 1500 metres.

The 1900 Olympics and 1904 Olympics showed improvements in times run, but it was not until the 1908 Olympics that a meeting of the top milers over the distance took place, and not until the 1912 Olympics that a true world-class race over the distance was run.[2]

The distance has now almost completely replaced the mile in major track meets.

Paavo Nurmi breaks the 1,500 m world record in Helsinki in 1924.

Men (outdoors) [edit]

Pre-IAAF [edit]

Time	Athlete	Date	Place
$4{:}24\frac{3}{5}$	J. Borel (FRA)	1892	
4:21	Fernand Meiers (FRA)	1893-05-28	Paris, France
$4{:}19\frac{4}{5}$	Felix Bourdier (FRA)	1894-07-22	Paris, France

Figure 14-3. Screenshot of the Wikipedia page with data on the world record for the 1,500-meter race

The client's computer sends this message over the internet to the Wikipedia server. The server processes the request and sends a response, which also consists of a header and body. The header for the response looks like this:

```
< HTTP/1.1 200 OK
< date: Fri, 24 Feb 2023 00:11:49 GMT
< server: mw1369.eqiad.wmnet
< x-content-type-options: nosniff
< content-language: en
< vary: Accept-Encoding,Cookie,Authorization
< last-modified: Tue, 21 Feb 2023 15:00:46 GMT
< content-type: text/html; charset=UTF-8
...
< content-length: 153912
{blank_line}
```

The first line states that the request completed successfully; the status code is 200. The next lines give additional information for the client. We shortened this header quite a bit to focus on just a few pieces of information that tell us the content of the body is HTML and uses UTF-8 encoding, and the content is 153,912 characters long. Finally, the blank line at the end of the header tells the client that the server has finished sending header information, and the response body follows.

HTTP is used in almost every application that interacts with the internet. For example, if you visit this same Wikipedia page in your web browser, the browser makes the same basic HTTP request as the one just shown. When it receives the response, it displays the body in your browser's window, which looks like the screenshot in Figure 14-3.

In practice, we do not write out full HTTP requests ourselves. Instead, we use tools like the `requests` Python library to construct requests for us. The following code constructs the HTTP request for the page in Figure 14-3 for us. We simply pass the URL to `requests.get`. The "get" in the name indicates the `GET` method is being used:

```
import requests

url_1500 = 'https://en.wikipedia.org/wiki/1500_metres_world_record_progression'

resp_1500 = requests.get(url_1500)
```

We can check our request's status to make sure the server completed it successfully:

```
resp_1500.status_code

200
```

We can thoroughly examine the request and response through the object's attributes. As an example, let's take a look at the key-value pairs in the header in our request:

```
for key in resp_1500.request.headers:
    print(f'{key}: {resp_1500.request.headers[key]}')

User-Agent: python-requests/2.25.1
Accept-Encoding: gzip, deflate
Accept: */*
Connection: keep-alive
```

Although we did not specify any header information in our function call, `request.get` provided some basic information for us. If we need to send special header information, we can specify them in our call.

Now let's examine the header of the response we received from the server:

```
len(resp_1500.headers)

20
```

As we saw earlier, there's a lot of header information in the response. We just display the `date`, `content-type`, and `content-length`:

```
keys = ['date', 'content-type', 'content-length' ]
for key in keys:
    print(f'{key}: {resp_1500.headers[key]}')

date: Fri, 10 Mar 2023 01:54:13 GMT
content-type: text/html; charset=UTF-8
content-length: 23064
```

Finally, we display the first several hundred characters of the response body (the entire content is too long to display nicely here):

```
resp_1500.text[:600]
```

```
'<!DOCTYPE html>\n<html class="client-nojs vector-feature-language-in-header-
enabled vector-feature-language-in-main-page-header-disabled vector-feature-
language-alert-in-sidebar-enabled vector-feature-sticky-header-disabled vector-
feature-page-tools-disabled vector-feature-page-tools-pinned-disabled vector-
feature-toc-pinned-enabled vector-feature-main-menu-pinned-disabled vector-
feature-limited-width-enabled vector-feature-limited-width-content-enabled"
lang="en" dir="ltr">\n<head>\n<meta charset="UTF-8"/>\n<title>1500 metres world
record progression - Wikipedia</title>\n<script>document.documentE'
```

We confirm that the response is an HTML document and that it contains the title `1500 metres world record progression - Wikipedia`. We have successfully retrieved the web page shown in Figure 14-3.

Our HTTP request has been successful, and the server has returned a status code of `200`. There are hundreds of other HTTP status codes. Thankfully, they are grouped into categories to make them easier to remember (see Table 14-1).

Table 14-1. Response status codes

Code	Type	Description
100s	Informational	More input is expected from the client or server (100 Continue, 102 Processing, etc.).
200s	Success	The client's request was successful (200 OK, 202 Accepted, etc.).
300s	The redirection	Requested URL is located elsewhere and may need further action from the user (300 Multiple Choices, 301 Moved Permanently, etc.).
400s	Client error	A client-side error occurred (400 Bad Request, 403 Forbidden, 404 Not Found, etc.).
500s	Server error	A server-side error occurred or the server is incapable of performing the request (500 Internal Server Error, 503 Service Unavailable, etc.).

One common error code that might look familiar is 404, which tells us we have requested a resource that doesn't exist. We send such a request here:

```
url = "https://www.youtube.com/404errorwow"
bad_loc = requests.get(url)
bad_loc.status_code
```

```
404
```

The request we made to retrieve the web page was a `GET` HTTP request. There are four main HTTP request types: `GET`, `POST`, `PUT`, and `DELETE`. The two most commonly used methods are `GET` and `POST`. We just used `GET` to retrieve the web page:

```
resp_1500.request.method
```

```
'GET'
```

The `POST` request is used to send specific information from the client to the server. In the next section, we use `POST` to retrieve data from Spotify.

REST

Web services are increasingly implementing the REST (REpresentational State Transfer) architecture for developers to access their data. These include social media platforms like Twitter and Instagram, music apps like Spotify, real estate apps like Zillow, scientific sources of data such as the Climate Data Store, government data at the World Bank, and many, many more. The basic idea behind REST is that every URL identifies a resource (data).

REST is *stateless*, meaning that the server does not remember the client from one request to the next. This aspect of REST has a few advantages: the server and the client can understand any message received without seeing previous messages, code can be changed on either the client or server side without impacting the operation of the service, and access is scalable, fast, modular, and independent.

In this section, we work through an example to retrieve data from Spotify.

Our example follows Steven Morse's blog post (*https://oreil.ly/zI-5z*), where we use both `POST` and `GET` methods in a series of requests to retrieve data on songs by The Clash (*https://www.theclash.com*).

In practice, we wouldn't write `GET` and `POST` requests ourselves for Spotify. Instead, we'd use the `spotipy` (*https://oreil.ly/fPQX0*) library, which has functions to interact with the Spotify web API (*https://oreil.ly/NH4ZO*). That said, data scientists can often find themselves in the position of wanting to access data available via REST that doesn't have a Python library available, so this section shows how to get data from a RESTful website like Spotify.

Typically, a REST application provides documentation with examples on how to request its data. Spotify has extensive documentation geared to developers who want to build an app, but we can also access the service just to explore data. To do that, we need to register as a developer and get a client ID and secret. We then use these to identify us to Spotify in our HTTP requests.

After we register, we can begin to request data. This process has two steps: authenticate and request resources.

To authenticate, we issue a POST request, where we give the web service our client ID and secret. We provide these in the header of the request. In return, we receive a token from the server that authorizes us to make requests.

We begin the process and authenticate:

```
AUTH_URL = 'https://accounts.spotify.com/api/token'

import requests
auth_response = requests.post(AUTH_URL, {
    'grant_type': 'client_credentials',
    'client_id': CLIENT_ID,
    'client_secret': CLIENT_SECRET,
})
```

We provided our ID and secret in key-value pairs in the header of our POST request. We can check the status of our request to see if it was successful:

```
auth_response.status_code
```

```
200
```

Now let's check the type of content in the body of the response:

```
auth_response.headers['content-type']
```

```
'application/json'
```

The body of the response contains the token that we need in the next step to get the data. Since this information is JSON-formatted, we can check the keys and retrieve the token:

```
auth_response_data = auth_response.json()
auth_response_data.keys()
```

```
dict_keys(['access_token', 'token_type', 'expires_in'])
```

```
access_token = auth_response_data['access_token']
token_type = auth_response_data['token_type']
```

Notice that we hid our ID and secret so that others reading this book can't imitate us. This request won't be successful without a valid ID and secret. For example, here we make up an ID and secret and try to authenticate:

```
bad_ID = '0123456789'
bad_SECRET = 'a1b2c3d4e5'

auth_bad = requests.post(AUTH_URL, {
    'grant_type': 'client_credentials',
    'client_id': bad_ID, 'client_secret': bad_SECRET,
})
```

We check the status of this "bad" request:

```
auth_bad.status_code
```

```
400
```

According to Table 14-1, a code of 400 means that we issued a bad request. For one more example, Spotify shuts us down if we take too much time making requests. We ran into this issue a couple of times when writing this section and received the following code, telling us our token had expired:

```
res_clash.status_code
```

```
401
```

Now for the second step, let's get some data.

Requests for resources can be made via `GET` for Spotify. Other services may require POSTs. Requests must include the token we received from the web service when we authenticated, which we can use over and over. We pass the access token in the header of our `GET` request. We construct the name-value pairs as a dictionary:

```
headers = {"Authorization": f"{token_type} {access_token}"}
```

The developer API tells us that an artist's albums are available at URLs that look like *https://api.spotify.com/v1/artists/3RGLhK1IP9jnYFH4BRFJBS/albums*, where the code between *artists/* and */albums* is an artist's ID. This particular code is for The Clash. Information about the tracks on an album is available at a URL that looks like *https://api.spotify.com/v1/albums/49kzgMsxHU5CTeb2XmFHjo/tracks*, where the identifier here is for the album.

If we know the ID for an artist, we can retrieve the IDs for its albums, and in turn, we can get data about the tracks on the albums. Our first step was to get the ID for The Clash from Spotify's site:

```
artist_id = '3RGLhK1IP9jnYFH4BRFJBS'
```

Our first data request retrieves the group's albums. We construct the URL using `artist_id` and pass our access token in the header:

```
BASE_URL = "https://api.spotify.com/v1/"

res_clash = requests.get(
    BASE_URL + "artists/" + artist_id + "/albums",
    headers=headers,
    params={"include_groups": "album"},
)
```

```
res_clash.status_code
```

```
200
```

Our request was successful. Now let's check the `content-type` of the response body:

```
res_clash.headers['content-type']
```

```
'application/json; charset=utf-8'
```

The resource returned is JSON, so we can load it into a Python dictionary:

```
clash_albums = res_clash.json()
```

After poking around a bit, we can find that album information is in the `items` element. The keys for the first album are:

```
clash_albums['items'][0].keys()
```

```
dict_keys(['album_group', 'album_type', 'artists', 'available_markets', 'exter-
nal_urls', 'href', 'id', 'images', 'name', 'release_date', 'release_date_preci-
sion', 'total_tracks', 'type', 'uri'])
```

Let's print the album IDs, names, and release dates for a few albums:

```
for album in clash_albums['items'][:4]:
    print('ID: ', album['id'], ' ', album['name'], '----', album['release_date'])
```

```
ID:  7nL9UERtRQCB5eWEQCINsh   Combat Rock + The People's Hall ---- 2022-05-20
ID:  3un5bLdxz0zKhiZXlmnxWE   Live At Shea Stadium ---- 2008-08-26
ID:  4dMWTj1OkiCKFN5yBMP1vS   Live at Shea Stadium (Remastered) ---- 2008
ID:  1Au9637RH9pXjBv5uS3JpQ   From Here To Eternity Live ---- 1999-10-04
```

We see that some albums are remastered and others are live performances. Next, we cycle through the albums, pick up their IDs, and for each album we request information about the tracks:

```
tracks = []

for album in clash_albums['items']:
    tracks_url = f"{BASE_URL}albums/{album['id']}/tracks"
    res_tracks = requests.get(tracks_url, headers=headers)
    album_tracks = res_tracks.json()['items']

    for track in album_tracks:
        features_url = f"{BASE_URL}audio-features/{track['id']}"
        res_feat = requests.get(features_url, headers=headers)
        features = res_feat.json()

        features.update({
            'track_name': track.get('name'),
            'album_name': album['name'],
            'release_date': album['release_date'],
            'album_id': album['id']
        })

        tracks.append(features)
```

Over a dozen features are available to explore on the tracks. Let's close the example with a plot of danceability and loudness of The Clash songs:

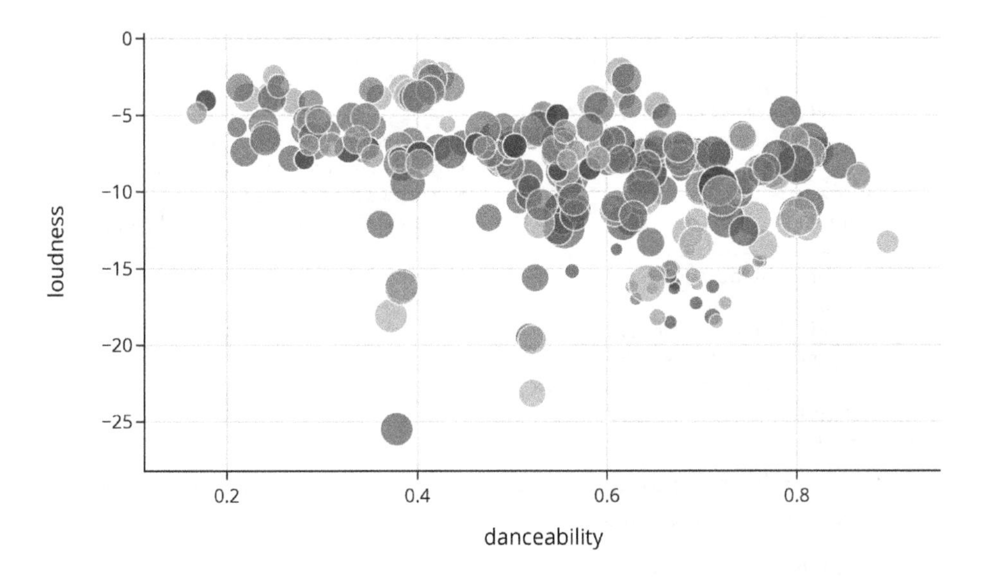

This section covered REST APIs, which provide standardized approaches for programs to download data. The example shown here downloaded JSON data. At other times, the data from a REST request may be in an XML format. And sometimes a REST API isn't available for the data we want, and we must extract the data from web pages themselves in HTML, a format similar to XML. We describe how to work with these formats next.

XML, HTML, and XPath

The eXtensible Markup Language (XML) can represent all types of information, such as data sent to and from web services, including web pages, spreadsheets, visual displays like SVG, social network structures, word processing documents like Microsoft's docx, databases, and much more. For a data scientist, knowing a little about XML can come in handy.

Despite its name, XML is not a language. Rather, it is a very general structure we can use to define formats to represent and organize data. XML provides a basic structure and syntax for these "dialects" or vocabularies. If you read or compose HTML, you will recognize the format of XML.

The basic unit in XML is the *element*, which is also referred to as a *node*. An element has a name and may have attributes, child elements, and text.

The following annotated snippet of an XML plant catalog provides an example of these pieces (this content is adapted from W3Schools (*https://oreil.ly/qPa6s*)):

```
<catalog>                                   The topmost node, aka root node.
    <plant>                                 The first child of the root node.
```

```
        <common>Bloodroot</common>      common is the first child of plant.
        <botanical>Sanguinaria canadensis</botanical>
        <zone>4</zone>                  This zone node has text content: 4.
        <light>Mostly Shady</light>
        <price curr="USD">$2.44</price> This node has an attribute.
        <availability date="0399"/>     Empty nodes can be collapsed.
    </plant>                            Nodes must be closed.
    <plant>                             The two plant nodes are siblings.
        <common>Columbine</common>
        <botanical>Aquilegia canadensis</botanical>
        <zone>3</zone>
        <light>Mostly Shady</light>
        <price curr="CAD">$9.37</price>
        <availability date="0199"/>
    </plant>
</catalog>
```

We added the indentation to this snippet of XML to make it easier to see the structure. It is not needed in the actual file.

XML documents are plain-text files with the following syntax rules:

- Each element begins with a start tag, like `<plant>`, and closes with an end tag of the same name, like `</plant>`.
- XML elements can contain other XML elements.
- XML elements can be plain-text, like "Columbine" in `<common>Columbine</common>`.
- XML elements can have optional attributes. The element `<price curr="CAD">` has an attribute `curr` with value `"CAD"`.
- In the special case when a node has no children, the end tag can be folded into the start tag. An example is `<availability date="0199"/>`.

We call an XML document well formed when it follows certain rules. The most important of these are:

- One root node contains all of the other elements in the document.
- Elements nest properly; an open node closes around all of its children and no more.
- Tag names are case-sensitive.
- Attribute values have a `name="value"` format with single or double quotes.

There are additional rules for a document to be well formed. These relate to whitespace, special characters, naming conventions, and repeated attributes.

The hierarchical nature of well-formed XML means it can be represented as a tree. Figure 14-4 shows a tree representation of the plant catalog XML.

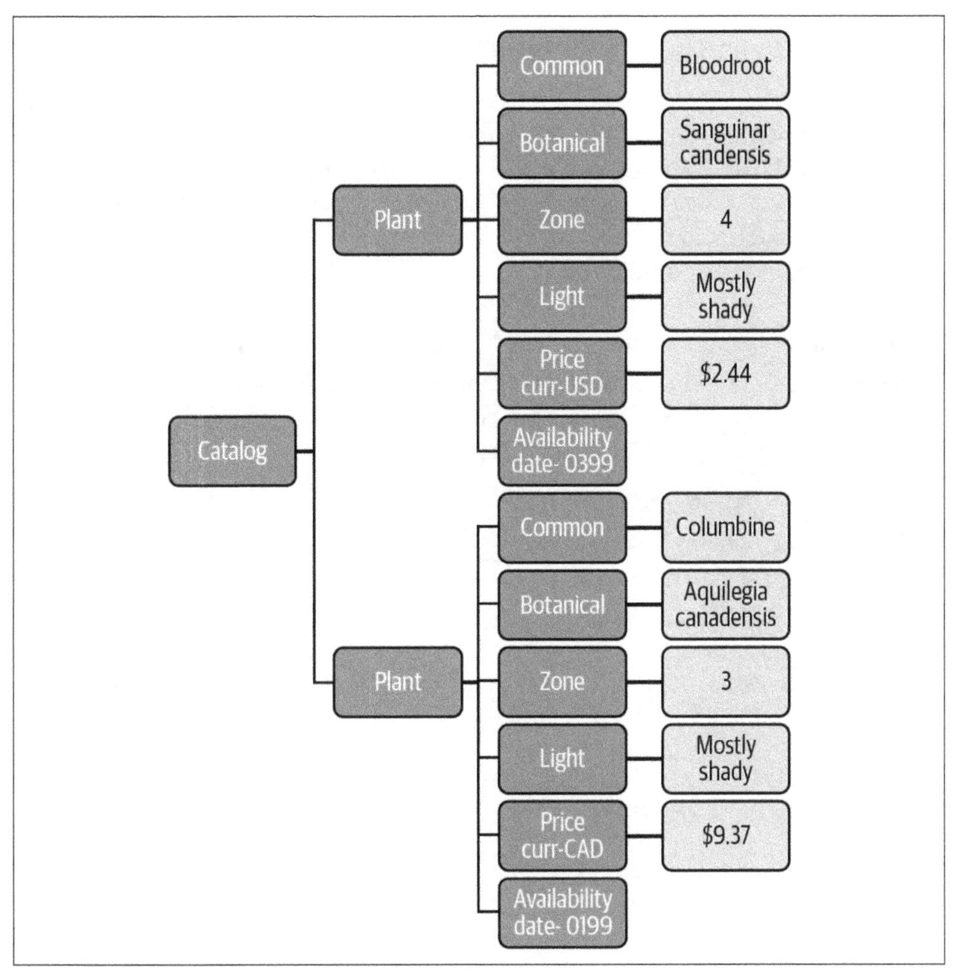

Figure 14-4. Hierarchy of an XML document; the lighter gray boxes represent text elements and, by design, these cannot have child nodes

Like with JSON, an XML document is plaintext. We can read it with a plain-text viewer, and it's easy for machines to read and create XML content. The extensible nature of XML allows content to be easily merged into higher-level container documents and easily exchanged with other applications. XML also supports binary data and arbitrary character sets.

As mentioned already, HTML looks a lot like XML. That's no accident, and indeed, XHTML is a subset of HTML that follows the rules of well-formed XML. Let's return to our earlier example of the Wikipedia page that we retrieved from the internet and show how to used XML tools to create a dataframe from the contents of one of its tables.

Example: Scraping Race Times from Wikipedia

Earlier in this chapter, we used an HTTP request to retrieve the HTML page from Wikipedia shown in Figure 14-3. The contents of this page are in HTML, which is essentially an XML vocabulary. We can use the hierarchical structure of the page and XML tools to access data in one of the tables and wrangle it into a dataframe. In particular, we are interested in the second table in the page, a portion of which appears in the screenshot in Figure 14-5.

Time ◆	Auto ◆	Athlete ◆	Date ▲	Place ◆
3:55.8		Abel Kiviat (USA)	1912-06-08	Cambridge, Massachusetts, United States
3:54.7		John Zander (SWE)	1917-08-05	Stockholm, Sweden
3:52.6		Paavo Nurmi (FIN)	1924-06-19	Helsinki, Finland
3:51.0		Otto Peltzer (GER)	1926-09-11	Berlin, Germany
3:49.2		Jules Ladoumegue (FRA)	1930-10-05	Paris, France
3:49.2		Luigi Beccali (ITA)	1933-09-09	Turin, Italy
3:49.0		Luigi Beccali (ITA)	1933-09-17	Milan, Italy

Figure 14-5. Screenshot of the second table in a web page that contains the data we want to extract

Before we work on this table, we provide a quick summary of the format for a basic HTML table. Here is the HTML for a table with a header and two rows of three columns:

```
<table>
 <tbody>
  <tr>
   <th>A</th><th>B</th><th>C</th>
  </tr>
  <tr>
   <td>1</td><td>2</td><td>3</td>
  </tr>
  <tr>
   <td>5</td><td>6</td><td>7</td>
  </tr>
 </tbody>
</table>
```

Notice how the table is laid out in rows with `<tr>` elements, and each cell in a row is a `<td>` element that contains the text to be displayed in the table.

Our first task is to create a tree structure from the content of the web page. To do this, we use the `lxml` library, which provides access to the C-library `libxml2` for handling XML content. Recall that `resp_1500` contains the response from our request, and the page is in the body of the response. We can parse the web page into a hierarchical structure with `fromstring` in the `lxml.html` module:

```
from lxml import html
```

```
tree_1500 = html.fromstring(resp_1500.content)
```

```
type(tree_1500)
```

```
lxml.html.HtmlElement
```

Now we can work with the document using its tree structure. We can find all the tables in the HTML document with the following search:

```
tables = tree_1500.xpath('//table')
type(tables)
```

```
list
```

```
len(tables)
```

```
7
```

This search uses the XPath `//table` expression, which we soon describe, to search for all table nodes anywhere in the document.

We found six tables in the document. If we examine the web page, including looking at its HTML source via the browser, we can figure out that the second table in the document contains the IAF-era times. This is the table we want. The screenshot in Figure 14-5 shows that the first column contains the race times, the third holds names, and the fourth has the dates of the races. We can extract each of these pieces of information in turn. We do this with the following XPath expressions:

```
times = tree_1500.xpath('//table[3]/tbody/tr/td[1]/b/text()')
names = tree_1500.xpath('//table[3]/tbody/tr/td[3]/a/text()')
dates = tree_1500.xpath('//table[3]/tbody/tr/td[4]/text()')
```

```
type(times[0])
```

```
lxml.etree._ElementUnicodeResult
```

These return values behave like a list, but each value is an element of the tree. We can convert them to strings:

```
date_str = [str(s) for s in dates]
name_str = [str(s) for s in names]
```

For the times, we want to transform them into seconds. The function `get_sec` does this conversion. And we want to extract the race year from the date string:

```
def get_sec(time):
    """convert time into seconds."""
```

```
        time = str(time)
        time = time.replace("+","")
        m, s = time.split(':')
        return float(m) * 60 + float(s)

    time_sec = [get_sec(rt) for rt in times]
    race_year = pd.to_datetime(date_str, format='%Y-%m-%d\n').year
```

We can create a dataframe and make a plot to show the progress in race times over the years:

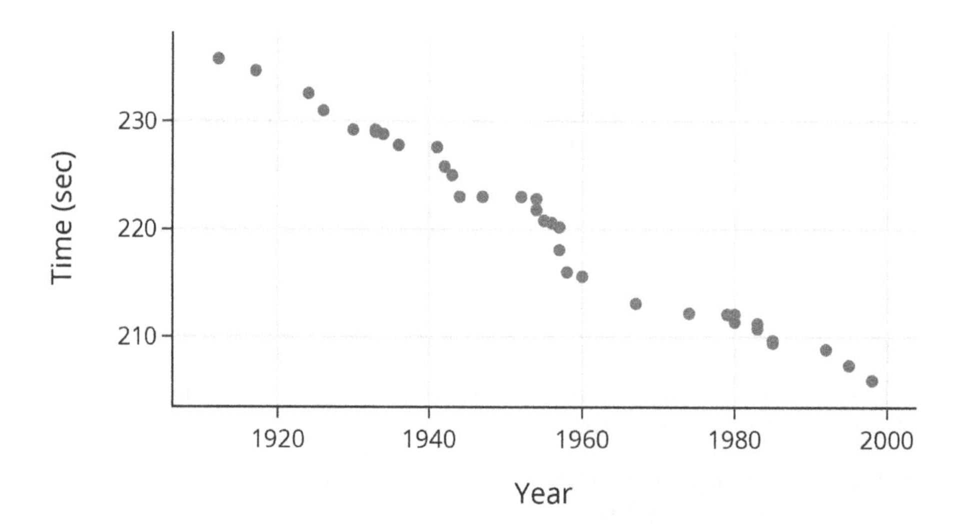

As you may have noticed, extracting data from an HTML page relies on careful examination of the source to find where in the document the numbers that we're after are. We relied heavily on the XPath tool to do the extraction. Its elegant language is quite powerful. We introduce it next.

XPath

When we work with XML documents, we typically want to extract data from them and bring it into a dataframe. XPath can help here. XPath can recursively traverse an XML tree to find elements. For example, we used the expression `//table` in the previous example to locate all table nodes in our web page.

XPath expressions operate on the hierarchy of well-formed XML. They are succinct and similar in format to the way files are located in a hierarchy of directories in a computer filesystem. But they're much more powerful. XPath is also similar to regular expressions in that we specify patterns to match content. Like with regular expressions, it takes experience to compose correct XPath expressions.

An XPath expression forms logical steps to identify and filter nodes in a tree. The result is a *node set* where each node occurs at most once. The node set also has an order that matches the order in which the nodes occur in the source; this can be quite handy.

Each XPath expression is made up of one or more *location steps*, separated by a "/". Each location step has three parts—the *axis*, *node test*, and optional *predicate*:

- The axis specifies the direction to look in, such as down, up, or across the tree. We exclusively use shortcuts for the axis. The default is to look down one step at children in the tree. `//` says to look down the tree as far as possible, and `..` indicates one step up to the parent.
- The node test identifies the name or the type of node to look for. This is typically just a tag name or `text()` for text elements.
- A predicate acts like a filter to further restrict the node set. This is given in square brackets, like `[2]`, which keeps the second node in the node set, and `[ @date ]`, which keeps all nodes with a date attribute.

We can tack together location steps to create powerful search instructions. Table 14-2 provides some examples that cover the most common expressions. Refer back to the tree in Figure 14-4 to follow along.

Table 14-2. XPath examples

Expression	Result	Description
'//common'	Two nodes	Look down the tree for any common nodes.
'/catalog/plant/common'	Two nodes	Travel the specific path from the root node *catalog* to all plant nodes to all common nodes within the plant nodes.
'//common/text()'	Bloodroot, Columbine	Locate the text content of all common nodes.
'//plant[2]/price/text()'	$9.37	Locate plant nodes anywhere in the tree, then filter to take only the second. From this plant node, travel to its price child and locate its text.
'//@date'	0399, 0199	Locate the attribute value of any attribute named "date" in the tree.
'//price[@curr="CAD"]/text()'	$9.37	The text content of any price node that has a currency attribute value of "CAD."

You can try out the XPath expressions in the table with the catalog file. We load the file into Python using the `etree` module. The `parse` method reads the file into an element tree:

```
from lxml import etree

catalog = etree.parse('data/catalog.xml')
```

The `lxml` library gives us access to XPath. Let's try it out.

This simple XPath expression locates all text content of any `<light>` node in the tree:

```
catalog.xpath('//light/text()')
```

```
['Mostly Shady', 'Mostly Shady']
```

Notice that two elements are returned. Although the text content is identical, we have two `<light>` nodes in our tree and so are given the text content of each. The following expression is a bit more challenging:

```
catalog.xpath('//price[@curr="CAD"]/../common/text()')
```

```
['Columbine']
```

The expression locates all `<price>` nodes in the tree, then filters them according to whether their `curr` attribute is `CAD`. Then, for the remaining nodes (there's only one in this case), travel up one step in the tree to the parent node and then back down to any child "common" nodes and on to their text content. Quite the trip!

Next, we provide an example that uses an HTTP request to retrieve XML-formatted data, and XPath to wrangle the content into a dataframe.

Example: Accessing Exchange Rates from the ECB

The European Central Bank (ECB) makes exchange rates available online in XML format. Let's begin by getting the most recent exchange rates from the ECB with an HTTP request:

```
url_base = 'https://www.ecb.europa.eu/stats/eurofxref/'
url2 = 'eurofxref-hist-90d.xml?d574942462c9e687c3235ce020466aae'
resECB = requests.get(url_base+url2)
```

```
resECB.status_code
```

```
200
```

Again, we can use the `lxml` library to parse the text document we received from the ECB, but this time the contents are in a string returned from the ECB, not in a file:

```
ecb_tree = etree.fromstring(resECB.content)
```

In order to extract the data we want, we need to know how it is organized. Here is a snippet of the content:

```
<gesmes:Envelope xmlns:gesmes="http://www.gesmes.org/xml/2002-08-01"
       xmlns="http://www.ecb.int/vocabulary/2002-08-01/eurofxref">
<gesmes:subject>Reference rates</gesmes:subject>
<gesmes:Sender>
<gesmes:name>European Central Bank</gesmes:name>
</gesmes:Sender>
<Cube>
<Cube time="2023-02-24">
<Cube currency="USD" rate="1.057"/>
<Cube currency="JPY" rate="143.55"/>
```

```
<Cube currency="BGN" rate="1.9558"/>
</Cube>
<Cube time="2023-02-23">
<Cube currency="USD" rate="1.0616"/>
<Cube currency="JPY" rate="143.32"/>
<Cube currency="BGN" rate="1.9558"/>
</Cube>
</Cube>
</gesmes:Envelope>
```

This document appears quite different in structure from the plant catalog. The snippet shows three levels of tags, all with the same name, and none have text content. All of the relevant information is contained in attribute values. Other new features are the `xmlns` in the root `<Envelope>` node, and the odd tag names, like `gesmes:Envelope`. These have to do with namespaces.

XML allows content creators to use their own vocabularies, called *namespaces*. The namespace gives the rules for a vocabulary, such as allowable tag names and attribute names, and restrictions on how nodes can be nested. And XML documents can merge vocabularies from different applications. To keep it all straight, information about the namespace(s) is provided in the document.

The root node in the ECB file is `<Envelope>`. The additional "gesmes:" in the tag name indicates that the tags belong to the gesmes vocabulary, which is an international standard for the exchange of time-series information. Another namespace is also in `<Envelope>`. It is the default namespace for the file because it doesn't have a prefix, like "`gesmes:`". Whenever a namespace is not provided in a tag name, the default is assumed.

The upshot of this is that we need to take into account these namespaces when we search for nodes. Let's see how this works when we extract the dates. From the snippet, we see that the dates reside in "time" attributes. These `<Cube>`s are children of the top `<Cube>`. We can give a very specific XPath expression to step from the root to its `<Cube>` child node and on to the next level of `<Cube>` nodes:

```
namespaceURI = 'http://www.ecb.int/vocabulary/2002-08-01/eurofxref'

date = ecb_tree.xpath('./x:Cube/x:Cube/@time', namespaces = {'x':namespaceURI})
date[:5]
```

```
['2023-07-18', '2023-07-17', '2023-07-14', '2023-07-13', '2023-07-12']
```

The `.` in the expression is a shortcut to signify "from here," and since we're at the top of the tree, it's equivalent to "from the root." We specified the namespace in our expression as "`x:`". Even though the `<Cube>` nodes are using the default namespace, we must specify it in our XPath expression. Fortunately, we can simply pass in the namespace as a parameter with our own label ("x" in this case) to keep our tag names short.

Like with the HTML table, we can convert the date values into strings and from strings into timestamps:

```
date_str = [str(s) for s in date]
timestamps = pd.to_datetime(date_str)
xrates = pd.DataFrame({"date":timestamps})
```

As for the exchange rates, they also appear in `<Cube>` nodes, but these have a "rate" attribute. For example, we can access all exchange rates for the British pound with the following XPath expression (we're ignoring the namespace for the moment):

```
//Cube[@currency = "GBP"]/@rate
```

This expression says look for all `<Cube>` nodes anywhere in the document, filter them according to whether the node has a currency attribute value of "GBP," and return their rate attribute values.

Since we want to extract exchange rates for multiple currencies, we generalize this XPath expression. We also want to convert the exchange rates to a numeric storage type, and make them relative to the first day's rate so that the different currencies are on the same scale, which makes them more amenable for plots:

```
currs = ['GBP', 'USD', 'CAD']

for ctry in currs:
    expr = './/x:Cube[@currency = "' + ctry + '"]/@rate'
    rates = ecb_tree.xpath(expr, namespaces = {'x':namespaceURI})
    rates_num = [float(rate) for rate in rates]
    first = rates_num[len(rates_num)-1]
    xrates[ctry] = [rate / first for rate in rates_num]
```

We wrap up this example with line plots of the exchange rates:

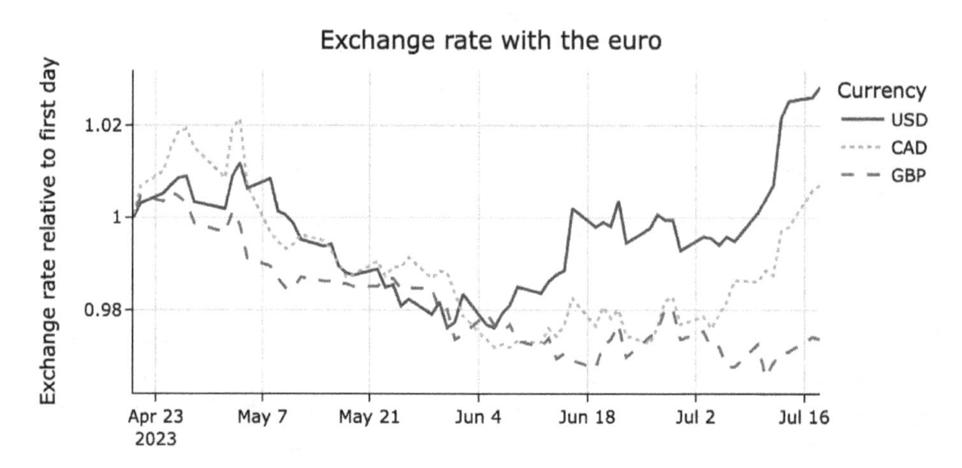

Combining knowledge of JSON, HTTP, REST, and HTML gives us access to a vast variety of data available on the web. For example, in this section we wrote code to scrape data from a Wikipedia page. One key advantage of this approach is that we can likely rerun this code in a few months to automatically update the data and the plots. One key drawback is that our approach is tightly coupled to the structure of the web page—if someone updates the Wikipedia page and the table is no longer the second table on the page, our code will also need some edits in order to work. That said, having the skills needed to scrape data from the web opens the door to a wide range of data and enables all kinds of useful analyses.

Summary

The internet abounds with data that are stored and exchanged in many different formats. In this chapter, our aim was to give you a taste of the variety of formats available and a basic understanding of how to acquire data from online sources and services. We also addressed the important goal of acquiring data in a reproducible fashion. Rather than copying and pasting from a web page or completing a form by hand, we demonstrated how to write code to acquire data. This code gives you a record of your workflow and of the data provenance.

With each format introduced, we described a model for its structure. A basic understanding of a dataset's organization helps you uncover issues with quality, mistakes in reading a source file, and how best to wrangle and analyze the data. In the longer run, as you continue to develop your data science skills, you will be exposed to other forms of data exchange, and we expect this approach of considering the organizational model and getting your hands dirty with some simple cases will serve you well.

We only touched the surface of web services. There are many other useful topics, like keeping connections to a server alive as you issue multiple requests or retrieve data in batches, using cookies, and making multiple connections. But understanding the basics presented here can get you a long way. For example, if you use a library to retrieve data from an API but run into an error, you can start looking at the HTTP requests to debug your code. And you will know what's possible when a new web service comes online.

Web etiquette is a topic that we must mention. If you plan to scrape data from a website, it's a good idea to check that you have permission to do so. When we sign up to be a client for a web app, we typically check a box indicating our agreement to the terms of service.

If you use a web service or scrape web pages, be careful not to overburden the site with your requests. If a site offers a version of the data in a format like CSV, JSON, or XML, it's better to download and use these than to scrape from a web page. Likewise, if there is a Python library that provides structured access to a web app, use it rather than writing your own code. When you make requests, start small to test your code, and consider saving the results so that you don't have to repeat requests unnecessarily.

The aim of this chapter wasn't to make you an expert in these specific data formats. Instead, we wanted to give you the confidence needed to learn more about a data format, to evaluate the pros and cons of different formats, and to participate in projects that might use formats that you haven't seen before.

Now that you have experience working with different data formats, we return to the topic of modeling that we introduced in Chapter 4, picking it back up in earnest.

PART V
Linear Modeling

CHAPTER 15

Linear Models

At this point in the book, we've covered the four stages of the data science lifecycle to different extents. We've talked about formulating questions and obtaining and cleaning data, and we've used exploratory data analysis to better understand the data. In this chapter, we extend the constant model introduced in Chapter 4 to the *linear model*. Linear models are a popular tool in the last stage of the lifecycle: understanding the world.

Knowing how to fit linear models opens the door to all kinds of useful data analyses. We can use these models to make *predictions*—for example, environmental scientists developed a linear model to predict air quality based on air sensor measurements and weather conditions (see Chapter 12). In that case study, understanding how measurements from two instruments varied enabled us to calibrate inexpensive sensors and improve their air quality readings. We can also use these models to make *inferences* about the form of a relationship between features—for example, in Chapter 18 we'll see how veterinarians used a linear model to infer the coefficients for length and girth for a donkey's weight: $Length + 2 \times Girth - 175$. In that case study, the model enables vets working in the field to prescribe medication for sick donkeys. Models can also help *describe relationships* and provide insights—for example, in this chapter we explore relationships between factors correlated with upward mobility, such as commute time, income inequality, and the quality of K–12 education. We carry out a descriptive analysis that follows an analysis social scientists have used to shape public conversation and inform policy recommendations.

We start by describing the simple linear model, which summarizes the relationship between two features with a line. We explain how to fit this line to data using the loss minimization approach introduced in Chapter 4. Then we introduce the multiple linear model, which models one feature using multiple other features. To fit such a model, we use linear algebra and reveal the geometry behind fitting a linear model with squared error loss. Finally, we cover feature engineering techniques that let us include categorical features and transformed features when building models.

Simple Linear Model

Like with the constant model, our goal is to approximate the signal in a feature by a constant. Now we have additional information from a second feature to help us. In short, we want to use information from a second feature to make a better model than the constant model. For example, we might describe the sale price of a house by its size or predict a donkey's weight from its length. In each of these examples, we have an *outcome* feature (sale price, weight) that we want to explain, describe, or predict with the help of an *explanatory* feature (house size, length).

We use *outcome* to refer to the feature that we are trying to model and *explanatory* for the feature that we are using to explain the outcome. Different fields have adopted conventions for describing this relationship. Some call the outcome the dependent variable and the explanatory the independent variable. Others use response and covariate; regress and regressor; explained and explanatory; endogenous and exogenous. In machine learning, *target* and *features* or *predicted* and *predictors* are common. Unfortunately, many of these pairs connote a causal relationship. The notion of explaining or predicting is not necessarily meant to imply that one causes the other. Particularly confusing is the independent-dependent usage, and we recommend avoiding it.

One possible model we might use is a line. Mathematically, that means we have an intercept, θ_0, and a slope, θ_1, and we use the explanatory feature x to approximate the outcome, y, by a point on the line:

$$y \approx \theta_0 + \theta_1 x$$

As x changes, the estimate for y changes but still falls on the line. Typically, the estimate isn't perfect, and there is some error in using the model; that's why we use the symbol $\approx$ to mean "approximately."

To find a line that does a good job of capturing the signal in the outcome, we use the same approach introduced in Chapter 4 and minimize the average squared loss. Specifically, we follow these steps:

1. Find the errors: $y_i - (\theta_0 + \theta_1 x_i),\ i = 1, \ldots, n$
2. Square the errors (i.e., use squared loss): $[y_i - (\theta_0 + \theta_1 x_i)]^2$

3. Calculate the average loss over the data:

$$\frac{1}{n}\sum_i [y_i - (\theta_0 + \theta_1 x_i)]^2$$

To fit the model, we find the slope and intercept that give us the smallest average loss; in other words, we minimize the *mean squared error*, or MSE for short. We call the minimizing values for the intercept and slope $\hat{\theta}_0$ and $\hat{\theta}_1$.

Notice that the errors we calculate in step 1 are measured in the vertical direction, meaning for a specific x, the error is the vertical distance between the data point (x, y) and the point on the line $(x, \theta_0 + \theta_1 x)$. Figure 15-1 shows this notion. On the left is a scatterplot of points with a line used to estimate y from x. We have marked two specific points by squares and their corresponding approximations on the line by diamonds. The dotted segment from the actual point to the line shows the error. The plot on the right is a scatterplot of all the errors; for reference, we marked the errors corresponding to the two square points in the left plot with squares in the right plot as well.

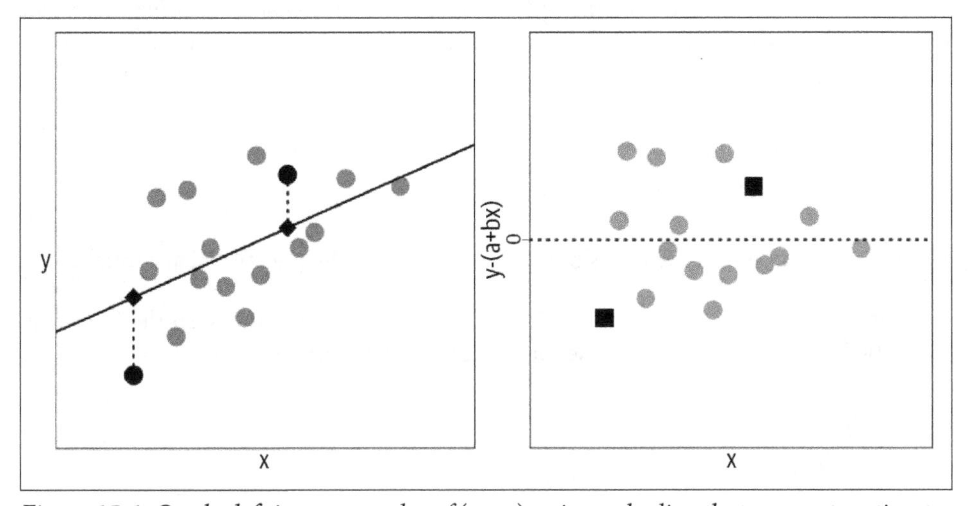

Figure 15-1. On the left is a scatterplot of (x_i, y_i) pairs and a line that we use to estimate y from x. Two specific points are represented by squares and their estimates by diamonds. On the right is a scatterplot of the errors: $y_i - (\theta_0 + \theta_1 x_i)$.

Later in this chapter, we derive the values $\hat{\theta}_0$ and $\hat{\theta}_1$ that minimize the mean squared error. We show that these are:

$$\hat{\theta}_0 = \bar{y} - \hat{\theta}_1 \bar{x}$$
$$\hat{\theta}_1 = r(\mathbf{x}, \mathbf{y}) \frac{SD(\mathbf{y})}{SD(\mathbf{x})}$$

Here, **x** represents the values $x_1, \ldots, x_n$ and **y** is similarly defined; $r(\mathbf{x}, \mathbf{y})$ is the correlation coefficient of the (x_i, y_i) pairs.

Putting the two together, the equation for the line becomes:

$$\begin{aligned} \hat{\theta}_0 + \hat{\theta}_1 x &= \bar{y} - \hat{\theta}_1 \bar{x} + \hat{\theta}_1 x \\ &= \bar{y} + r(\mathbf{x}, \mathbf{y}) SD(\mathbf{y}) \frac{(x - \bar{x})}{SD(\mathbf{x})} \end{aligned}$$

This equation has a nice interpretation: for a given x value, we find how many standard deviations above (or below) average it is, and then we predict (or explain, depending on the setting) y to be r times as many standard deviations above (or below) its average.

We see from the expression for the optimal line that the *sample correlation coefficient* plays an important role. Recall that r measures the strength of the linear association and is defined as:

$$r(\mathbf{x}, \mathbf{y}) = \sum_i \frac{(x_i - \bar{x})}{SD(\mathbf{x})} \frac{(y_i - \bar{y})}{SD(\mathbf{y})}$$

Here are a few important features of the correlation that help us fit linear models:

- r is unitless. Notice that x, $\bar{x}$, and $SD(\mathbf{x})$ all have the same units, so the following ratio has no units (and likewise for the terms involving y_i):

  $$\frac{(x_i - \bar{x})}{SD(\mathbf{x})}$$

- r is between -1 and $+1$. Only when all of the points fall exactly along a line is the correlation either $+1$ or -1, depending on whether the slope of the line is positive or negative.
- r measures the strength of a linear association, not whether or not the data have a linear association. The four scatterplots in Figure 15-2 all have the same correlation coefficient of about 0.8 (as well as the same averages and standard

deviations), but only one plot, the one on the top left, has what we think of as a linear association with random errors about the line.

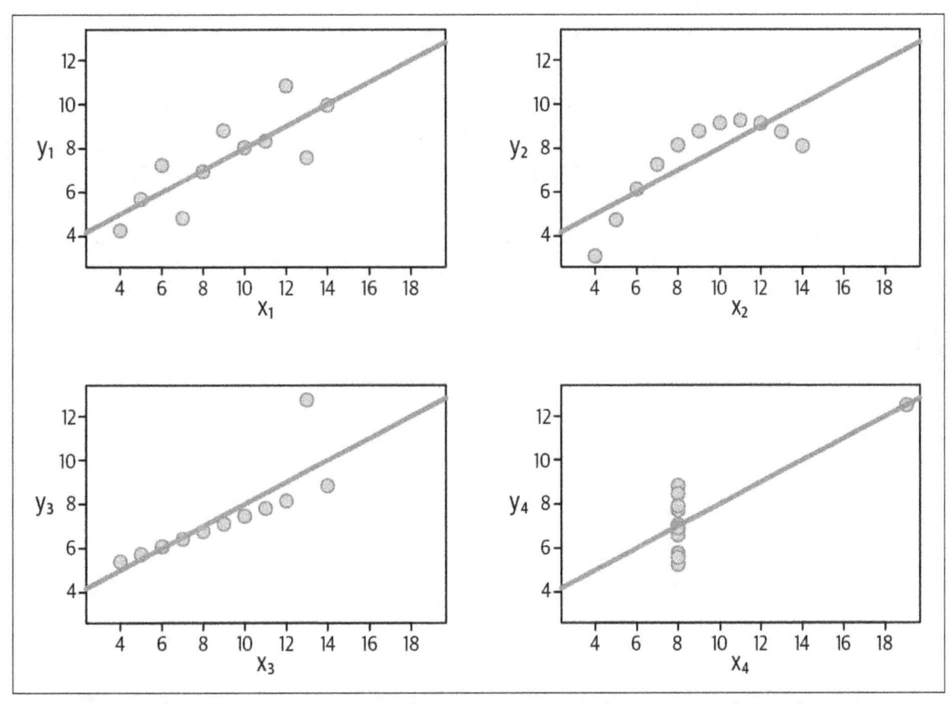

Figure 15-2. These four sets of points, known as Anscombe's quartet, have the same correlation of 0.8, and the same means and standard deviations. The plot in the top left exhibits a linear association; top right shows a perfect nonlinear association; bottom left, with the exception of one point, is a perfect linear association; and bottom right, with the exception of one point, has no association.

Again, we do not expect the pairs of data points to fall exactly along a line, but we do expect the scatter of points to be reasonably described by the line, and we expect the deviations between y_i and the estimate $\hat{\theta}_0 + \hat{\theta}_1 x_i$ to be roughly symmetrically distributed about the line with no apparent patterns.

Linear models were introduced in Chapter 12, where we used the relationship between measurements from high-quality air monitors operated by the Environmental Protection Agency and neighboring inexpensive air quality monitors to calibrate the inexpensive monitors for more accurate predictions. We revisit that example to make the notion of a simple linear model more concrete.

Example: A Simple Linear Model for Air Quality

Recall from Chapter 12 that our aim is to use air quality measurements from the accurate Air Quality System (AQS) sensors operated by the US government to predict the measurements made by PurpleAir (PA) sensors. The pairs of data values come from neighboring instruments that measure the average daily concentration of particulate matter in the air on the same day. (The unit of measurement is an average count of particles under 2.5 mm in size per cubic liter of air in a 24-hour period.) In this section, we focus on air quality measurements at one location in Georgia. These are a subset of the data we examined in the case study in Chapter 12. The measurements are daily averages from August 2019 to mid-November 2019:

	date	id	region	pm25aqs	pm25pa
5258	2019-08-02	GA1	Southeast	8.65	16.19
5259	2019-08-03	GA1	Southeast	7.70	13.59
5260	2019-08-04	GA1	Southeast	6.30	10.30
...	...	...	...	...	...
5439	2019-10-18	GA1	Southeast	6.30	12.94
5440	2019-10-21	GA1	Southeast	7.50	13.62
5441	2019-10-30	GA1	Southeast	5.20	14.55

```
184 rows × 5 columns
```

The feature `pm25aqs` contains measurements from the AQS sensor and `pm25pa` from the PurpleAir monitor. Since we are interested in studying how well the AQS measurements predict the PurpleAir measurements, our scatterplot places PurpleAir readings on the y-axis and AQS readings on the x-axis. We also add a trend line:

```
px.scatter(GA, x="pm25aqs", y="pm25pa", trendline='ols',
           trendline_color_override="darkorange",
           labels={'pm25aqs':'AQS PM2.5', 'pm25pa':'PurpleAir PM2.5'},
           width=350, height=250)
```

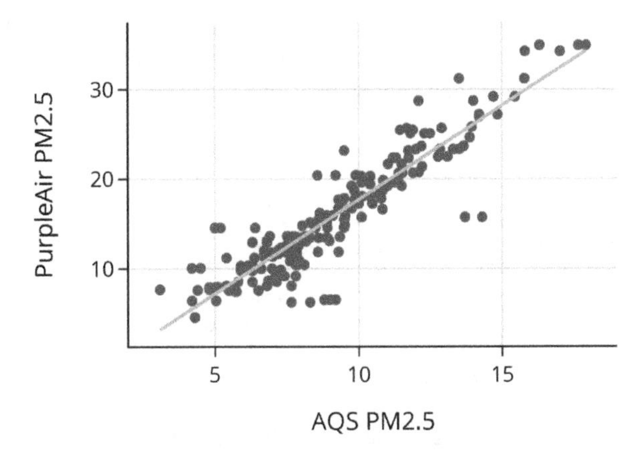

This scatterplot shows a linear relationship between the measurements from these two kinds of instruments. The model that we want to fit has the following form:

$$PA \approx \theta_0 + \theta_1 AQ$$

where *PA* refers to the PurpleAir average daily measurement and *AQ* to its partner AQS measurement.

Since `pandas.Series` objects have built-in methods to compute standard deviations (SDs) and correlation coefficients, we can quickly define functions that calculate the best-fitting line:

```
def theta_1(x, y):
    r = x.corr(y)
    return r * y.std() / x.std()

def theta_0(x, y):
    return y.mean() - theta_1(x, y) * x.mean()
```

Now we can fit the model by computing $\hat{\theta}_0$ and $\hat{\theta}_1$ for these data:

```
t1 = theta_1(GA['pm25aqs'], GA['pm25pa'])
t0 = theta_0(GA['pm25aqs'], GA['pm25pa'])
```

```
Model: -3.36 + 2.10AQ
```

This model matches the trend line shown in the scatterplot. That's not by accident. The parameter value for `trendline` in the call to `scatter()` is `"ols"`, which stands for *ordinary least squares*, another name for fitting a linear model by minimizing squared error.

Let's examine the errors. First, we find the predictions for PA measurements given the AQS measurements, and then we calculate the errors—the difference between the actual PA measurements and the predictions:

```
prediction = t0 + t1 * GA["pm25aqs"]
error = GA["pm25pa"] - prediction
fit = pd.DataFrame(dict(prediction=prediction, error=error))
```

Let's plot these errors against the predicted values:

```
fig = px.scatter(fit, y='error', x='prediction',
                 labels={"prediction": "Prediction",
                         "error": "Error"},
                 width=350, height=250)

fig.add_hline(0, line_width=2, line_dash='dash', opacity=1)
fig.update_yaxes(range=[-12, 12])
```

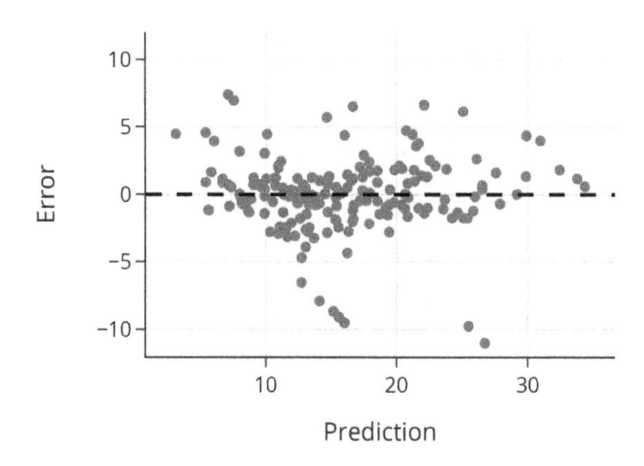

An error of 0 means that the actual measurement falls on the fitted line; we also call this line the *least squares line* or the *regression line*. A positive value means it is above the line, and negative means it's below. You might be wondering how good this model is and what it says about our data. We consider these topics next.

Interpreting Linear Models

The original scatterplot of paired measurements shows that the PurpleAir recordings are often quite a bit higher than the more accurate AQS measurements. Indeed, the equation for our simple line model has a slope of about 2.1. We interpret the slope to mean that a change of 1 ppm measured by the AQS monitor is associated with a change of 2 ppm in the PA measurement, on average. So, if on one day the AQS sensor measures 10 ppm and on the next day it is 5 ppm higher, namely 15 ppm, then our prediction for the PA measurement increases from one day to the next by $2 \times 5 = 10$ ppm.

Any change in the PurpleAir reading is not caused by the change in the AQS reading. Rather, they both reflect the air quality, and our model captures the relationship between the two devices. Oftentimes, the term *prediction* is taken to mean *causation*, but that is not the case here. Instead, the prediction just refers to our use of the *linear association* between PA and AQS measurements.

As for the intercept in the model, we might expect it to be 0, since when there is no particulate matter in the air we would think that both instruments would measure 0 ppm. But for an AQS of 0, the model predicts −3.4 ppm for PurpleAir, which doesn't make sense. There can't be negative amounts of particles in the air. This highlights the problem of using the model outside the range where measurements were taken. We observed AQS recordings between 3 and 18 ppm, and in this range the model fits well. While it makes sense for the line to have an intercept of 0, such a model doesn't fit well in a practical sense and the predictions tend to be much worse.

George Box, a renowned statistician, famously said, "All models are wrong, but some are useful." Here is a case where despite the intercept of the line not passing through 0, the simple linear model is useful in predicting air quality measurements for a PurpleAir sensor. Indeed, the correlation between our two features is very high:

```
GA[['pm25aqs', 'pm25pa']].corr()
```

	pm25aqs	pm25pa
pm25aqs	1.00	0.92
pm25pa	0.92	1.00

Aside from looking at correlation coefficients, there are other ways to assess the quality of a linear model.

Assessing the Fit

The earlier plot of the errors against the fitted values gives a visual assessment of the quality of the fit. (This plot is called a *residual plot* because the errors are sometimes referred to as *residuals*.) A good fit should show a cloud of points around the horizontal line at 0 with no clear pattern. When there is a pattern, we can usually conclude that the simple linear model doesn't entirely capture the signal. We saw earlier that there are no apparent patterns in the residual plot.

Another type of residual plot that can be useful is a plot of the residuals against a feature that is not in the model. If we see a pattern, then we may want to include this feature in the model, in addition to the feature(s) already in the model. Also, when the data have a time component, we want to check for patterns in the residuals over time. For these particular data, since the measurements are daily averages over a four-month period, we plot the error against the date the measurement is recorded:

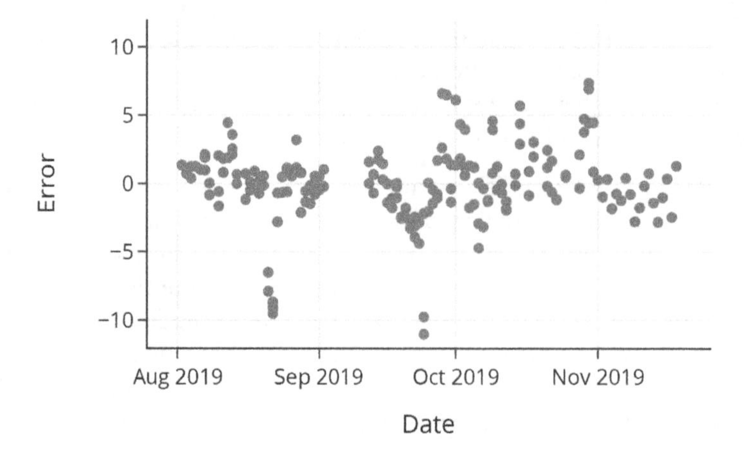

It looks like there are a few consecutive days near the end of August and again near the end of September where the data are far below what is expected. Looking back at the original scatterplot (and the first residual plot), we can see two small clusters of horizontal points below the main point cloud. The plot we just made indicates that we should check the original data and any available information about the equipment to determine whether it was properly functioning on those days.

The residual plot can also give us a general sense of how accurate the model is in its predictions. Most of the errors lie between ±6 ppm of the line. And we find the standard deviation of the errors to be about 2.8 ppm:

```
error.std()
```

```
2.796095864304746
```

In comparison, the standard deviation of the PurpleAir measurements is quite a bit larger:

```
GA['pm25pa'].std()
```

```
6.947418231019876
```

The model error may be further reduced if we find the monitor wasn't working on those days in late August and September and so exclude them from the dataset. In any event, for situations where the air is quite clean, the error is relatively large, but in absolute terms it is inconsequential. We are typically more concerned about the case when there is air pollution, and in that case, an error of 2.8 ppm seems reasonable.

Let's return to the process of how to find this line, the process of *model fitting*. In the next section, we derive the intercept and slope by minimizing the mean squared error.

Fitting the Simple Linear Model

We stated earlier in this chapter that when we minimize the average loss over the data:

$$\frac{1}{n}\sum_i [y_i - (\theta_0 + \theta_1 x_i)]^2$$

the best-fitting line has intercept and slope:

$$\hat{\theta}_0 = \bar{y} - \hat{\theta}_1 \bar{x}$$
$$\hat{\theta}_1 = r(\mathbf{x}, \mathbf{y}) \frac{SD(\mathbf{y})}{SD(\mathbf{x})}$$

In this section, we use calculus to derive these results.

With the simple linear model, the mean squared error is a function of two model parameters, the intercept and slope. This means that if we use calculus to find the minimizing parameter values, we need to find the partial derivatives of the MSE with respect to θ_0 and θ_1. We can also find these minimizing values through other techniques:

Gradient descent
: We can use numerical optimization techniques, such as gradient descent, when the loss function is more complex and it's faster to find an approximate solution that's pretty accurate (see Chapter 20).

Quadratic formula
: Since the average loss is a quadratic function of θ_0 and θ_1, we can use the quadratic formula (along with some algebra) to solve for the minimizing parameter values.

Geometric argument
: Later in this chapter, we use a geometric interpretation of least squares to fit multiple linear models. This approach relates to the Pythagorean theorem and has several intuitive benefits.

We choose calculus to optimize the simple linear model since it is quick and straightforward. To begin, we take the partial derivatives of the sum of squared errors with respect to each parameter (we can ignore the e$1/n$ in the MSE because it doesn't affect the location of the minimum):

$$\frac{\partial}{\partial \theta_0} \sum_i [y_i - (\theta_0 + \theta_1 x_i)]^2 = \sum_i 2(y_i - \theta_0 - \theta_1 x_i)(-1)$$

$$\frac{\partial}{\partial \theta_1} \sum_i [y_i - (\theta_0 + \theta_1 x_i)]^2, = \sum_i 2(y_i - \theta_0 - \theta_1 x_i)(-x_i)$$

Then we set the partial derivatives equal to 0 and simplify a bit by multiplying both sides of the equations by $-1/2$ to get:

$$0 = \sum_i (y_i - \hat{\theta}_0 - \hat{\theta}_1 x_i)$$

$$0 = \sum_i (y_i - \hat{\theta}_0 - \hat{\theta}_1 x_i) x_i$$

These equations are called the *normal equations*. In the first equation, we see that $\hat{\theta}_0$ can be represented as a function of $\hat{\theta}_1$:

$$\hat{\theta}_0 = \bar{y} - \hat{\theta}_1 \bar{x}$$

Plugging this value into the second equation gives us:

$$\begin{aligned} 0 &= \sum_i (y_i - \bar{y} + \hat{\theta}_1 \bar{x} - \hat{\theta}_1 x_i) x_i \\ &= \sum_i [(y_i - \bar{y}) - \hat{\theta}_1 (x_i - \bar{x})] x_i \\ \hat{\theta}_1 &= \frac{\Sigma_i (y_i - \bar{y}) x_i}{\Sigma_i (x_i - \bar{x}) x_i} \end{aligned}$$

After some algebra, we can represent $\hat{\theta}_1$ in terms of quantities that we are familiar with:

$$\hat{\theta}_1 = r(\mathbf{x}, \mathbf{y}) \frac{SD(\mathbf{y})}{SD(\mathbf{x})}$$

As shown earlier in this chapter, this representation says that a point on the fitted line at x can be written as follows:

$$\hat{\theta}_0 + \hat{\theta}_1 x = \bar{y} + r(\mathbf{x}, \mathbf{y}) SD(\mathbf{y}) \frac{(x - \bar{x})}{SD(\mathbf{x})}$$

We have derived the equation for the least squares line that we used in the previous section. There, we used the pandas built-in methods to compute $SD(\mathbf{x})$, $SD(\mathbf{y})$, and $r(\mathbf{x}, \mathbf{y})$, to easily calculate the equation for this line. However, in practice we recommend using the functionality provided in `scikit-learn` to do the model fitting:

```
from sklearn.linear_model import LinearRegression

y = GA['pm25pa']
x = GA[['pm25aqs']]
reg = LinearRegression().fit(x, y)
```

Our fitted model is:

```
Model: PA estimate = -3.36 + 2.10AQS
```

Notice that we provided `y` as an array and `x` as a dataframe to `LinearRegression`. We will soon see why when we fit multiple explanatory features in a model.

The `LinearRegression` method offers numerically stable algorithms to fit linear models by least squares. This is especially important when fitting multiple variables, which we introduce next.

Multiple Linear Model

So far in this chapter, we've used a single input variable to predict an outcome variable. Now we introduce the *multiple linear model* that uses more than one feature to predict (or describe or explain) the outcome. Having multiple explanatory features can improve our model's fit to the data and improve predictions.

We start by generalizing from a simple linear model to one that includes a second explanatory variable, called v. This model is linear in both x and v, meaning that for a pair of values for x and v, we can describe, explain, or predict y by the linear combination:

$$y \approx \theta_0 + \theta_1 x + \theta_2 v$$

Notice that for a particular value of v, say $v^\star$, we could express the preceding equation as:

$$y \approx (\theta_0 + \theta_2 v^\star) \ + \ \theta_1 x$$

In other words, when we hold v constant at $v^\star$, we have a simple linear relation between x and y with slope θ_1 and intercept $\theta_0 + \theta_2 v^\star$. For a different value of v, say

$v^\dagger$, we again have a simple linear relationship between x and y. The slope for x remains the same and the only change is the intercept, which is now $\theta_0 + \theta_2 v^\dagger$.

With multiple linear regression, we need to remember to interpret the coefficient θ_1 of x in the presence of the other variables in the model. Holding fixed the values of the other variables in the model (that's just v in this case), an increase of 1 unit in x corresponds to a θ_1 change in y, on average. One way to visualize this kind of multiple linear relationship is to create facets of scatterplots of (x, y) where in each plot the values of v are roughly the same. We make such a scatterplot for the air quality measurements next, and provide examples of additional visualizations and statistics to examine when fitting a multiple linear model.

The scientists who studied the air quality monitors (see Chapter 12) were looking for an improved model that incorporated weather factors. One weather variable they examined was a daily measurement for relative humidity. Let's consider a two-variable linear model to explain the PurpleAir measurements based on the AQS sensor measurements and relative humidity. This model has the following form:

$$PA \approx \theta_0 + \theta_1 AQ + \theta_2 RH$$

where PA, AQ, and RH refer to the variables: the PurpleAir average daily measurement, AQS measurement, and relative humidity, respectively.

For a first step, we make a facet plot to compare the relationship between the two air quality measurements for fixed values of humidity. To do this, we transform relative humidity to a categorical variable so that each facet consists of observations with similar humidity:

```
rh_cat = pd.cut(GA['rh'], bins=[43,50,55,60,78],
                labels=['<50','50-55','55-60','>60'])
```

Then we use this qualitative feature to subdivide the data into a two-by-two panel of scatterplots:

```
fig = px.scatter(GA, x='pm25aqs', y='pm25pa',
                 facet_col=rh_cat, facet_col_wrap=2,
                 facet_row_spacing=0.15,
                 labels={'pm25aqs':'AQS PM2.5', 'pm25pa':'PurpleAir PM2.5'},
                 width=550, height=350)

fig.update_layout(margin=dict(t=30))
fig.show()
```

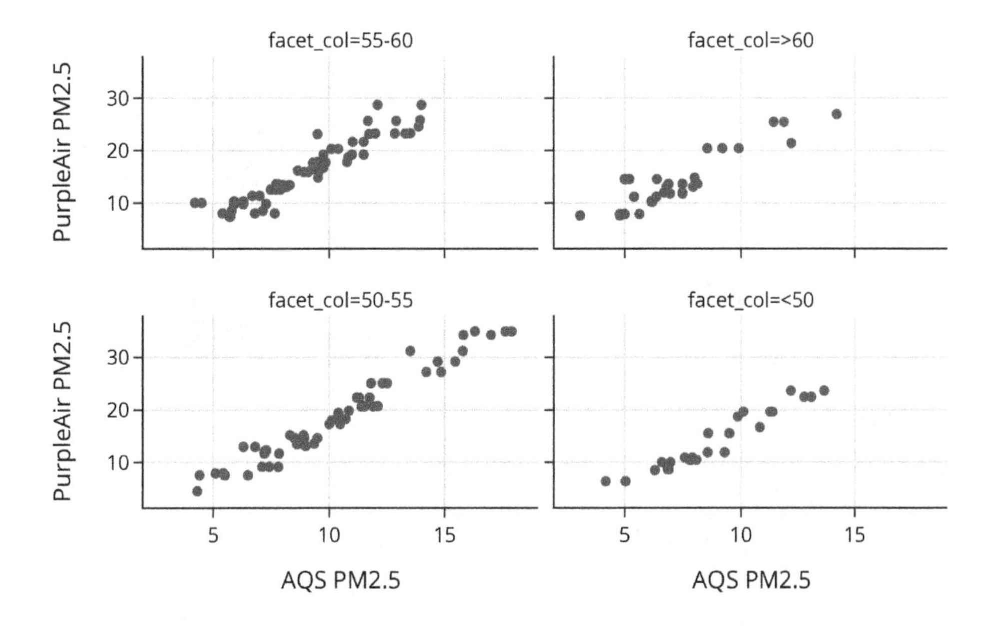

These four plots show a linear relationship between the two sources of air quality measurements. And the slopes appear to be similar, which means that a multiple linear model may fit well. It's difficult to see from these plots if the relative humidity affects the intercept much.

We also want to examine the pairwise scatterplots between the three features. When two explanatory features are highly correlated, their coefficients in the model may be unstable. While linear relationships between three or more features may not show up in pairwise plots, it's still a good idea to check:

```
fig = px.scatter_matrix(
    GA[['pm25pa', 'pm25aqs', 'rh']],
    labels={'pm25aqs':'AQS', 'pm25pa':'PurpleAir', 'rh':'Humidity'},
    width=550, height=400)

fig.update_traces(diagonal_visible=False)
```

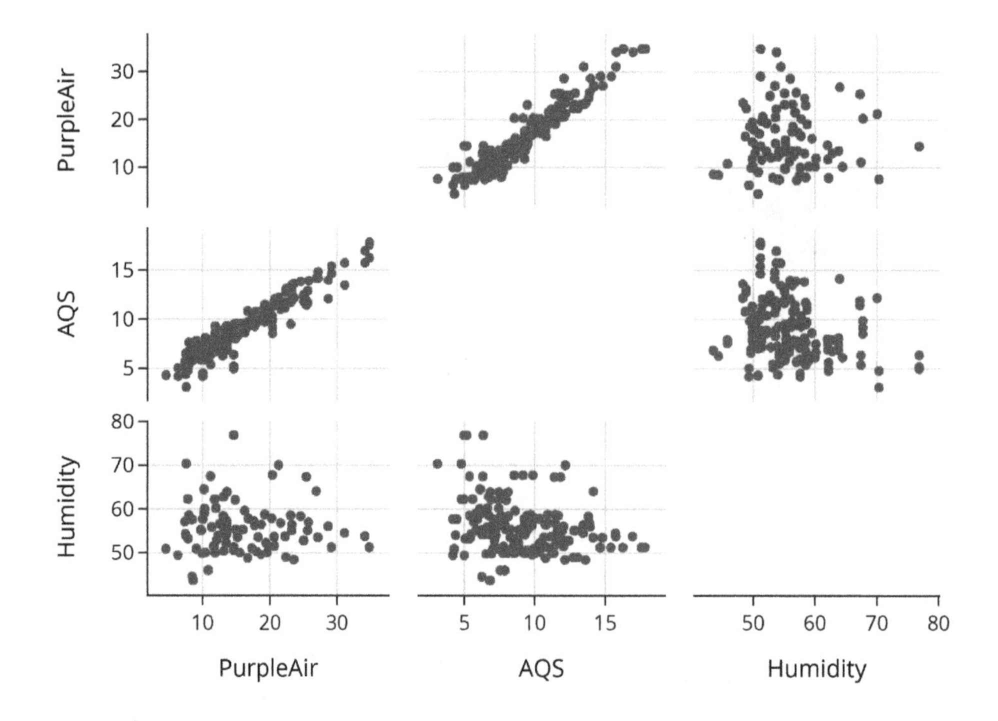

The relationship between humidity and air quality does not appear to be particularly strong. Another pairwise measure we should examine is the correlations between features:

	pm25pa	pm25aqs	rh
pm25pa	1.00	0.95	-0.06
pm25aqs	0.95	1.00	-0.24
rh	-0.06	-0.24	1.00

One small surprise is that relative humidity has a small negative correlation with the AQS measurement of air quality. This suggests that humidity might be helpful in the model.

In the next section, we derive the equation for the fit. But for now, we use the functionality in `LinearRegression` to fit the model. The only change from earlier is that we provide two columns for the explanatory variables (that's why the `x` input is a dataframe):

```
from sklearn.linear_model import LinearRegression

y = GA['pm25pa']
X2 = GA[['pm25aqs', 'rh']]
```

```
model2 = LinearRegression().fit(X2, y)
```

The fitted multiple linear model, including the coefficient units, is:

```
PA estimate = -15.8 ppm + 2.25 ppm/ppm x AQS +  0.21 ppm/percent x RH
```

The coefficient for humidity in the model adjusts the air quality prediction by 0.21 ppm for each percentage point of relative humidity. Notice that the coefficient for AQS differs from the simple linear model that we fitted earlier. This happens because the coefficient reflects the additional information coming from relative humidity.

Lastly, to check the quality of the fit, we make residual plots of the predicted values and the errors. This time, we use `LinearRegression` to compute the predictions for us:

```
predicted_2var = model2.predict(X2)
error_2var = y - predicted_2var

fig = px.scatter(y = error_2var, x=predicted_2var,
                 labels={"y": "Error", "x": "Predicted PurpleAir measurement"},
                 width=350, height=250)

fig.update_yaxes(range=[-12, 12])
fig.add_hline(0, line_width=3, line_dash='dash', opacity=1)

fig.show()
```

The residual plot appears to have no clear patterns, which indicates that the model fits pretty well. Notice also that the errors nearly all fall within –4 and +4 ppm, a smaller range than in the simple linear model. And we find the standard deviation of the residuals is quite a bit smaller:

```
error_2var.std()

1.8211427707294048
```

The residual standard deviation has been reduced from 2.8 ppm in the one variable model to 1.8 ppm, a good size reduction.

The correlation coefficient can't capture the strength of a linear association model when we have more than one explanatory variable. Instead, we adapt the MSE to give us a sense of model fit. In the next section, we describe how to fit a multiple linear model and use the MSE to assess fit.

Fitting the Multiple Linear Model

In the previous section, we considered the case of two explanatory variables; one of these we called x and the other v. Now we want to generalize the approach to p explanatory variables. The idea of choosing different letters to represent variables quickly fails us. Instead, we use a more formal and general approach that represents multiple predictors as a matrix, as depicted in Figure 15-3. We call **X** the *design matrix*. Notice that **X** has shape $n \times (p+1)$. Each column of **X** represents a feature, and each row represents an observation. That is, $x_{i,j}$ is the measurement taken on observation i for feature j.

$$\mathrm{X}=\begin{bmatrix} 1 & x_{1,1} & \dots & x_{1,p} \\ 1 & x_{2,1} & \dots & x_{2,p} \\ & & \vdots & \\ 1 & x_{n,1} & \dots & x_{n,p} \end{bmatrix}$$

Figure 15-3. In this design matrix X, each row represents an observation/record and each column a feature/variable

One technicality: the design matrix is defined as a mathematical matrix, not a dataframe, so you might notice that a matrix doesn't include the column or row labels that a dataframe has.

That said, we usually don't have to worry about converting dataframes into matrices since most Python libraries for modeling treat dataframes of numbers as if they were matrices.

For a given observation, say, the second row in **X**, we approximate the outcome y_2 by the linear combination:

$$y_2 \approx \theta_0 + \theta_1 x_{2,1} + \ldots + \theta_p x_{2,p}$$

It's more convenient to express the linear approximation in matrix notation. To do this, we write the model parameters as a $p+1$ column vector $\boldsymbol{\theta}$:

$$\theta = \begin{bmatrix} \theta_0 \\ \theta_1 \\ \vdots \\ \theta_p \end{bmatrix}$$

Putting these notational definitions together, we can write the vector of predictions for the entire dataset using matrix multiplication:

$$\mathbf{X}\boldsymbol{\theta}$$

If we check the dimensions of $\mathbf{X}$ and $\boldsymbol{\theta}$, we can confirm that $\mathbf{X}\boldsymbol{\theta}$ is an n-dimensional column vector. So the error in using this linear prediction can be expressed as the vector:

$$\mathbf{e} = \mathbf{y} - \mathbf{X}\boldsymbol{\theta}$$

where the outcome variable is also represented as a column vector:

$$\mathbf{y} = \begin{bmatrix} y_1 \\ y_2 \\ \vdots \\ y_n \end{bmatrix}$$

This matrix representation of the multiple linear model can help us find the model that minimizes mean squared error. Our goal is to find the model parameters $(\theta_0, \theta_1, \ldots, \theta_p)$ that minimize the mean squared error:

$$\frac{1}{n}\sum_i [y_i - (\theta_0 + \theta_1 x_{i,1} + \cdots + \theta_p x_{i,p})]^2 = \frac{1}{n}\|\mathbf{y} - \mathbf{X}\boldsymbol{\theta}\|^2$$

Here, we use the notation $\|\mathbf{v}\|^2$ for a vector $\mathbf{v}$ as a shorthand for the sum of each vector element squared: $\|\mathbf{v}\|^2 = \sum_i v_i^2$. The square root, $\sqrt{\|\mathbf{v}\|^2}$, corresponds to the length of the vector $\mathbf{v}$ and is also called the ℓ_2 norm of $\mathbf{v}$. So, minimizing the mean squared error is the same thing as finding the shortest error vector.

We can fit our model using calculus as we did for the simple linear model. However, this approach gets cumbersome, and instead we use a geometric argument that is more intuitive and easily leads to useful properties of the design matrix, errors, and predicted values.

Our goal is to find the parameter vector, which we call $\hat{\boldsymbol{\theta}}$, that minimizes our average squared loss—we want to make $\|\mathbf{y} - \mathbf{X}\boldsymbol{\theta}\|^2$ as small as possible for a given $\mathbf{X}$ and $\mathbf{y}$. The key insight is that we can restate this goal in a geometric way. Since the model predictions and the true outcomes are both vectors, we can think of them as vectors in a *vector space*. When we change our model parameters $\boldsymbol{\theta}$, the model makes different predictions, but any prediction must be a linear combination of the column vectors of $\mathbf{X}$; that is, the prediction must be in what is called span($\mathbf{X}$). This notion is illustrated in Figure 15-4, where the shaded region consists of the possible linear models. Notice that $\mathbf{y}$ is not entirely captured in span($\mathbf{X}$); this is typically the case.

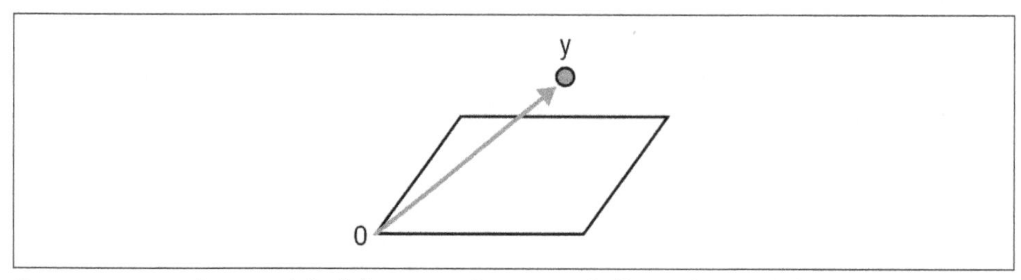

Figure 15-4. In this simplified diagram, the space of all possible model prediction vectors span($\mathbf{X}$) *is illustrated as a plane in three-dimensional space, and the observed* $\mathbf{y}$ *as a vector*

Although the squared loss can't be exactly zero because $\mathbf{y}$ isn't in the span($\mathbf{X}$), we can find the vector that lies as close to $\mathbf{y}$ as possible while still being in span($\mathbf{X}$). This vector is called $\hat{\mathbf{y}}$.

The error is the vector $\mathbf{e} = \mathbf{y} - \hat{\mathbf{y}}$. Its length $\|\mathbf{e}\|$ represents the distance between the true outcome and our model's prediction. Visually, $\mathbf{e}$ has the smallest magnitude when it is *perpendicular* to the span($\mathbf{X}$), as shown in Figure 15-5. The proof of this fact is omitted, and we rely on the figures to convince you of it.

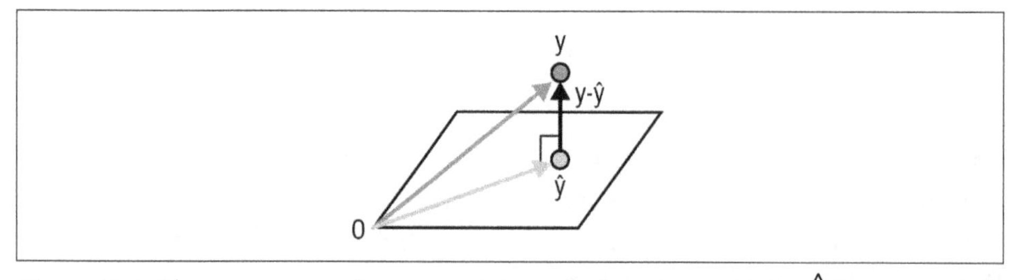

Figure 15-5. The mean squared error is minimized when the prediction $\hat{\mathbf{y}}$ *lies in* span($\mathbf{X}$) *perpendicular to* $\mathbf{y}$

The fact that the smallest error, $\mathbf{e}$, must be perpendicular to $\hat{\mathbf{y}}$ lets us derive a formula for $\hat{\boldsymbol{\theta}}$ as follows:

$$
\begin{aligned}
\mathbf{X}\hat{\boldsymbol{\theta}} + \mathbf{e} &= \mathbf{y} && \text{(the definition of } \mathbf{y}, \hat{\mathbf{y}}, \mathbf{e}\text{)} \\
\mathbf{X}^\top\mathbf{X}\hat{\boldsymbol{\theta}} + \mathbf{X}^\top\mathbf{e} &= \mathbf{X}^\top\mathbf{y} && \text{(left-multiply by } \mathbf{X}^\top\text{)} \\
\mathbf{X}^\top\mathbf{X}\hat{\boldsymbol{\theta}} &= \mathbf{X}^\top\mathbf{y} && (\mathbf{e} \perp \text{span}(\mathbf{X})) \\
\hat{\boldsymbol{\theta}} &= (\mathbf{X}^\top\mathbf{X})^{-1}\mathbf{X}^\top\mathbf{y} && \text{(left-multiply by } (\mathbf{X}^\top\mathbf{X})^{-1}\text{)}
\end{aligned}
$$

This general approach to derive $\hat{\boldsymbol{\theta}}$ for the multiple linear model also gives us $\hat{\theta}_0$ and $\hat{\theta}_1$ for the simple linear model. If we set $\mathbf{X}$ to be the two-column matrix that contains the intercept column and one feature column, this formula for $\hat{\boldsymbol{\theta}}$ and some linear algebra gets the intercept and slope of the least-squares-fitted simple linear model. In fact, if $\mathbf{X}$ is simply a single column of 1s, then we can use this formula to show that $\hat{\theta}$ is just the mean of $\mathbf{y}$. This nicely ties back to the constant model that we introduced in Chapter 4.

While we can write a simple function to derive the $\hat{\boldsymbol{\theta}}$ based on the formula

$$\hat{\boldsymbol{\theta}} = (\mathbf{X}^\top\mathbf{X})^{-1}\mathbf{X}^\top\mathbf{y}$$

we recommend leaving the calculation of $\hat{\boldsymbol{\theta}}$ to the optimally tuned methods provided in the `scikit-learn` and `statsmodels` libraries. They handle cases where the design matrix is sparse, highly co-linear, and not invertible.

This solution for $\hat{\boldsymbol{\theta}}$ (along with the pictures) reveals some useful properties of the fitted coefficients and the predictions:

- The residuals, $\mathbf{e}$, are orthogonal to the predicted values, $\hat{\mathbf{y}}$.
- The average of the residuals is 0 if the model has an intercept term.
- The variance of the residuals is just the MSE.

These properties explain why we examine plots of the residuals against the predictions. When we fit a multiple linear model, we also plot the residuals against variables that we are considering adding to the model. If they showed a linear pattern, then we would consider adding them to the model.

In addition to examining the SD of the errors, the ratio of the MSE for a multiple linear model to the MSE for the constant model gives a measure of the model fit. This is called the *multiple* R^2 and is defined as:

$$R^2 = 1 - \frac{\|\mathbf{y} - \mathbf{X}\hat{\theta}\|^2}{\|\mathbf{y} - \bar{y}\|^2}$$

As the model fits the data closer and closer, the multiple R^2 gets nearer to 1. That might seem like a good thing, but there can be problems with this approach because R^2 continues to grow even as we add meaningless features to our model, as long as the features expand the span($\mathbf{X}$). To account for the size of a model, we often adjust the numerator and denominator in R^2 by the number of fitted coefficients in the models. That is, we normalize the numerator by $1/[n - (p + 1)]$ and the denominator by $1/(n - 1)$. Better approaches to selecting a model are covered in Chapter 16.

Next, we consider a social science example where there are many variables available to us for modeling.

Example: Where Is the Land of Opportunity?

The US is called "the land of opportunity" because people believe that even those with few resources can end up wealthy in the US—economists call this notion "economic mobility." In one study, economist Raj Chetty and colleagues did a large-scale data analysis on economic mobility in the US (*https://doi.org/10.1093/qje/qju022*). His basic question was whether the US is a land of opportunity. To answer this somewhat vague question, Chetty needed a way to measure economic mobility.

Chetty had access to 2011–2012 federal income tax records for everyone born in the US between 1980 and 1982, along with their parents' tax records filed in their birth year. They matched the 30-year-olds to their parents by finding the parents' 1980–1982 tax records that listed them as dependents. In total, his dataset had about 10 million people. To measure economic mobility, Chetty grouped people born in a particular geographic region whose parents' income was in the 25th income percentile in 1980–1982. He then found the group's average income percentile in 2011. Chetty calls this average *absolute upward mobility* (AUM). If a region's AUM is 25, then people born into the 25th percentile generally stay in the 25th percentile—they remain where their parents were when they were born. High AUM values mean that the region has more upward mobility. Those born into the 25th income percentile in these regions generally wind up in a higher income bracket than their parents. For reference, the US average AUM is about 41 at the time of this writing. Chetty calculated the AUM for regions called commuting zones (CZs), which are roughly on the same scale as counties.

While the granularity of the original data is at an individual level, the data Chetty analyzed has a granularity at the CZ level. Income records can't be publicly available because of privacy laws, but the AUM for a commuting zone can be made available. However, even with the granularity of a commuting zone, not all commuting zones are included in the data set because with 40 features in the data, it might be possible to identify individuals in small CZs. This limitation points to a potential coverage bias. Measurement bias is another potential problem. For example, children born into the 25th income percentile who become extremely wealthy may not file income tax.

We also point out the limitations of working with data that are regional averages rather than individual measurements. The relationships found among features are often more highly correlated at the aggregate level than at the individual level. This phenomenon is called *ecological regression*, and interpretations of findings from aggregated data need to be made with care.

Chetty had a hunch that some places in the US have higher economic mobility than others. His analysis found this to be true. He found that some cities—such as San Jose, Calif.; Washington, DC; and Seattle—have higher mobility than others, such as Charlotte, N.C.; Milwaukee; and Atlanta. This means that, for example, people move from low to high income brackets in San Jose at a higher rate compared to Charlotte. Chetty used linear models to find that social and economic factors like segregation, income inequality, and local school systems are related to economic mobility.

In this analysis, our outcome variable is the AUM for a commuting zone, since we are interested in finding features that correlate with AUM. There are many possible such features in Chetty's data, but we first investigate one in particular: the fraction of people in a CZ who have a 15-minute or shorter commute to work.

Explaining Upward Mobility Using Commute Time

We begin our investigation by loading the data into a dataframe called `cz_df`:

	aum	travel_lt15	gini	rel_tot	...	taxrate	worked_14	foreign	region
0	38.39	0.33	0.47	0.51	...	0.02	3.75e-03	1.18e-02	South
1	37.78	0.28	0.43	0.54	...	0.02	4.78e-03	2.31e-02	South
2	39.05	0.36	0.44	0.67	...	0.01	2.89e-03	7.08e-03	South
...	...	...	...	...	...	...	...	...	...
702	44.12	0.42	0.42	0.29	...	0.02	4.82e-03	9.85e-02	West
703	41.41	0.49	0.41	0.26	...	0.01	4.39e-03	4.33e-02	West
704	43.20	0.24	0.42	0.32	...	0.02	3.67e-03	1.13e-01	West

```
705 rows × 9 columns
```

Each row represents one commuting zone. The column `aum` has the average AUM for people born in the commuting zone in 1980–1982 to parents in the 25th income percentile. There are many columns in this dataframe, but for now we focus on the fraction of people in a CZ that have a 15-minute or shorter commute time, which is called `travel_lt15`. We plot AUM against this fraction to look at the relationship between the two variables:

```
px.scatter(cz_df, x='travel_lt15', y='aum', width=350, height=250,
           labels={'travel_lt15':'Commute time under 15 min',
                   'aum':'Upward mobility'})
```

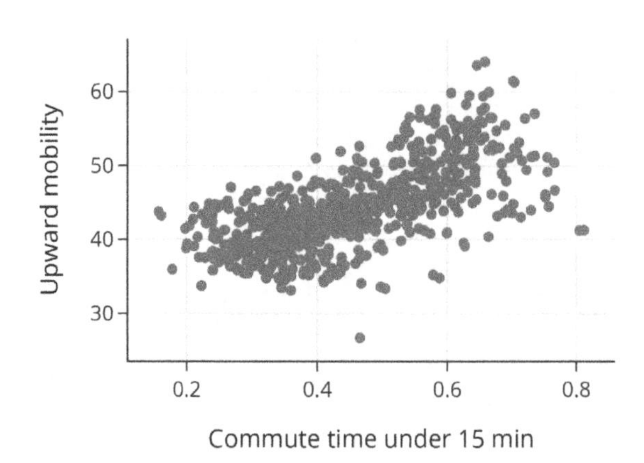

The scatterplot shows a rough linear association between AUM and commute time. Indeed, we find the correlation to be quite strong:

```
cz_df[['aum', 'travel_lt15']].corr()
```

	aum	travel_lt15
aum	1.00	0.68
travel_lt15	0.68	1.00

Let's fit a simple linear model to explain AUM with commute time:

```
from sklearn.linear_model import LinearRegression

y = cz_df['aum']
X = cz_df[['travel_lt15']]

model_ct = LinearRegression().fit(X, y)
```

The coefficients from the MSE minimization are:

```
Intercept: 31.3
    Slope: 28.7
```

Interestingly, an increase in upward mobility of a CZ is associated with an increase in the fraction of people with a short commute time.

We can compare the SD of the AUM measurements to the SD of the residuals. This comparison gives us a sense of how useful the model is in explaining the AUM:

```
prediction = model_ct.predict(X)
error = y - prediction

print(f"SD(errors): {np.std(error):.2f}")
print(f"   SD(AUM): {np.std(cz_df['aum']):.2f}")
```

```
SD(errors): 4.14
   SD(AUM): 5.61
```

The size of the errors about the regression line has decreased from the constant model by about 25%.

Next, we examine the residuals for lack of fit since it can be easier to see potential problems with the fit in a residual plot:

```
fig = px.scatter(x=prediction, y=error,
                 labels=dict(x='Prediction for AUM', y='Error'),
                 width=350, height=250)

fig.add_hline(0, line_width=2, line_dash='dash', opacity=1)
fig.update_yaxes(range=[-20, 15])

fig.show()
```

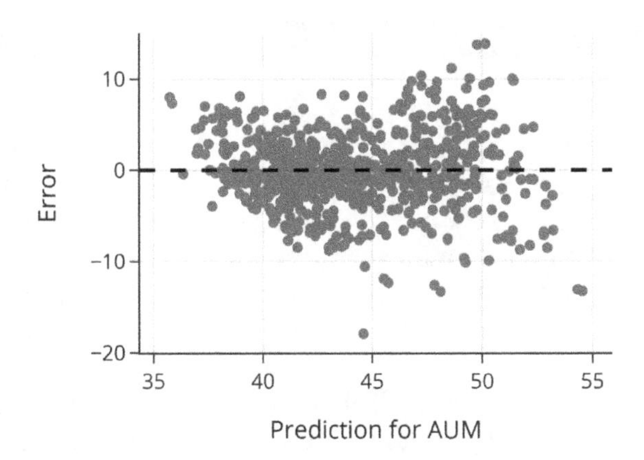

It appears that the errors grow with AUM. We might try a transformation of the response variable, or fitting a model that is quadratic in the commute time fraction. We consider transformations and polynomials in the next section. First we see whether including additional variables offers a more accurate prediction of AUM.

Relating Upward Mobility Using Multiple Variables

In his original analysis, Chetty created several high-level features related to factors such as segregation, income, and K–12 education. We consider seven of Chetty's predictors as we aim to build a more informative model for explaining AUM. These are described in Table 15-1.

Table 15-1. Potential explanation for modeling AUM

Column name	Description
`travel_lt15`	Fraction of people with a ≤15-minute commute to work.
`gini`	Gini coefficient, a measure of wealth inequality. Values are between 0 and 1, where small values mean wealth is evenly distributed and large values mean more inequality.
`rel_tot`	Fraction of people who self-reported as religious.
`single_mom`	Fraction of children with a single mother.
`taxrate`	Local tax rate.
`worked_14`	Fraction of 14- to 16-year-olds who work.
`foreign`	Fraction of people born outside the US.

Let's first examine the correlations between AUM and the explanatory features and between the explanatory features themselves:

	aum	travel_lt15	gini	rel_tot	single_mom	taxrate	worked_14	foreign
aum	1.00	0.68	-0.60	0.52	-0.77	0.35	0.65	-0.03
travel_lt15	0.68	1.00	-0.56	0.40	-0.42	0.34	0.60	-0.19
gini	-0.60	-0.56	1.00	-0.29	0.57	-0.15	-0.58	0.31
rel_tot	0.52	0.40	-0.29	1.00	-0.31	0.08	0.28	-0.11
single_mom	-0.77	-0.42	0.57	-0.31	1.00	-0.26	-0.60	-0.04
taxrate	0.35	0.34	-0.15	0.08	-0.26	1.00	0.35	0.26
worked_14	0.65	0.60	-0.58	0.28	-0.60	0.35	1.00	-0.15
foreign	-0.03	-0.19	0.31	-0.11	-0.04	0.26	-0.15	1.00

We see that the fraction of single mothers in the commuting zone has the strongest correlation with AUM, which implies that it is also the single best feature to explain AUM. In addition, we see that several explanatory variables are highly correlated with each other; the Gini coefficient is highly correlated with the fraction of teenagers who work, the fraction of single mothers, and the fraction with less than a 15-minute commute. With such highly correlated features, we need to take care in interpreting the coefficients because several different models might equally explain AUM with the covariates standing in for one another.

The vector geometry perspective that we introduced earlier in this chapter can help us understand the problem. Recall that a feature corresponds to a column vector in n-dimensions, like $\mathbf{x}$. With two highly correlated features, $\mathbf{x}_1$ and $\mathbf{x}_2$, these vectors are nearly in alignment. So the projection of the response vector $\mathbf{y}$ onto one of these vectors is nearly the same as the projection onto the other. The situation gets even murkier when several features are correlated with one another.

To begin, we can consider all possible two-feature models to see which one has the smallest prediction error. Chetty derived 40 potential variables to use as predictors, which would have us checking $(40 \times 39)/2 = 780$ models. Fitting models, with all pairs, triples, and so on, of variables quickly grows out of control. And it can lead to finding spurious correlations (see Chapter 17).

Here, we keep things a bit simpler and examine just one two-variable model that includes the travel time and single-mother features. After that, we look at the model that has all seven numeric explanatory features in our dataframe:

```
X2 = cz_df[['travel_lt15', 'single_mom']]
y = cz_df['aum']

model_ct_sm = LinearRegression().fit(X2, y)
```

```
Intercept: 49.0
Fraction with under 15 minute commute coefficient: 18.10
Fraction of single moms coefficient: 18.10
```

Notice that the coefficient for travel time is quite different than the coefficient for this variable in the simple linear model. That's because the two features in our model are highly correlated.

Next we compare the errors from the two fits:

```
prediction_ct_sm = model_ct_sm.predict(X2)
error_ct_sm = y - prediction_ct_sm
```

```
 SD(errors in model 1): 4.14
 SD(errors in model 2): 2.85
```

The SD of the residuals have been reduced by another 30%. Adding a second variable to the model seems worth the extra complexity.

Let's again visually examine the residuals. We use the same scale on the y-axis to make it easier to compare this residual plot with the plot for the one-variable model:

```
fig = px.scatter(x=prediction_ct_sm, y=error_ct_sm,
          labels=dict(x='Two-variable prediction for AUM', y='Error'),
          width=350, height=250)

fig.add_hline(0, line_width=2, line_dash='dash', opacity=1)
```

```
fig.update_yaxes(range=[-20, 15])

fig.show()
```

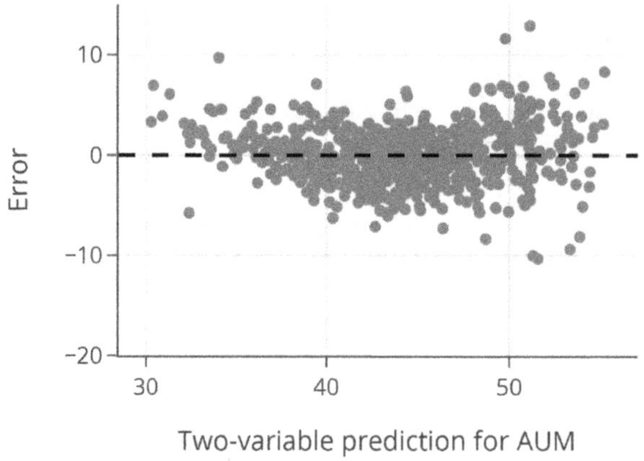

The larger variability in the errors for higher AUM is even more evident. The implications are that the estimates, $\hat{y}$, are unaffected, but their accuracy depends on AUM. This problem can be addressed with *weighted regression*.

Once again, we point out that data scientists from different backgrounds use different terminology to refer to the same concept. For example, the terminology that calls each row in the design matrix **X** an observation and each column a variable is more common among people with backgrounds in statistics. Others say that each column of the design matrix represents a *feature* or that each row represents a *record*. Also, we say that our overall process of fitting and interpreting models is called *modeling*, while others call it *machine learning*.

Now let's fit a multiple linear model that uses all seven variables to explain upward mobility. After fitting the model, we again plot the errors using the same y-axis scale as in the previous two residual plots:

```
X7 = cz_df[predictors]
model_7var = LinearRegression().fit(X7, y)

prediction_7var = model_7var.predict(X7)
error_7var = y - prediction_7var

fig = px.scatter(
    x=prediction_7var, y=error_7var,
    labels=dict(x='Seven-variable prediction for AUM', y='Error'),
    width=350, height=250)
```

```
fig.add_hline(0, line_width=2, line_dash='dash', opacity=1)
fig.update_yaxes(range=[-20, 15])

fig.show()
```

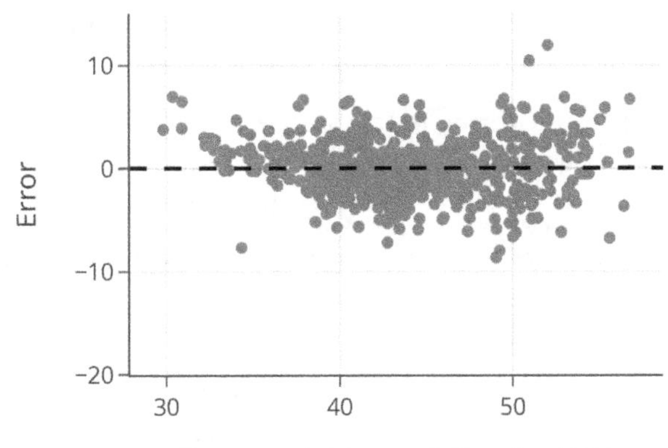

The model with seven features does not appear to be much better than the two-variable model. In fact, the standard deviation of the residuals has only decreased by 8%:

```
error_7var.std()
```

```
2.588739233574256
```

We can compare the multiple R^2 for these three models:

```
R² for 7-variable model: 0.79
R² for 2-variable model: 0.74
R² for 1-variable model: 0.46
```

The adjustment for the number of features in the model makes little difference for us since we have over 700 observations. Now we have confirmed our earlier findings that using two variables greatly improves the explanatory capability of the model, and the seven-variable model offers little improvement over the two-variable model. The small gain is likely not worth the added complexity of the model.

So far, our models have used only numeric predictor variables. But categorical data is often useful for model fitting as well. Additionally, in Chapter 10 we transformed variables and created new variables from combinations of variables. We address how to incorporate these variables into linear models next.

Feature Engineering for Numeric Measurements

All of the models that we have fit so far in this chapter have used numeric features that were originally provided in the dataframe. In this section, we look at variables that are created from transformations of numeric features. Transforming variables to use in modeling is called *feature engineering*.

We introduced feature engineering in Chapters 9 and 10. There, we transformed features so that they had symmetric distributions. Transformations can capture more kinds of patterns in the data and lead to better and more accurate models.

Let's return to the dataset we used as an example in Chapter 10: house sale prices in the San Francisco Bay Area. We restrict the data to houses sold in 2006, when sale prices were relatively stable, so we don't need to account for trends in price.

We wish to model sale price. Recall that visualizations in Chapter 10 showed us that sale price was related to several features, like the size of the house, size of the lot, number of bedrooms, and location. We log-transformed both sale price and the size of the house to improve their relationship, and we saw that box plots of sale price by the number of bedrooms and box plots by city revealed interesting relationships too. In this section, we include transformed numeric features in a linear model. In the next section, we also add an ordinal feature (the number of bedrooms) and a nominal feature (the city) to the model.

To begin, we'll model sale price on house size. The correlation matrix tell us which of our numeric explanatory variables (original and transformed) is most strongly correlated with sale price:

	price	br	lsqft	bsqft	log_price	log_bsqft	log_lsqft	ppsf	log_ppsf
price	1.00	0.45	0.59	0.79	0.94	0.74	0.62	0.49	0.47
br	0.45	1.00	0.29	0.67	0.47	0.71	0.38	-0.18	-0.21
lsqft	0.59	0.29	1.00	0.46	0.55	0.44	0.85	0.29	0.27
bsqft	0.79	0.67	0.46	1.00	0.76	0.96	0.52	-0.08	-0.10
log_price	0.94	0.47	0.55	0.76	1.00	0.78	0.62	0.51	0.52
log_bsqft	0.74	0.71	0.44	0.96	0.78	1.00	0.52	-0.11	-0.14
log_lsqft	0.62	0.38	0.85	0.52	0.62	0.52	1.00	0.29	0.27
ppsf	0.49	-0.18	0.29	-0.08	0.51	-0.11	0.29	1.00	0.96
log_ppsf	0.47	-0.21	0.27	-0.10	0.52	-0.14	0.27	0.96	1.00

Sale price correlates most highly with house size, called `bsqft` for building square feet. We make a scatterplot of sale price against house size to confirm the association is linear:

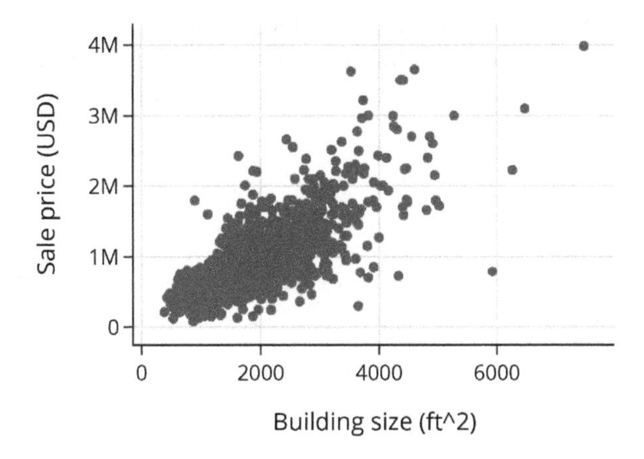

The relationship does look roughly linear, but the very large and expensive houses are far from the center of the distribution and can overly influence the model. As shown in Chapter 10, the log transformation makes the distributions of price and size more symmetric (both are log base 10 to make it easier to convert the values into the original units):

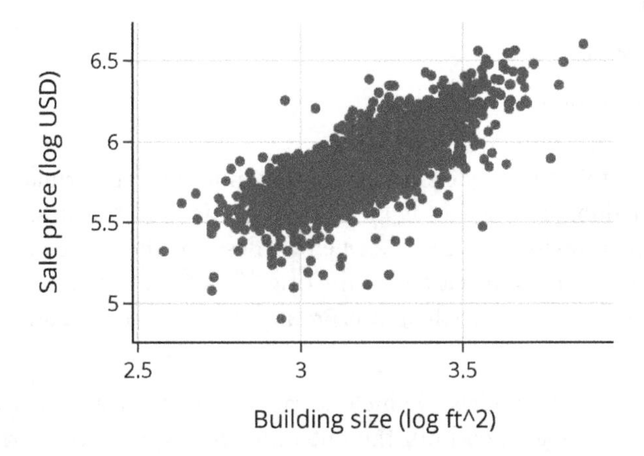

Ideally, a model that uses transformations should make sense in the context of the data. If we fit a simple linear model based on log(size), then when we examine the coefficient, we think in terms of a percentage increase. For example, a doubling of x increases the prediction by $\theta\log(2)$, since $\theta\log(2x) = \theta\log(2) + \theta\log(x)$.

Let's begin by fitting a model that explains log-transformed price by the house's log-transformed size. But first, we note that this model is still considered a linear model. If we represent sale price by y and house size by x, then the model is:

$$\log(y) = \theta_0 + \theta_1 \log(x)$$

(Note that we have ignored the approximation in this equation to make the linear relationship clearer.) This equation may not seem linear, but if we rename $\log(y)$ to w and $\log(x)$ to v, then we can express this "log–log" relationship as a linear model in w and v:

$$w = \theta_0 + \theta_1 v$$

Other examples of models that can be expressed as linear combinations of transformed features are:

$$\begin{aligned}
\log(y) &= \theta_0 + \theta_1 x \\
y &= \theta_0 + \theta_1 x + \theta_2 x^2 \\
y &= \theta_0 + \theta_1 x + \theta_2 z + \theta_3 xz
\end{aligned}$$

Again, if we rename $\log(y)$ to w, x^2 to u, and xz as t, then we can express each of these models as linear in these renamed features. In order, the preceding models are now:

$$\begin{aligned}
w &= \theta_0 + \theta_1 x \\
y &= \theta_0 + \theta_1 x + \theta_2 u \\
y &= \theta_0 + \theta_1 x + \theta_2 z + \theta_3 t
\end{aligned}$$

In short, we can think of models that include nonlinear transformations of features and/or combinations of features as linear in their derived features. In practice, we don't rename the transformed features when we describe the model; instead, we write the model using the transformations of the original features because it's important to keep track of them, especially when interpreting the coefficients and checking residual plots.

When we refer to these models, we include mention of the transformations. That is, we call a model *log–log* when both the outcome and explanatory variables are log-transformed; we say it's *log–linear* when the outcome is log-transformed but not the explanatory variable; we describe a model as having *polynomial features* of, say, degree two when the first and second power transformations of the explanatory variable are included; and we say a model includes an *interaction term* between two explanatory features when the product of these two features is included in the model.

Let's fit a log–log model of price on size:

```
X1_log = sfh[['log_bsqft']]
y_log = sfh['log_price']

model1_log_log = LinearRegression().fit(X1_log, y_log)
```

The coefficients and predicted values from this model cannot be directly compared to a model fitted using linear features because the units are the log of dollars and log of square feet, not dollars and square feet.

Next, we examine the residuals and predicted values with a plot:

```
prediction = model1_log_log.predict(X1_log)
error = y_log - prediction

fig = px.scatter(x=prediction, y=error,
                 labels=dict(x='Predicted sale price (log USD)', y='Error'),
                 width=350, height=250)

fig.add_hline(0, line_width=2, line_dash='dash', opacity=1)
fig.show()
```

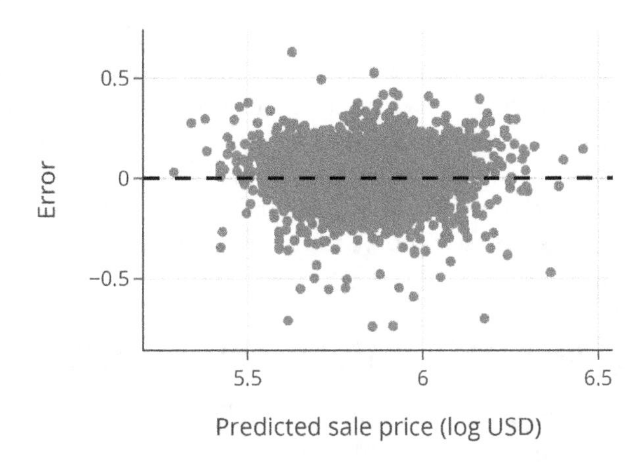

The residual plot looks reasonable, but it contains thousands of points, which makes it hard to see curvature.

To see if additional variables might be helpful, we can plot the residuals from the fitted model against a variable that is not in the model. If we see patterns, that indicates we might want to include this additional feature or a transformation of it. Earlier, we found that the distribution of price was related to the city where the house is located, so let's examine the relationship between the residuals and city:

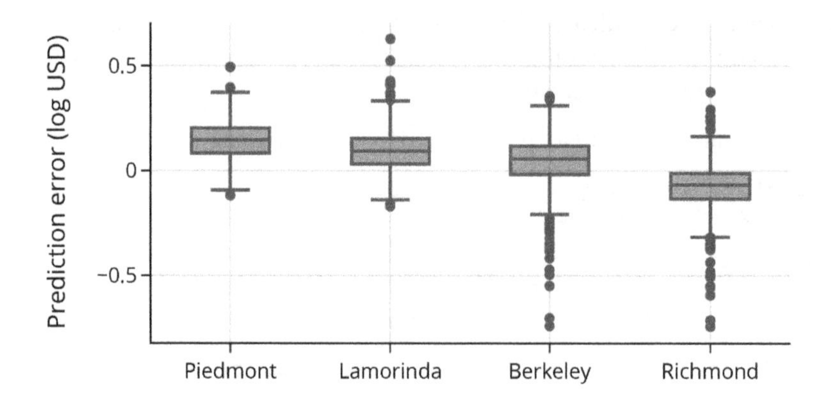

This plot shows us that the distribution of errors appears shifted by city. Ideally, the median of each city's box plot lines up with 0 on the y-axis. Instead, more than 75% of the houses sold in Piedmont have positive errors, meaning the actual sale price is above the predicted value. And at the other extreme, more than 75% of sale prices in Richmond fall below their predicted values. These patterns suggest that we should include city in the model. From a context point of view, it makes sense for location to impact sale price. In the next section, we show how to incorporate a nominal variable into a linear model.

Feature Engineering for Categorical Measurements

The first model we ever fit was the constant model in Chapter 4. There, we minimized squared loss to find the best-fitting constant:

$$\min_{c} \sum_{i} (y_i - c)^2$$

We can think of including a nominal feature in a model in a similar fashion. That is, we find the best-fitting constant to each subgroup of the data corresponding to a category:

$$\min_{c_B} \sum_{i \in \text{Berkeley}} (y_i - c_B)^2 \quad \min_{c_L} \sum_{i \in \text{Lamorinda}} (y_i - c_L)^2$$
$$\min_{c_P} \sum_{i \in \text{Piedmont}} (y_i - c_P)^2 \quad \min_{c_R} \sum_{i \in \text{Richmond}} (y_i - c_R)^2$$

Another way to describe this model is with *one-hot encoding*.

One-hot encoding takes a categorical feature and creates multiple numeric features that have only the values 0 or 1. To one-hot encode a feature, we create new features, one for each unique category. In this case, since we have four cities—Berkeley, Lamorinda, Piedmont, and Richmond—we create four new features in a design matrix, called X_{city}. Each row in X_{city} contains one value of 1, and it appears in the column that corresponds to the city. All other columns contain 0 for that row. Figure 15-6 illustrates this notion.

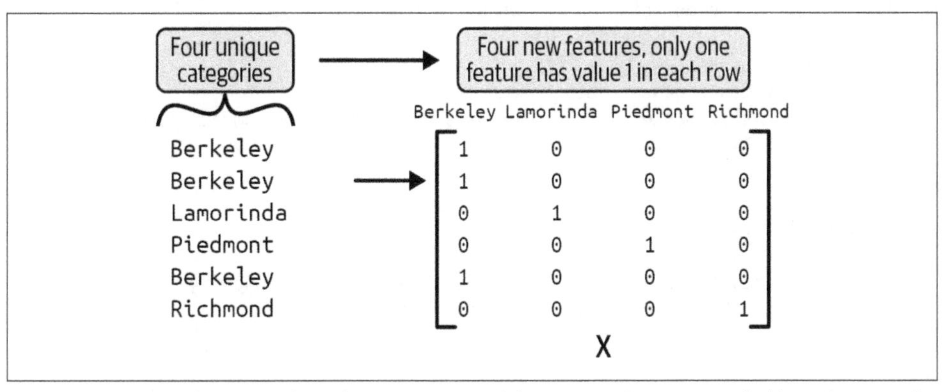

Figure 15-6. One-hot encoding for a categorical feature with six rows (left) and its resulting design matrix (right)

Now we can concisely represent the model as follows:

$$\theta_B x_{i,B} + \theta_L x_{i,L} + \theta_P x_{i,P} + \theta_R x_{i,R}$$

Here, we have indexed the columns of the design matrix by B, L, P, and R, rather than j, to make it clear that each column represents a column of 0s and 1s where, say, a 1 appears for $x_{i,P}$ if the ith house is located in Piedmont.

One-hot encoding creates features that have only 0-1 values. These features are also known as *dummy variable* or *indicator variable*. The term "dummy variable" is more common in econometrics, and the usage of "indicator variable" is more common in statistics.

Our goal is to minimize least square loss over $\boldsymbol{\theta}$:

$$\begin{aligned}
\|\mathbf{y} - \mathbf{X}\boldsymbol{\theta}\|^2 &= \sum_i (y_i - \theta_B x_{i,B} + \theta_L x_{i,L} + \theta_P x_{i,P} + \theta_R x_{i,R})^2 \\
&= \sum_{i \in Berkeley} (y_i - \theta_B x_{i,B})^2 + \sum_{i \in Lamorinda} (y_i - \theta_L x_{i,L})^2 \\
&+ \sum_{i \in Piedmont} (y_i - \theta_P x_{i,P})^2 + \sum_{i \in Richmond} (y_i - \theta_R x_{i,R})^2
\end{aligned}$$

where $\boldsymbol{\theta}$ is the column vector $[\theta_B, \theta_L, \theta_P, \theta_R]$. Notice that this minimization reduces to four minimizations, one for each city. That's the idea that we started with at the beginning of this section.

We can use `OneHotEncoder` to create this design matrix:

```
from sklearn.preprocessing import OneHotEncoder

enc = OneHotEncoder(
    # categories argument sets column order
    categories=[["Berkeley", "Lamorinda", "Piedmont", "Richmond"]],
    sparse=False,
)

X_city = enc.fit_transform(sfh[['city']])

categories_city=["Berkeley","Lamorinda", "Piedmont", "Richmond"]
X_city_df = pd.DataFrame(X_city, columns=categories_city)

X_city_df
```

	Berkeley	Lamorinda	Piedmont	Richmond
0	1.0	0.0	0.0	0.0
1	1.0	0.0	0.0	0.0
2	1.0	0.0	0.0	0.0
...	...	...	...	...
2664	0.0	0.0	0.0	1.0
2665	0.0	0.0	0.0	1.0
2666	0.0	0.0	0.0	1.0

```
2667 rows × 4 columns
```

Let's fit a model using these one-hot encoded features:

```
y_log = sfh['log_price']

model_city = LinearRegression(fit_intercept=False).fit(X_city_df, y_log)
```

And examine the multiple R^2:

```
R-square for city model: 0.57
```

If we only know the city where a house is located, the model does a reasonably good job of estimating its sale price. Here are the coefficients from the fit:

```
model_city.coef_

array([5.87, 6.03, 6.1 , 5.67])
```

As expected from the box plots, the estimated sale price (in log $) depends on the city. But if we know the size of the house as well as the city, we should have an even better model. We saw earlier that the simple log–log model that explains sale price by house size fits reasonably well, so we expect that the city feature (as one-hot encoded variables) should further improve the model.

Such a model looks like this:

$$y_i \approx \theta_1 x_i + \theta_B x_{i,B} + \theta_L x_{i,L} + \theta_P x_{i,P} + \theta_R x_{i,R}$$

Notice that this model describes the relationship between log(price), which is represented as y, and log(size), which is represented as x, as linear with the same coefficient for log(size) for each city. But the intercept term depends on the city:

$$\begin{aligned}
y_i &\approx \theta_1 x_i + \theta_B & \text{for houses in Berkeley} \\
y_i &\approx \theta_1 x_i + \theta_L & \text{for houses in Lamorinda} \\
y_i &\approx \theta_1 x_i + \theta_P & \text{for houses in Piedmont} \\
y_i &\approx \theta_1 x_i + \theta_R & \text{for houses in Richmond}
\end{aligned}$$

We next make a facet of scatterplots, one for each city, to see if this relationship roughly holds:

```
fig = px.scatter(sfh, x='log_bsqft', y='log_price',
                 facet_col='city', facet_col_wrap=2,
                 labels={'log_bsqft':'Building size (log ft^2)',
                         'log_price':'Sale price (log USD)'},
                 width=500, height=400)

fig.update_layout(margin=dict(t=30))
fig
```

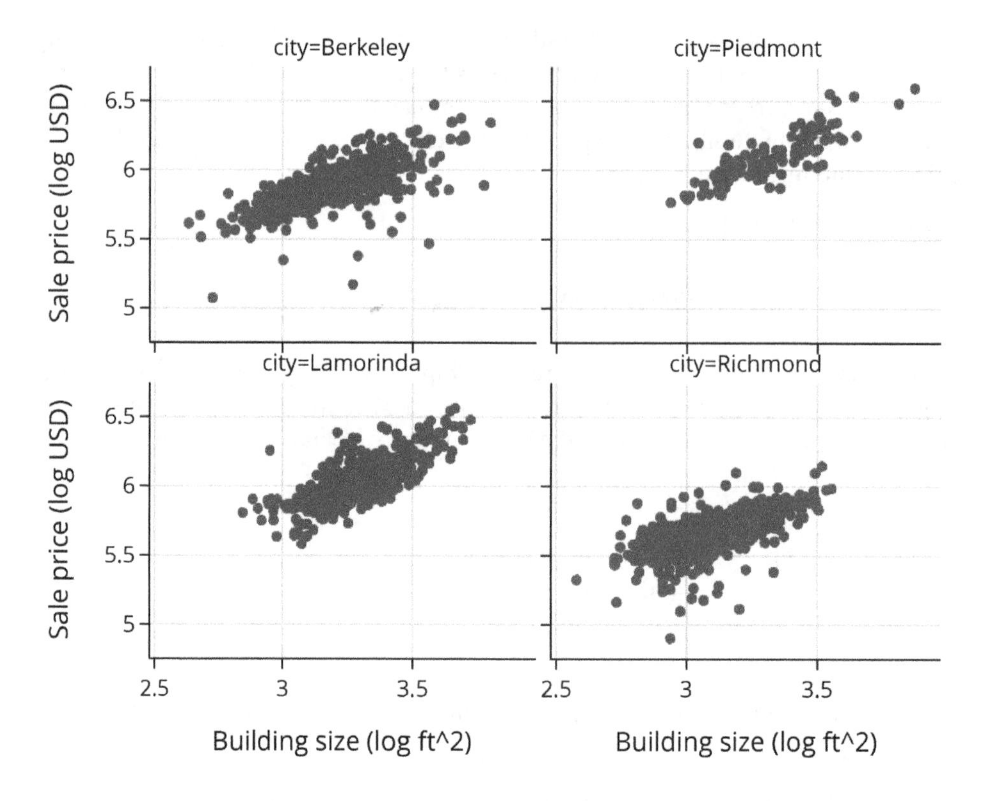

The shift is evident in the scatterplot. We concatenate our two design matrices together to fit the model that includes size and city:

```
X_size = sfh['log_bsqft']

X_city_size = pd.concat([X_size.reset_index(drop=True), X_city_df], axis=1)
X_city_size.drop(0)
```

	log_bsqft	Berkeley	Lamorinda	Piedmont	Richmond
1	3.14	1.0	0.0	0.0	0.0
2	3.31	1.0	0.0	0.0	0.0
3	2.96	1.0	0.0	0.0	0.0
...	...	...	...	...	...
2664	3.16	0.0	0.0	0.0	1.0
2665	3.47	0.0	0.0	0.0	1.0
2666	3.44	0.0	0.0	0.0	1.0

```
2666 rows × 5 columns
```

Now let's fit a model that incorporates the quantitative feature, the house size, and the qualitative feature, location (city):

```
model_city_size = LinearRegression(fit_intercept=False).fit(X_city_size, y_log)
```

The intercepts reflect which cities have more expensive houses, even taking into account the size of the house:

```
model_city_size.coef_
```

```
array([0.62, 3.89, 3.98, 4.03, 3.75])
```

```
R-square for city and log(size):  0.79
```

This fit, which includes the nominal variable `city` and the log-transformed house size, is better than both the simple log–log model with house size and the model that fits constants for each city.

Notice that we dropped the intercept from the model so that each subgroup has its own intercept. However, a common practice is to remove one of the one-hot encoded features from the design matrix and keep the intercept. For example, if we drop the feature for Berkeley houses and add the intercept, then the model is:

$$\theta_0 + \theta_1 x_i + \theta_L x_{i,L} + \theta_P x_{i,P} + \theta_R x_{i,R}$$

The meaning of the coefficients for the dummy variables has changed in this representation. For example, consider this equation for a house in Berkeley and a house in Piedmont:

$$\begin{aligned} &\theta_0 + \theta_1 x_i && \text{for a house in Berkeley} \\ &\theta_0 + \theta_1 x_i + \theta_P && \text{for a house in Piedmont} \end{aligned}$$

In this representation, the intercept θ_0 is for Berkeley houses, and the coefficient θ_P measures the typical difference between a Piedmont house and a Berkeley house. In this representation, we can more easily compare θ_P to 0 to see if these two cities have essentially the same average price.

If we include the intercept and all of the city variables, then the columns of the design matrix are linearly dependent, which means that we can't solve for the coefficients. Our predictions will be the same in either case, but there will not be a unique solution to the minimization.

We also prefer the representation of the model that drops one dummy variable and includes an intercept term when we include one-hot encodings of two categorical variables. This practice maintains consistency in the interpretation of the coefficients.

We demonstrate how to build a model with two sets of dummy variables, using the `statsmodels` library. This library uses a formula language to describe the model to fit, so we don't need to create the design matrix ourselves. We import the formula API:

```
import statsmodels.formula.api as smf
```

Let's first repeat our fit of the model with the nominal variable `city` and house size to show how to use the formula language and compare the results:

```
model_size_city = smf.ols(formula='log_price ~ log_bsqft + city',
                          data=sfh).fit()
```

The string provided for the `formula` parameter describes the model to fit. The model has `log_price` as the outcome and fits a linear combination of `log_bsqft` and `city` as explanatory variables. Notice that we do not need to create dummy variables to fit the model. Conveniently, `smf.ols` does the one-hot encoding of the city feature for us. The fitted coefficients of the following model include an intercept term and drop the Berkeley indicator variable:

```
print(model_size_city.params)
```

```
Intercept              3.89
city[T.Lamorinda]      0.09
city[T.Piedmont]       0.14
city[T.Richmond]      -0.15
log_bsqft              0.62
dtype: float64
```

If we want to drop the intercept, we can add –1 to the formula, which is a convention that indicates dropping the column of ones from the design matrix. In this particular example, the space spanned by all of the one-hot encoded features is equivalent to the space spanned by the 1 vector and all but one of the dummy variables, so the fit is the same. However, the coefficients are different as they reflect the different parameterization of the design matrix:

```
smf.ols(formula='log_price ~ log_bsqft + city - 1', data=sfh).fit().params
```

```
city[Berkeley]      3.89
city[Lamorinda]     3.98
city[Piedmont]      4.03
city[Richmond]      3.75
log_bsqft           0.62
dtype: float64
```

Additionally, we can add interaction terms between the city and size variables to allow each city to have a different coefficient for size. We specify this in the formula by adding the term `log_bsqft:city`. We don't go into details here.

Now let's fit a model with two categorical variables: the number of bedrooms and the city. Recall that we earlier reassigned the count of bedrooms that were above 6 to 6, which essentially collapses 6, 7, 8, ... into the category 6+. We can see this relationship in the box plots of price (log $) by the number of bedrooms:

```
px.box(sfh, x="br", y="log_price", width=450, height=250,
       labels={'br':'Number of bedrooms','log_price':'Sale price (log USD)'})
```

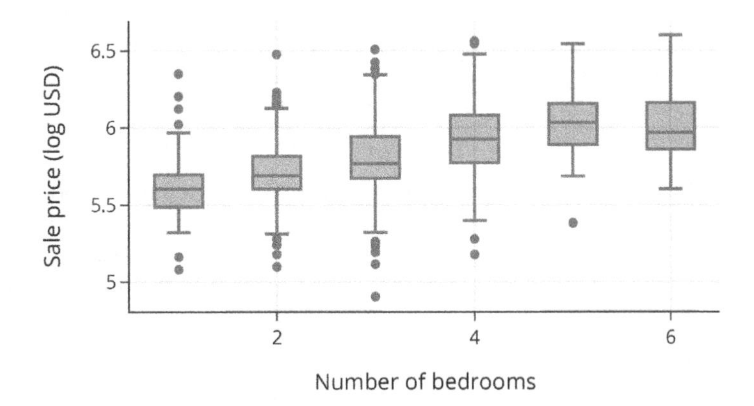

The relationship does not appear linear: for each additional bedroom, the sale price does not increase by the same amount. Given that the number of bedrooms is discrete, we can treat this feature as categorical, which allows each bedroom encoding to contribute a different amount to the cost:

```
model_size_city_br = smf.ols(formula='log_price ~ log_bsqft + city + C(br)',
                             data=sfh).fit()
```

We have used the term `C(br)` in the formula to indicate that we want the number of bedrooms, which is numeric, to be treated like a categorical variable.

Let's examine the multiple R^2 from the fit:

```
model_size_city_br.rsquared.round(2)
```

```
0.79
```

The multiple R^2 has not increased even though we have added five more one-hot encoded features. The R^2 is adjusted for the number of parameters in the model and by this measure is no better than the earlier one that included only city and size.

In this section, we introduced feature engineering for qualitative features. We saw how the one-hot encoding technique lets us include categorical data in linear models and gives a natural interpretation for model parameters.

Summary

Linear models help us describe relationships between features. We discussed the simple linear model and extended it to linear models in multiple variables. Along the way, we applied mathematical techniques that are widely useful in modeling—calculus to minimize loss for the simple linear model and matrix geometry for the multiple linear model.

Linear models may seem basic, but they are used for all sorts of tasks today. And they are flexible enough to allow us to include categorical features as well as nonlinear transformations of variables, such as log transformations, polynomials, and ratios. Linear models have the advantage of being broadly interpretable for nontechnical people, yet sophisticated enough to capture many common patterns in data.

It can be tempting to throw all of the variables available to us into a model to get the "best fit possible." But we should keep in mind the geometry of least squares when fitting models. Recall that p explanatory variables can be thought of as p vectors in n-dimensional space, and if these vectors are highly correlated, then the projections onto this space will be similar to projections onto smaller spaces made up of fewer vectors. This implies that:

- Adding more variables may not provide a large improvement in the model.
- Interpretation of the coefficients can be difficult.
- Several models can be equally effective in predicting/explaining the response variable.

If we are concerned with making inferences, where we want to interpret/understand the model, then we should err on the side of simpler models. On the other hand, if our primary concern is the predictive ability of a model, then we tend not to concern ourselves with the number of coefficients and their interpretation. But this "black box" approach can lead to models that, say, overly depend on anomalous values in the data or models that are inadequate in other ways. So be careful with this approach, especially when the predictions may be harmful to people.

In this chapter, we used linear models in a descriptive way. We introduced a few notions for deciding when to include a feature in a model by examining residuals for patterns, comparing the size of standard errors and the change in the multiple R^2. Oftentimes, we settled for a simpler model that was easier to interpret. In the next chapter, we look at other, more formal tools for choosing the features to include in a model.

CHAPTER 16

Model Selection

So far when we fit models, we have used a few strategies to decide which features to include:

- Assess model fit with residual plots.
- Connect the statistical model to a physical model.
- Keep the model simple.
- Compare improvements in the standard deviation of the residuals and in the MSE between increasingly complex models.

For example, when we examined the one-variable model of upward mobility in Chapter 15, we found curvature in the residual plot. Adding a second variable greatly improved the fit in terms of average loss (MSE and, relatedly, multiple R^2), but some curvature remained in the residuals. A seven-variable model made little improvement over the two-variable model in terms of a decrease in MSE, so although the two-variable model still showed some patterns in the residuals, we opted for this simpler model.

As another example, when we model the weight of a donkey in Chapter 18, we will take guidance from a physical model. We'll ignore the donkey's appendages and draw on the similarity between a barrel and a donkey's body to begin fitting a model that explains weight by its length and girth (comparable to a barrel's height and circumference). We'll then continue to adjust that model by adding categorical features related to the donkey's physical condition and age, collapsing categories, and excluding other possible features to keep the model simple.

The decisions we make in building these models are based on judgment calls, and in this chapter we augment these with more formal criteria. To begin, we provide an example that shows why it's typically not a good idea to include too many features in a model. This phenomenon, called *overfitting*, often leads to models that follow the data too closely and capture some of the noise in the data. Then, when new observations come along, the predictions are worse than those from a simpler model. The remainder of the chapter provides techniques, such as the train-test split, cross-validation, and regularization, for limiting the impact of overfitting. These techniques are especially helpful when there are a large number of potential features to include in a model. We also provide a synthetic example, where we know the true model, to explain the concepts of model variance and bias and how they relate to over- and underfitting.

Overfitting

When we have many features available to include in a model, choosing which ones to include or exclude rapidly gets complicated. In the upward mobility example in Chapter 15, we chose two of the seven variables to fit the model, but there are 21 pairs of features that we could have examined and fitted for a two-variable model. And there are over one hundred models to choose from if we consider all one-, two-, ..., seven-variable models. It can be hard to examine hundreds of residual plots to decide how simple is simple enough, and to settle on a model. Unfortunately, the notion of minimizing MSE isn't entirely helpful either. With each variable that we add to a model, the MSE typically gets smaller. Recall from the geometric perspective of model fitting (Chapter 15) that adding a feature to a model adds an n-dimensional vector to the feature space, and the error between the outcome vector and its projection into the space spanned by the explanatory variables shrinks. We might view this as a good thing because our model fits the data more closely, but there is a danger in overfitting.

Overfitting happens when the model follows the data too closely and picks up the variability in the random noise in the outcome. When this happens, new observations are not well-predicted. An example helps clarify this idea.

Example: Energy Consumption

In this example, we examine a dataset you can download (*https://oreil.ly/ngD4G*) that contains information from utility bills for a private residence in Minnesota. We have records of the monthly gas usage in a home (cubic feet) and the average temperature that month (degrees Fahrenheit).[1] We first read in the data:

```
heat_df = pd.read_csv("data/utilities.csv", usecols=["temp", "ccf"])
heat_df
```

1 These data are from Daniel T. Kaplan (CreateSpace Independent Publishing Platform, 2009).

	temp	ccf
0	29	166
1	31	179
2	15	224
...	...	...
96	76	11
97	55	32
98	39	91

```
99 rows × 2 columns
```

We will begin by looking at a scatterplot of gas consumption as a function of temperature:

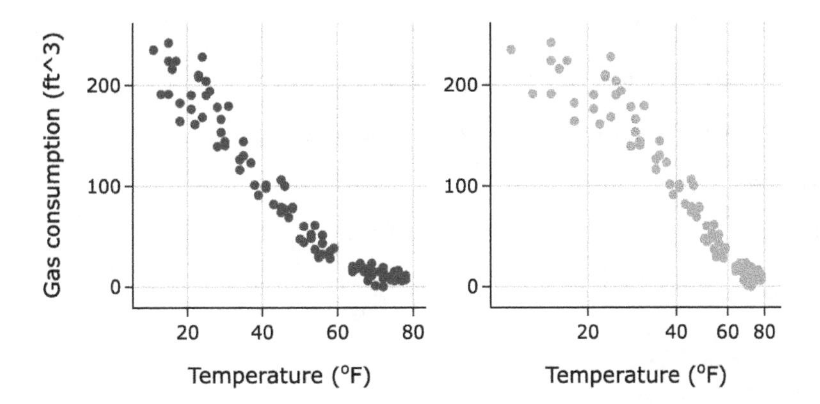

The relationship shows curvature (left plot), but when we try to straighten it with a log transformation (right plot), a different curvature arises in the low-temperature region. Additionally, there are two unusual points. When we refer back to the documentation, we find that these points represent recording errors, so we remove them.

Let's see if a quadratic curve can capture the relationship between gas usage and temperature. Polynomials are still considered linear models. They are linear in their polynomial features. For example, we can express a quadratic model as:

$$\theta_0 + \theta_1 x + \theta_2 x^2$$

This model is linear in the features x and x^2, and in matrix notation we can write this model as $\mathbf{X}\boldsymbol{\theta}$, where $\mathbf{X}$ is the design matrix:

$$\begin{bmatrix} 1 & x_1 & x_1^2 \\ 1 & x_2 & x_2^2 \\ \vdots & \vdots & \vdots \\ 1 & x_n & x_n^2 \end{bmatrix}$$

We can create the polynomial features of the design matrix with the `PolynomialFeatures` tool in `scikit-learn`:

```
y = heat_df['ccf']
X = heat_df[['temp']]

from sklearn.preprocessing import PolynomialFeatures

poly = PolynomialFeatures(degree=2, include_bias=False)
poly_features = poly.fit_transform(X)
poly_features
```

```
array([[  29.,  841.],
       [  31.,  961.],
       [  15.,  225.],
       ...,
       [  76., 5776.],
       [  55., 3025.],
       [  39., 1521.]])
```

We set the parameter `include_bias` to `False` because we plan to fit the polynomial with the `LinearRegression` method in `scikit-learn`, and by default it includes the constant term in the model. We fit the polynomial with:

```
from sklearn.linear_model import LinearRegression

model_deg2 = LinearRegression().fit(poly_features, y)
```

To get a quick idea as to the quality of the fit, let's overlay the fitted quadratic on the scatterplot and also look at the residuals:

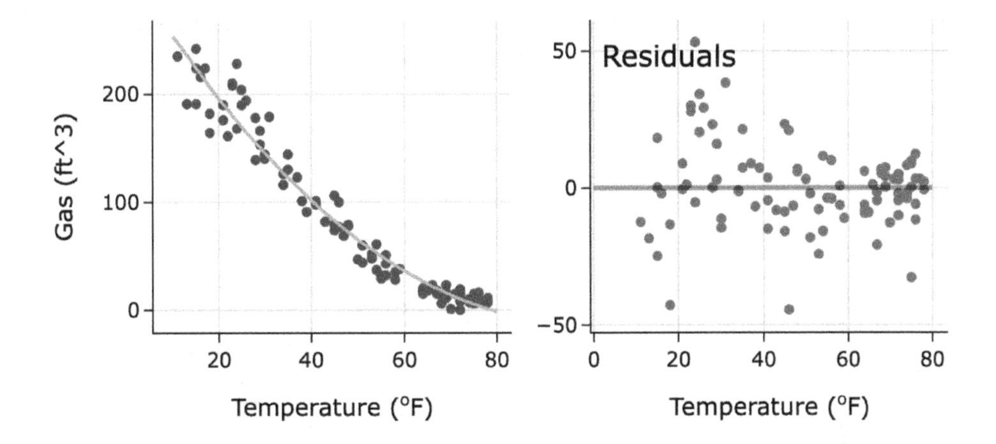

The quadratic captures the curve in the data quite well, but the residuals show a slight upward trend in the temperature range of 70°F to 80°F, which indicates some lack of fit. There is also some funneling in the residuals, where the variability in gas consumption is greater in the colder months. We might expect this behavior since we have only the monthly average temperature.

For comparison, we fit a few more models with higher-degree polynomials and collectively examine the fitted curves:

```
poly12 = PolynomialFeatures(degree=12, include_bias=False)
poly_features12 = poly12.fit_transform(X)

degrees = [1, 2, 3, 6, 8, 12]

mods = [LinearRegression().fit(poly_features12[:, :deg], y)
        for deg in degrees]
```

We use the polynomial features in this section to demonstrate overfitting, but directly fitting the $x, x^2, x^3, \ldots$ polynomials is not advisable in practice. Unfortunately, these polynomial features tend to be highly correlated. For example, the correlation between x and x^2 for the energy data is 0.98. Highly correlated features give unstable coefficients, where a small change in an x-value can lead to a large change in the coefficients of the polynomial. Also, when the x-values are large, the normal equations are poorly conditioned and the coefficients can be difficult to interpret and compare.

A better practice is to use polynomials that have been constructed to be orthogonal to one another. These polynomials fill the same space as the original polynomials, but they are uncorrelated with one another and give a more stable fit.

Let's place all of the polynomial fits on the same graph so that we can see how the higher-degree polynomials bend more and more strangely:

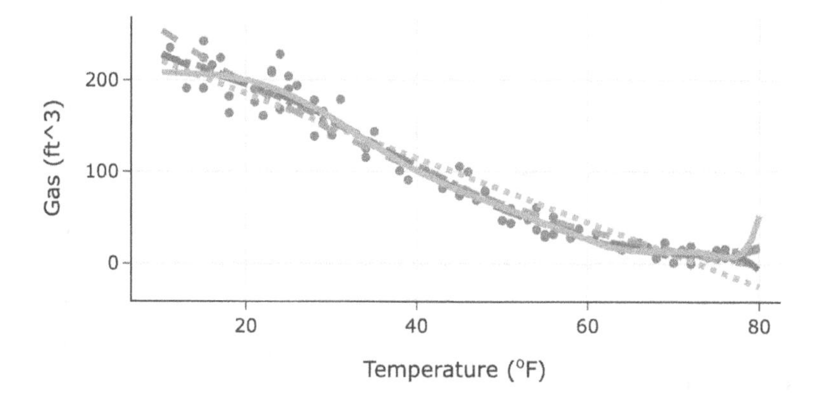

We can also visualize the different polynomial fits in separate facets:

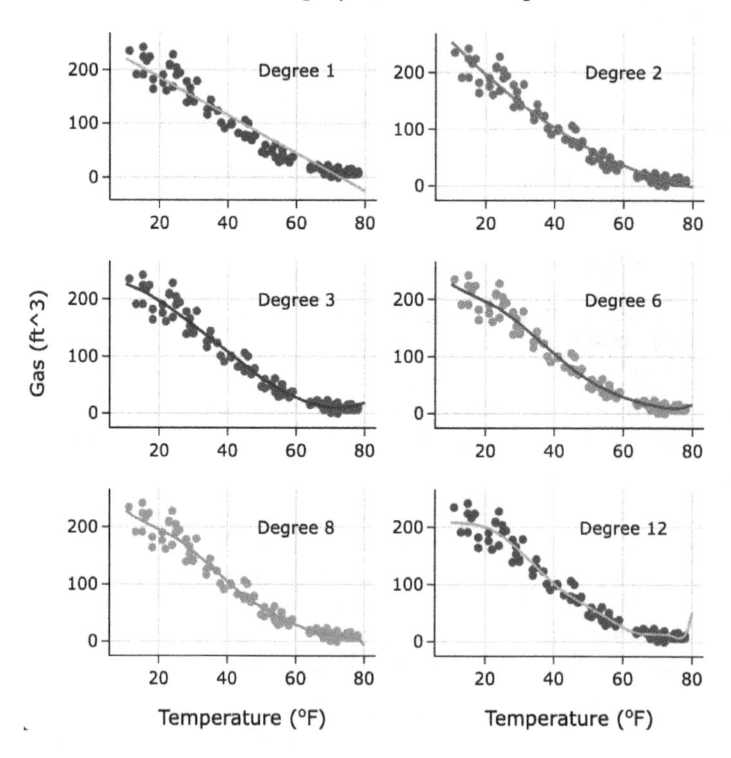

The degree 1 curve (the straight line) in the upper-left facet misses the curved pattern in the data. The degree 2 curve begins to capture it, and the degree 3 curve looks like an improvement, but notice the upward bend at the right side of the plot. The polynomials of degrees 6, 8, and 12 follow the data increasingly closely, as they get increasingly curvy. These polynomials seem to fit spurious bumps in the data. Altogether, these six curves illustrate under- and overfitting. The fitted line in the upper left underfits and misses the curvature entirely. And the degree 12 polynomial in the bottom right definitely overfits with a wiggly pattern that we don't think makes sense in this context.

In general, as we add more features, models get more complex and the MSE drops, but at the same time, the fitted model grows increasingly erratic and sensitive to the data. When we overfit, the model follows the data too closely, and predictions are poor for new observations. One simple technique to assess a fitted model is to compute the MSE on new data, data that were not used in building the model. Since we don't typically have the capacity to acquire more data, we set aside some of the original data to evaluate the fitted model. This technique is the topic of the next section.

Train-Test Split

Although we want to use all of our data in building a model, we also want to get a sense of how the model behaves with new data. We often do not have the luxury of collecting additional data to assess a model, so instead we set aside a portion of our data, called the *test set*, to stand in for new data. The remainder of the data is called the *train set*, and we use this portion to build the model. Then, after we have chosen a model, we pull out the test set and see how well the model (fitted on the train set) predicts the outcomes in the test set. Figure 16-1 demonstrates this idea.

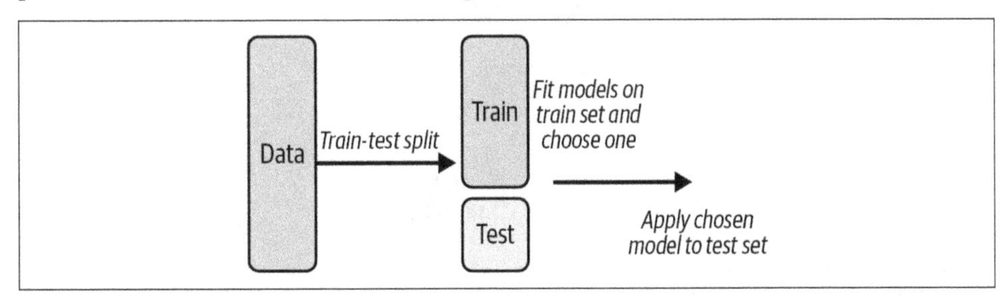

Figure 16-1. The train-test split divides the data into two parts: the train set is used to build the model and the test set evaluates that model

Typically, the test set consists of 10% to 25% of the data. What might not be clear from the diagram is that this division into two parts is often made at random, so the train and test sets are similar to each other.

We can describe this process using the notion introduced in Chapter 15. The design matrix, **X**, and outcome, **y**, are each divided into two parts; the design matrix, labeled

$\mathbf{X}_T$, and corresponding outcomes, $\mathbf{y}_T$, form the train set. We minimize the average squared loss over $\boldsymbol{\theta}$ with these data:

$$\min_{\boldsymbol{\theta}} \|\mathbf{y}_T - \mathbf{X}_T\boldsymbol{\theta}\|^2$$

The coefficient, $\hat{\boldsymbol{\theta}}_T$, that minimizes the training error is used to predict outcomes for the test set, which is labeled $\mathbf{X}_S$ and $\mathbf{y}_S$:

$$\|\mathbf{y}_S - \mathbf{X}_S\hat{\boldsymbol{\theta}}_T\|^2$$

Since $\mathbf{X}_S$ and $\mathbf{y}_S$ are not used to build the model, they give a reasonable estimate of the loss we might expect for a new observation.

We demonstrate the train-test split with our polynomial model for gas consumption from the previous section. To do this, we carry out the following steps:

1. Split the data at random into two parts, the train and test sets.
2. Fit several polynomial models to the train set and choose one.
3. Compute the MSE on the test set for the chosen polynomial (with coefficients fitted on the train set).

For the first step, we divide the data with the `train_test_split` method in `scikit-learn` and set aside 22 observations for model evaluation:

```
from sklearn.model_selection import train_test_split

test_size = 22

X_train, X_test, y_train, y_test = train_test_split(
    X, y, test_size=test_size, random_state=42)


print(f'Training set size: {len(X_train)}')
print(f'Test set size: {len(X_test)}')
```

```
Training set size: 75
Test set size: 22
```

As in the previous section, we fit models of gas consumption to various polynomials in temperature. But this time, we use only the training data:

```
poly = PolynomialFeatures(degree=12, include_bias=False)
poly_train = poly.fit_transform(X_train)

degree = np.arange(1,13)
```

```
mods = [LinearRegression().fit(poly_train[:, :j], y_train)
        for j in degree]
```

We find the MSE for each of these models:

```
from sklearn.metrics import mean_squared_error

error_train = [
    mean_squared_error(y_train, mods[j].predict(poly_train[:, : (j + 1)]))
    for j in range(12)
]
```

To visualize the change in MSE, we plot MSE for each fitted polynomial against its degree:

```
px.line(x=degree, y=error_train, markers=True,
        labels=dict(x='Degree of polynomial', y='Train set MSE'),
        width=350, height=250)
```

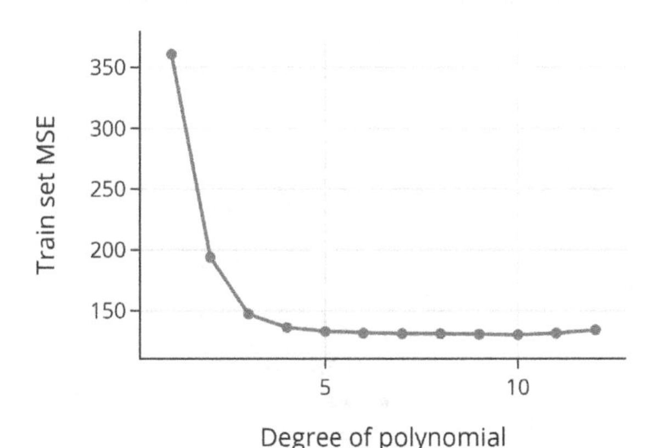

Notice that the training error decreases with the additional model complexity. We saw earlier that the higher-order polynomials showed a wiggly behavior that we don't think reflects the underlying structure in the data. With this in mind, we might choose a model that is simpler but shows a large reduction in MSE. That could be degree 3, 4, or 5. Let's go with degree 3 since the difference between these three models in terms of MSE is quite small and it's the simplest.

Now that we have chosen our model, we provide an independent assessment of its MSE using the test set. We prepare the design matrix for the test set and use the degree 3 polynomial fitted on the train set to predict the outcome for each row in the test set. Lastly, we compute the MSE for the test set:

```
poly_test = poly.fit_transform(X_test)
y_hat = mods[2].predict(poly_test[:, :3])
```

```
mean_squared_error(y_test, y_hat)
```

```
307.44460133992294
```

The MSE for this model is quite a bit larger than the MSE computed on the training data. This demonstrates the problem with using the same data to fit and evaluate a model: the MSE doesn't adequately reflect the MSE for a new observation. To further demonstrate the problem with overfitting, we compute the error for the test for each of these models:

```
error_test = [
    mean_squared_error(y_test, mods[j].predict(poly_test[:, : (j + 1)]))
    for j in range(12)
]
```

In practice, we do not look at the test set until we have committed to a model. Alternating between fitting a model on the train set and evaluating it on the test set can lead to overfitting. But for demonstration purposes, we plot the MSE on the test set for all of the polynomial models we fitted:

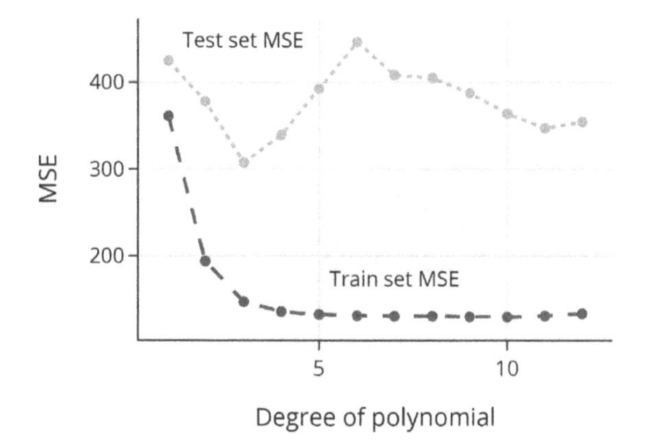

Notice how the MSE for the test set is larger than the MSE for the train set for all models, not just the model that we selected. More importantly, notice how the MSE for the test set initially decreases as the model goes from underfitting to one that follows the curvature in the data a bit better. Then, as the model grows in complexity, the MSE for the test set increases. These more complex models overfit the training data and lead to large errors in predicting the test set. An idealization of this phenomenon is captured in the diagram in Figure 16-2.

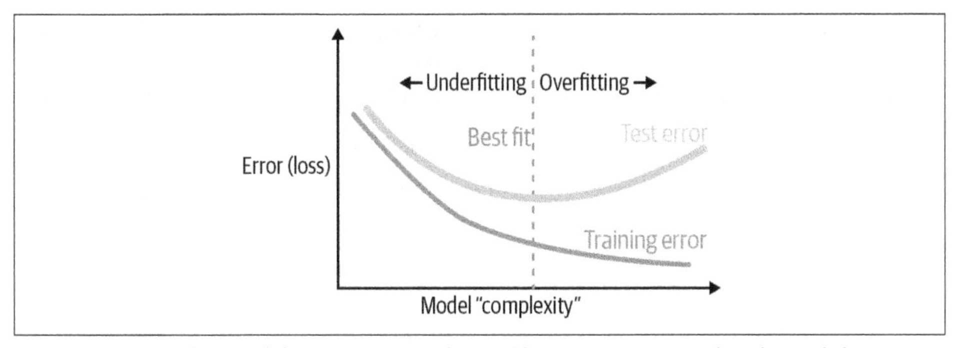

Figure 16-2. As the model grows in complexity, the train set error shrinks and the test set error increases

The test data provides an assessment of the prediction error for new observations. It is crucial to use the test set only once, after we have committed to a model. Otherwise, we fall into the trap of using the same data to choose and evaluate the model. When choosing the model, we fell back on the simplicity argument because we were aware that increasingly complex models tend to overfit. However, we can extend the train-test method to help select the model as well. This is the topic of the next section.

Cross-Validation

We can use the train-test paradigm to help us choose a model. The idea is to further divide the train set into separate parts where we fit the model on one part and evaluate it on another. This approach is called *cross-validation*. We describe one version, called *k-fold cross-validation*. Figure 16-3 shows the idea behind this division of the data.

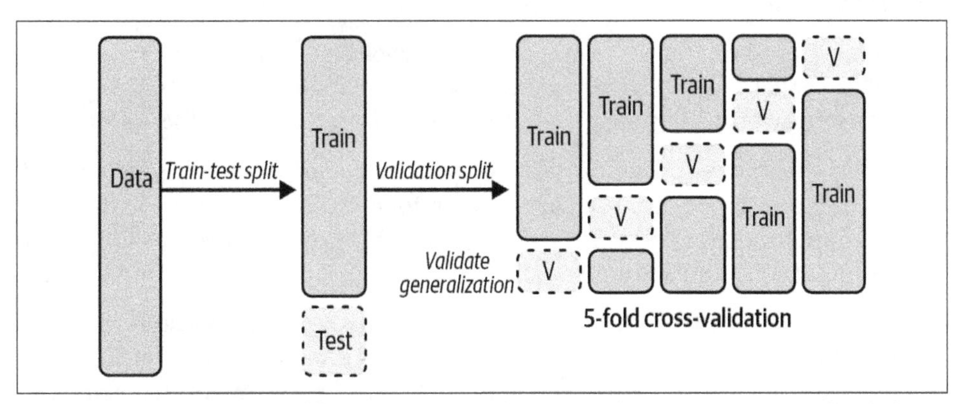

Figure 16-3. An example of fivefold cross-validation in which the train set is divided into five parts that are used in turn to validate models built on the remainder of the data

Cross-validation can help select the general form of a model. By this we mean the degree of the polynomial, the number of features in the model, or a cutoff for a regularization penalty (covered in the next section). The basic steps behind k-fold cross-validation are as follows:

1. Divide the train set into k parts of roughly the same size; each part is called a *fold*. Use the same technique that was used to create the train and test sets to make the folds. Typically, we divide the data at random.
2. Set one fold aside to act as a test set:
   - Fit all models on the remainder of the training data (the training data less the particular fold).
   - Use the fold you set aside to evaluate all of these models.
3. Repeat this process for a total of k times, where each time you set aside one fold, use the rest of the train set to fit the models, and evaluate them on the fold that was set aside.
4. Combine the error in fitting each model across the folds, and choose the model with the smallest error.

These fitted models will not have identical coefficients across folds. As an example, when we fit a polynomial of, say, degree 3, we average the MSE across the k folds to get an average MSE for the k fitted polynomials of degree 3. We then compare the MSEs and choose the degree of the polynomial with the lowest MSE. The actual coefficients for the x, x^2, and x^3 terms in the cubic polynomial are not the same in each of the k fits. Once the polynomial degree is selected, we refit the model using all of the training data and evaluate it on the test set. (We haven't used the test set in any of the earlier steps to select the model.)

Typically, we use 5 or 10 folds. Another popular choice puts one observation in each fold. This special case is called *leave-one-out cross-validation*. Its popularity stems from the simplicity in adjusting a least squares fit to have one fewer observation.

Generally, k-fold cross-validation takes some computation time since we typically have to refit each model from scratch for each fold. The `scikit-learn` library provides a convenient `sklearn.model_selection.KFold` (*https://oreil.ly/tnHTv*) class to implement k-fold cross-validation.

To give you an idea of how k-fold cross-validation works, we'll demonstrate the technique on the gas consumption example. However, this time we'll fit a different type of model. In the original scatterplot of the data, it looks like the points fall along two connected line segments. In cold temperatures, the relationship between gas consumption and temperature looks roughly linear with a negative slope of about −4 cubic ft/degree, and in warmer months, the relationship appears nearly flat. So, rather than fitting a polynomial, we can fit a bent line to the data.

Let's start by fitting a line with a bend at 65 degrees. To do this, we create a feature that enables the points with temperatures above 65°F to have a different slope. The model is:

$$y = \theta_0 + \theta_1 x + \theta_2 (x - 65)^+$$

Here, $( )^+$ stands for "positive part," so when x is less than 65 it evaluates to 0, and when x is 65 or greater it is just $x - 65$. We create this new feature and add it to the design matrix:

```
y = heat_df["ccf"]
X = heat_df[["temp"]]
X["temp65p"] = (X["temp"] - 65) * (X["temp"] >= 65)
```

Then we fit the model with these two features:

```
bend_index = LinearRegression().fit(X, y)
```

Let's overlay this fitted "curve" on the scatterplot to see how well it captures the shape of the data:

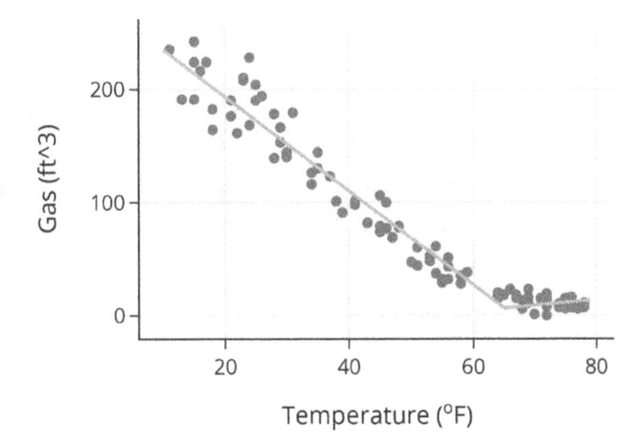

This model appears to fit the data much better than a polynomial. But many bent line models are possible. The line might bend at 55 degrees or 60 degrees, and so on. We can use k-fold cross-validation to choose the temperature value at which the line bends. Let's consider models with bends at $40, 41, 42, \ldots, 68, 69$ degrees. For each of these, we need to create the additional feature to enable the line to bend there:

```
bends = np.arange(40, 70, 1)

for i in bends:
    col = "temp" + i.astype("str") + "p"
    heat_df[col] = (heat_df["temp"] - i) * (heat_df["temp"] >= i)
heat_df
```

	temp	ccf	temp40p	temp41p	...	temp66p	temp67p	temp68p	temp69p
0	29	166	0	0	...	0	0	0	0
1	31	179	0	0	...	0	0	0	0
2	15	224	0	0	...	0	0	0	0
...	...	...	...	...	...	...	...	...	...
96	76	11	36	35	...	10	9	8	7
97	55	32	15	14	...	0	0	0	0
98	39	91	0	0	...	0	0	0	0

```
97 rows × 32 columns
```

The first step in cross-validation is to create our train and test sets. Like before, we choose 22 observations at random to be placed in the test set. That leaves 75 for the train set:

```
y = heat_df['ccf']
X = heat_df.drop(columns=['ccf'])

test_size = 22

X_train, X_test, y_train, y_test = train_test_split(
    X, y, test_size=test_size, random_state=0)
```

Now we can divide the train set into folds. We use three folds so that we have 25 observations in each fold. For each fold, we fit 30 models, one for each bend in the line. For this step, we divide the data with the `KFold` method in `scikit-learn`:

```
from sklearn.model_selection import KFold

kf = KFold(n_splits=3, shuffle=True, random_state=42)

validation_errors = np.zeros((3, 30))


def validate_bend_model(X, y, X_valid, y_valid, bend_index):
    model = LinearRegression().fit(X.iloc[:, [0, bend_index]], y)
    predictions = model.predict(X_valid.iloc[:, [0, bend_index]])
    return mean_squared_error(y_valid, predictions)


for fold, (train_idx, valid_idx) in enumerate(kf.split(X_train)):
    cv_X_train, cv_X_valid = (X_train.iloc[train_idx, :],
                              X_train.iloc[valid_idx, :])
```

```
    cv_Y_train, cv_Y_valid = (y_train.iloc[train_idx],
                              y_train.iloc[valid_idx])

    error_bend = [
        validate_bend_model(
            cv_X_train, cv_Y_train, cv_X_valid, cv_Y_valid, bend_index
        )
        for bend_index in range(1, 31)
    ]

    validation_errors[fold][:] = error_bend
```

Then we find the mean validation error across the three folds and plot them against the location of the bend:

```
totals = validation_errors.mean(axis=0)
```

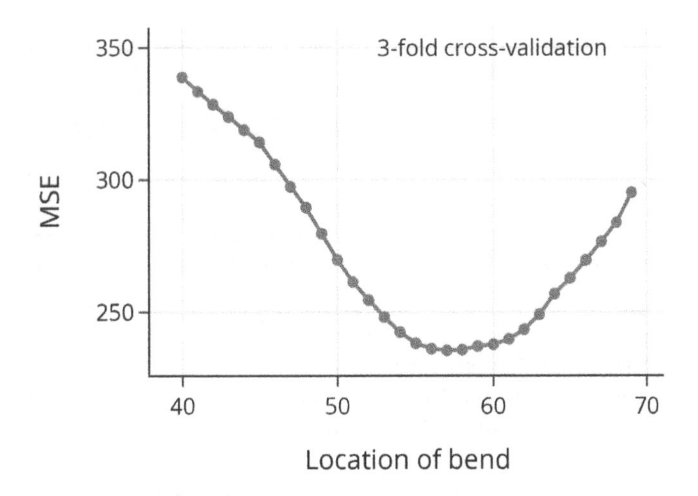

The MSE looks quite flat for 57 to 60 degrees. The minimum occurs at 58, so we choose that model. To assess this model on the test set, we first fit the bent line model at 58 degrees on the entire train set:

```
bent_final = LinearRegression().fit(
    X_train.loc[:, ["temp", "temp58p"]], y_train
)
```

Then we use the fitted model to predict gas consumption for the test set:

```
y_pred_test = bent_final.predict(X_test.loc[:, ["temp", "temp58p"]])

mean_squared_error(y_test, y_pred_test)

71.40781435952441
```

Let's overlay the bent-line fit on the scatterplot and examine the residuals to get an idea as to the quality of the fit:

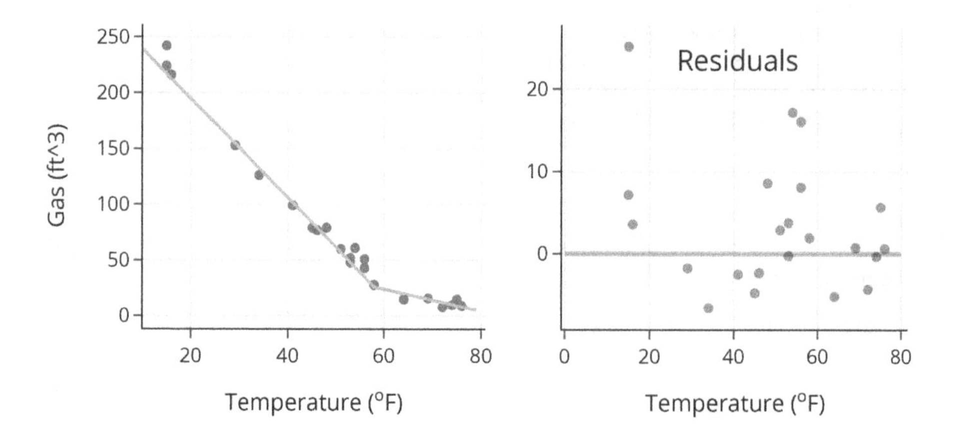

The fitted curve looks reasonable, and the residuals are much smaller than those from the polynomial fit.

For teaching purposes in this section, we use `KFold` to manually split up the training data into three folds, then find the model validation errors using a loop. In practice, we suggest using `sklearn.model_selection.GridSearchCV` with an `sklearn.pipeline.Pipeline` object, which can automatically break the data into training and validation sets and find the model that has the lowest average validation error across the folds.

Using cross-validation to manage model complexity has a couple of critical limitations: typically it requires the complexity to vary discretely, and there may not be a natural way to order the models. Rather than changing the dimensions of a sequence of models, we can fit a large model and apply constraints on the size of the coefficients. This notion is called regularization and is the topic of the next section.

Regularization

We just saw how cross-validation can help find a dimension for a fitted model that balances under- and overfitting. Rather than selecting the dimension of the model, we can build a model with all of the features, but restrict the size of the coefficients. We keep from overfitting by adding to the MSE a penalty term on the size of the coefficients. The penalty, called a *regularization term*, is $\lambda \Sigma_{j=1}^{p} \theta_j^2$. We fit the model by minimizing the combination of mean squared error plus this penalty:

$$\frac{1}{n}\sum_{i=1}^{n}(y_i - \mathbf{x}_i\boldsymbol{\theta})^2 + \lambda\sum_{j=1}^{p}\theta_j^2$$

When the *regularization parameter*, λ, is large, it penalizes large coefficients. (We typically choose it by cross-validation.)

Penalizing the square of the coefficients is called L_2 regularization, or *ridge regression*. Another popular regularization penalizes the absolute size of the coefficients:

$$\frac{1}{n}\sum_{i=1}^{n}(y_i - \mathbf{x}_i\boldsymbol{\theta})^2 + \lambda\sum_{j=1}^{p}|\theta_j|$$

This L_1 regularized linear model is also called *lasso regression* (lasso stands for Least Absolute Shrinkage and Selection Operator).

To get an idea about how regularization works, let's think about the extreme cases: when λ is really large and when it's close to 0 (λ is never negative). With a big regularization parameter, the coefficients are heavily penalized, so they shrink. On the other hand, when λ is tiny, the coefficients aren't restricted. In fact, when λ is 0, we're back in the world of ordinary least squares. A couple of issues crop up when we think about controlling the size of the coefficients through regularization:

- We do not want to regularize the intercept term. This way, a large penalty fits a constant model.
- When features have very different scales, the penalty can impact them differently, with large-valued features being penalized more than others. To avoid this, we standardize all of the features to have mean 0 and variance 1 before fitting the model.

Let's look at an example with 35 features.

Model Bias and Variance

In this section, we provide a different way to think about the problem of over- and underfitting. We carry out a simulation study where we generate synthetic data from a model of our design. This way, we know the true model, and we can see how close we get to the truth when we fit models to the data.

We concoct a general model of data as follows:

$$y = g(\mathbf{x}) + \epsilon$$

This expression makes it easy to see the two components of the model: the signal $g(x)$ and the noise ϵ. In our model, we assume the noise has no trend or pattern, constant variance, and each observation's noise is independent of the others.

As an example, let's take $g(x) = \sin(x) + 0.3x$ and the noise from a normal curve with center 0 and SD = 0.2. We can generate data from this model with the following functions:

```
def g(x):
    return np.sin(x) + 0.3 * x

def gen_noise(n):
    return np.random.normal(scale=0.2, size=n)

def draw(n):
    points = np.random.choice(np.arange(0, 10, 0.05), size=n)
    return points, g(points) + gen_noise(n)
```

Let's generate 50 data points (x_i, y_i), $i = 1, \ldots, 50$, from this model:

```
np.random.seed(42)

xs, ys = draw(50)
noise = ys - g(xs)
```

We can plot our data, and since we know the true signal, we can find the errors and plot them too:

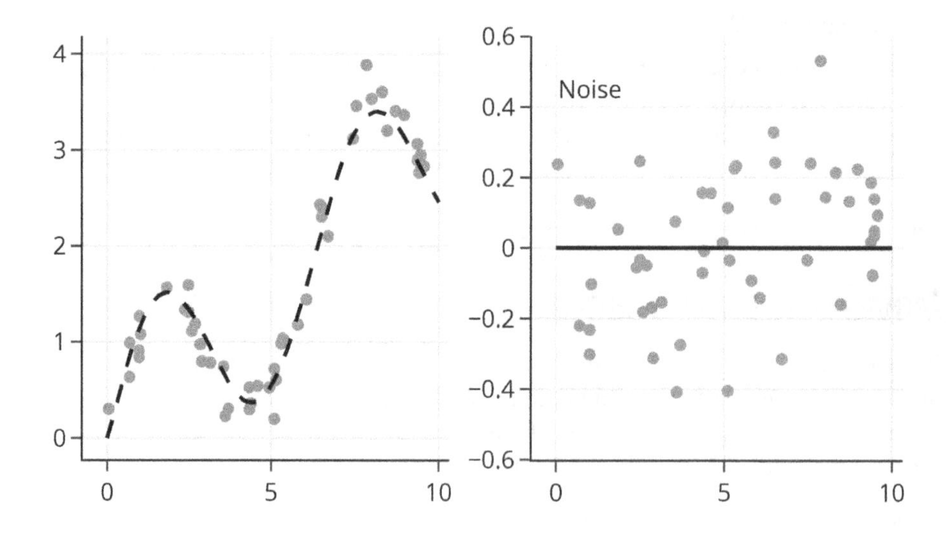

The plot on the left shows g as a dashed curve. We can also see that the (x, y) pairs form a scatter of dots about this curve. The righthand plot shows the errors, $y - g(x)$, for the 50 points. Notice that they do not form a pattern.

When we fit a model to the data, we minimize the mean squared error. Let's write this minimization in generality:

$$\min_{f \in \mathscr{F}} \frac{1}{n} \sum_{i=1}^{n} [y_i - f(\mathbf{x}_i)]^2$$

The minimization is over the collection of functions $\mathscr{F}$. We have seen in this chapter that this collection of functions might be polynomials of 12 degrees, or simply bent lines. An important point is that the true model, g, doesn't have to be one of the functions in the collection.

Let's take $\mathscr{F}$ to be the collection of second-degree polynomials; in other words, functions that can be expressed as $\theta_0 + \theta_1 x + \theta_2 x^2$. Since $g(x) = \sin(x) + 0.3x$, it doesn't belong to the collection of functions that we are optimizing over.

Let's fit a polynomial to our 50 data points:

```
poly = PolynomialFeatures(degree=2, include_bias=False)
poly_features = poly.fit_transform(xs.reshape(-1, 1))

model_deg2 = LinearRegression().fit(poly_features, ys)
```

```
Fitted Model: 0.98 + -0.19x + 0.05x^2
```

Again, we know the true model is not quadratic (because we built it). Let's plot the data and the fitted curve:

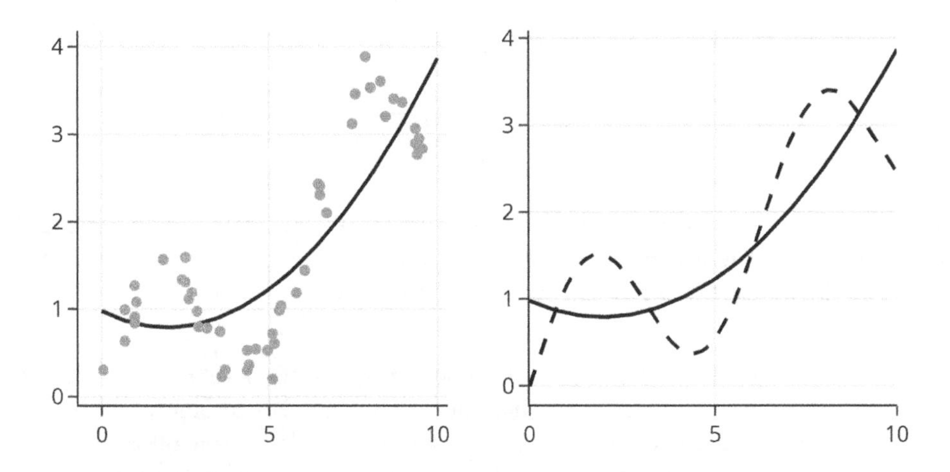

The quadratic doesn't fit the data well, and it doesn't represent the underlying curve well either because the set of models that we are choosing from (second-order polynomials) can't capture the curvature in g.

If we repeat this process and generate another 50 points from the true model and fit a second-degree polynomial to these data, then the fitted coefficients of the quadratic will change because it depends on the new set of data. We can repeat this process many times, and average the fitted curves. This average curve will resemble the typical best fit of a second-degree polynomial to 50 points from our true model. To demonstrate this notion, let's generate 25 sets of 50 data points and fit a quadratic to each dataset:

```
def fit(n):
    xs_new = np.random.choice(np.arange(0, 10, 0.05), size=n)
    ys_new = g(xs_new) + gen_noise(n)
    X_new = xs_new.reshape(-1, 1)
    mod_new = LinearRegression().fit(poly.fit_transform(X_new), ys_new)
    return mod_new.predict(poly_features_x_full).flatten()

fits = [fit(50) for j in range(25)]
```

We can show on a plot all 25 fitted models along with the true function, g, and the average of the fitted curves, $\bar{f}$. To do this, we use transparency for the 25 fitted models to distinguish overlapping curves:

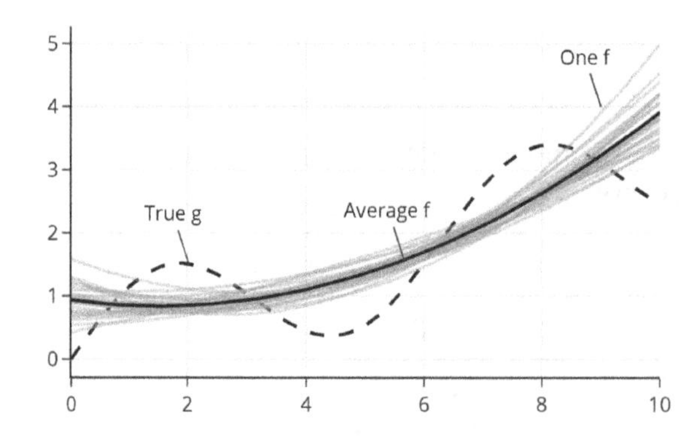

We can see that the 25 fitted quadratics vary with the data. This concept is called *model variation*. The average of the 25 quadratics is represented by the solid black line. The difference between the average quadratic and the true curve is called the *model bias*.

When the signal, g, does not belong to the model space, $\mathscr{F}$, we have model bias. If the model space can approximate g well, then the bias is small. For instance, a 10-degree polynomial can get pretty close to the g used in our example. On the other hand, we have seen earlier in this chapter that higher-degree polynomials can overfit the data and vary a lot trying to get close to the data. The more complex the model space, the greater the variability in the fitted model. Underfitting with too simple a model can lead to high model bias (the difference between g and $\bar{f}$), and overfitting with too complex a model can result in high model variance (the fluctuations of $\hat{f}$ around $\bar{f}$). This notion is called the *bias-variance trade-off*. Model selection aims to balance these competing sources of a lack of fit.

Summary

In this chapter, we saw that problems arise when we minimize mean squared error to both fit a model and evaluate it. The train-test split helps us get around this problem, where we fit a model with the train set and evaluate our fitted model on test data that have been set aside.

It's important to not "overuse" the test set, so we keep it separate until we have committed to a model. To help us commit, we might use cross-validation, which imitates the division of data into test and train sets. Again, it's important to cross-validate using only the train set and keep the original test set away from any model selection process.

Regularization takes a different approach and penalizes the mean squared error to keep the model from fitting the data too closely. In regularization, we use all of the data available to fit the model, but shrink the size of the coefficients.

The bias-variance trade-off allows us to more precisely describe the modeling phenomena that we have seen in this chapter: underfitting relates to model bias; overfitting results in model variance. In Figure 16-4, the x-axis measures model complexity and the y-axis measures these two components of model misfit: model bias and model variance. Notice that as the complexity of the model being fit increases, model bias decreases and model variance increases. Thinking in terms of test error, we have seen this error first decrease and then increase as the model variance outweighs the decrease in model bias. To select a useful model, we must strike a balance between model bias and variance.

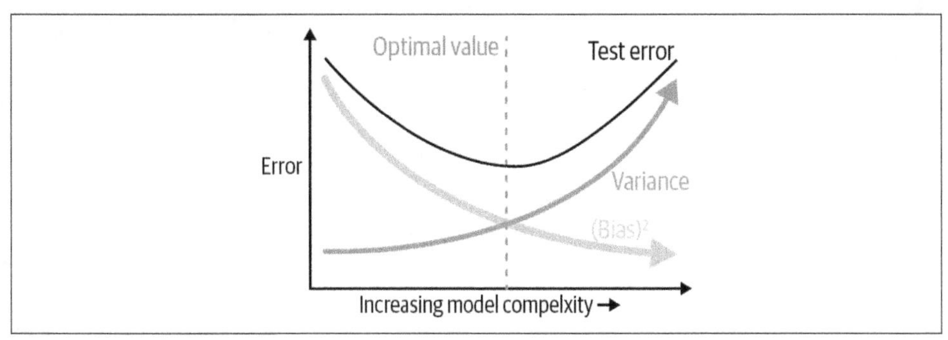

Figure 16-4. Bias-variance trade-off

Collecting more observations reduces bias if the model can fit the population process exactly. If the model is inherently incapable of modeling the population (as in our synthetic example), even infinite data cannot get rid of model bias. In terms of variance, collecting more data also reduces variance. One recent trend in data science is to select a model with low bias and high intrinsic variance (such as a neural network) but to collect many data points so that the model variance is low enough to make accurate predictions. While effective in practice, collecting enough data for these models tends to require large amounts of time and money.

Creating more features, whether useful or not, typically increases model variance. Models with many parameters have many possible combinations of parameters and therefore have higher variance than models with few parameters. On the other hand, adding a useful feature to the model, such as a quadratic feature when the underlying process is quadratic, reduces bias. But even adding a useless feature rarely increases bias.

Being aware of the bias-variance trade-off can help you do a better job of fitting models. And using techniques like the train-test split, cross-validation, and regularization can ameliorate this issue.

Another part of modeling considers the variation in the fitted coefficients and curve. We might want to provide a confidence interval for a coefficient or a prediction band for a future observation. These intervals and bands give a sense of the accuracy of the fitted model. We discuss this notion next.

CHAPTER 17

Theory for Inference and Prediction

When you want to generalize your findings beyond descriptions for your collection of data to a larger setting, the data needs to be representative of that larger world. For example, you may want to predict air quality at a future time based on a sensor reading (Chapter 12), test whether an incentive improves the productivity of contributors based on experimental findings (Chapter 3), or construct an interval estimate for the amount of time you might spend waiting for a bus (Chapter 5). We touched on all of these scenarios in earlier chapters. In this chapter, we'll formalize the framework for making predictions and inferences.

At the core of this framework is the notion of a distribution, be it a population, empirical (aka sample), or probability distribution. Understanding the connections between these distributions is central to the basics of hypothesis testing, confidence intervals, prediction bands, and risk. We begin with a brief review of the urn model, introduced in Chapter 3, then we introduce formal definitions of hypothesis tests, confidence intervals, and prediction bands. We use simulation in our examples, including the bootstrap as a special case. We wrap up the chapter with formal definitions of expectation, variance, and standard error—essential concepts in the theory of testing, inference, and prediction.

Distributions: Population, Empirical, Sampling

The population, sampling, and empirical distributions are important concepts that guide us when we make inferences about a model or predictions for new observations. Figure 17-1 provides a diagram that can help distinguish between them. The diagram uses the notions of population and access frame from Chapter 2 and the urn model from Chapter 3. On the left is the population that we are studying, represented as marbles in an urn with one marble for each unit. We have simplified the situation to where the access frame and the population are the same; that is, we can access every unit in the population. (The problems that arise when this is not the case are covered in Chapters 2 and 3.) The arrow from the urn to the sample represents the design, meaning the protocol for selecting the sample from the frame. The diagram shows this selection process as a chance mechanism, represented by draws from an urn filled with indistinguishable marbles. On the right side of the diagram, the collection of marbles constitutes our sample (the data we got).

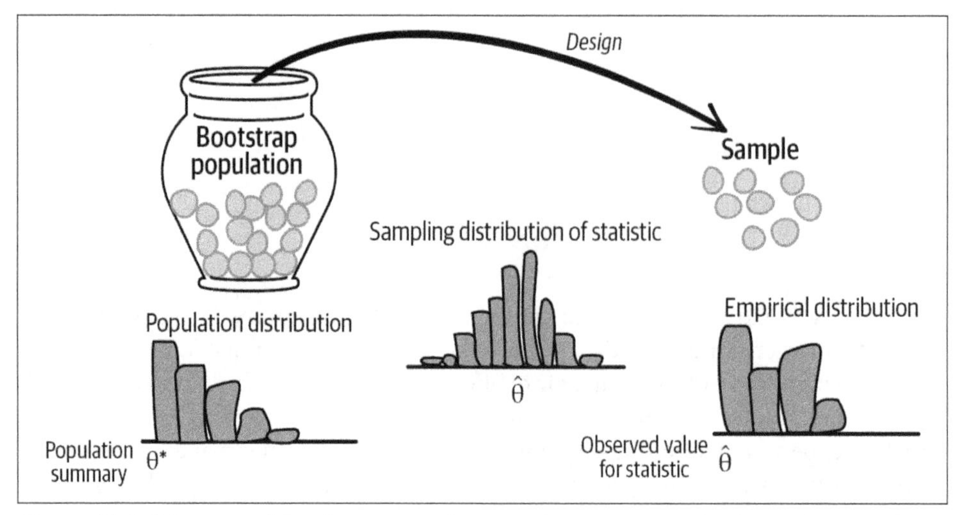

Figure 17-1. Diagram of the data generation process

We have kept the diagram simple by considering measurements for just one feature. Below the urn in the diagram is the *population histogram* for that feature. The population histogram represents the distribution of values across the entire population. On the far right, the *empirical histogram* shows the distribution of values for our actual sample. Notice that these two distributions are similar in shape. This happens when our sampling mechanism produces representative samples.

We are often interested in a summary of the sample measurements, such as the mean, median, slope from a simple linear model, and so on. Typically, this summary statistic is an estimate for a population parameter, such as the population mean or median. The population parameter is shown as θ^* on the left of the diagram; on the right, the summary statistic, calculated from the sample, is $\hat{\theta}$.

The chance mechanism that generates our sample might well produce a different set of data if we were to conduct our investigation over again. But if the protocols are well designed, we expect the sample to still resemble the population. In other words, we can infer the population parameter from the summary statistic calculated from the sample. The *sampling distribution* in the middle of the diagram is a *probability distribution* for the statistic. It shows the possible values that the statistic might take for different samples and their chances. In Chapter 3, we used simulation to estimate the sampling distribution in several examples. In this chapter, we revisit these and other examples from earlier chapters to formalize the analyses.

One last point about these three histograms: as introduced in Chapter 10, the rectangles provide the fraction of observations in any bin. In the case of the population histogram, this is the fraction of the entire population; for the empirical histogram, the area represents the fraction in the sample; and for the sampling distribution, the area represents the chance the data generation mechanism would produce a sample statistic in this bin.

Finally, we typically don't know the population distribution or parameter, and we try to infer the parameter or predict values for unseen units in the population. At other times, a conjecture about the population can be tested using the sample. Testing is the topic of the next section.

Basics of Hypothesis Testing

In our experience, hypothesis testing is one of the more challenging areas of data science—challenging to learn and challenging to apply. This is not necessarily because hypothesis testing is deeply technical; rather, hypothesis testing can be counterintuitive because it makes use of contradictions. As the name suggests, we often start hypothesis testing with a *hypothesis*: a statement about the world that we would like to verify.

In an ideal world, we would directly prove our hypothesis is true. Unfortunately, we often don't have access to all the information needed to determine the truth. For example, we might hypothesize that a new vaccine is effective, but contemporary medicine doesn't yet understand all the details of the biology that govern vaccine efficacy. Instead, we turn to the tools of probability, random sampling, and data design.

One reason hypothesis testing can be confusing is that it's a lot like "proof by contradiction," where we assume the opposite of our hypothesis is true and try to show that the data we observe is inconsistent with that assumption. We approach the problem this way because often, something can be true for many reasons, but we only need a single example to contradict an assumption. We call this "opposite hypothesis" the *null hypothesis* and our original hypothesis the *alternative hypothesis*.

To make matters a bit more confusing, the tools of probability don't directly prove or disprove things. Instead, they tell us how likely or unlikely something we observe is under assumptions, like the assumptions of the null hypothesis. That's why it's so important to design the data collection well.

Recall the randomized clinical trial of the J&J vaccine (Chapter 3), where 43,738 people enrolled in the trial were randomly split into two equal groups. The treatment group was given the vaccine and the control was given a fake vaccine, called a placebo. This random assignment created two groups that were similar in every way except for the vaccine.

In this trial, 117 people in the treatment group fell ill and 351 in the control group got sick. Since we want to provide convincing evidence that the vaccine works, we start with a null hypothesis that it doesn't work, meaning it was just by chance that the random assignment led to so few illnesses in the treatment group. We can then use probability to calculate the chance of observing so few sick people in the treatment group. The probability calculations are based on the urn that has 43,738 marbles in it, with 468 marked 1 to denote a sick person. We then found that the probability of at most 117 marbles being drawn in 21,869 draws with replacement from the urn was nearly zero. We take this as evidence to reject the null hypothesis in favor of the alternative hypothesis that the vaccine works. Because the J&J experiment was well designed, a rejection of the null leads us to conclude that the vaccine works. In other words, the truth of the hypothesis is left to us and how willing we are to be potentially wrong.

In the rest of this section, we go over the four basic steps of a hypothesis test. We then provide two examples that continue two of the examples from Chapter 3, and delve deeper into the formalities for testing.

There are four basic steps to hypothesis testing:

Step 1: Set up
: You have your data, and you want to test whether a particular model is reasonably consistent with the data. So you specify a statistic, $\hat{\theta}$, such as the sample average, fraction of zeros in a sample, or fitted regression coefficient, with the goal of comparing your data's statistic to what might have been produced under the model.

Step 2: Model
: You spell out the model that you want to test in the form of a data generation mechanism, along with any specific assumptions about the population. This model typically includes specifying θ^*, which may be the population mean, the proportion of zeros, or a regression coefficient. The sampling distribution of the statistic under this model is referred to as the *null distribution*, and the model itself is called the *null hypothesis*.

Step 3: Compute
: How likely, according to the null model in step 2, is it to get data (and the resulting statistic) at least as extreme as what you actually got in step 1? In formal inference, this probability is called the *p-value*. To approximate the p-value, we often use the computer to generate a large number of repeated random trials using the assumptions in the model and find the fraction of samples that give a value of the statistic at least as extreme as our observed value. Other times, we can instead use mathematical theory to find the p-value.

Step 4: Interpret

The *p*-value is used as a measure of surprise. If the model that you spelled out in step 2 is believable, how surprised should you be to get the data (and summary statistic) that you actually got? A moderately sized *p*-value means that the observed statistic is pretty much what you would expect to get for data generated by the null model. A tiny *p*-value raises doubts about the null model. In other words, if the model is correct (or approximately correct), then it would be very unusual to get such an extreme value of the test statistic from data generated by the model. In this case, either the null model is wrong or a very unlikely outcome has occurred. Statistical logic says to conclude that the pattern is real, that it is more than just coincidence. Then it's up to you to explain why the data generation process led to such an unusual value. This is when a careful consideration of the scope is important.

Let's demonstrate these steps in the testing process with a couple of examples.

Example: A Rank Test to Compare Productivity of Wikipedia Contributors

Recall the Wikipedia example from Chapter 2, where a randomly selected set of 200 contributors were chosen from among the top 1% of contributors who were active in the past 30 days on the English-language Wikipedia and who had never received an award. These 200 contributors were divided at random into two groups of 100. The contributors in one group, the treatment group, were each given an informal award, while no one in the other group was given one. All 200 contributors were followed for 90 days and their activity on Wikipedia recorded.

It has been conjectured that informal awards have a reinforcing effect on volunteer work, and this experiment was designed to formally study this conjecture. We carry out a hypothesis test based on the rankings of the data.

First, we read the data into a dataframe:

```
wiki = pd.read_csv("data/Wikipedia.csv")
wiki.shape
```

```
(200, 2)
```

```
wiki.describe()[3:]
```

	experiment	postproductivity
min	0.0	0.0
25%	0.0	57.5
50%	0.5	250.5
75%	1.0	608.0
max	1.0	2344.0

The dataframe has 200 rows, one for each contributor. The feature `experiment` is either 0 or 1, depending on whether the contributor was in the control or treatment group, respectively, and `postproductivity` is a count of the edits made by the contributor in the 90 days after the awards were made. The gap between the quartiles (lower, middle, and upper) suggests the distribution of productivity is skewed. We make a histogram to confirm:

```
px.histogram(
        wiki, x='postproductivity', nbins=50,
        labels={'postproductivity': 'Number of actions in 90 days post award'},
        width=350, height=250)
```

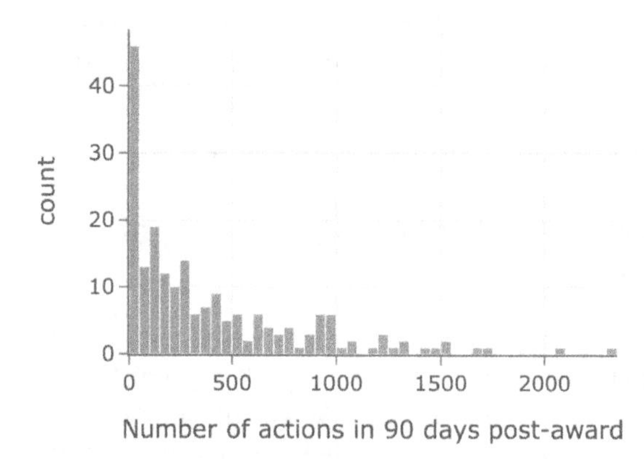

Indeed, the histogram of post-award productivity is highly skewed, with a spike near zero. The skewness suggests a statistic based on the ordering of the values from the two samples.

To compute our statistic, we order all productivity values (from both groups) from smallest to largest. The smallest value has rank 1, the second smallest rank 2, and so on, up to the largest value, which has a rank of 200. We use these ranks to compute our statistic, $\hat{\theta}$, which is the average rank of the treatment group. We chose this statistic because it is insensitive to highly skewed distributions. For example, whether the largest value is 700 or 700,000, it still receives the same rank, namely 200. If the informal award incentivizes contributors, then we would expect the average rank of the treatment group to be typically higher than the control.

The null model assumes that an informal award has *no* effect on productivity, and any difference observed between the treatment and control groups is due to the chance process in assigning contributors to groups. The null hypothesis is set up for the status quo to be rejected; that is, we hope to find a surprise in assuming no effect.

The null hypothesis can be represented by 100 draws from an urn with 200 marbles, marked 1, 2, 3, ..., 200. In this case, the average rank would be $(1 + 200)/2 = 100.5$.

We use the `rankdata` method in `scipy.stats` to rank the 200 values and compute the sum of ranks in the treatment group:

```
from scipy.stats import rankdata
ranks = rankdata(wiki['postproductivity'], 'average')
```

Let's confirm that the average rank of the 200 values is 100.5:

```
np.average(ranks)
```

```
100.5
```

And find the average rank of the 100 productivity scores in the treatment group:

```
observed = np.average(ranks[100:])
observed
```

```
113.68
```

The average rank in the treatment group is higher than expected, but we want to figure out if it is an unusually high value. We can use simulation to find the sampling distribution for this statistic to see if 113 is a routine value or a surprising one.

To carry out this simulation, we set up the urn as the `ranks` array from the data. Shuffling the 200 values in the array and taking the first 100 represents a randomly sampled treatment group. We write a function to shuffle the array of ranks and find the average of the first 100.

```
rng = np.random.default_rng(42)
def rank_avg(ranks, n):
    rng.shuffle(ranks)
    return np.average(ranks[n:])
```

Our simulation mixes the marbles in the urn, draws 100 times, computes the average rank for the 100 draws, and repeats this 100,000 times.

```
rank_avg_simulation = [rank_avg(ranks, 100) for _ in range(100_000)]
```

Here is a histogram of the simulated averages:

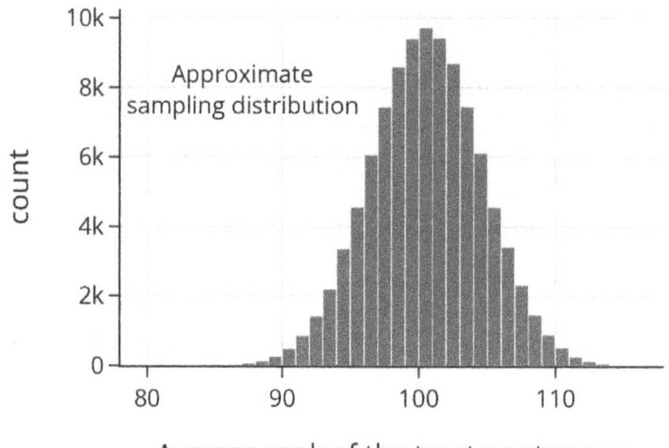

As we expected, the sampling distribution of the average rank is centered on 100 (100.5 actually) and is bell-shaped. The center of this distribution reflects the assumptions of the treatment having no effect. Our observed statistic is well outside the typical range of simulated average ranks, and we use this simulated sampling distribution to find the approximate *p*-value for observing a statistic at least as big as ours:

```
np.mean(rank_avg_simulation > observed)
```

```
0.00058
```

This is a big surprise. Under the null, the chance of seeing an average rank at least as large as ours is about 5 in 10,000.

This test raises doubt about the null model. Statistical logic has us conclude that the pattern is real. How do we interpret this? The experiment was well designed. The 200 contributors were selected at random from the top 1%, and then they were divided at random into two groups. These chance processes say that we can rely on the sample of 200 being representative of top contributors, and on the treatment and control groups being similar to each other in every way except for the application of the treatment (the award). Given the careful design, we conclude that informal awards have a positive effect on productivity for top contributors.

Earlier, we implemented a simulation to find the *p*-value for our observed statistic. In practice, rank tests are commonly used and made available in most statistical software:

```
from scipy.stats import ranksums

ranksums(x=wiki.loc[wiki.experiment == 1, 'postproductivity'],
         y=wiki.loc[wiki.experiment == 0, 'postproductivity'])
```

```
RanksumsResult(statistic=3.220386553232206, pvalue=0.0012801785007519996)
```

The p-value here is twice the p-value we computed because we considered only values greater than the observed, whereas the `ranksums` test computed the the p-value for both sides of the distribution. In our example, we are only interested in an increase in productivity, and so use a one-sided p-value, which is half the reported value (0.0006) and close to our simulated value.

This somewhat unusual test statistic that uses ranks rather than the actual data values was developed in the 1950s and 1960s, before today's era of powerful laptop computers. The mathematical properties of rank statistics is well developed and the sampling distribution is well behaved (it is symmetric and shaped like the bell curve even for small datasets). Rank tests remain popular for A/B testing where samples tend to be highly skewed, and it is common to carry out many, many tests where p-values can be computed rapidly from the normal distribution.

The next example revisits the vaccine efficacy example from Chapter 3. There, we encountered a hypothesis test without actually calling it that.

Example: A Test of Proportions for Vaccine Efficacy

The approval of a vaccine is subject to stricter requirements than the simple test we performed earlier where we compared the disease counts in the treatment group to those of the control group. The CDC requires stronger evidence of success based on a comparison of the proportion of sick individuals in each group. To explain, we express the sample proportion of sick people in the control and treatment groups as $\hat{p}_C$ and $\hat{p}_T$, respectively, and use these proportions to compute vaccine efficacy:

$$\hat{\theta} = \frac{\hat{p}_C - \hat{p}_T}{\hat{p}_C} = 1 - \frac{\hat{p}_T}{\hat{p}_C}$$

The observed value of vaccine efficacy in the J&J trial is:

$$1 - \frac{117/21869}{351/21869} = 1 - \frac{117}{351} = 0.667$$

If the treatment doesn't work, the efficacy would be near 0. The CDC sets a standard of 50% for vaccine efficacy, meaning that the efficacy has to exceed 50% to be approved for distribution. In this situation, the null model assumes that vaccine efficacy is 50% ($\theta^* = 0.5$), and any difference of the observed value from the expected is due to the chance process in assigning people to groups. Again, we set the null hypothesis to be the status quo that the vaccine isn't effective enough to warrant approval, and we hope to find a surprise and reject the null.

With a little algebra, the null model $0.5 = 1 - p_T/p_C$ reduces to $p_T = 0.5p_C$. That is, the null hypothesis implies that the proportion of ill people among those receiving the treatment is at most half that of the control. Notice that the actual values for the two risks (p_T and p_C) are not assumed in the null. That is, the model doesn't assume the treatment doesn't work, but rather, that its efficacy is no larger than 0.5.

Our urn model in this situation is a bit different from what we set up in Chapter 3. The urn still has 43,738 marbles in it, corresponding to the enrollees in the experiment. But now each marble has two numbers on it, which for simplicity appear in a pair, such as (0, 1). The number on the left is the response if the person receives the treatment, and the number on the right corresponds to the response to no treatment (the control). As usual, 1 means they become ill and 0 means they stay healthy.

The null model assumes that the proportion of ones on the left of the pair is half the proportion on the right. Since we don't know these two proportions, we can use the data to estimate them. There are three types of marbles in the urn $(0, 0)$, $(0, 1)$, and $(1, 1)$. We assume that $(1, 0)$, which corresponds to a person getting ill under treatment and not under control, is not possible. We observed 351 people getting sick in control and 117 in treatment. With the assumption that the treatment rate of illness is half that of the control, we can tray a scenario for the makeup of the urn. For example, we can study the case where 117 people in treatment didn't get sick but would have if they were in the control group, so combined, all 585 people (351 + 117 + 117) would get the virus if they didn't receive the vaccine and half of them would not get the virus if they received treatment. Table 17-1 shows these counts.

Table 17-1. Vaccine trial urn

Label	Count
(0, 0)	43,152
(0, 1)	293
(1, 0)	0
(1, 1)	293
Total	43,738

We can use these counts to carry out a simulation of the clinical trial and compute vaccine efficacy. As shown in Chapter 3, the multivariate hypergeometric function simulates draws from an urn when there are more than two kinds of marbles. We set up this urn and sampling process:

```
N = 43738
n_samp = 21869
N_groups = np.array([293, 293, (N - 586)])

from scipy.stats import multivariate_hypergeom
```

```
def vacc_eff(N_groups, n_samp):
    treat = multivariate_hypergeom.rvs(N_groups, n_samp)
    ill_t = treat[1]
    ill_c = N_groups[0] - treat[0] + N_groups[1] - treat[1]
    return (ill_c - ill_t) / ill_c
```

Now we can simulate the clinical trial 100,000 times and calculate the vaccine efficacy for each trial:

```
np.random.seed(42)
sim_vacc_eff = np.array([vacc_eff(N_groups, n_samp) for _ in range(100_000)])
```

```
px.histogram(x=sim_vacc_eff, nbins=50,
             labels=dict(x='Simulated vaccine efficacy'),
             width=350, height=250)
```

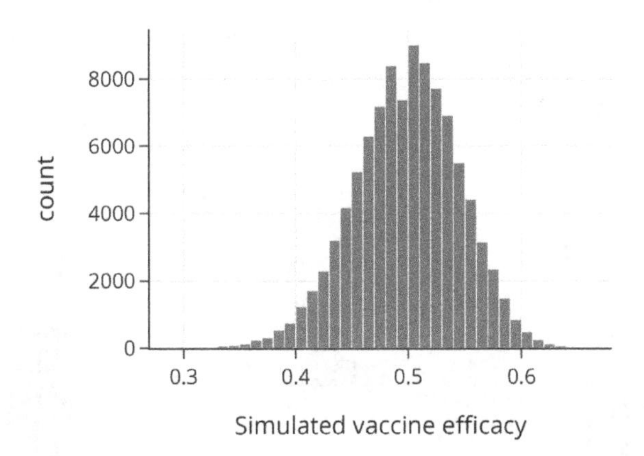

The sampling distribution is centered at 0.5, which agrees with our model assumptions. We see that 0.667 is far out in the tail of this distribution:

```
np.mean(sim_vacc_eff > 0.667)
```

```
1e-05
```

Only a tiny handful of the 100,000 simulations have a vaccine efficacy as large as the observed 0.667. This is a rare event, and that's why the CDC approved the Johnson & Johnson vaccine for distribution.

In this example of hypothesis testing, we were not able to completely specify the model, and we had to provide approximate values for p_C and p_T based on our observed values of $\hat{p}_C$ and $\hat{p}_T$. At times, the null model isn't entirely specified, and we must rely on the data to set up the model. The next section introduces a general approach, called the bootstrap, to approximate the model using the data.

Bootstrapping for Inference

In many hypothesis tests the assumptions of the null hypothesis lead to a complete specification of a hypothetical population and data design (see Figure 17-1), and we use this specification to simulate the sampling distribution of a statistic. For example, the rank test for the Wikipedia experiment led us to sample the integers 1, ..., 200, which we easily simulated. Unfortunately, we can't always specify the population and model completely. To remedy the situation, we substitute the data for the population. This substitution is at the heart of the notion of the bootstrap. Figure 17-2 updates Figure 17-1 to reflect this idea; here the population distribution is replaced by the empirical distribution to create what is called the *bootstrap population*.

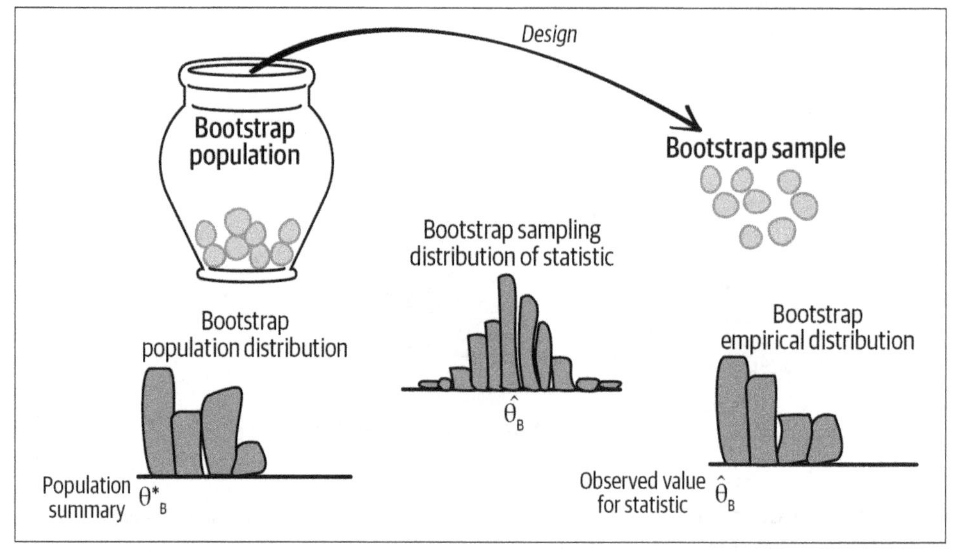

Figure 17-2. Diagram of bootstrapping the data generation process

The rationale for the bootstrap goes like this:

- Your sample looks like the population because it is a representative sample, so we replace the population with the sample and call it the bootstrap population.
- Use the same data generation process that produced the original sample to get a new sample, which is called a *bootstrap sample*, to reflect the change in the population. Calculate the statistic on the bootstrap sample in the same manner as before and call it the *bootstrap statistic*. The *bootstrap sampling distribution* of the bootstrap statistic should be similar in shape and spread to the true sampling distribution of the statistic.
- Simulate the data generation process many times, using the bootstrap population, to get bootstrap samples and their bootstrap statistics. The distribution of the

simulated bootstrap statistics approximates the bootstrap sampling distribution of the bootstrap statistic, which itself approximates the original sampling distribution.

Take a close look at Figure 17-2 and compare it to Figure 17-1. Essentially, the bootstrap simulation involves two approximations: the original sample approximates the population, and the simulation approximates the sampling distribution. We have been using the second approximation in our examples so far; the approximation of the population by the sample is the core notion behind bootstrapping. Notice that in Figure 17-2, the distribution of the bootstrap population (on the left) looks like the original sample histogram; the sampling distribution (in the middle) is still a probability distribution based on the same data generation process as in the original study, but it now uses the bootstrap population; and the sample distribution (on the right) is a histogram of one sample taken from the bootstrap population.

You might be wondering how to take a simple random sample from your bootstrap population and not wind up with the exact same sample each time. After all, if your sample has 100 units in it and you use it as your bootstrap population, then 100 draws from the bootstrap population without replacement will take all of the units and give you the same bootstrap sample every time. There are two approaches to solving this problem:

- When sampling from the bootstrap population, draw units from the bootstrap population with replacement. Essentially, if the original population is very large, then there is little difference between sampling with and without replacement. This is the more common approach by far.
- "Blow up the sample" to be the same size as the original population. That is, tally the fraction of each unique value in the sample, and add units to the bootstrap population so that it is the same size as the original population, while maintaining the proportions. For example, if the sample is size 30 and 1/3 of the sample values are 0, then a bootstrap population of 750 should include 250 zeros. Once you have this bootstrap population, use the original data generation procedure to take the bootstrap samples.

The example of vaccine efficacy used a bootstrap-like process, called the *parameterized bootstrap*. Our null model specified 0-1 urns, but we didn't know how many 0s and 1s to put in the urn. We used the sample to determine the proportions of 0s and 1s; that is, the sample specified the parameters of the multivariate hypergeometric. Next, we use the example of calibrating air quality monitors to show how bootstrapping could be used to test a hypothesis.

It's a common mistake to think that the center of the bootstrap sampling distribution is the same as the center of the true sampling distribution. If the mean of the sample is not 0, then the mean of the bootstrap population is also not 0. That's why we use the spread of the bootstrap distribution, and not its center, in hypothesis testing. The next example shows how we might use the bootstrap to test a hypothesis.

The case study on calibrating air quality monitors (see Chapter 12) fit a model to adjust the measurements from an inexpensive monitor to more accurately reflect true air quality. This adjustment included a term in the model related to humidity. The fitted coefficient was about 0.2 so that on days of high humidity the measurement is adjusted upward more than on days of low humidity. However, this coefficient is close to 0, and we might wonder whether including humidity in the model is really needed. In other words, we want to test the hypothesis that the coefficient for humidity in the linear model is 0. Unfortunately, we can't fully specify the model, because it is based on measurements taken over a particular time period from a set of air monitors (both PurpleAir and those maintained by the EPA). This is where the bootstrap can help.

Our model makes the assumption that the air quality measurements taken resemble the population of measurements. Note that weather conditions, the time of year, and the location of the monitors make this statement a bit hand-wavy; what we mean here is that the measurements are similar to others taken under the same conditions as those when the original measurements were taken. Also, since we can imagine a virtually infinite supply of air quality measurements, we think of the procedure for generating measurements as draws with replacement from the urn. Recall that in Chapter 2 we modeled the urn as repeated draws with replacement from an urn of measurement errors. This situation is a bit different because we are also including the other factors mentioned already (weather, season, location).

Our model is focused on the coefficient for humidity in the linear model:

$$\text{PA} \approx \theta_0 + \theta_1\text{AQ} + \theta_2\text{RH}$$

Here, PA refers to the PurpleAir PM2.5 measurement, RH is the relative humidity, and AQ stands for the more exact measurement of PM2.5 made by the more accurate AQS monitors. The null hypothesis is $\theta_2 = 0$; that is, the null model is the simpler model:

$$\text{PA} \approx \theta_0 + \theta_1\text{AQ}$$

To estimate θ_2, we use the linear model fitting procedure from Chapter 15.

Our bootstrap population consists of the measurements from Georgia that we used in Chapter 15. Now we sample rows from the dataframe (which is equivalent to our urn) with replacement using the chance mechanism `randint`. This function takes random samples with replacement from a set of integers. We use the random sample of indices to create the bootstrap sample from the dataframe. Then we fit the linear model and get the coefficient for humidity (our bootstrap statistic). The following `boot_stat` function performs this simulation process:

```
from scipy.stats import randint

def boot_stat(X, y):
    n = len(X)
    bootstrap_indexes = randint.rvs(low=0, high=(n - 1), size=n)
    theta2 = (
        LinearRegression()
        .fit(X.iloc[bootstrap_indexes, :], y.iloc[bootstrap_indexes])
        .coef_[1]
    )
    return theta2
```

We set up the design matrix and the outcome variable and check our `boot_stat` function once to test it:

```
X = GA[['pm25aqs', 'rh']]
y = GA['pm25pa']

boot_stat(X, y)
```

```
0.21572251745549495
```

When we repeat this process 10,000 times, we get an approximation to the bootstrap sampling distribution of the bootstrap statistic (the fitted humidity coefficient):

```
np.random.seed(42)
boot_theta_hat = np.array([boot_stat(X, y) for _ in range(10_000)])
```

We are interested in the shape and spread of this bootstrap sampling distribution (we know that the center will be close to the original coefficient of 0.21):

```
px.histogram(x=boot_theta_hat, nbins=50,
             labels=dict(x='Bootstrapped humidity coefficient'),
             width=350, height=250)
```

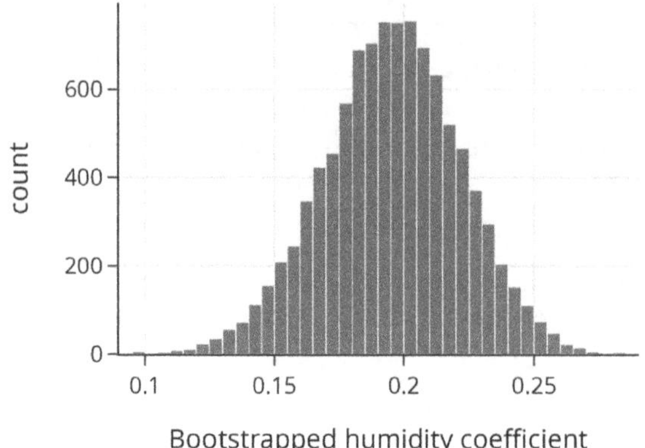

By design, the center of the bootstrap sampling distribution will be near $\hat{\theta}$ because the bootstrap population consists of the observed data. So, rather than compute the chance of a value at least as large as the observed statistic, we find the chance of a value at least as small as 0. The hypothesized value of 0 is far from the sampling distribution.

None of the 10,000 simulated regression coefficients are as small as the hypothesized coefficient. Statistical logic leads us to reject the null hypothesis that we do not need to adjust the model for humidity.

The form of the hypothesis test we performed here looks different than the earlier tests because the sampling distribution of the statistic is not centered on the null. That is because we are using the bootstrap to create the sampling distribution. We are, in effect, using a confidence interval for the coefficient to test the hypothesis. In the next section we introduce interval estimates more generally, including those based on the bootstrap, and we connect the concepts of hypothesis testing and confidence intervals.

Basics of Confidence Intervals

We have seen that modeling leads to estimates, such as the typical time that a bus is late (Chapter 4), a humidity adjustment to an air quality measurement (Chapter 15), and an estimate of vaccine efficacy (Chapter 2). These examples are point estimates for unknown values, called *parameters*: the median lateness of the bus is 0.74 minutes; the humidity adjustment to air quality is 0.21 PM2.5 per humidity percentage point; and the ratio of COVID infection rates in vaccine efficacy is 0.67. However, a different sample would have produced a different estimate. Simply providing a point estimate doesn't give a sense of the estimate's precision. Alternatively, an interval estimate can reflect the estimate's accuracy. These intervals typically take one of two forms:

- A *bootstrap confidence interval* created from the percentiles of the bootstrap sampling distribution
- A *normal confidence interval* constructed using the standard error (SE) of the sampling distribution and additional assumptions about the distribution having the shape of a normal curve

We describe these two types of intervals and then give an example. Recall that the sampling distribution (see Figure 17-1) is a probability distribution that reflects the chance of observing different values of $\hat{\theta}$. Confidence intervals are constructed from the spread of the sampling distribution of $\hat{\theta}$, so the endpoints of the interval are random because they are based on $\hat{\theta}$. These intervals are designed so that 95% of the time the interval covers θ^*.

As its name suggests, the percentile-based bootstrap confidence interval is created from the percentiles of the bootstrap sampling distribution. Specifically, we compute the quantiles of the sampling distribution of $\hat{\theta}_B$, where $\hat{\theta}_B$ is the bootstrapped statistic. For a 95th percentile interval, we identify the 2.5 and 97.5 quantiles, called $q_{2.5,B}$ and $q_{97.5,B}$, respectively, where 95% of the time the bootstrapped statistic is in the interval:

$$q_{2.5,B} \le \hat{\theta}_B \le q_{97.5,B}$$

This bootstrap percentile confidence interval is considered a quick-and-dirty interval. There are many alternatives that adjust for bias, take into consideration the shape of the distribution, and are better suited for small samples.

The percentile confidence interval does not rely on the sampling distribution having a particular shape or the center of the distribution being θ^*. In contrast, the normal confidence interval often doesn't require bootstrapping to compute, but it does make additional assumptions about the shape of the sampling distribution of $\hat{\theta}$.

We use the normal confidence interval when the sampling distribution is well approximated by a normal curve. For a normal probability distribution, with center μ and spread σ, there is a 95% chance that a random value from this distribution is in the interval $\mu \pm 1.96\sigma$. Since the center of the sampling distribution is typically θ^*, the chance is 95% that for a randomly generated $\hat{\theta}$:

$$|\hat{\theta} - \theta^*| \le 1.96 SE(\hat{\theta})$$

where $SE(\hat{\theta})$ is the spread of the sampling distribution of $\hat{\theta}$. We use this inequality to make a 95% confidence interval for θ^*:

$$[\hat{\theta} - 1.96SE(\hat{\theta}), \quad \hat{\theta} + 1.96SE(\hat{\theta})]$$

Confidence intervals of other sizes can be formed with different multiples of $SE(\hat{\theta})$, all based on the normal curve. For example, a 99% confidence interval is $\pm 2.58SE$, and a one-sided upper 95% confidence interval is $[\hat{\theta} - 1.64SE(\hat{\theta}), \quad \infty]$.

The SD of a parameter estimate is often called the *standard error*, or SE, to distinguish it from the SD of a sample, population, or one draw from an urn. In this book, we don't differentiate between them. We call them SDs.

We provide an example of each type of interval next.

Earlier in this chapter we tested the hypothesis that the coefficient for humidity in a linear model for air quality is 0. The fitted coefficient for these data was 0.21. Since the null model did not completely specify the data generation mechanism, we resorted to bootstrapping. That is, we used the data as the population, took a sample of 11,226 records with replacement from the bootstrap population, and fitted the model to find the bootstrap sample coefficient for humidity. Our simulation repeated this process 10,000 times to get an approximate bootstrap sampling distribution.

We can use the percentiles of this bootstrap sampling distribution to create a 99% confidence interval for θ^*. To do this, we find the quantiles, $q_{0.5}$ and $q_{99.5}$, of the bootstrap sampling distribution:

```
q_995 = np.percentile(boot_theta_hat, 99.5, method='lower')
q_005 = np.percentile(boot_theta_hat, 0.05, method='lower')

print(f"Lower 0.05th percentile: {q_005:.3f}")
print(f"Upper 99.5th percentile: {q_995:.3f}")
```

```
Lower 0.05th percentile: 0.099
Upper 99.5th percentile: 0.260
```

Alternatively, since the histogram of the sampling distribution looks roughly normal in shape, we can create a 99% confidence interval based on the normal distribution. First, we find the standard error of $\hat{\theta}$, which is just the standard deviation of the sampling distribution of $\hat{\theta}$:

```
standard_error = np.std(boot_theta_hat)
standard_error
```

```
0.026534986093303445
```

Then, a 99% confidence interval for θ^* is 2.58 SEs away from the observed $\hat{\theta}$ in either direction:

```
print(f"Lower 0.05th endpoint: {theta2_hat - (2.58 * standard_error):.3f}")
print(f"Upper 99.5th endpoint: {theta2_hat + (2.58 * standard_error):.3f}")
```

```
Lower 0.05th endpoint: 0.138
Upper 99.5th endpoint: 0.275
```

These two intervals (bootstrap percentile and normal) are close but clearly not identical. We might expect this given the slight asymmetry in the bootstrapped sampling distribution.

There are other versions of the normal-based confidence interval that reflect the variability in estimating the standard error of the sampling distribution using the SD of the data. And there are still other confidence intervals for statistics that are percentiles, rather than averages. (Also note that for permutation tests, the bootstrap tends not to be as accurate as normal approximations.)

Confidence intervals can be easily misinterpreted as the chance that the parameter θ^* is in the interval. However, the confidence interval is created from one realization of the sampling distribution. The sampling distribution gives us a different probability statement; 95% of the time, an interval constructed in this way will contain θ^*. Unfortunately, we don't know whether this particular time is one of those that happens 95 times in 100 or not. That is why the term *confidence* is used rather than *probability* or *chance*, and we say that we are 95% confident that the parameter is in our interval.

Confidence intervals and hypothesis tests are related in the following way. If, say, a 95% confidence interval contains the hypothesized value θ^*, then the p-value for the test is less than 5%. That is, we can invert a confidence interval to create a hypothesis test. We used this technique in the previous section when we carried out the test that the coefficient for humidity in the air quality model is 0. In this section, we have created a 99% confidence interval for the coefficient (based on the bootstrap percentiles), and since 0 does not belong to the interval, the p-value is less than 1% and statistical logic would lead us to conclude that the coefficient is not 0.

Another kind of interval estimate is the prediction interval. Prediction intervals focus on the variation in observations rather than the variation in an estimator. We explore these next.

Basics of Prediction Intervals

Confidence intervals convey the accuracy of an estimator, but sometimes we want the accuracy of a prediction for a future observation. For example, someone might say: half the time my bus arrives three-quarters of a minute late at most, but how late might it get? As another example, the California Department of Fish and Wildlife sets the minimum catch size for Dungeness crabs at 146 mm, and a recreational fishing company might wonder how much bigger than 146 mm their customer's catch might be when they bring them fishing. And for another example, a vet estimates the weight of a donkey to be 169 kg based on its length and girth and uses this estimate to administer medication. For the donkey's safety, the vet is keen to know how different the donkey's real weight might be from this estimate.

What these examples have in common is an interest in the prediction of a future observation and the desire to quantify how far that future observation might be from this prediction. Just like with confidence intervals, we compute the statistic (the estimator) and use it in making the prediction, but now we're interested in typical deviations of future observations from the prediction. In the following sections, we work through examples of prediction intervals based on quantiles, standard deviations, and those conditional on covariates. Along the way, we provide additional information about the typical variation of observations about a prediction.

Example: Predicting Bus Lateness

Chapter 4 models the lateness of a Seattle bus in arriving at a particular stop. We observed that the distribution was highly skewed and chose to estimate the typical lateness by the median, which was 0.74 minutes. We reproduce the sample histogram from that chapter here:

```
times = pd.read_csv("data/seattle_bus_times_NC.csv")
fig = px.histogram(times, x="minutes_late", width=350, height=250)
fig.update_xaxes(range=[-12, 60], title_text="Minutes late")
fig
```

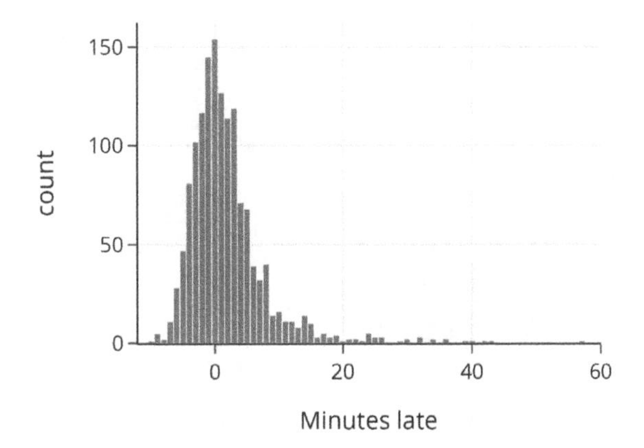

The prediction problem addresses how late a bus might be. While the median is informative, it doesn't provide information about the skewness of the distribution. That is, we don't know how late the bus might be. The 75th percentile, or even the 95th percentile, would add useful information to consider. We compute those percentiles here:

```
median:          0.74 mins late
75th percentile: 3.78 mins late
95th percentile: 13.02 mins late
```

From these statistics, we learn that while more than half the time the bus is not even a minute late, one-quarter of the time it's almost four minutes late, and with some regularity it can happen that the bus is nearly 15 minutes late. These three values together help us make plans.

Example: Predicting Crab Size

Fishing for Dungeness crabs is highly regulated, including limiting the shell size to 146 mm in width for crabs caught for recreation. To better understand the distribution of shell size of Dungeness crabs, the California Department of Fish and Wildlife worked with commercial crab fishers from Northern California and Southern Oregon to capture, measure, and release crabs. Here is a histogram of crab shell sizes for the approximately 450 crabs caught:

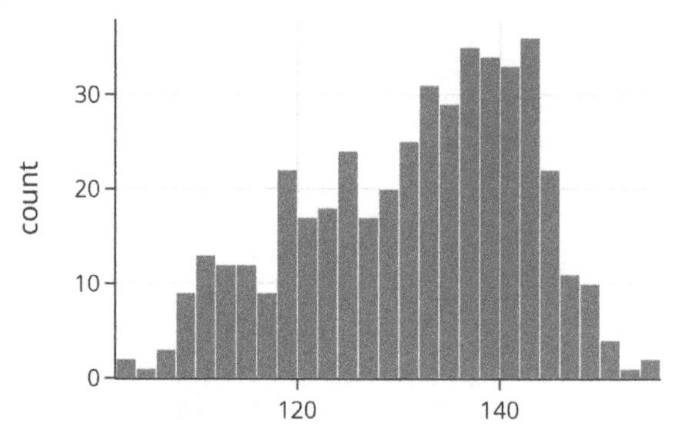

The distribution is somewhat skewed left, but the average and standard deviations are reasonable summary statistics of the distribution:

```
crabs['shell'].describe()[:3]
```

```
count    452.00
mean     131.53
std       11.07
Name: shell, dtype: float64
```

The average, 132 mm, is a good prediction for the typical size of a crab. However, it lacks information about how far an individual crab may vary from the average. The standard deviation can fill in this gap.

In addition to the variability of individual observations about the center of the distribution, we also take into account the variability in our estimate of the mean shell size. We can use the bootstrap to estimate this variability, or we can use probability theory (we do this in the next section) to show that the standard deviation of the estimator is $SD(pop)/\sqrt{n}$. We also show, in the next section, that these two sources of variation combine as follows:

$$\sqrt{SD(pop)^2 + \frac{SD(pop)^2}{n}} = SD(pop)\sqrt{1 + \frac{1}{n}}$$

We substitute $SD(sample)$ for $SD(pop)$ and apply this formula to our crabs:

```
np.std(crabs['shell']) * np.sqrt(1 + 1/len(crabs))
```

```
11.073329460297957
```

We see that including the SE of the sample average essentially doesn't change the prediction error because the sample is so large. We conclude that crabs routinely differ from the typical size of 132 mm by 11 to 22 mm. This information is helpful in developing policies around crab fishing to maintain the health of the crab population and to set expectations for the recreational fisher.

Example: Predicting the Incremental Growth of a Crab

After Dungeness crabs mature, they continue to grow by casting off their shell and building a new, larger one to grow into each year; this process is called *molting*. The California Department of Fish and Wildlife wanted a better understanding of crab growth so that it could set better limits on fishing that would protect the crab population. The crabs caught in the study mentioned in the previous example were about to molt, and in addition to their size, the change in shell size from before to after molting was also recorded:

```
crabs.corr()
```

	shell	inc
shell	1.0	-0.6
inc	-0.6	1.0

These two measurements are negatively correlated, meaning that the larger the crab, the less they grow when they molt. We plot the growth increment against the shell size to determine whether the relationship between these variables is roughly linear:

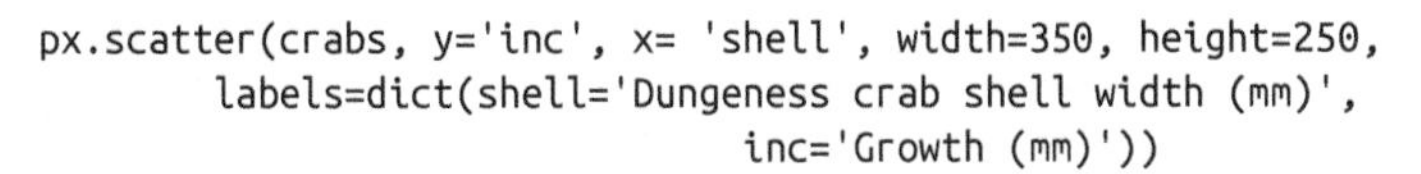

```
px.scatter(crabs, y='inc', x= 'shell', width=350, height=250,
        labels=dict(shell='Dungeness crab shell width (mm)',
                                    inc='Growth (mm)'))
```

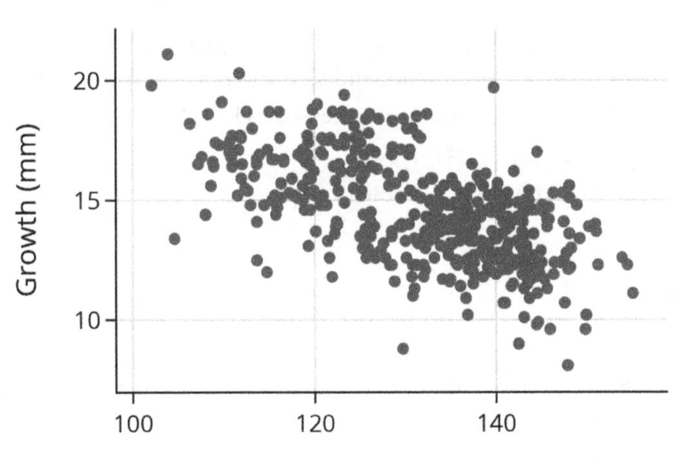

The relationship appears linear, and we can fit a simple linear model to explain the growth increment by the pre-molt size of the shell. For this example, we use the statsmodels library, which provides prediction intervals with get_prediction. We first set up the design matrix and response variable, and then we use least squares to fit the model:

```
import statsmodels.api as sm

X = sm.add_constant(crabs[['shell']])
y = crabs['inc']

inc_model = sm.OLS(y, X).fit()

print(f"Increment estimate = {inc_model.params[0]:0.2f} + ",
      f"{inc_model.params[1]:0.2f} x Shell Width")
```

```
Increment estimate = 29.80 +  -0.12 x Shell Width
```

When modeling, we create prediction intervals for given values of the explanatory variable. For example, if a newly caught crab is 120 mm across, then we use our fitted model to predict its shell's growth.

As in the previous example, the variability of our prediction for an individual observation includes the variability in our estimate of the crab's growth and the crab-to-crab variation in shell size. Again, we can use the bootstrap to estimate this variation, or we can use probability theory to show that these two sources of variation combine as follows:

$$SD(\mathbf{e})\sqrt{1 + \mathbf{x}_0(\mathbf{X}^\top\mathbf{X})^{-1}\mathbf{x}_0^\top}$$

Here $\mathbf{X}$ is the design matrix that consists of the original data, $\mathbf{e}$ is the $n \times 1$ column vector of residuals from the regression, and $\mathbf{x}_0$ is the $1 \times (p + 1)$ row vector of features for the new observation (in this example, these are $[1, 120]$):

```
new_data = dict(const=1, shell=120)
new_X = pd.DataFrame(new_data, index=[0])
new_X
```

	const	shell
0	1	120

We use the get_prediction method in statsmodels to find a 95% prediction interval for a crab with a 120 mm shell:

```
pred = inc_model.get_prediction(new_X)
pred.summary_frame(alpha=0.05)
```

	mean	mean_se	mean_ci_lower	mean_ci_upper	obs_ci_lower	obs_ci_upper
0	15.86	0.12	15.63	16.08	12.48	19.24

Here we have both a confidence interval for the average growth increment for a crab with a 120 mm shell, [15.6, 16.1] and a prediction interval for the growth increment, [12.5, 19.2]. The prediction interval is quite a bit wider because it takes into account the variation in individual crabs. This variation is seen in the spread of the points about the regression line, which we approximate by the SD of the residuals. The correlation between shell size and growth increment means that the variation in a growth increment prediction for a particular shell size is smaller than the overall SD of the growth increment:

```
print(f"Residual SD:    {np.std(inc_model.resid):0.2f}")
print(f"Crab growth SD: {np.std(crabs['inc']):0.2f}")
```

```
Residual SD:    1.71
Crab growth SD: 2.14
```

The intervals provided by `get_prediction` rely on the normal approximation to the distribution of growth increment. That's why the 95% prediction interval endpoints are roughly twice the residual SD away from the prediction. In the next section, we dive deeper into these calculations of standard deviations, estimators, and predictions. We also discuss some of the assumptions that we make in calculating them.

Probability for Inference and Prediction

Hypothesis testing, confidence intervals, and prediction intervals rely on probability calculations computed from the sampling distribution and the data generation process. These probability frameworks also enable us to run simulation and bootstrap studies for a hypothetical survey, an experiment, or some other chance process in order to study its random behavior. For example, we found the sampling distribution for an average of ranks under the assumption that the treatment in a Wikipedia experiment was not effective. Using simulation, we quantified the typical deviations from the expected outcome and the distribution of the possible values for the summary statistic. The triptych in Figure 17-1 provided a diagram to guide us in the process; it helped keep straight the differences between the population, probability, and sample and also showed their connections. In this section, we bring more mathematical rigor to these concepts.

We formally introduce the notions of expected value, standard deviation, and random variable, and we connect them to the concepts we have been using in this chapter for testing hypotheses and making confidence and prediction intervals. We begin with the specific example from the Wikipedia experiment, and then we generalize. Along the way, we connect this formalism to the triptych that we have used as our guide throughout the chapter.

Formalizing the Theory for Average Rank Statistics

Recall in the Wikipedia experiment that we pooled the post-award productivity values from the treatment and control groups and converted them into ranks, $1, 2, 3, \ldots, 200$, so the population is simply made up of the integers from 1 to 200. Figure 17-3 is a diagram that represents this specific situation. Notice that the population distribution is flat and ranges from 1 to 200 (left side of Figure 17-3). Also, the population summary (called *population parameter*) we use is the average rank:

$$\theta^* = \text{Avg(pop)} = \frac{1}{200}\Sigma_{k=1}^{200} k = 100.5$$

Another relevant summary is the spread about θ^*, defined as the population standard deviation:

$$\text{SD(pop)} = \sqrt{\frac{1}{200}\Sigma_{k=1}^{200}(k - \theta^*)^2} = \sqrt{\frac{1}{200}\Sigma_{k=1}^{200}(k - 100.5)^2} \approx 57.7$$

The SD(pop) represents the typical deviation of a rank from the population average. To calculate SD(pop) for this example takes some mathematical handiwork:

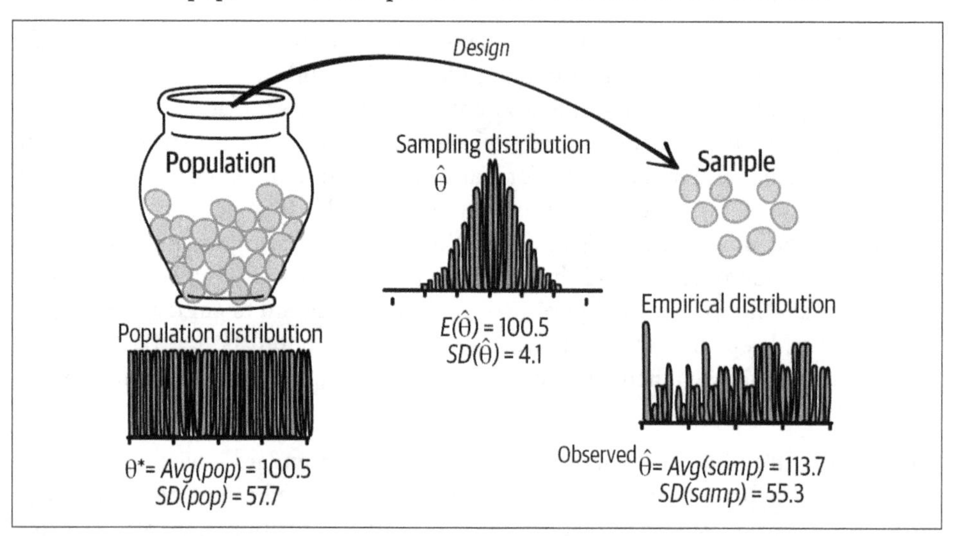

Figure 17-3. Diagram of the data generation process for the Wikipedia experiment; this is a special case where we know the population

The observed sample consists of the integer ranks of the treatment group; we refer to these values as $k_1, k_2, \ldots, k_{100}$. The sample distribution appears on the right in Figure 17-3 (each of the 100 integers appears once).

The parallel to the population average is the sample average, which is our statistic of interest:

$$\text{Avg(sample)} = \frac{1}{100}\Sigma_{i=1}^{100} k_i = \bar{k} = 113.7$$

The Avg(sample) is the observed value for $\hat{\theta}$. Similarly, the spread about Avg(sample), called the standard deviation of the sample, represents the typical deviation of a rank in the sample from the sample average:

$$\text{SD(sample)} = \sqrt{\frac{1}{100}\Sigma_{i=1}^{100}(k_i - \bar{k})^2} = 553.$$

Notice the parallel between the definitions of the sample statistic and the population parameter in the case where they are averages. The parallel between the two SDs is also noteworthy.

Next we turn to the data generation process: draw 100 marbles from the urn (with values $1, 2, \ldots, 200$), without replacement, to create the treatment ranks. We represent the action of drawing the first marble from the urn, and the integer that we get, by the capital letter Z_1. This Z_1 is called a *random variable*. It has a probability distribution determined by the urn model. That is, we can list all of the values that Z_1 might take and the probability associated with each:

$$\mathbb{P}(Z_1 = k) = \frac{1}{200} \quad \text{for } k = 1, \ldots, 200$$

In this example, the probability distribution of Z_1 is determined by a simple formula because all of the integers are equally likely to be drawn from the urn.

We often summarize the distribution of a random variable by its *expected value* and *standard deviation*. Like with the population and sample, these two quantities give us a sense of what to expect as an outcome and how far the actual value might be from what is expected.

For our example, the expected value of Z_1 is simply:

$$\begin{aligned}
\mathbb{E}[Z_1] &= 1\mathbb{P}(Z_1 = 1) + 2\mathbb{P}(Z_1 = 2) + \cdots + 200\mathbb{P}(Z_1 = 200) \\
&= 1 \times \frac{1}{200} + 2 \times \frac{1}{200} + \cdots + 200 \times \frac{1}{200} \\
&= 100.5
\end{aligned}$$

Notice that $\mathbb{E}[Z_1] = \theta^*$, the population average from the urn. The average value in a population and the expected value of a random variable that represents one draw at random from an urn that contains the population are always the same. This is more easily seen by expressing the population average as an average of the unique values in the population, weighted by the fraction of units that have that value. The expected value of a random variable of a draw at random from the population urn uses the exact same weights because they match the chance of selecting the particular value.

The term *expected value* can be a bit confusing because it need not be a possible value of the random variable. For example, $\mathbb{E}[Z_1] = 100.5$, but only integers are possible values for Z_1.

Next, the variance of Z_1 is defined as follows:

$$\begin{aligned}\mathbb{V}(Z_1) &= \mathbb{E}[Z_1 - \mathbb{E}(Z_1)]^2 \\ &= [1 - \mathbb{E}(Z_1)]^2\mathbb{P}(Z_1 = 1) + \cdots + [200 - \mathbb{E}(Z_1)]^2\mathbb{P}(Z_1 = 200) \\ &= (1 - 100.5)^2 \times \frac{1}{200} + \cdots + (200 - 100.5)^2 \times \frac{1}{200} \\ &= 3333.25\end{aligned}$$

Additionally, we define the standard deviation of Z_1 as follows:

$$SD(Z_1) = \sqrt{\mathbb{V}(Z_1)} = 57.7$$

We again point out that the standard deviation of Z_1 matches the SD(pop).

To describe the entire data generation process in Figure 17-3, we also define $Z_2, Z_3, \ldots, Z_{100}$ as the result of the remaining 99 draws from the urn. By symmetry, these random variables should all have the same probability distribution. That is, for any $k = 1, \ldots, 200$:

$$\mathbb{P}(Z_1 = k) \;=\; \mathbb{P}(Z_2 = k) \;=\; \cdots \;=\; \mathbb{P}(Z_{100} = k) \;=\; \frac{1}{200}$$

This implies that each Z_i has the same expected value, 100.5, and standard deviation, 57.7. However, these random variables are not independent. For example, if you know that $Z_1 = 17$, then it is not possible for $Z_2 = 17$.

To complete the middle portion of Figure 17-3, which involves the sampling distribution of $\hat{\theta}$, we express the average rank statistic as follows:

$$\hat{\theta} = \frac{1}{100}\Sigma_{i=1}^{100} Z_i$$

We can use the expected value and SD of Z_1 and our knowledge of the data generation process to find the expected value and SD of $\hat{\theta}$. We first find the expected value of $\hat{\theta}$:

$$\begin{aligned} \mathbb{E}(\hat{\theta}) &= \mathbb{E}\left[\frac{1}{100}\Sigma_{i=1}^{100} Z_i\right] \\ &= \frac{1}{100}\Sigma_{i=1}^{100} \mathbb{E}[Z_i] \\ &= 100.5 \\ &= \theta^* \end{aligned}$$

In other words, the expected value of the average of random draws from the population equals the population average. Here we provide formulas for the variance of the average in terms of the population variance, as well as the SD:

$$\begin{aligned} \mathbb{V}(\hat{\theta}) &= \mathbb{V}\left[\frac{1}{100}\Sigma_{i=1}^{100} Z_i\right] \\ &= \frac{200-100}{100-1} \times \frac{\mathbb{V}(Z_i)}{100} \\ &= 16.75 \end{aligned}$$

$$\begin{aligned} SD(\hat{\theta}) &= \sqrt{\frac{100}{199}}\frac{SD(Z_1)}{10} \\ &= 4.1 \end{aligned}$$

These computations relied on several properties of expected value and variance of a random variable and sums of random variables. Next, we provide properties of sums and averages of random variables that can be used to derive the formulas we just presented.

General Properties of Random Variables

In general, a *random variable* represents a numeric outcome of a chance event. In this book, we use capital letters like X or Y or Z to denote a random variable. The probability distribution for X is the specification $\mathbb{P}(X = x) = p_x$ for all values x that the random variable takes on.

Then, the expected value of X is defined as:

$$\mathbb{E}[X] = \sum_x x p_x$$

The variance X is defined as:

$$\begin{aligned} \mathbb{V}(X) &= \mathbb{E}[(X - \mathbb{E}[X])^2] \\ &= \sum_x [x - \mathbb{E}(X)]^2 p_x \end{aligned}$$

And the SD(X) is the square root of $\mathbb{V}(X)$.

Although random variables can represent quantities that are either discrete (such as the number of children in a family drawn at random from a population) or continuous (such as the air quality measured by an air monitor), we address only random variables with discrete outcomes in this book. Since most measurements are made to a certain degree of precision, this simplification doesn't limit us too much.

Simple formulas provide the expected value, variance, and standard deviation when we make scale and shift changes to random variables, such as $a + bX$ for constants a and b:

$$\begin{aligned} \mathbb{E}(a + bX) &= a + b\mathbb{E}(X) \\ \mathbb{V}(a + bX) &= b^2\mathbb{V}(X) \\ SD(a + bX) &= |b| SD(X) \end{aligned}$$

To convince yourself that these formulas make sense, think about how a distribution changes if you add a constant a to each value or scale each value by b. Adding a to each value would simply shift the distribution, which in turn would shift the expected value but not change the size of the deviations about the expected value. On the other hand, scaling the values by, say, 2 would spread the distribution out and essentially double both the expected value and the deviations from the expected value.

We are also interested in the properties of the sum of two or more random variables. Let's consider two random variables, X and Y. Then:

$$\mathbb{E}(a + bX + cY) = a + b\mathbb{E}(X) + c\mathbb{E}(Y)$$

But to find the variance of $a + bX + cY$, we need to know how X and Y vary together, which is called the *joint distribution* of X and Y. The joint distribution of X and Y assigns probabilities to combinations of their outcomes:

$$\mathbb{P}(X = x, Y = y) = p_{x,y}$$

A summary of how X and Y vary together, called the *covariance*, is defined as:

$$\begin{aligned} Cov(X, Y) &= \mathbb{E}[(X - \mathbb{E}[X])(Y - \mathbb{E}[Y])] \\ &= \mathbb{E}[(XY) - \mathbb{E}(X)\mathbb{E}(Y)] \\ &= \Sigma_{x,y}[(xy) - \mathbb{E}(X)\mathbb{E}(Y)]p_{x,y} \end{aligned}$$

The covariance enters into the calculation of $(a + bX + cY)$, as shown here:

$$\mathbb{V}(a + bX + cY) = b^2\mathbb{V}(X) + 2bcCov(X, Y) + c^2\mathbb{V}(Y)$$

In the special case where X and Y are independent, their joint distribution is simplified to $p_{x,y} = p_x p_y$. And in this case, $Cov(X, Y) = 0$, so:

$$\mathbb{V}(a + bX + cY) = b^2\mathbb{V}(X) + c^2\mathbb{V}(Y)$$

These properties can be used to show that for random variables $X_1, X_2, \ldots, X_n$ that are independent with expected value μ and standard deviation σ, the average, $\bar{X}$, has the following expected value, variance, and standard deviation:

$$\begin{aligned} \mathbb{E}(\bar{X}) &= \mu \\ \mathbb{V}(\bar{X}) &= \sigma^2 / n \\ SD(\bar{X}) &= \sigma / \sqrt{n} \end{aligned}$$

This situation arises with the urn model where $X_1, \ldots, X_n$ are the result of random draws with replacement. In this case, μ represents the average of the urn and σ the standard deviation.

However, when we make random draws from the urn without replacement, the X_i are not independent. In this situation, $\bar{X}$ has the following expected value and variance:

$$\begin{aligned} \mathbb{E}(\bar{X}) &= \mu \\ \mathbb{V}(\bar{X}) &= \frac{N - n}{N - 1} \times \frac{\sigma^2}{n} \end{aligned}$$

Notice that while the expected value is the same as when the draws are without replacement, the variance and SD are smaller. These quantities are adjusted by $(N-n)/(N-1)$, which is called the *finite population correction factor*. We used this formula earlier to compute the $SD(\hat{\theta})$ in our Wikipedia example.

Returning to Figure 17-3, we see that the sampling distribution for $\bar{X}$ in the center of the diagram has an expectation that matches the population average; the SD decreases like $1/\sqrt{n}$ but even more quickly because we are drawing without replacement; and the distribution is shaped like a normal curve. We saw these properties earlier in our simulation study.

Now that we have outlined the general properties of random variables and their sums, we connect these ideas to testing, confidence, and prediction intervals.

Probability Behind Testing and Intervals

As mentioned at the beginning of this chapter, probability is the underpinning of conducting a hypothesis test, providing a confidence interval for an estimator and a prediction interval for a future observation.

We now have the technical machinery to explain these concepts, which we have carefully defined in this chapter without the use of formal technicalities. This time we present the results in terms of random variables and their distributions.

Recall that a hypothesis test relies on a null model that provides the probability distribution for the statistic, $\hat{\theta}$. The tests we carried out were essentially computing (sometimes approximately) the following probability. Given the assumptions of the null distribution:

$$\mathbb{P}(\hat{\theta} \geq \text{observed statistic})$$

Oftentimes, the random variable is normalized to make these computations easier and standard:

$$\mathbb{P}\left(\frac{\hat{\theta} - \theta^*}{SD(\hat{\theta})} \geq \frac{\text{observed stat} - \theta^*}{SD(\hat{\theta})}\right)$$

When $SD(\hat{\theta})$ is not known, we have approximated it via simulation, and when we have a formula for $SD(\hat{\theta})$ in terms of $SD(pop)$, we substitute $SD(samp)$ for $SD(pop)$. This normalization is popular because it simplifies the null distribution. For example, if $\hat{\theta}$ has an approximate normal distribution, then the normalized version will have a standard normal distribution with center 0 and SD 1. These approximations are useful when a lot of hypothesis tests are being carried out, such as with A/B testing, since there is no need to simulate for every statistic because we can just use the normal curve probabilities.

The probability statement behind a confidence interval is quite similar to the probability calculations used in testing. In particular, to create a 95% confidence interval where the sampling distribution of the estimator is roughly normal, we standardize and use the probability:

$$\begin{aligned}\mathbb{P}\left(\frac{|\hat{\theta} - \theta^*|}{SD(\hat{\theta})} \leq 1.96\right) &= \mathbb{P}\left(\hat{\theta} - 1.96SD(\hat{\theta}) \leq \theta^* \leq \hat{\theta} + 1.96SD(\hat{\theta})\right) \\ &\approx 0.95\end{aligned}$$

Note that $\hat{\theta}$ is a random variable in the preceding probability statement and θ^* is considered a fixed unknown parameter value. The confidence interval is created by substituting the observed statistic for $\hat{\theta}$ and calling it a 95% confidence interval:

$$\left[\text{observed stat} - 1.96SD(\hat{\theta}),\ \text{observed stat} + 1.96SD(\hat{\theta})\right]$$

Once the observed statistic is substituted for the random variable, then we say that we are 95% confident that the interval we have created contains the true value θ^*. In other words, in 100 cases where we compute an interval in this way, we expect 95 of them to cover the population parameter that we are estimating.

Next, we consider prediction intervals. The basic notion is to provide an interval that denotes the expected variation of a future observation about the estimator. In the simple case where the statistic is $\bar{X}$ and we have a hypothetical new observation X_0 that has the same expected value, say μ, and standard deviation, say σ, of the X_i, then we find the expected variation of the squared loss:

$$\begin{aligned}\mathbb{E}[(X_0 - \bar{X})^2] &= \mathbb{E}\{[(X_0 - \mu) - (\bar{X} - \mu)]^2\} \\ &= \mathbb{V}(X_0) + \mathbb{V}(\bar{X}) \\ &= \sigma^2 + \sigma^2/n \\ &= \sigma\sqrt{1 + 1/n}\end{aligned}$$

Notice there are two parts to the variation: one due to the variation of X_0 and the other due to the approximation of $\mathbb{E}(X_0)$ by $\bar{X}$.

In the case of more complex models, the variation in prediction also breaks down into two components: the inherent variation in the data about the model plus the variation in the sampling distribution due to the estimation of the model. Assuming the model is roughly correct, we can express it as follows:

$$\mathbf{Y} = \mathbf{X}\boldsymbol{\theta}^* + \boldsymbol{\epsilon}$$

where $\boldsymbol{\theta}^*$ is a $(p+1) \times 1$ column vector, $\mathbf{X}$ is an $n \times (p+1)$ design matrix, and $\boldsymbol{\epsilon}$ consists of n independent random variables that each have expected value 0 and variance σ^2. In this equation, $\mathbf{Y}$ is a vector of random variables, where the expected value of each variable is determined by the design matrix and the variance is σ^2. That is, the variation about the line is constant in that it does not change with $\mathbf{x}$.

When we create prediction intervals in regression, they are given a $1 \times (p+1)$ row vector of covariates, called $\mathbf{x}_0$. Then the prediction is $\mathbf{x}_0\hat{\boldsymbol{\theta}}$, where $\hat{\boldsymbol{\theta}}$ is the estimated parameter vector based on the original $\mathbf{y}$ and design matrix $\mathbf{X}$. The expected squared error in this prediction is:

$$\begin{aligned}
\mathbb{E}[(Y_0 - \mathbf{x}_0\hat{\boldsymbol{\theta}})^2] &= \mathbb{E}\{[(Y_0 - \mathbf{x}_0\boldsymbol{\theta}^*) - (\mathbf{x}_0\hat{\boldsymbol{\theta}} - \mathbf{x}_0\boldsymbol{\theta}^*)]^2\} \\
&= \mathbb{V}(\epsilon_0) + \mathbb{V}(\mathbf{x}_0\hat{\boldsymbol{\theta}}) \\
&= \sigma^2[1 + \mathbf{x}_0(\mathbf{X}^\top\mathbf{X})^{-1}\mathbf{x}_0^\top]
\end{aligned}$$

We approximate the variance of ϵ with the variance of the residuals from the least squares fit.

The prediction intervals we create using the normal curve rely on the additional assumption that the distribution of the errors is approximately normal. This is a stronger assumption than we make for the confidence intervals. With confidence intervals, the probability distribution of X_i need not look normal for $\bar{X}$ to have an approximate normal distribution. Similarly, the probability distribution of $\boldsymbol{\epsilon}$ in the linear model need not look normal for the estimator $\hat{\theta}$ to have an approximate normal distribution.

We also assume that the linear model is approximately correct when making these prediction intervals. In Chapter 16, we considered the case where the fitted model doesn't match the model that has produced the data. We now have the technical machinery to derive the model bias-variance trade-off introduced in that chapter. It's very similar to the prediction interval derivation with a couple of small twists.

Probability Behind Model Selection

In Chapter 16, we introduced model under- and overfitting with mean squared error (MSE). We described a general setup where the data might be expressed as follows:

$$y = g(\mathbf{x}) + \epsilon$$

The ϵ are assumed to behave like random errors that have no trends or patterns, have constant variance, and are independent of one another. The *signal* in the model is the function $g()$. The data are the $(\mathbf{x}_i, y_i)$ pairs, and we fit models by minimizing the MSE:

$$\min_{f \in \mathcal{F}} \frac{1}{n} \sum_{i=1}^{n} (y_i - f(\mathbf{x}_i)^2$$

Here $\mathcal{F}$ is the collection of models over which we are minimizing. This collection might be all polynomials of degree m or less, bent lines with a bend at point k, and so on. Note that g doesn't have to be in the collection of functions that we are using to fit a model.

Our goal in model selection is to land on a model that predicts a new observation well. For a new observation, we would like the expected loss to be small:

$$\mathbb{E}[y_0 - f(\mathbf{x}_0)]^2$$

This expectation is with respect to the distribution of possible $(\mathbf{x}_0, y_0)$ and is called *risk*. Since we don't know the population distribution of $(\mathbf{x}_0, y_0)$, we can't calculate the risk, but we can approximate it by the average loss over the data we have collected:

$$\mathbb{E}[y_0 - f(\mathbf{x}_0)]^2 \approx \frac{1}{n} \sum_{i=1}^{n} (y_i - f(\mathbf{x}_i))^2$$

This approximation goes by the name of *empirical risk*. But hopefully you recognize it as the MSE:

We fit models by minimizing the empirical risk (or MSE) over all possible models, $\mathcal{F} = \{f\}$:

$$\min_{f \in \mathcal{F}} \frac{1}{n} \sum_{i=1}^{n} (y_i - f(\mathbf{x}_i))^2$$

The fitted model is called $\hat{f}$, a slightly more general representation of the linear model $\mathbf{X}\hat{\boldsymbol{\theta}}$. This technique is aptly called *empirical risk minimization*.

In Chapter 16, we saw problems arise when we used the empirical risk to both fit a model and evaluate the risk for a new observation. Ideally, we want to estimate the risk (expected loss):

$$\mathbb{E}[(y_0 - \hat{f}(\mathbf{x}_0))^2]$$

where the expected value is over the new observation $(\mathbf{x}_0, y_0)$ and over $\hat{f}$ (which involves the original data $(\mathbf{x}_i, y_i)$, $i = 1, \ldots, n$).

To understand the problem, we decompose this risk into three parts representing the model bias, the model variance, and the irreducible error from ϵ:

$$\begin{aligned}
&\mathbb{E}[y_0 - \hat{f}(x_0)]^2 \\
&= \mathbb{E}[g(x_0) + \epsilon_0 - \hat{f}(x_0)]^2 && \text{definition of } y_0 \\
&= \mathbb{E}[g(x_0) + \epsilon_0 - \mathbb{E}[\hat{f}(x_0)] + \mathbb{E}[\hat{f}(x_0)] - \hat{f}(x_0)]^2 && \text{adding } \pm\mathbb{E}[\hat{f}(x_0)] \\
&= \mathbb{E}[g(x_0) - \mathbb{E}[\hat{f}(x_0)] - (\hat{f}(x_0) - \mathbb{E}[\hat{f}(x_0)]) + \epsilon_0]^2 && \text{rearranging terms} \\
&= [g(x_0) - \mathbb{E}[\hat{f}(x_0)]]^2 + \mathbb{E}[\hat{f}(x_0) - \mathbb{E}[\hat{f}(x_0)]]^2 + \sigma^2 && \text{expanding the square} \\
&= \quad \text{model bias}^2 \quad + \quad \text{model variance} \quad + \quad \text{error}
\end{aligned}$$

To derive the equality labeled "expanding the square," we need to formally prove that the cross-product terms in the expansion are all 0. This takes a bit of algebra and we don't present it here. But the main idea is that the terms ϵ_0 and $(\hat{f}(x_0) - \mathbb{E}[\hat{f}(x_0])$ are independent and both have the expected value 0. The remaining three terms in the final equation—model bias, model variance, and irreducible error—are described as follows:

Model bias

The first of the three terms in the final equation is model bias (squared). When the signal, g, does not belong to the model space, we have model bias. If the model space can approximate g well, then the bias is small. Note that this term is not present in our prediction intervals because we assumed that there is no (or minimal) model bias.

Model variance

The second term represents the variability in the fitted model that comes from the data. We have seen in earlier examples that high-degree polynomials can overfit, and so vary a lot from one set of data to the next. The more complex the model space, the greater the variability in the fitted model.

Irreducible error

Finally, the last term is the variability in the error, the ϵ_0, which is dubbed the "irreducible error." This error sticks around whether we have underfit with a simple model (high bias) or overfit with a complex model (high variance).

This representation of the expected loss shows the bias-variance decomposition of a fitted model. Model selection aims to balance these two competing sources of error. The train-test split, cross-validation, and regularization introduced in Chapter 16 are techniques to either mimic the expected loss for a new observation or penalize a model from overfitting.

While we have covered a lot of theory in this chapter, we have attempted to tie it to the basics of the urn model and the three distributions: population, sample, and sampling. We wrap up the chapter with a few cautions to keep in mind when performing hypothesis tests and when making confidence or prediction intervals.

Summary

Throughout this chapter, we based our development of the theory behind inference and prediction on the urn model. The urn induced a probability distribution on the estimator, such as the sample mean and the least squares regression coefficients. We end this chapter with some cautions about these statistical procedures.

We saw how the SD of an estimator has a factor of the square root of the sample size in the denominator. When samples are large, the SD can be quite small and can lead to rejecting a hypothesis or very narrow confidence intervals. When this happens it's good to consider the following:

- Is the difference that you have detected an important difference? That is, a *p*-value may be quite small, indicating a surprising result, but the actual effect observed may be unimportant. *Statistical significance* does not imply *practical significance*.
- Keep in mind that these calculations do not incorporate bias, such as nonresponse bias and measurement bias. The bias might well be larger than any difference due to chance variation in the sampling distribution.

At times, we know the sample is not from a chance mechanism, but it can still be useful to carry out a hypothesis test. In this case, the null model would test whether the sample (and estimator) are as if they were at random. When this test is rejected, we confirm that something nonrandom has led to the observed data. This can be a useful conclusion: that the difference between what we expect and what we observed is not explained by chance.

At other times, the sample consists of the complete population. When this happens, we might not need to make confidence intervals or hypothesis tests because we have observed all values in the population. That is, inference is not required. However, we can instead place a different interpretation on hypothesis tests: we can suppose that any relation observed between two features was randomly distributed without relation to each other.

We also saw how the bootstrap can be used when we don't have enough information about the population. The bootstrap is a powerful technique, but it has limitations:

- Make sure that the original sample is large and random so that the sample resembles the population.
- Repeat the bootstrap process many times. Typically 10,000 replications is a reasonable number.
- The bootstrap tends to have difficulties when:
  - — The estimator is influenced by outliers.
  - — The parameter is based on extreme values of the distribution.
  - — The sampling distribution of the statistic is far from bell shaped.

Alternatively, we rely on the sampling distribution being approximately normal in shape. At times, the sampling distribution looks roughly normal but has thicker tails. In these situations, the family of t-distributions might be appropriate to use instead of the normal.

A model is usually only an approximation of underlying reality, and the precision of the statement that θ^* exactly equals 0 is at odds with this notion of a model. The inference depends on the correctness of our model. We can partially check the model assumptions, but some amount of doubt goes with any model. In fact, it often happens that the data suggest more than one possible model, and these models may even be contradictory.

Lastly, at times, the number of hypothesis tests or confidence intervals can be quite large, and we need to exercise caution to avoid spurious results. This problem is called p-hacking and is another example of the reproducibility crisis in science described in Chapter 10. P-hacking is based on the notion that if we test, say, 100 hypotheses, all of which are true, then we would expect to get a few surprise results and reject a few of these hypotheses. This phenomenon can happen in multiple linear regression when we have a large number of features in a model, and techniques have been developed to limit the dangers of these false discoveries.

We next recap the modeling process with a case study.

CHAPTER 18

Case Study: How to Weigh a Donkey

Donkeys play important roles in rural Kenya. People need them to transport crops, water, and people and to plow fields. When a donkey gets sick, the veterinarian needs to figure out how much the donkey weighs in order to prescribe the right amount of medicine. But many vets in rural Kenya don't have access to a scale, so they need to guess the donkey's weight. Too little medicine can allow an infection to reemerge; too much medicine can cause a harmful overdose. There are over 1.8 million donkeys in Kenya, so it's important to have a simple, accurate way to estimate the weight of a donkey.

In this case study, we follow the work of Kate Milner and Jonathan Rougier (*https://doi.org/10.1111/j.1740-9713.2014.00768.x*) to create a model that veterinarians in the Kenyan countryside can use to make accurate estimates of a donkey's weight. As usual, we walk through the steps of the data science lifecycle, but this time our work departs from the basics covered so far in this book. You can think of this case study as an opportunity to reflect on many of the core principles of working with data and to understand how they can be extended to address the context of the situation. We directly evaluate sources of measurement error, design a special loss function that reflects the concern about an overdose, build a model while keeping applicability top of mind, and evaluate model predictions using special criteria that are relative to the donkey's size.

We begin with the scope of the data.

Donkey Study Question and Scope

Our motivating question is: how can a vet accurately estimate the weight of a donkey when they're out in the countryside without a scale? Let's think about the information that is more readily available to them. They can carry a tape measure and find the size of the donkey in other dimensions, like its height. They can observe the animal's sex, assess its general condition, and inquire about the donkey's age. So we can refine our question to: how can a vet accurately predict the weight of a donkey from easy-to-get measurements?

To address this more precise question, The Donkey Sanctuary (*https://oreil.ly/uUyZj*) carried out a study at 17 mobile deworming sites in rural Kenya.

In terms of scope (Chapter 2), the target population is the donkey population in rural Kenya. The access frame is the set of all donkeys that were brought into the deworming sites. The sample consists of all donkeys brought to these sites between July 23 and August 11, 2010, with a few caveats: if there were too many donkeys to measure at a site, the scientists selected a subset of donkeys to measure, and any pregnant or visibly diseased donkeys were excluded from the study.

To avoid accidentally weighing a donkey twice, each donkey was marked after weighing. To quantify the measurement error and assess repeatability of the weighing process, 31 donkeys were measured twice, without the staff knowing that a donkey was being reweighed.

With this sampling process in mind, potential sources of bias for this data include:

Coverage bias
: The 17 sites were located in the regions surrounding the Yatta district in eastern Kenya and the Naivasha district in the Rift Valley.

Selection bias
: Only donkeys brought to the sanctuary were enrolled in the study, and when there were too many donkeys at a site, a nonrandom sample was selected.

Measurement bias
: In addition to measurement error, the scale might have a bias. Ideally, the scale(s) would be calibrated before and after use at a site (Chapter 12).

Despite these potential sources of bias, the access frame seems reasonable for accessing donkeys from rural areas in Kenya that have owners looking after the health of their animals.

Our next step is to clean the data.

Wrangling and Transforming

We begin by taking a peek at the contents of our datafile. To do this, we open the file and examine the first few rows (Chapter 8):

```
from pathlib import Path

# Create a Path pointing to our datafile
```

```
insp_path = Path('data/donkeys.csv')

with insp_path.open() as f:
    # Display first five lines of file
    for _ in range(5):
        print(f.readline(), end='')
```

```
BCS,Age,Sex,Length,Girth,Height,Weight,WeightAlt
3,<2,stallion,78,90,90,77,NA
2.5,<2,stallion,91,97,94,100,NA
1.5,<2,stallion,74,93,95,74,NA
3,<2,female,87,109,96,116,NA
```

Since the file is CSV-formatted, we can easily read it into a dataframe:

```
donkeys = pd.read_csv("data/donkeys.csv")
donkeys
```

	BCS	Age	Sex	Length	Girth	Height	Weight	WeightAlt
0	3.0	<2	stallion	78	90	90	77	NaN
1	2.5	<2	stallion	91	97	94	100	NaN
2	1.5	<2	stallion	74	93	95	74	NaN
...	...	...	...	...	...	...	...	...
541	2.5	10-15	stallion	103	118	103	174	NaN
542	3.0	2-5	stallion	91	112	100	139	NaN
543	3.0	5-10	stallion	104	124	110	189	NaN

```
544 rows × 8 columns
```

Over five hundred donkeys participated in the survey, and eight measurements were made on each donkey. According to the documentation, the granularity is a single donkey (Chapter 9). Table 18-1 provides descriptions of the eight features.

Table 18-1. Donkey study codebook

Feature	Data type	Feature type	Description
BCS	float64	Ordinal	Body condition score: from 1 (emaciated) to 3 (healthy) to 5 (obese) in increments of 0.5.
Age	string	Ordinal	Age in years, under 2, 2–5, 5–10, 10–15, 15–20, and over 20 years
Sex	string	Nominal	Sex categories: stallion, gelding, female
Length	int64	Numeric	Body length (cm) from front leg elbow to back of pelvis
Girth	int64	Numeric	Body circumference (cm), measured just behind front legs
Height	int64	Numeric	Body height (cm) up to point where neck connects to back
Weight	int64	Numeric	Weight (kilogram)
WeightAlt	float64	Numeric	Second weight measurement taken on a subset of donkeys

Figure 18-1 is a stylized representation of a donkey as a cylinder with neck and legs appended. Height is measured from the ground to the base of the neck above the shoulders; girth is around the body, just behind the legs; and length is from the front elbow to the back of the pelvis.

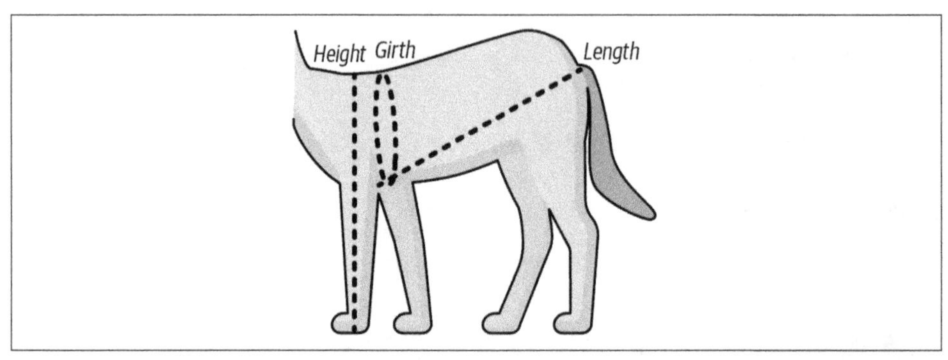

Figure 18-1. Diagram of a donkey's girth, length, and height, characterized as measurements on a cylinder

Our next step is to perform some quality checks on the data. In the previous section, we listed a few potential quality concerns based on scope. Next, we check the quality of the measurements and their distributions.

Let's start by comparing the two weight measurements made on the subset of donkeys to check on the consistency of the scale. We make a histogram of the difference between these two measurements for the 31 donkeys that were weighed twice:

```
donkeys = donkeys.assign(difference=donkeys["WeightAlt"] - donkeys["Weight"])

px.histogram(donkeys, x="difference", nbins=15,
    labels=dict(
        difference="Differences of two weighings (kg)<br>on the same donkey"
    ),
    width=350, height=250,
)
```

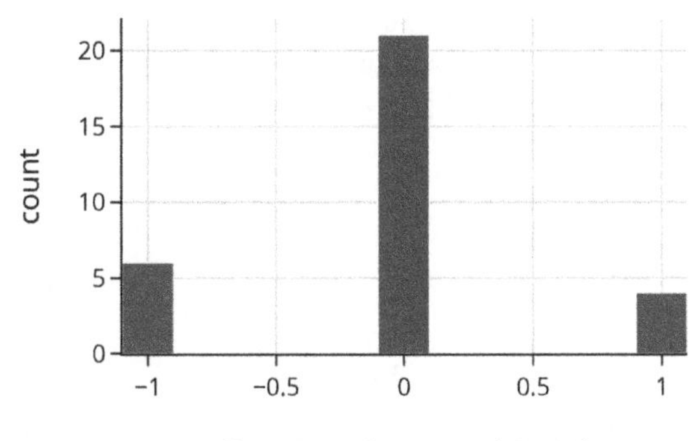

The measurements are all within 1 kg of each other, and the majority are exactly the same (to the nearest kilogram). This gives us confidence in the accuracy of the measurements.

Next, we look for unusual values in the body condition score:

```
donkeys['BCS'].value_counts()
```

```
BCS
3.0    307
2.5    135
3.5     55
       ...
1.5      5
4.5      1
1.0      1
Name: count, Length: 8, dtype: int64
```

From this output, we see that there's only one emaciated (BCS = 1) and one obese (BCS = 4.5) donkey. Let's look at the complete records for these two donkeys:

```
donkeys[(donkeys['BCS'] == 1.0) | (donkeys['BCS'] == 4.5)]
```

	BCS	Age	Sex	Length	Girth	Height	Weight	WeightAlt
291	4.5	10-15	female	107	130	106	227	NaN
445	1.0	>20	female	97	109	102	115	NaN

Since these BCS values are extreme, we want to be cautious about including these two donkeys in our analysis. We have only one donkey in each of these extreme categories, so our model might well not extend to donkeys with a BCS of 1 or 4.5. We remove these two records from the dataframe and note that our analysis may not extend to emaciated or obese donkeys. In general, we exercise caution in dropping records from a dataframe. Later, we may also decide to remove the five donkeys with a score of 1.5 if they appear anomalous in our analysis, but for now, we keep them in our dataframe. In general, we need a good reason to exclude data, and we should document these actions since they can impact our findings. Removing data can lead to over fitting, if we drop any record that disagrees with the model.

We remove these two outliers next:

```
def remove_bcs_outliers(donkeys):
    return donkeys[(donkeys['BCS'] >= 1.5) & (donkeys['BCS'] <= 4)]

donkeys = (pd.read_csv('data/donkeys.csv')
           .pipe(remove_bcs_outliers))
```

Now, we examine the distribution of values for weight to see if there are any issues with quality:

```
px.histogram(donkeys, x='Weight', nbins=40, width=350, height=250,
             labels={'Weight':'Weight (kg)'})
```

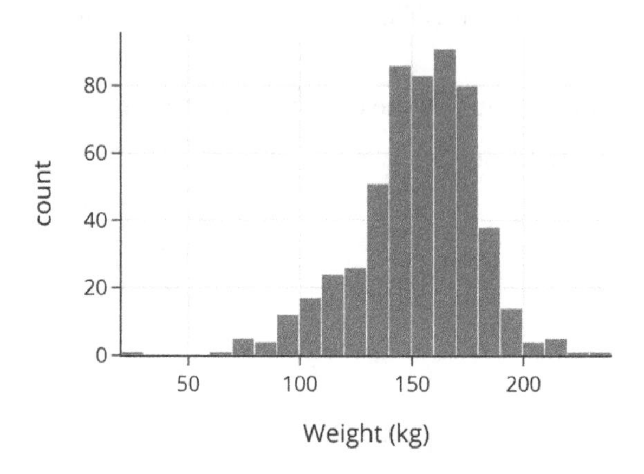

It appears there is one very light donkey weighing less than 30 kg. Next, we check the relationship between weight and height to assess the quality of the data for analysis:

```
px.scatter(donkeys, x='Height', y='Weight', width=350, height=250,
           labels={'Weight':'Weight (kg)', 'Height':'Height (cm)'})
```

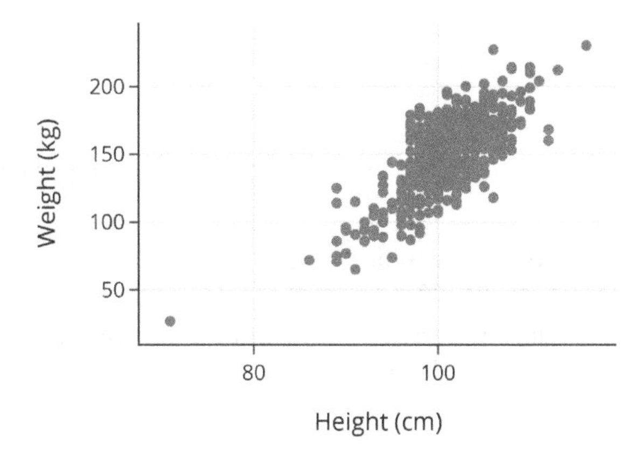

The small donkey is far from the main concentration of donkeys and would overly influence our models. For this reason, we exclude it. Again, we keep in mind that we may also want to exclude the one or two heavy donkeys if they appear to overly influence our future model fitting:

```
def remove_weight_outliers(donkeys):
    return donkeys[(donkeys['Weight'] >= 40)]

donkeys = (pd.read_csv('data/donkeys.csv')
           .pipe(remove_bcs_outliers)
           .pipe(remove_weight_outliers))

donkeys.shape
```

```
(541, 8)
```

In summary, based on our cleaning and quality checks, we removed three anomalous observations from the dataframe. Now we're nearly ready to begin our exploratory analysis. Before we proceed, we set aside some of our data as a test set.

We talked about why it's important to separate out a test set from the train set in Chapter 16. A best practice is to separate out a test set early in the analysis, before we explore the data in detail, because in EDA, we begin to make decisions about what kinds of models to fit and what variables to use in the model. It's important that our test set isn't involved in these decisions so that it imitates how our model would perform with entirely new data.

We divide our data into an 80/20 split, where we use 80% of the data to explore and build a model. Then we evaluate the model with the 20% that has been set aside. We use a simple random sample to split the dataframe into the test and train sets. To begin, we randomly shuffle the indices of the dataframe:

```
np.random.seed(42)
n = len(donkeys)
indices = np.arange(n)
np.random.shuffle(indices)
n_train = int(np.round((0.8 * n)))
```

Next, we assign the first 80% of the dataframe to the train set and the remaining 20% to the test set:

```
train_set = donkeys.iloc[indices[:n_train]]
test_set = donkeys.iloc[indices[n_train:]]
```

Now we're ready to explore the training data and look for useful relationships and distributions that inform our modeling.

Exploring

Let's look at the features in our dataframe for shapes and relationships that will help us make transformations and models (Chapter 10). We start by looking at how the categorical features of age, sex, and body condition relate to weight:

```
f1 = px.box(train_set, x="Age", y="Weight",
            category_orders = {"Age":['<2', '2-5', '5-10',
                                      '10-15', '15-20', '>20']})
f2 = px.box(train_set, x="Sex", y="Weight")

# We wrote the left_right function as a shorthand for plotly's make_subplots
fig = left_right(f1, f2, column_widths=[0.7, 0.3])

fig.update_xaxes(title='Age (yr)', row=1, col=1)
fig.update_xaxes(title='Sex', row=1, col=2)
fig.update_yaxes(title='Weight (kg)', row=1, col=1)
```

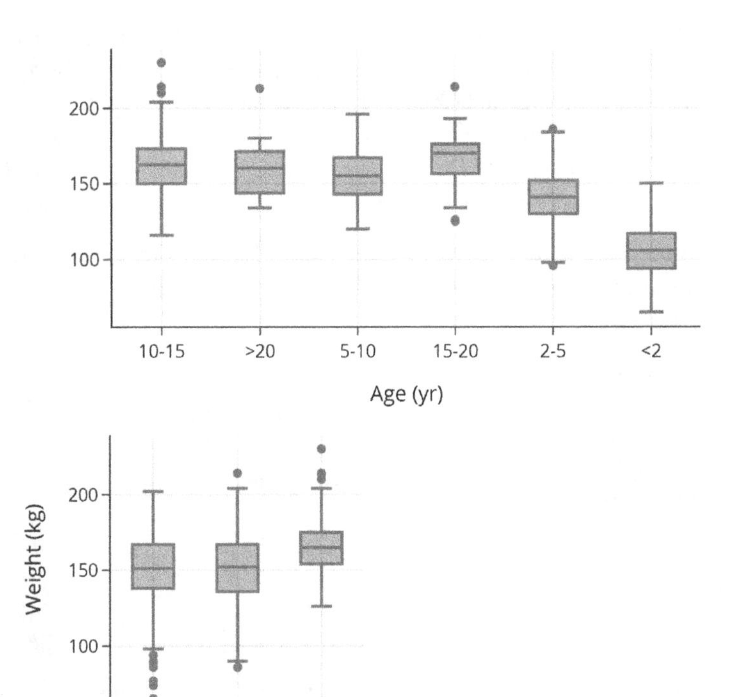

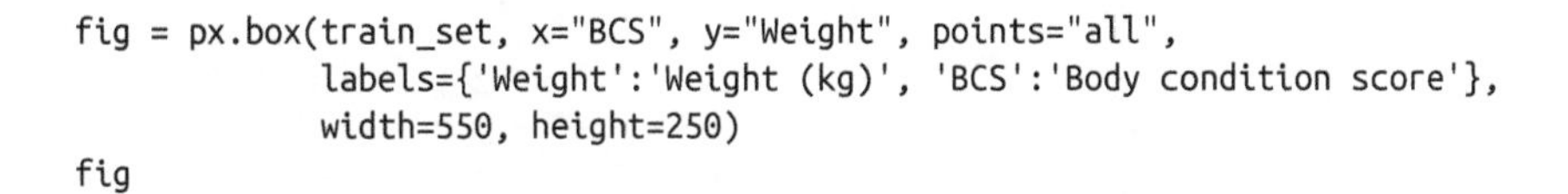

```
fig = px.box(train_set, x="BCS", y="Weight", points="all",
             labels={'Weight':'Weight (kg)', 'BCS':'Body condition score'},
             width=550, height=250)
fig
```

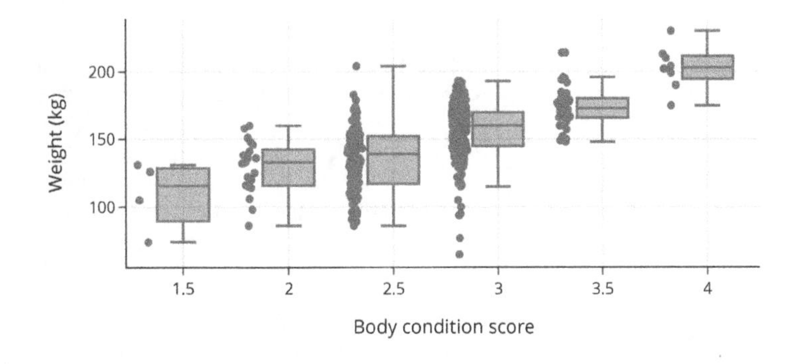

Notice that we plotted the points as well as the boxes for the body condition score because we saw earlier that there are only a handful of observations with a score of

1.5, so we don't want to read too much into a box plot with only a few data points (Chapter 11). It appears that the median weight increases with the body condition score, but not in a simple linear fashion. On the other hand, weight distributions for the three sex categories appear roughly the same. As for age, once a donkey reaches five years, the distribution of weight doesn't seem to change much. But donkeys under age 2 and donkeys from 2 to 5 years of age have lower weights in general.

Next, let's examine the quantitative variables. We plot all pairs of quantitative variables in the scatterplot matrix:

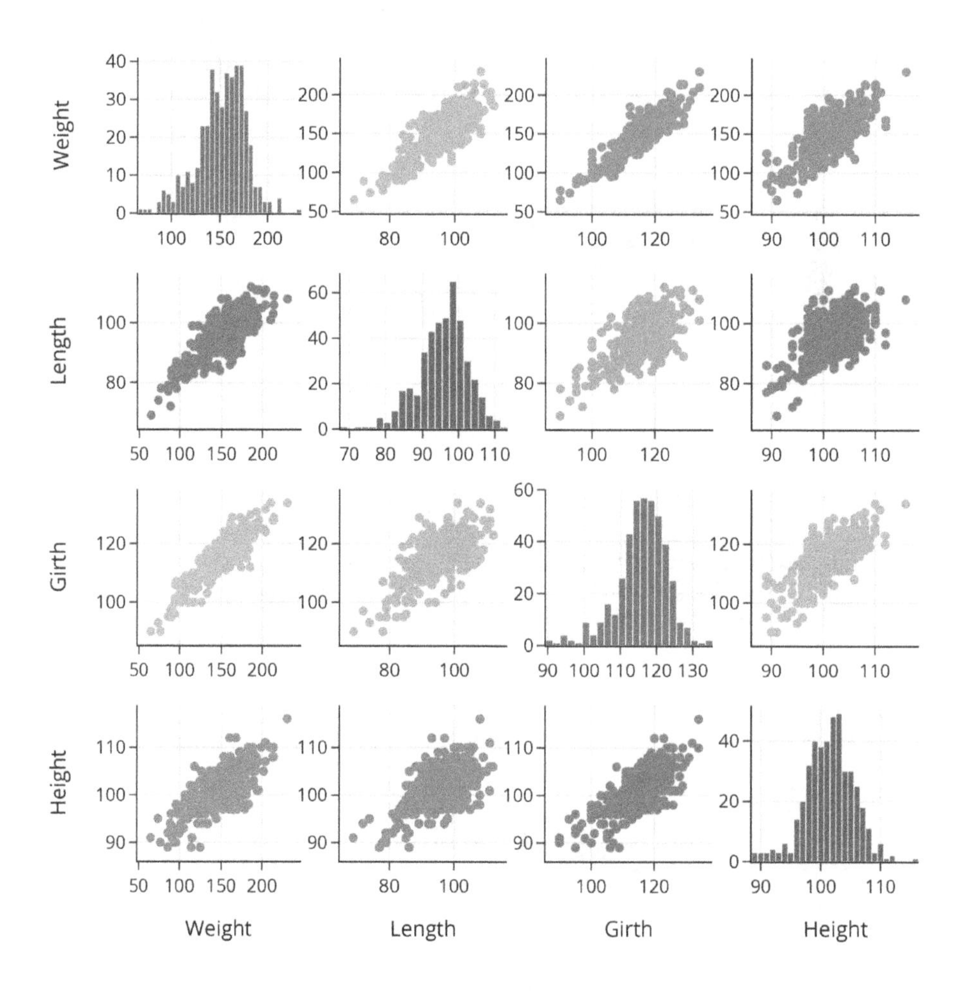

The height, length, and girth of donkeys all appear linearly associated with weight and with each other. This is not too surprising; given one of the donkey's dimensions, we should have a good guess about the other dimensions. Girth appears most highly correlated with weight, and this is confirmed by the correlation coefficient matrix:

```
train_numeric.corr()
```

	Weight	Length	Girth	Height
Weight	1.00	0.78	0.90	0.71
Length	0.78	1.00	0.66	0.58
Girth	0.90	0.66	1.00	0.70
Height	0.71	0.58	0.70	1.00

Our explorations uncovered several aspects of the data that may be relevant for modeling. We found that the donkey's girth, length, and height all have linear associations with weight and with each other, and girth has the strongest linear relationship with weight. We also observed that the body condition score has a positive association with weight; the sex of the donkey does not appear related to weight; and neither does age for those donkeys over 5 years. In the next section, we use these findings to build our model.

Modeling a Donkey's Weight

We want to build a simple model for predicting the weight of a donkey. The model should be easy for a vet to implement in the field with only a hand calculator. The model should also be easy to interpret.

We also want the model to depend on the vet's situation—for example, whether they're prescribing an antibiotic or an anesthetic. For brevity, we only consider the case of prescribing an anesthetic. Our first step is to choose a loss function that reflects this situation.

A Loss Function for Prescribing Anesthetics

An overdose of an anesthetic can be much worse than an underdose. It's not hard for a vet to see when a donkey has too little anesthetic (it'll complain), and the vet can give the donkey a bit more. On the other hand, too much anesthetic can have serious consequences and can even be fatal. Because of this, we want an asymmetric loss function: it should have a bigger loss for an overestimate of weight compared to an underestimate. This is in contrast to the other loss functions that we have used so far in this book, which have all been symmetric.

We've created a loss function `anes_loss(x)` with this in mind:

```
def anes_loss(x):
    w = (x >= 0) + 3 * (x < 0)
    return np.square(x) * w
```

The relative error is $100(y - \hat{y}) / \hat{y}$, where y is the true value and $\hat{y}$ is the prediction. We can demonstrate the asymmetry of the loss function with a plot:

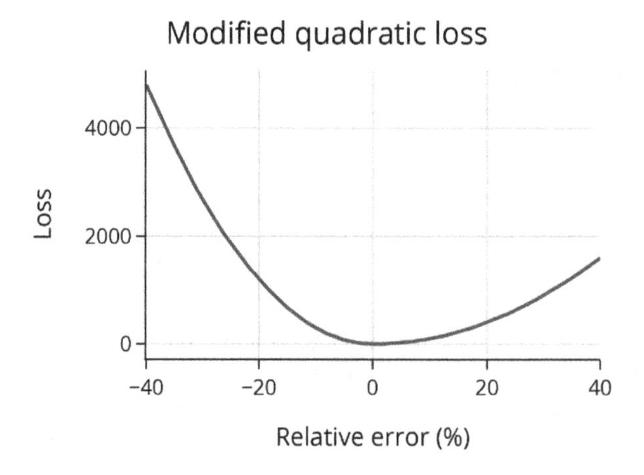

Note that a value of –10 on the x-axis reflects an overestimate of 10%.

Next, let's fit a simple linear model using this loss function.

Fitting a Simple Linear Model

We saw that girth has the highest correlation with weight among the donkeys in our train set. So we fit a model of the form:

$$\theta_0 + \theta_1 \text{Girth}$$

To find the best fit θ_0 and θ_1 to the data, we first create a design matrix that has girth and an intercept term. We also create the y vector of observed donkey weights:

```
X = train_set.assign(intr=1)[['intr', 'Girth']]
y = train_set['Weight']
X
```

	intr	Girth
230	1	116
74	1	117
354	1	123
...	...	...
157	1	123
41	1	103
381	1	106

```
433 rows × 2 columns
```

Now we want to find the θ_0 and θ_1 that minimize the average anesthetic loss over the data. To do this, we could use calculus as we did in Chapter 15, but here we'll instead use the `minimize` method from the `scipy` package, which performs a numerical optimization (see Chapter 20):

```
from scipy.optimize import minimize

def training_loss(X, y):
    def loss(theta):
        predicted = X @ theta
        return np.mean(anes_loss(100 * (y - predicted) / predicted))
    return loss

results = minimize(training_loss(X, y), np.ones(2))
theta_hat = results['x']
```

```
After fitting:
θ₀ = -218.51
θ₁ =    3.16
```

Let's see how this simple model does. We can use the model to predict the donkey weights on our train set, then find the errors in the predictions. The residual plot that follows shows the model error as a percentage of the predicted value. It's more important for the prediction errors to be small relative to the size of the donkey, since a 10 kg error is much worse for a 100 kg donkey than a 200 kg one. Thus, we find the relative error of each prediction:

```
predicted = X @ theta_hat
resids = 100 * (y - predicted) / predicted
```

Let's examine a scatterplot of the relative errors:

```
resid = pd.DataFrame({
    'Predicted weight (kg)': predicted, 'Percent error': resids})
px.scatter(resid, x='Predicted weight (kg)', y='Percent error',
           width=350, height=250)
```

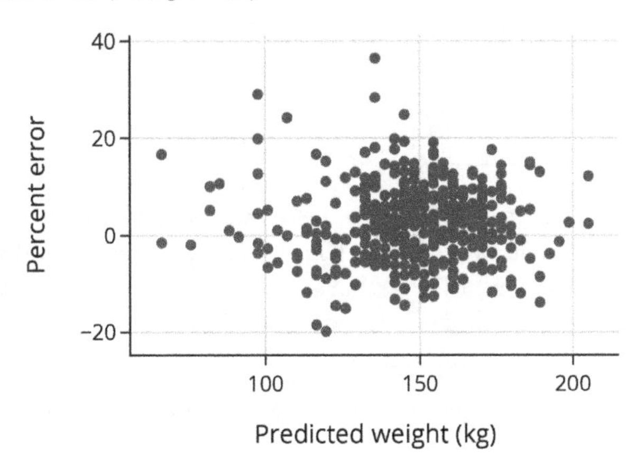

With the simplest model, some of the predictions are off by 20% to 30%. Let's see if a slightly more complicated model improves the predictions.

Fitting a Multiple Linear Model

Let's consider additional models that incorporate the other numeric variables. We have three numeric variables that measure the donkey's girth, length, and height, and there are seven total ways to combine these variables in a model:

```
[['Girth'],
 ['Length'],
 ['Height'],
 ['Girth', 'Length'],
 ['Girth', 'Height'],
 ['Length', 'Height'],
 ['Girth', 'Length', 'Height']]
```

For each of these variable combinations, we can fit a model with our special loss function. Then we can look at how well each model does on the train set:

```
def training_error(model):
    X = train_set.assign(intr=1)[['intr', *model]]
    theta_hat = minimize(training_loss(X, y), np.ones(X.shape[1]))['x']
    predicted = X @ theta_hat
    return np.mean(anes_loss(100 * (y - predicted)/ predicted))

model_risks = [
    training_error(model)
    for model in models
]
```

	model	mean_training_error
0	[Girth]	94.36
1	[Length]	200.55
2	[Height]	268.88
3	[Girth, Length]	65.65
4	[Girth, Height]	86.18
5	[Length, Height]	151.15
6	[Girth, Length, Height]	63.44

As we stated earlier, the girth of the donkey is the single best predictor for weight. However, the combination of girth and length has an average loss that is quite a bit smaller than girth alone, and this particular two-variable model is nearly as good as the model that includes all three. Since we want a simple model, we select the two-variable model over the three-variable one.

Next, we use feature engineering to incorporate categorical variables into the model, which improves our model.

Bringing Qualitative Features into the Model

In our exploratory analysis, we found that the box plots of weight for a donkey's body condition and age could contain useful information in predicting weight. Since these are categorical features, we can transform them into 0-1 variables with one-hot encoding, as explained in Chapter 15.

One-hot encoding lets us adjust the intercept term in the model for each combination of categories. Our current model includes the numeric variables girth and length:

$$\theta_0 + \theta_1 \text{Girth} + \theta_2 \text{Length}$$

If we cleaned up the age feature to consist of three categories—`Age<2`, `Age2-5`, and `Age>5`—a one-hot encoding of age creates three 0-1 features, one for each category. Including the one-hot-encoded feature in the model gives:

$$\begin{aligned} &\theta_0 + \theta_1 \text{Girth} \ + \theta_2 \text{Length} \\ &\quad + \theta_3 \text{Age<2} \ + \theta_4 \text{Age2-5} \end{aligned}$$

In this model, `Age<2` is 1 for a donkey younger than 2 and 0 otherwise. Similarly, `Age2-5` is 1 for a donkey between 2 and 5 years old and 0 otherwise.

We can think of this model as fitting three linear models that are identical except for the size of the constant, since the model is equivalent to:

$$\begin{aligned} (\theta_0 + \theta_3) + \theta_1 \text{Girth} + \theta_2 \text{Length} &\qquad \text{for a donkey under 2} \\ (\theta_0 + \theta_4) + \theta_1 \text{Girth} + \theta_2 \text{Length} &\qquad \text{for a donkey between 2 and 4} \\ \theta_0 + \theta_1 \text{Girth} + \theta_2 \text{Length} &\qquad \text{for a donkey over 5} \end{aligned}$$

Now let's apply a one-hot encoding to all three of our categorical variables (body condition, age, and sex):

```
X_one_hot = (
    train_set.assign(intr=1)
    [['intr', 'Length', 'Girth', 'BCS', 'Age', 'Sex']]
    .pipe(pd.get_dummies, columns=['BCS', 'Age', 'Sex'])
    .drop(columns=['BCS_3.0', 'Age_5-10', 'Sex_female'])
)
X_one_hot
```

	intr	Length	Girth	BCS_1.5	...	Age_<2	Age_>20	Sex_gelding	Sex_stallion
230	1	101	116	0	...	0	0	0	1
74	1	92	117	0	...	0	0	0	1
354	1	103	123	0	...	0	1	0	0
...	...	...	...	...	...	...	...	...	...
157	1	93	123	0	...	0	0	0	1
41	1	89	103	0	...	1	0	0	0
381	1	86	106	0	...	0	0	0	0

```
433 rows × 15 columns
```

We dropped one dummy variable for each categorical feature. Since `BCS`, `Age`, and `Sex` have six, six, and three categories, respectively, we have added 12 dummy variables to the design matrix for a total of 15 columns, including the intercept term, girth, and length.

Let's see which categorical variables, if any, improve on our two-variable model. To do this, we can fit the model that includes the dummies from all three categorical features, along with girth and length:

```
results = minimize(training_loss(X_one_hot, y), np.ones(X_one_hot.shape[1]))

theta_hat = results['x']

y_pred = X_one_hot @ theta_hat
training_error = (np.mean(anes_loss(100 * (y - y_pred)/ y_pred)))

print(f'Training error: {training_error:.2f}')
```

```
Training error: 51.47
```

According to average loss, this model does better than the previous model with only `Girth` and `Length`. But let's try to make this model simpler while keeping its accuracy. To do this, we look at the coefficients for each of the dummy variables to see how close they are to 0 and to one another. In other words, we want to see how much the intercept might change if we include the coefficients in the model. A plot of the coefficients will make this comparison easy:

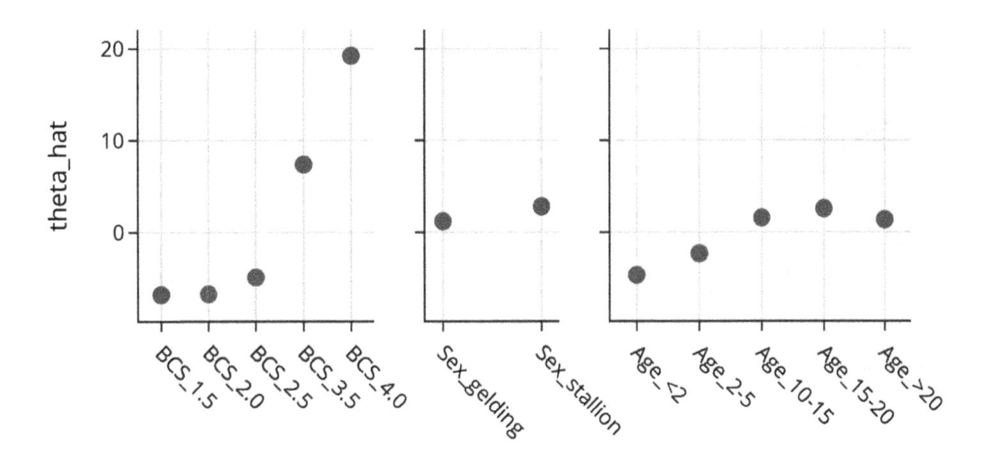

The coefficients confirm what we saw in the box plots. The coefficients for the sex of the donkey are close to zero, meaning that knowing the sex doesn't really change the weight prediction. We also see that combining the age categories for donkeys over 5 years will simplify the model without losing much. Lastly, since there are so few donkeys with a body condition score of 1.5 and its coefficient is close to that of a BCS of 2, we are inclined to combine these two categories.

We update the design matrix in view of these findings:

```
def combine_bcs(X):
    new_bcs_2 = X['BCS_2.0'] + X['BCS_1.5']
    return X.assign(**{'BCS_2.0': new_bcs_2}).drop(columns=['BCS_1.5'])

def combine_age_and_sex(X):
    return X.drop(columns=['Age_10-15', 'Age_15-20', 'Age_>20',
                           'Sex_gelding', 'Sex_stallion'])

X_one_hot_simple = (
    X_one_hot.pipe(combine_bcs)
    .pipe(combine_age_and_sex)
)
```

And then we fit the simpler model:

```
results = minimize(training_loss(X_one_hot_simple, y),
                   np.ones(X_one_hot_simple.shape[1]))
theta_hat = results['x']
y_pred = X_one_hot_simple @ theta_hat
training_error = (np.mean(anes_loss(100 * (y - y_pred)/ y_pred)))
print(f'Training error: {training_error:.2f}')
```

```
Training error: 53.20
```

The average error is close enough to that of the more complex model for us to settle on this simpler one. Let's display the coefficients and summarize the model:

	var	theta_hat
0	intr	-175.25
1	Length	1.01
2	Girth	1.97
3	BCS_2.0	-6.33
4	BCS_2.5	-5.11
5	BCS_3.5	7.36
6	BCS_4.0	20.05
7	Age_2-5	-3.47
8	Age_<2	-6.49

Our model is roughly:

$$\text{Weight} \approx -175 + \text{Length} + 2\text{Girth}$$

After this initial approximation, we use the categorical features to make some adjustments:

- BCS 2 or less? Subtract 6.5 kg.
- BCS 2.5? Subtract 5.1 kg.
- BCS 3.5? Add 7.4 kg.
- BCS 4? Add 20 kg.
- Age under 2 years? Subtract 6.5 kg.
- Age between 2 and 5 years? Subtract 3.5 kg.

This model seems quite simple to implement because after our initial estimate based on the length and girth of the donkey, we add or subtract a few numbers based on answers to a few yes/no questions. Let's see how well the model does in predicting the weights of the donkeys in the test set.

Model Assessment

Remember that we put aside 20% of our data before exploring and modeling with the remaining 80%. We are now ready to apply what we have learned from the training set to the test set. That is, we take our fitted model and use it to predict the weights of the donkeys in the test set. To do this, we need to prepare the test set. Our model uses the girth and length of the donkey, as well as dummy variables for the donkey's age and body condition score. We apply all of our transformations on the train set to our test set:

```
y_test = test_set['Weight']

X_test = (
    test_set.assign(intr=1)
    [['intr', 'Length', 'Girth', 'BCS', 'Age', 'Sex']]
    .pipe(pd.get_dummies, columns=['BCS', 'Age', 'Sex'])
    .drop(columns=['BCS_3.0', 'Age_5-10', 'Sex_female'])
    .pipe(combine_bcs)
    .pipe(combine_age_and_sex)
)
```

We consolidate all of our manipulations of the design matrix to create the final version that we settled on in our modeling with the train set. Now we are ready to use the θs that we fitted with the train set to make weight predictions for those donkeys in the test set:

```
y_pred_test = X_test @ theta_hat
test_set_error = 100 * (y_test - y_pred_test) / y_pred_test
```

Then we can plot the relative prediction errors:

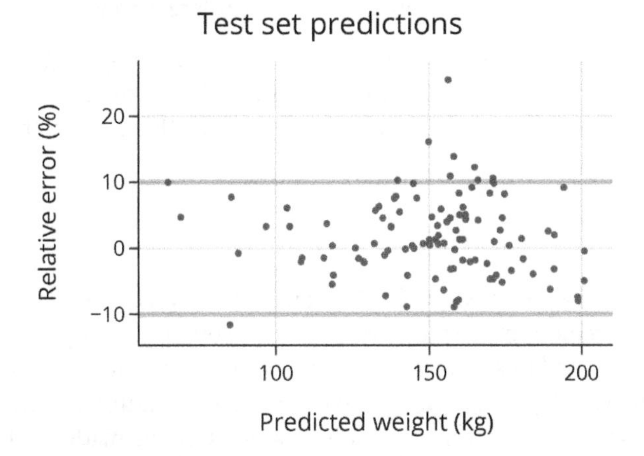

Remember that positive relative error means underestimating weight, which is not as critical as overestimating weight. From this residual plot, we see that nearly all of the test set weights are within 10% of the predictions, and only one error that exceeds 10% errs on the side of overestimation. This makes sense given that our loss function penalized overestimation more.

An alternative scatterplot that shows the actual and predicted values along with lines marking 10% error gives a different view:

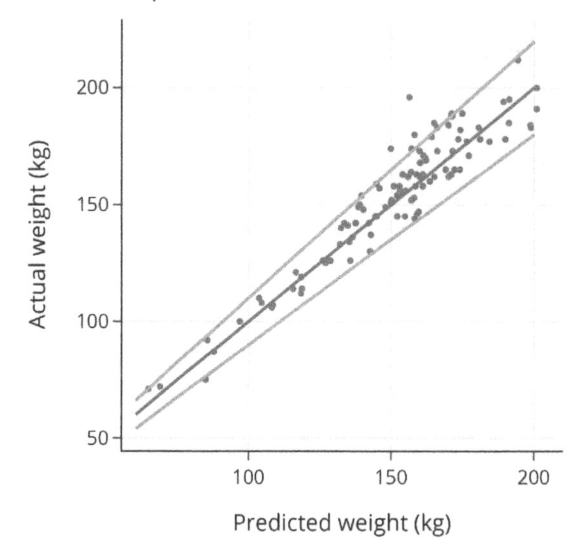

The 10% lines lie farther from the prediction line for larger weights.

We've accomplished our goal! We have a model that uses easy-to-get measurements, is simple enough to explain on an instruction sheet, and makes predictions within 10% of the actual donkey weight. Next, we summarize this case study and reflect on our model.

Summary

In this case study, we demonstrated the different purposes of modeling: description, inference, and prediction. For description, we sought a simple, understandable model. We handcrafted this model, beginning with our findings from the exploratory phase of the analysis. Every action we took to include a feature in the model, collapse categories, or transform a feature amounts to a decision we made while investigating the data.

In modeling a natural phenomenon such as the weight of a donkey, we would ideally make use of physical and statistical models. In this case, the physical model is the representation of a donkey by a cylinder. An inquisitive reader might have pointed out that we could have used this representation directly to estimate the weight of a donkey (cylinder) from its length and girth (since girth is $2\pi r$):

$$weight \propto girth^2 \times length$$

This physical model suggests that the log-transformed weight is approximately linear in girth and length:

$$\log(weight) \propto 2\log(girth) + \log(length)$$

Given this physical model, you might wonder why we did not use logarithmic or square transformations in fitting our model. We leave you to investigate such a model in greater detail. But generally, if the range of values measured is small, then the log function is roughly linear. To keep our model simple, we chose not to make these transformations given the strength of the statistical model seen by the high correlation between girth and weight.

We did a lot of *data dredging* in this modeling exercise. We examined all possible models built from linear combinations of the numeric features, and we examined coefficients of dummy variables to decide whether to collapse categories. When we create models using an iterative approach like this, it is extremely important that we set aside data to assess the model. Evaluating the model on new data reassures us that the model we chose works well. The data that we set aside did not enter into any decision making when building the model, so it gives us a good sense of how well the model works for making predictions.

We should keep the data scope and its potential biases described earlier in mind. Our model has done well on the test set, but the test and train sets come from the same data collection process. We expect our model to work well in practice as long as the scope remains the same for new data.

Finally, this case study shows how fitting models is often a balance between simplicity and complexity and between physical and statistical models. A physical model can be a good starting point in modeling, and a statistical model can inform a physical model. As data scientists, we needed to make judgment calls at each step in the analysis. Modeling is both an art and a science.

This case study and several chapters preceding it have focused on fitting linear models. Next, we consider a different kind of modeling for the situation when the response variable we are explaining or predicting is qualitative, not quantitative.

PART VI

Classification

CHAPTER 19
Classification

This chapter continues our foray into the fourth stage of the data science lifecycle: fitting and evaluating models to understand the world. So far, we've described how to fit a constant model using absolute error (Chapter 4) and simple and multiple linear models using squared error (Chapter 15). We've also fit linear models with an asymmetric loss function (Chapter 18) and with regularized loss (Chapter 16). In all of these cases, we aimed to predict or explain the behavior of a numeric outcome—bus wait times, smoke particles in the air, and donkey weights are all numeric variables.

In this chapter we expand our view of modeling. Instead of predicting numeric outcomes, we build models to predict nominal outcomes. These sorts of models enable banks to predict whether a credit card transaction is fraudulent or not, doctors to classify tumors as benign or malignant, and your email service to identify spam and set it aside from your usual emails. This type of modeling is called *classification* and occurs widely in data science.

Just as with linear regression, we formulate a model, choose a loss function, fit the model by minimizing average loss for our data, and assess the fitted model. But unlike linear regression, our model is not linear, the loss function is not squared error, and our assessment compares different kinds of classification errors. Despite these differences, the overall structure of model fitting carries over to this setting. Together, regression and classification compose the primary approaches for *supervised learning*, the general task of fitting models based on observed outcomes and covariates.

We begin by introducing an example that we use throughout this chapter.

Example: Wind-Damaged Trees

In 1999, a huge storm with winds over 90 mph damaged millions of trees in the Boundary Waters Canoe Area Wilderness (*https://oreil.ly/O2qOL*) (BWCAW), which has the largest tract of virgin forest in the eastern US. In an effort to understand the susceptibility of trees to wind damage, a researcher named Roy Lawrence Rich (*https://oreil.ly/plX02*) carried out a ground survey of the BWCAW. In the years following this study, other researchers have used this dataset to model *windthrow*, or the uprooting of trees in strong winds.

The population under study are the trees in the BWCAW. The access frame are *transects*: straight lines that cut through the natural landscape. These particular transects begin close to a lake and travel orthogonally to the gradient of the land for 250–400 meters. Along these transects, surveyors stop every 25 meters and examine a 5-by-5-meter plot. At each plot, trees are counted, categorized as blown down or standing, measured in diameter at 6 ft from the ground, and their species recorded.

Sampling protocols like this are common for studying natural resources. In the BWCAW, over 80% of the land in the region is within 500 meters of a lake, so the access frame nearly covers the population. The study took place over the summers of 2000 and 2001, and no other natural disasters happened between the 1999 storm and when the data were collected.

Measurements were collected on over 3,600 trees, but in this example, we examine just the black spruce. There are over 650 of them. We read in these data:

```
trees = pd.read_csv('data/black_spruce.csv')
trees
```

	diameter	storm	status
0	9.0	0.02	standing
1	11.0	0.03	standing
2	9.0	0.03	standing
...	...	...	...
656	9.0	0.94	fallen
657	17.0	0.94	fallen
658	8.0	0.98	fallen

```
659 rows × 3 columns
```

Each row corresponds to a single tree and has the following attributes:

`diameter`
: Diameter of the tree in cm, measured at 6 ft above the ground

`storm`
: Severity of the storm (fraction of trees that fell in a 25-meter-wide area containing the tree)

`status`
: Tree has "fallen" or is "standing"

Let's begin with some exploratory analysis before we turn to modeling. First, we calculate some simple summary statistics:

```
trees.describe()[3:]
```

	diameter	storm
min	5.0	0.02
25%	6.0	0.21
50%	8.0	0.36
75%	12.0	0.55
max	32.0	0.98

Based on the quartiles, the distribution of tree diameter seems skewed right. Let's compare the distribution of diameters for the standing and fallen trees with histograms:

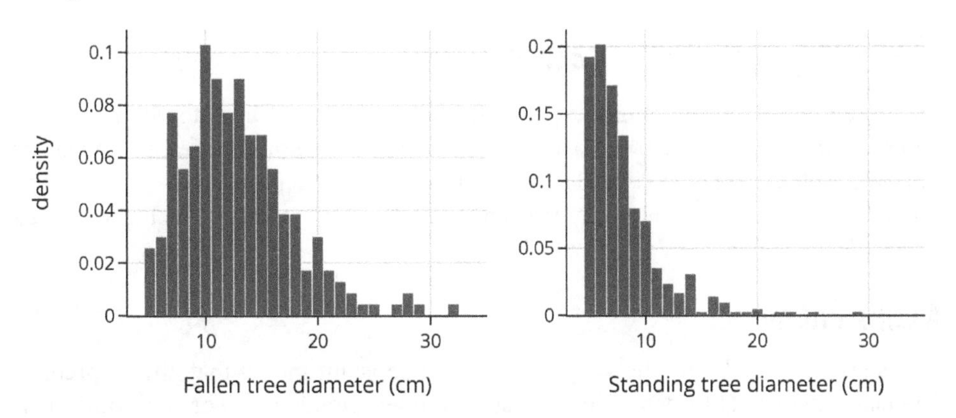

The distribution of the diameter of the trees that fell in the storm is centered at 12 cm with a right skew. In comparison, the standing trees were nearly all under 10 cm in diameter with a mode at about 6 cm (only trees with a diameter of at least 5 cm are included in the study).

Another feature to investigate is the strength of the storm. We plot the storm strength against the tree diameter using the symbol and marker color to distinguish the standing trees from the fallen ones. Since the diameter is essentially measured to the nearest cm, many trees have the same diameter, so we jitter the values by adding a bit of noise to the diameter values to help reduce overplotting (see Chapter 11). We also adjust the opacity of the marker colors to reveal the denser regions on the plot:

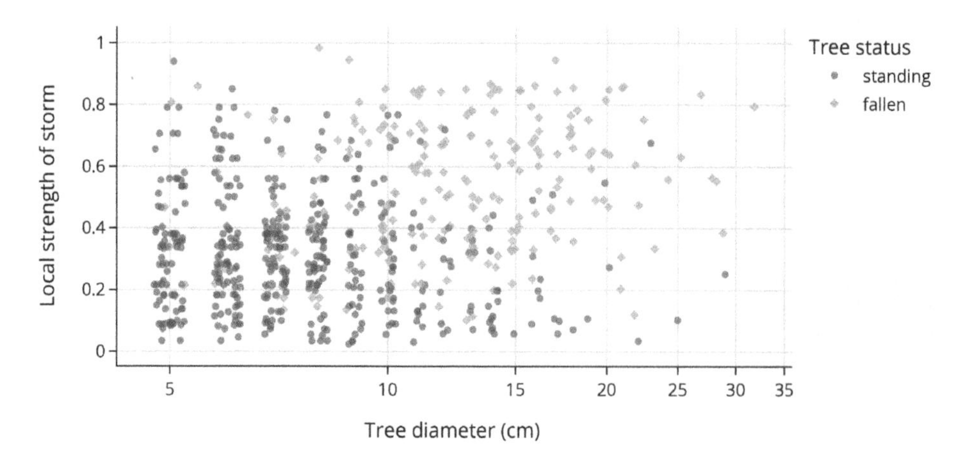

From this plot, it looks like both the tree diameter and the strength of the storm are related to windthrow: whether the tree was uprooted or left standing. Notice that windthrow, the feature we want to predict, is a nominal variable. In the next section, we consider how this impacts the prediction problem.

Modeling and Classification

We'd like to create a model that explains the susceptibility of trees to windthrow. In other words, we need to build a model for a two-level nominal feature: fallen or standing. When the response variable is nominal, this modeling task is called *classification*. In this case there are only two levels, so this task is more specifically called *binary classification*.

A Constant Model

Let's start by considering the simplest model: a constant model that always predicts one class. We use C to denote the constant model's prediction. For our windthrow dataset, this model will predict either C = standing or C = fallen for every input.

In classification, we want to track how often our model predicts the correct category. For now, we simply use a count of the correct predictions. This is sometimes called the *zero-one error* because the loss function takes on one of two possible values: 1 when an incorrect prediction is made and 0 for a correct prediction. For a given observed outcome y_i and prediction C, we can express this loss function as follows:

$$\ell(C, y) = \begin{cases} 0 & \text{when } C \text{ matches } y \\ 1 & \text{when } C \text{ is a mismatch for } y \end{cases}$$

When we have collected data, $\mathbf{y} = [y_1, \ldots, y_n]$, then the average loss is:

$$L(C, \mathbf{y}) = \frac{1}{n} \sum_{i=1}^{n} \ell(C, y) \\ = \frac{\# \text{ mismatches}}{n}$$

For the constant model (see Chapter 4), the model minimizes the loss when C is set to the most prevalent category.

In the case of the black spruce, we have the following proportions of standing and fallen trees:

```
trees['status'].value_counts() / len(trees)
```

```
status
standing    0.65
fallen      0.35
Name: count, dtype: float64
```

So our prediction is that a tree stands, and the average loss for our dataset is 0.35.

That said, this prediction is not particularly helpful or insightful. For example, in our EDA of the trees dataset, we saw that the size of the tree is correlated with whether the tree stands or falls. Ideally, we could incorporate this information into the model, but the constant model doesn't let us do this. Let's build some intuition for how we can incorporate predictors into our model.

Examining the Relationship Between Size and Windthrow

We want to take a closer look at how tree size is related to windthrow. For convenience, we transform the nominal windthrow feature into a 0-1 numeric feature where 1 stands for a fallen tree and 0 for standing:

```
trees['status_0_1'] = (trees['status'] == 'fallen').astype(int)
trees
```

	diameter	storm	status	status_0_1
0	9.0	0.02	standing	0
1	11.0	0.03	standing	0
2	9.0	0.03	standing	0
...	...	...	...	...
656	9.0	0.94	fallen	1
657	17.0	0.94	fallen	1
658	8.0	0.98	fallen	1

```
659 rows × 4 columns
```

This representation is useful in many ways. For example, the average of `status_0_1` is the proportion of fallen trees in the dataset:

```
pr_fallen = np.mean(trees['status_0_1'])
print(f"Proportion of fallen black spruce: {pr_fallen:0.2f}")
```

```
Proportion of fallen black spruce: 0.35
```

Having this 0-1 feature also lets us make a plot to show the relationship between tree diameter and windthrow. This is analogous to our process for linear regression, where we make scatterplots of the outcome variable against explanatory variable(s) (see Chapter 15).

Here we plot the tree status against the diameter, but we add a small amount of random noise to the status to help us see the density of 0 and 1 values at each diameter. As before, we jitter the diameter values too and adjust the opacity of the markers to reduce overplotting. We also add a horizontal line at the proportion of fallen trees:

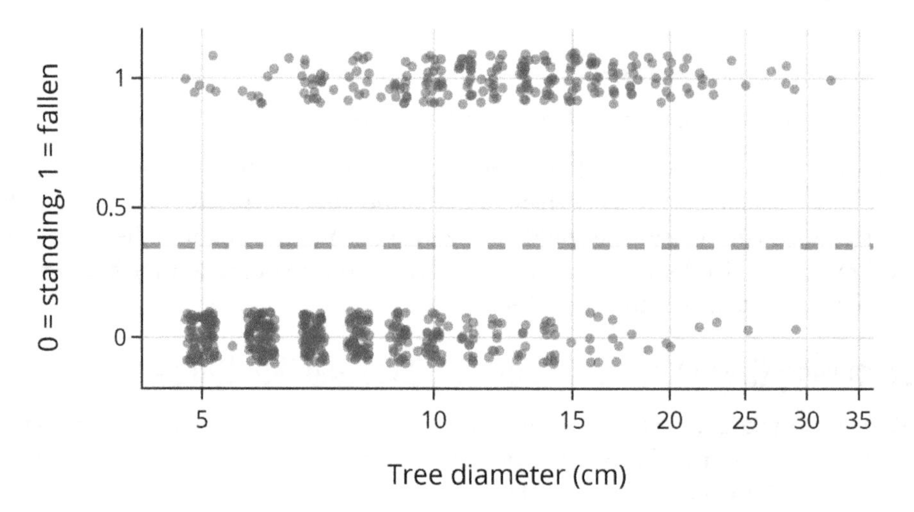

This scatterplot shows that the smaller trees are more likely to be standing than the larger trees. Notice that the average status for trees (0.35) essentially fits a constant model to the response variable. If we consider tree diameter as an explanatory feature, we should be able to improve the model.

A starting place might be to compute the proportion of fallen trees for different diameters. The following block of code divides tree diameter into intervals and computes the proportion of fallen trees in each bin:

```
splits = [4, 5, 6, 7, 8, 9, 10, 12, 14, 17, 20, 25, 32]
tree_bins = (
    trees["status_0_1"]
    .groupby(pd.cut(trees["diameter"], splits))
    .agg(["mean", "count"])
```

```
    .rename(columns={"mean": "proportion"})
    .assign(diameter=lambda df: [i.right for i in df.index])
)
```

We can plot these proportions against tree diameter:

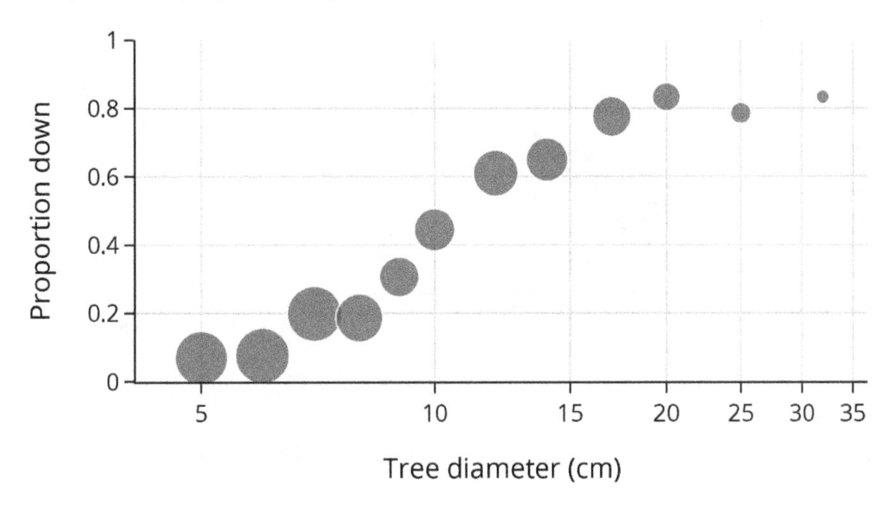

The size of the markers reflects the number of trees in the diameter bin. We can use these proportions to improve our model. For example, for a tree that is 6 cm in diameter, we would classify it as standing, whereas for a 20 cm tree, our classification would be fallen. A natural starting place for binary classification is to model the observed proportions and then use these proportions to classify. Next, we develop a model for these proportions.

Modeling Proportions (and Probabilities)

Recall that when we model, we need to choose three things: a model, a loss function, and a method to minimize the average loss on our train set. In the previous section, we chose a constant model, the 0-1 loss, and a proof to fit the model. However, the constant model doesn't incorporate predictor variables. In this section, we address this issue by introducing a new model called the *logistic* model.

To motivate these models, notice that the relationship between tree diameter and the proportion of downed trees does not appear linear. For demonstration, let's fit a simple linear model to these data to show that it has several undesirable features. Using the techniques from Chapter 15, we fit a linear model of tree status to diameter:

```
from sklearn.linear_model import LinearRegression
X = trees[['diameter']]
y = trees['status_0_1']

lin_reg = LinearRegression().fit(X, y)
```

Then we add this fitted line to our scatterplot of proportions:

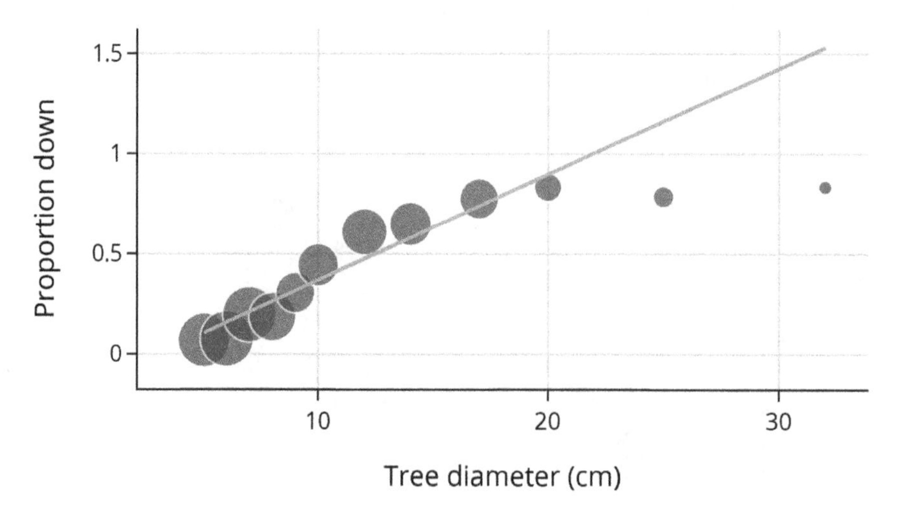

Clearly, the model doesn't fit the proportions well at all. There are several problems:

- The model gives proportions greater than 1 for large trees.
- The model doesn't pick up the curvature in the proportions.
- An extreme point (such as a tree that's 30 cm across) shifts the fitted line to the right, away from the bulk of the data.

To address these issues, we introduce the *logistic model*.

A Logistic Model

The logistic model is one of the most widely used basic models for classification and a simple extension of the linear model. The *logistic function*, often called the *sigmoid function*, is defined as:

$$\mathbf{logistic}(t) = \frac{1}{1 + \exp(-t)}$$

The *sigmoid* function is typically denoted by $\sigma(t)$. Sadly, the Greek letter σ is widely used to mean a lot of things in data science and statistics, like the standard deviation, logistic function, and a permutation. You'll have to be careful when seeing σ and use context to understand its meaning.

We can plot the logistic function to reveal its s-shape (sigmoid-shape) and confirm that it outputs numbers between 0 and 1. The function monotonically increases with *t*, and large values of *t* get close to 1:

```
def logistic(t):
    return 1. / (1. + np.exp(-t))
```

Since the logistic function maps to the interval between 0 and 1, it is commonly used when modeling proportions and probabilities. Also, we can write the logistic as a function of a line, $\theta_0 + \theta_1 x$:

$$\sigma(\theta_0 + \theta_1 x) = \frac{1}{1 + \exp(-\theta_0 - \theta_1 x)}$$

To help build your intuition for the shape of this function, the following plot shows the logistic function as we vary θ_0 and θ_1:

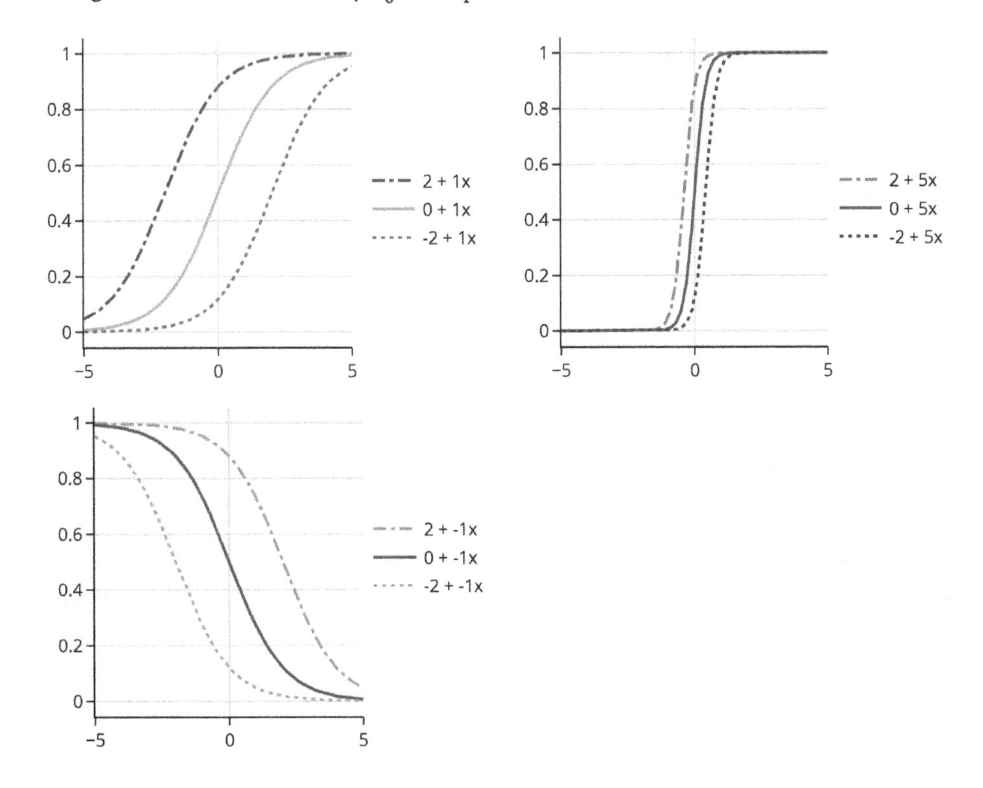

We can see that changing the magnitude of θ_1 changes the sharpness of the curve; the farther away from 0, the steeper the curve. Flipping the sign of θ_1 reflects the curve about the vertical line $x = 0$. Changing θ_0 shifts the curve left and right.

The logistic function can be seen as a transformation: it transforms a linear function into a nonlinear smooth curve, and the output always lies between 0 and 1. In fact, the output of a logistic function has a deeper probabilistic interpretation, which we describe next.

Log Odds

Recall that the odds are the ratio $p/(1-p)$ for a probability p. For example, when we toss a fair coin, the odds of getting heads are 1; for a coin that's twice as likely to land heads as tails ($p = 2/3$), the odds of getting heads are 2. The logistic model is also called the *log odds* model because the logistic function coincides with a linear function of the log odds.

We can see this in the following equations. To show this, we multiply the numerator and denominator of the sigmoid function by $\exp(t)$:

$$\begin{aligned}\sigma(t) &= \frac{1}{1+\exp(-t)} = \frac{\exp(t)}{1+\exp(t)} \\ (1-\sigma(t)) &= 1 - \frac{\exp(t)}{1+\exp(t)} = \frac{1}{1+\exp(t)}\end{aligned}$$

Then we take the logarithm of the odds and simplify:

$$\log\left(\frac{\sigma(t)}{1-\sigma(t)}\right) = \log(\exp(t)) = t$$

So, for $\sigma(\theta_0 + \theta_1 x)$, we find the log odds are a linear function of x:

$$\log\left(\frac{\sigma(\theta_0 + \theta_1 x)}{1-\sigma(\theta_0 + \theta_1 x)}\right) = \log(\exp(\theta_0 + \theta_1 x)) = \theta_0 + \theta_1 x$$

This representation of the logistic in terms of log odds gives a useful interpretation for the coefficient θ_1. Suppose the explanatory variable increases by 1. Then the odds change as follows:

$$\begin{aligned}\text{odds} &= \exp\left(\theta_0 + \theta_1(x+1)\right) \\ &= \exp(\theta_1) \times \exp(\theta_0 + \theta_1 x)\end{aligned}$$

We see that the odds increase or decrease by a factor of $\exp(\theta_1)$.

Here, the log function is the natural logarithm. Since the natural log is the default in data science, we typically don't bother to write it as ln.

Next, let's add a logistic curve to our plot of proportions to get a sense of how well it might fit the data.

Using a Logistic Curve

In the following plot, we've added a logistic curve on top of the plot of proportions of fallen trees:

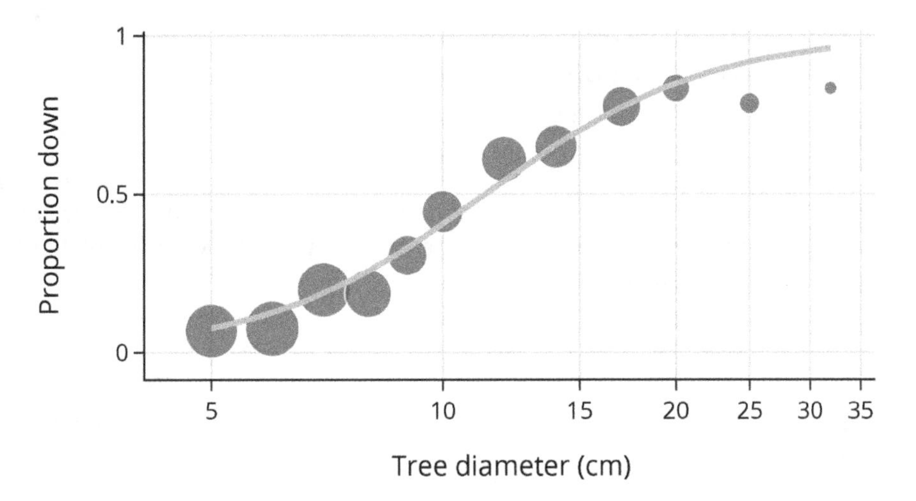

We can see that the curve follows the proportions reasonably well. In fact, we selected this particular logistic by fitting it to the data. The fitted logistic regression is:

```
σ(-7.4 + 3.0x)
```

Now that we've seen that logistic curves can model probabilities well, we turn to the process of fitting logistic curves to data. In the next section, we proceed to our second step in modeling: selecting an appropriate loss function.

A Loss Function for the Logistic Model

The logistic model gives us probabilities (or empirical proportions), so we write our loss function as $\ell(p, y)$, where p is between 0 and 1. The response takes on one of two values because our outcome feature is a binary classification. Thus, any loss function reduces to:

$$\ell(p, y) = \begin{cases} \ell(p, 0) & \text{if } y \text{ is } 0 \\ \ell(p, 1) & \text{if } y \text{ is } 1 \end{cases}$$

Once again, using 0 and 1 to represent the categories has an advantage because we can conveniently write the loss as:

$$\ell(p, y) = \; y\ell(p, y) + (1 - y)\ell(p, 1 - y)$$

We encourage you to confirm this equivalence by considering the two cases $y = 1$ and $y = 0$.

The logistic model pairs well with *log loss*:

$$\begin{aligned} \ell(p, y) &= \begin{cases} -\log(p) & \text{if } y \text{ is } 1 \\ -\log(1 - p) & \text{if } y \text{ is } 0 \end{cases} \\ &= -\, y\log(p) - (1 - y)\log(1 - p) \end{aligned}$$

Note that the log loss is not defined at 0 and 1 because $-\log(p)$ tends to ∞ as p approaches 0, and similarly for $-\log(1 - p)$ as p tends to 1. We need to be careful to avoid the end points in our minimization. We can see this in the following plot of the two forms of the loss function:

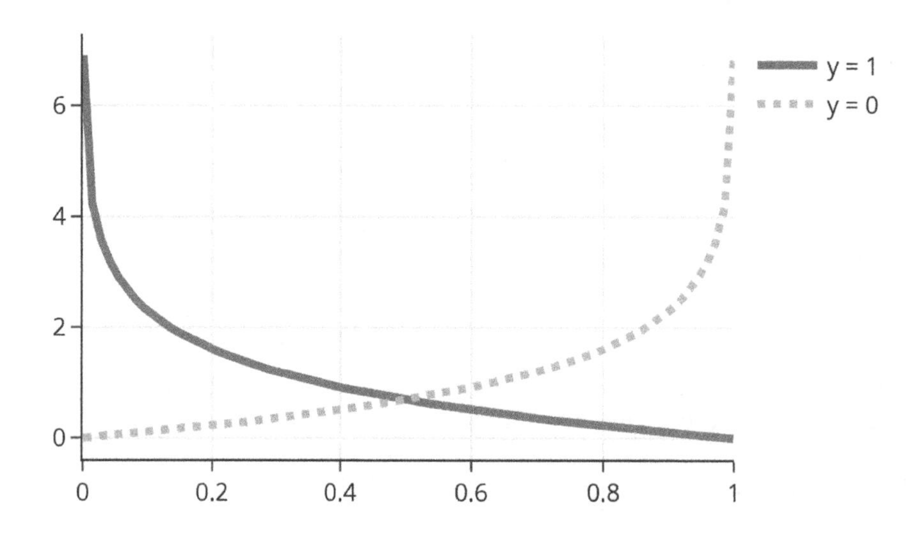

When y is 1 (solid line), the loss is small for p near 1, and when y is 0 (dotted line), the loss is small near 0.

If our goal is to fit a constant to the data using log loss, then the average loss is:

$$
\begin{aligned}
L(p, \mathbf{y}) &= \frac{1}{n}\sum_i [-y_i \log(p) - (1 - y_i)\log(1 - p)] \\
&= -\frac{n_1}{n}\log(p) - \frac{n_0}{n}\log(1 - p))
\end{aligned}
$$

Here n_0 and n_1 are the number of y_i that are 0 and 1, respectively. We can differentiate with respect to p to find the minimizer:

$$
\frac{\partial L(p, \mathbf{y})}{\partial p} = -\frac{n_1}{np} + \frac{n_0}{n(1 - p)}
$$

Then we set the derivative to 0 and solve for the minimizing value $\hat{p}$:

$$
\begin{aligned}
0 &= -\frac{n_1}{n\hat{p}} + \frac{n_0}{n(1 - \hat{p})} \\
0 &= -\hat{p}(1 - \hat{p})\frac{n_1}{\hat{p}} + \hat{p}(1 - \hat{p})\frac{n_0}{(1 - \hat{p})} \\
n_1(1 - \hat{p}) &= n_0\hat{p} \\
\hat{p} &= \frac{n_1}{n}
\end{aligned}
$$

(The final equation results from noting that $n_0 + n_1 = n$.)

To fit a more complex model based on the logistic function, we can substitute $\sigma(\theta_0 + \theta_1 x)$ for p. And the loss for the logistic model becomes:

$$
\begin{aligned}
\ell(\sigma(\theta_0 + \theta_1 x), y) &= y\ell(\sigma(\theta_0 + \theta_1 x), y) + (1 - y)\ell(\sigma(\theta_0 + \theta_1 x), 1 - y) \\
&= y\log(\sigma(\theta_0 + \theta_1 x)) + (1 - y)\log(\sigma(\theta_0 + \theta_1 x))
\end{aligned}
$$

Averaging the loss over the data, we arrive at:

$$
\begin{aligned}
L(\theta_0, \theta_1, \mathbf{x}, \mathbf{y}) = \frac{1}{n}\sum_i & -y_i \log(\sigma(\theta_0 + \theta_1 x_i)) \\
& -(1 - y_i)\log(1 - \sigma(\theta_0 + \theta_1 x_i))
\end{aligned}
$$

Unlike with squared loss, there is no closed-form solution to this loss function. Instead, we use iterative methods like gradient descent (see Chapter 20) to minimize the average loss. This is also one of the reasons we don't use squared error loss for logistic models—the average squared error is nonconvex, which makes it hard to optimize. The notion of convexity is covered in greater detail in Chapter 20, and Figure 20-4 gives a picture for intuition.

Log loss is also called *logistic loss* and *cross-entropy loss*. Another name for it is the *negative log likelihood*. This name refers to the technique of fitting models using the likelihood that a probability distribution produced our data. We do not go any further into the background of these alternative approaches here.

Fitting the logistic model (with the log loss) is called *logistic regression*. Logistic regression is an example of a generalized linear model, a linear model with a nonlinear transformation.

We can fit logistic models with `scikit-learn`. The package designers made the API very similar to fitting linear models by least squares (see Chapter 15). First, we import the logistic regression module:

```
from sklearn.linear_model import LogisticRegression
```

Then we set up the regression problem with outcome `y`, the status of the tree, and covariate `X`, the diameter (which we have log-transformed):

```
trees['log_diam'] = np.log(trees['diameter'])
X = trees[['log_diam']]
y = trees['status_0_1']
```

Then we fit the logistic regression and examine the intercept and coefficient for diameter:

```
lr_model = LogisticRegression()
lr_model.fit(X, y)

[intercept] = lr_model.intercept_
[[coef]] = lr_model.coef_
print(f'Intercept:           {intercept:.1f}')
print(f'Diameter coefficient: {coef:.1f}')
```

```
Intercept:           -7.4
Diameter coefficient: 3.0
```

When making a prediction, the `predict` function returns the predicted (most likely) class, and `predict_proba` returns the predicted probability. For a tree with diameter 6, we expect the prediction to be 0 (meaning `standing`) with a high probability. Let's check:

```
diameter6 = pd.DataFrame({'log_diam': [np.log(6)]})
[pred_prof] = lr_model.predict_proba(diameter6)
print(f'Predicted probabilities: {pred_prof}')
```

```
Predicted probabilities: [0.87 0.13]
```

Thus, the model predicts that a tree with a diameter of 6 has a 0.87 probability for the class `standing` and a 0.13 probability for `fallen`.

Now that we've fit a model with one feature, we might want to see if including another feature like the strength of the storm can improve the model. To do this, we can fit a multiple logistic regression by adding a feature to `X` and fitting the model again.

Notice that the logistic regression fits a model to predict probabilities—the model predicts that a tree with diameter 6 has a 0.87 probability of class `standing` and a 0.13 probability of class `fallen`. Since probabilities can be any number between 0 and 1, we need to convert the probabilities back to categories to perform classification. We address this classification problem in the next section.

From Probabilities to Classification

We started this chapter by presenting a binary classification problem where we want to model a nominal response variable. At this point, we have used logistic regression to model proportions or probabilities, and we're now ready to return to the original problem: we use the predicted probabilities to classify records. For our example, this means that for a tree of a particular diameter, we use the fitted coefficients from the logistic regression to estimate the chance it is fallen. If the chance is high, we classify a tree as fallen; otherwise, we classify it as standing. But we need to choose a threshold for making this *decision rule*.

The `sklearn` logistic regression model's `predict` function implements the basic decision rule: predict `1` if the predicted probability $p > 0.5$. Otherwise, predict 0. We've overlaid this decision rule on top of the model predictions as a dotted line:

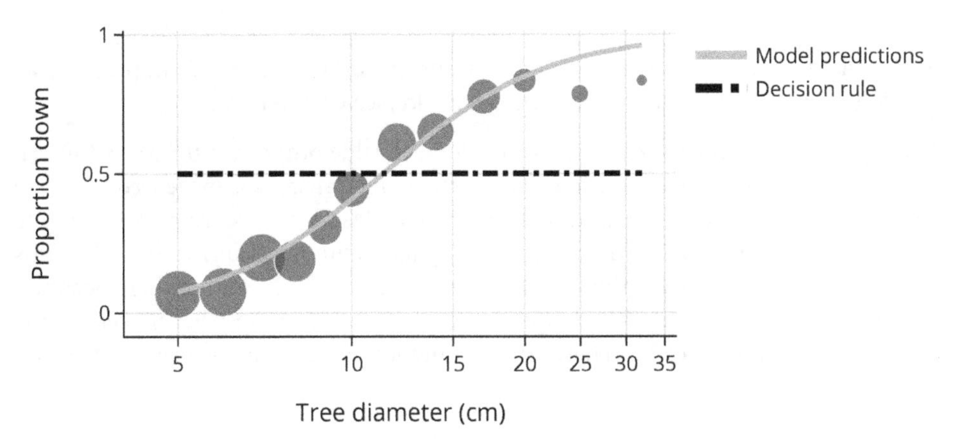

In this section, we consider a more general decision rule. For some choice of τ, predict 1 if the model's predicted probability $p > \tau$, otherwise predict 0. By default, `sklearn` sets $\tau = 0.5$. Let's explore what happens when τ is set to other values.

Choosing an appropriate value for τ depends on our goals. Suppose we want to maximize accuracy. The *accuracy* of a classifier is the fraction of correct predictions. We can compute the accuracy for different thresholds, meaning different τ values:

```
def threshold_predict(model, X, threshold):
    return np.where(model.predict_proba(X)[:, 1] > threshold, 1.0, 0.0)

def accuracy(threshold, X, y):
    return np.mean(threshold_predict(lr_model, X, threshold) == y)

thresholds = np.linspace(0, 1, 200)
accs = [accuracy(t, X, y) for t in thresholds]
```

To understand how accuracy changes with respect to τ, we make a plot:

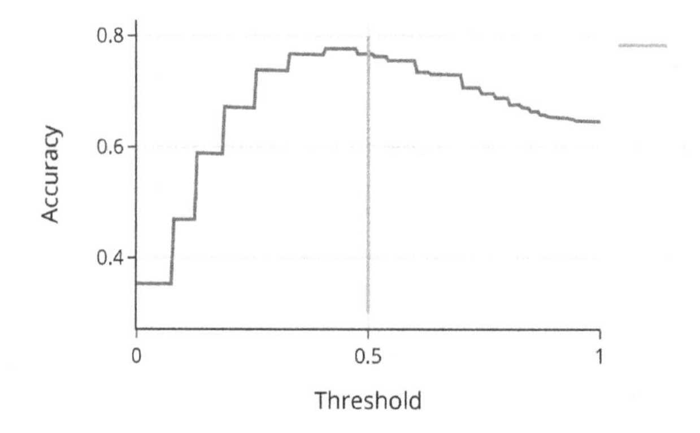

Notice that the threshold with the highest accuracy isn't exactly at 0.5. In practice, we should use cross-validation to select the threshold (see Chapter 16).

The threshold that maximizes accuracy could be a value other than 0.5 for many reasons, but a common one is *class imbalance*, where one category is more frequent than another. Class imbalance can lead to a model that classifies a record as belonging to the more common category. In extreme cases (like fraud detection) when only a tiny fraction of the data contain a particular class, our models can achieve high accuracy by simply always predicting the frequent class without learning what makes a good classifier for the rare class. There are techniques for managing class imbalance, such as:

- Resampling the data to reduce or eliminate the class imbalance
- Adjusting the loss function to put a larger penalty on the smaller class

In our example, the class imbalance is not that extreme, so we continue without these adjustments.

The problem of class imbalance explains why accuracy alone is often not how we want to judge a model. Instead, we want to differentiate between the types of correct and incorrect classifications. We describe these next.

The Confusion Matrix

A convenient way to visualize errors in a binary classification is to look at the confusion matrix. The confusion matrix compares what the model predicts with the actual outcomes. There are two types of error in this situation:

False positives
: When the actual class is 0 (false) but the model predicts 1 (true)

False negatives
: When the actual class is 1 (true) but the model predicts 0 (false)

Ideally, we would like to minimize both kinds of errors, but we often need to manage the balance between these two sources.

The terms *positive* and *negative* come from disease testing, where a test indicating the presence of a disease is called a positive result. This can be a bit confusing because having a disease doesn't seem like something positive at all. And $y = 1$ denotes the "positive" case. To keep things straight, it's a good idea to confirm your understanding of what $y = 1$ stands for in the context of your data.

`scikit-learn` has a function to compute and plot the confusion matrix:

```
from sklearn.metrics import confusion_matrix
mat = confusion_matrix(y, lr_model.predict(X))
mat
```

```
array([[377,  49],
       [104, 129]])
```

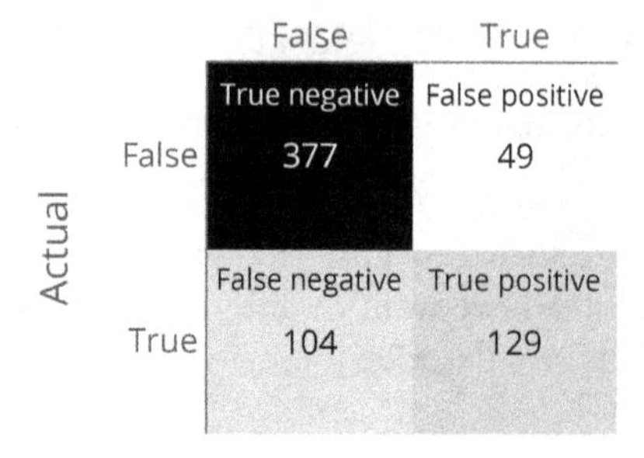

Ideally, we want to see all of the counts in the diagonal squares True negative and True positive. That means we have correctly classified everything. But this is rarely the case, and we need to assess the size of the errors. For this, it's easier to compare rates than counts. Next, we describe different rates and when we might prefer to prioritize one or the other.

Precision Versus Recall

In some settings, there might be a much higher cost to missing positive cases. For example, if we are building a classifier to identify tumors, we want to make sure that we don't miss any malignant tumors. Conversely, we're less concerned about classifying a benign tumor as malignant because a pathologist would still need to take a closer look to verify the malignant classification. In this case, we want to have a high true positive rate among the records that are actually positive. The rate is called *sensitivity*, or *recall*:

$$\text{Recall} = \frac{\text{True Positives}}{\text{True Positives+False Negatives}} = \frac{\text{True Positives}}{\text{Actually True}}$$

Higher recall runs the risk of predicting true on false records (false positives).

On the other hand, when classifying email as spam (positive) or ham (negative), we might be annoyed if an important email gets thrown into our spam folder. In this setting, we want high *precision*, the accuracy of the model for positive predictions:

$$\text{Precision} = \frac{\text{True Positives}}{\text{True Positives+False Positives}} = \frac{\text{True Positives}}{\text{Predicted True}}$$

Higher-precision models are often more likely to predict that true observations are negative (higher false-negative rate).

A common analysis compares the precision and recall at different thresholds:

```
from sklearn import metrics
precision, recall, threshold = (
    metrics.precision_recall_curve(y, lr_model.predict_proba(X)[:, 1]))

tpr_df = pd.DataFrame({"threshold":threshold,
                       "precision":precision[:-1], "recall": recall[:-1], })
```

To see how precision and recall relate, we plot them both against the threshold τ:

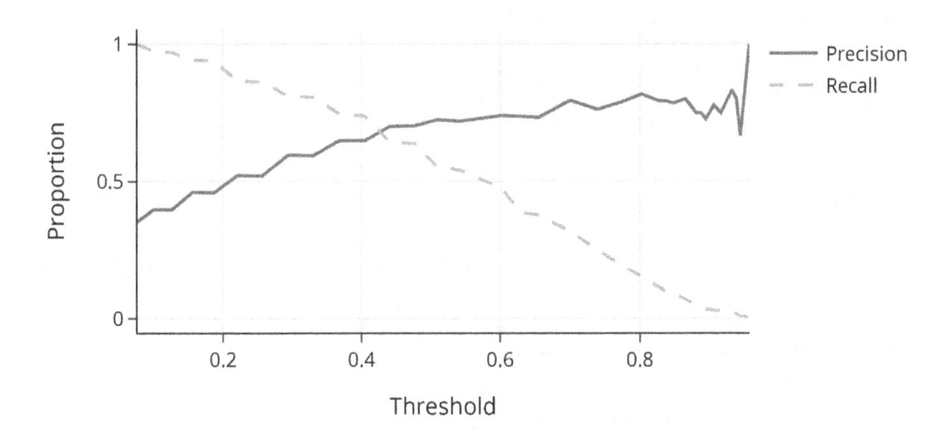

Another common plot used to evaluate the performance of a classifier is the *precision-recall curve*, or PR curve for short. It plots the precision-recall pairs for each threshold:

```
fig = px.line(tpr_df, x="recall", y="precision",
              labels={"recall":"Recall","precision":"Precision"})
fig.update_layout(width=450, height=250, yaxis_range=[0, 1])
fig
```

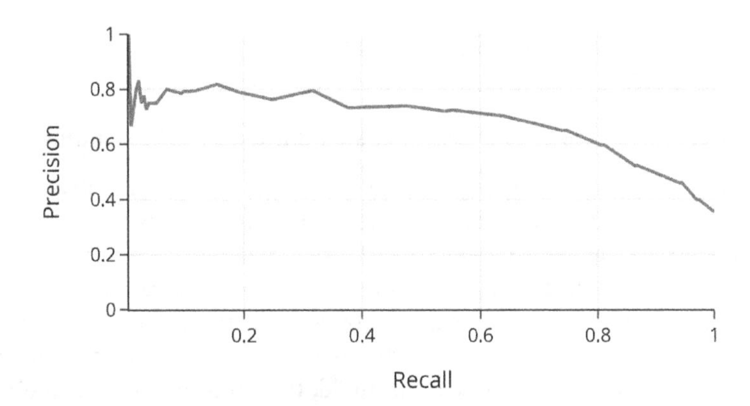

Notice that the righthand end of the curve reflects the imbalance in the sample. The precision matches the fraction of fallen trees in the sample, 0.35. Plotting multiple PR curves for different models can be particularly useful for comparing models.

Using precision and recall gives us more control over what kinds of errors matter. As an example, let's suppose we want to ensure that at least 75% of the fallen trees are classified as fallen. We can find the threshold where this occurs:

```
fall75_ind = np.argmin(recall >= 0.75) - 1

fall75_threshold = threshold[fall75_ind]
fall75_precision = precision[fall75_ind]
fall75_recall = recall[fall75_ind]
```

```
Threshold: 0.33
Precision: 0.59
Recall:    0.81
```

We find that about 41% (1 – precision) of the trees that we classify as fallen are actually standing. In addition, we find the fraction of trees below this threshold to be:

```
print("Proportion of samples below threshold:",
      f"{np.mean(lr_model.predict_proba(X)[:,1] < fall75_threshold):0.2f}")
```

```
Proportion of samples below threshold: 0.52
```

So, we have classified 52% of the samples as standing (negative). *Specificity* (also called *true negative rate*) measures the proportion of data belonging to the negative class that the classifier labels as negative:

$$\text{Specificity} = \frac{\text{True Negatives}}{\text{True Negatives} + \text{False Positives}} = \frac{\text{True Negatives}}{\text{Predicted False}}$$

The specificity for our threshold is:

```
act_neg = (y == 0)
true_neg = (lr_model.predict_proba(X)[:,1] < fall75_threshold) & act_neg
```

```
Specificity: 0.70
```

In other words, 70% of the trees classified as standing are actually standing.

As we've seen, there are several ways to use the 2-by-2 confusion matrix. Ideally, we want accuracy, precision, and recall to all be high. This happens when most predictions fall along the diagonal for the table, so our predictions are nearly all correct–true negatives and true positives. Unfortunately, in most scenarios our models will have some amount of error. In our example, trees of the same diameter include a mix of fallen and standing, so we can't perfectly classify trees based on their diameter. In practice, when data scientists choose a threshold, they need to consider their context to decide whether to prioritize precision, recall, or specificity.

Summary

In this chapter, we fit simple logistic regressions with one explanatory variable, but we can easily include other variables in the model by adding more features to our design matrix. For example, if some predictors are categorical, we can include them as one-hot encoded features. These ideas carry over directly from Chapter 15. The technique of regularization (Chapter 16) also applies to logistic regression. We will integrate all of these modeling techniques—including using a train-test split to assess the model and cross-validation to choose the threshold—in the case study in Chapter 21 that develops a model to classify fake news.

Logistic regression is a cornerstone in machine learning since it naturally extends to more complex models. For example, logistic regression is one of the basic components of a neural network. When the response variable has more than two categories, logistic regression can be extended to multinomial logistic regression. Another extension of logistic regression for modeling counts is called Poisson regression. These different forms of regression are related to maximum likelihood, where the underlying model for the response is binomial, multinomial, or Poisson, respectively, and the goal is to optimize the likelihood of the data over the parameters of the respective distribution. This family of models is also known as generalized linear models. In all of these scenarios, closed-form solutions for minimizing loss don't exist, so optimization of the average loss relies on numerical methods, which we cover in the next chapter.

CHAPTER 20

Numerical Optimization

At this point in the book, our modeling procedure should feel familiar: we define a model, choose a loss function, and fit the model by minimizing the average loss over our training data. We've seen several techniques to minimize loss. For example, we used both calculus and a geometric argument in Chapter 15 to find a simple expression for fitting linear models using squared loss.

But empirical loss minimization isn't always so straightforward. Lasso regression, with the addition of the L_1 penalty to the average squared loss, no longer has a closed-form solution, and logistic regression uses cross-entropy loss to fit a nonlinear model. In these cases, we use *numerical optimization* to fit the model, where we systematically choose parameter values to evaluate the average loss in search of the minimizing value.

When we introduced loss functions in Chapter 4, we performed a simple numerical optimization to find the minimizer of the average loss. We created a grid of θ values and evaluated the average loss at all points in the grid (see Figure 20-1). The grid point with the smallest average loss we took as the best fit. Unfortunately, this sort of grid search quickly becomes impractical, for the following reasons:

- For complex models with many features, the grid becomes unwieldy. With only four features and a grid of 100 values for each feature, we must evaluate the average loss at 100^4 = 100,000,000 grid points.
- The range of parameter values to search over must be specified in advance to create the grid, and when we don't have a good sense of the range, we need to start with a wide grid and possibly repeat the grid search over narrower ranges.
- With a large number of observations, the evaluation of the average loss over the grid points can be slow.

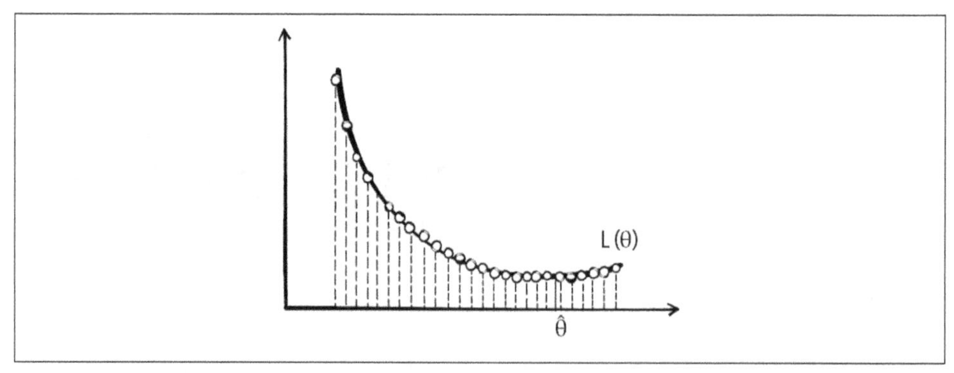

Figure 20-1. Searching over a grid of points can be computationally slow or inexact

In this chapter, we introduce numerical optimization techniques that take advantage of the shape and smoothness of the loss function in the search for the minimizing parameter values. We first introduce the basic idea behind the technique of gradient descent, then we give an example and describe the properties of the loss function that make gradient descent work, and finally, we provide a few extensions of gradient descent.

Gradient Descent Basics

Gradient descent is based on the notion that for many loss functions, the function is roughly linear in small neighborhoods of the parameter. Figure 20-2 gives a diagram of the basic idea.

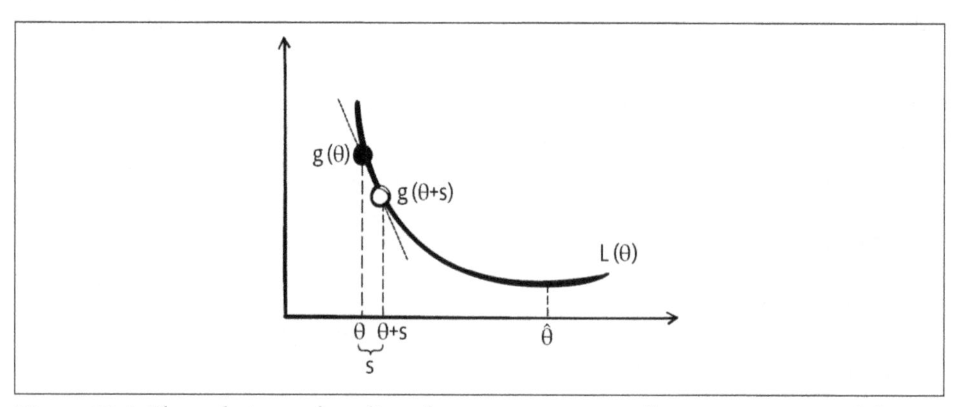

Figure 20-2. The technique of gradient descent moves in small increments toward the minimizing parameter value

In the diagram, we have drawn the tangent line to the loss curve L at some point θ to the left of the minimizing value, $\hat{\theta}$. Notice that the slope of the tangent line is negative. A short step to the right of θ to $\theta + s$, for some small amount s, gives a point on the tangent line close to the loss at $\theta + s$, and this loss is smaller than $L(\tilde{\theta})$. That is, since the slope, b, is negative, and the tangent line approximates the loss function in a neighborhood of θ, we have:

$$L(\theta + s) \approx L(\theta) + b \times s < L(\theta)$$

So, taking a small step to the right of this θ decreases the loss. On the other hand, on the other side of $\hat{\theta}$ in the diagram in Figure 20-2, the slope is positive, and taking a small step to the left decreases the loss.

When we take repeated small steps in the direction indicated by whether the slope of the tangent line is positive or negative at each new step, this leads to smaller and smaller values of the average loss and eventually brings us to the minimizing value $\hat{\theta}$ (or very close to it). This is the basic idea behind gradient descent.

More formally, to minimize $L(\boldsymbol{\theta})$ for a general vector of parameters, $\boldsymbol{\theta}$, the gradient (first-order partial derivative) determines the direction and size of the step to take. If we write the gradient, $\nabla_{\theta} L(\boldsymbol{\theta})$, as simply $g(\boldsymbol{\theta})$, then gradient descent says the increment or step is $-\alpha g(\boldsymbol{\theta})$ for some small positive α. Then the average loss at the new position is:

$$\begin{aligned} L(\theta + (-\alpha g(\boldsymbol{\theta}))) &\approx L(\theta) - \alpha g(\boldsymbol{\theta})^T g(\boldsymbol{\theta}) \\ &< L(\theta) \end{aligned}$$

Note that $g(\boldsymbol{\theta})$ is a $p \times 1$ vector and $g(\boldsymbol{\theta})^T g(\boldsymbol{\theta})$ is positive.

The steps in the gradient descent algorithm go as follows:

1. Choose a starting value, called $\boldsymbol{\theta}^{(0)}$ (a common choice is $\boldsymbol{\theta}^{(0)} = 0$).
2. Compute $\boldsymbol{\theta}^{(t+1)} = \boldsymbol{\theta}^{(t)} - \alpha g(\boldsymbol{\theta})$.
3. Repeat step 2 until $\boldsymbol{\theta}^{(t+1)}$ doesn't change (or changes little) between iterations.

The quantity α is called the *learning rate*. Setting α can be tricky. It needs to be small enough to not overshoot the minimum but large enough to arrive at the minimum in reasonably few steps (see Figure 20-3). There are many strategies for setting α. For example, it can be useful to decrease α over time. When α changes between iterations, we use the notation $\alpha^{(t)}$ to indicate that the learning rate varies during the search.

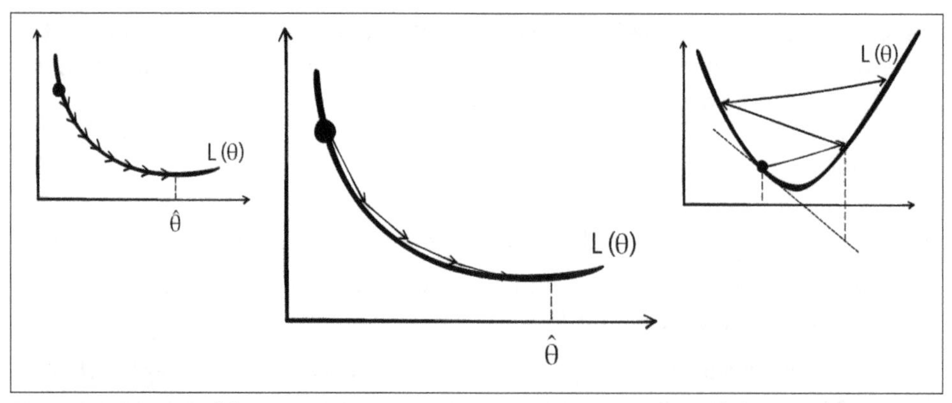

Figure 20-3. A small learning rate requires many steps to converge (left), and a large learning rate can diverge (right); choosing the learning rate well leads to fast convergence on the minimizing value (middle)

The gradient descent algorithm is simple yet powerful since we can use it for many types of models and many types of loss functions. It is the computational tool of choice for fitting many models, including linear regression on large datasets and logistic regression. We demonstrate the algorithm to fit a constant to the bus delay data (from Chapter 4) next.

Minimizing Huber Loss

Huber loss combines absolute loss and squared loss to get a function that is differentiable (like squared loss) and less sensitive to outliers (like absolute loss):

$$L(\theta, \mathbf{y}) = \frac{1}{n} \sum_{i=1}^{n} \begin{cases} \frac{1}{2}(y_i - \theta)^2 & |y_i - \theta| \le \gamma \\ \gamma(|y_i - \theta| - \frac{1}{2}\gamma) & \text{otherwise} \end{cases}$$

Since Huber loss is differentiable, we can use gradient descent. We first find the gradient of the average Huber loss:

$$\nabla_{\theta} L(\theta, \mathbf{y}) = \frac{1}{n} \sum_{i=1}^{n} \begin{cases} -(y_i - \theta) & |y_i - \theta| \le \gamma \\ -\gamma \cdot \text{sign}(y_i - \theta) & \text{otherwise} \end{cases}$$

We create the functions `huber_loss` and `grad_huber_loss` to compute the average loss and its gradient. We write these functions to have signatures that enable us to specify the parameter as well as the observed data that we average over and the transition point of the loss function:

```
def huber_loss(theta, dataset, gamma=1):
    d = np.abs(theta - dataset)
    return np.mean(
        np.where(d <= gamma,
                 (theta - dataset)**2 / 2.0,
                 gamma * (d - gamma / 2.0))
    )

def grad_huber_loss(theta, dataset, gamma=1):
    d = np.abs(theta - dataset)
    return np.mean(
        np.where(d <= gamma,
                 -(dataset - theta),
                 -gamma * np.sign(dataset - theta))
    )
```

Next, we write a simple implementation of gradient descent. The signature of our function includes the loss function, its gradient, and the data to average over. We also supply the learning rate.

```
def minimize(loss_fn, grad_loss_fn, dataset, alpha=0.2, progress=False):
    '''
    Uses gradient descent to minimize loss_fn. Returns the minimizing value of
    theta_hat once theta_hat changes less than 0.001 between iterations.
    '''
    theta = 0
    while True:
        if progress:
            print(f'theta: {theta:.2f} | loss: {loss_fn(theta, dataset):.3f}')
        gradient = grad_loss_fn(theta, dataset)
        new_theta = theta - alpha * gradient

        if abs(new_theta - theta) < 0.001:
            return new_theta

        theta = new_theta
```

Recall that the bus delays dataset consists of over 1,000 measurements of how many minutes late the northbound C-line buses are in arriving at the stop at 3rd Avenue and Pike Street in Seattle:

```
delays = pd.read_csv('data/seattle_bus_times_NC.csv')
```

In Chapter 4, we fit a constant model to these data for absolute loss and squared loss. We found that absolute loss yielded the median and square the mean of the data:

```
print(f"Mean:   {np.mean(delays['minutes_late']):.3f}")
print(f"Median: {np.median(delays['minutes_late']):.3f}")
```

```
Mean:   1.920
Median: 0.742
```

Now we use the gradient descent algorithm to find the minimizing constant model for Huber loss:

```
%%time
theta_hat = minimize(huber_loss, grad_huber_loss, delays['minutes_late'])
print(f'Minimizing theta: {theta_hat:.3f}')
print()
```

```
Minimizing theta: 0.701

CPU times: user 93 ms, sys: 4.24 ms, total: 97.3 ms
Wall time: 140 ms
```

The optimizing constant for Huber loss is close to the value that minimizes absolute loss. This comes from the shape of the Huber loss function. It is linear in the tails and so is not affected by outliers like with absolute loss and unlike with squared loss.

We wrote our `minimize` function to demonstrate the idea behind the algorithm. In practice, you will want to use well-tested, numerically sound implementations of an optimization algorithm. For example, the `scipy` package has a `minimize` method that we can use to find the minimizer of average loss, and we don't even need to compute the gradient. This algorithm is likely to be much faster than any one that we might write. In fact, we used it in Chapter 18 when we created our own asymmetric modification of quadratic loss for the special case where we wanted the loss to be greater for errors on one side of the minimum than the other.

More generally, we typically stop the algorithm when $\theta^{(t)}$ doesn't change much between iterations. In our function, we stop when $\theta^{(t+1)} - \theta^{(t)}$ is less than 0.001. It is also common to stop the search after a large number of steps, such as 1,000. If the algorithm has not arrived at the minimizing value after 1,000 iterations, then the algorithm might be diverging because the learning rate is too large or the minimum might exist in the limit at $\pm\infty$.

Gradient descent gives us a general way to minimize average loss when we cannot easily solve for the minimizing value analytically or when the minimization is computationally expensive. The algorithm relies on two important properties of the average loss function: it is both convex and differentiable in $\boldsymbol{\theta}$. We discuss how the algorithm relies on these properties next.

Convex and Differentiable Loss Functions

As its name suggests, the gradient descent algorithm requires the function being minimized to be differentiable. The gradient, $\nabla_{\theta} L(\boldsymbol{\theta})$, allows us to make a linear approximation to the average loss in small neighborhoods of $\boldsymbol{\theta}$. This approximation gives us the direction (and size) of the step, and as long as we don't overshoot the minimum, $\hat{\theta}$, we are bound to eventually reach it. Well, as long as the loss function is also convex.

The step-by-step search for the minimum also relies on the loss function being convex. The function in the diagram on the left in Figure 20-4 is convex, but the function on the right is not. The function on the right has a local minimum, and depending on where the algorithm starts, it might converge to this local minimum and miss the real minimum entirely. The property of convexity avoids this problem. A *convex function* avoids the problem of local minima. So, with an appropriate step size, gradient descent finds the globally optimal θ for any convex, differentiable function.

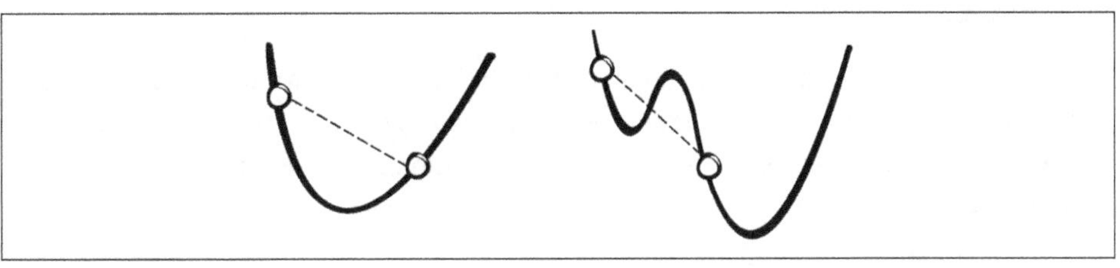

Figure 20-4. With nonconvex functions (right), gradient descent might locate a local minimum rather than a global minimum, which is not possible with convex functions (left)

Formally, a function f is convex if for any two input values, $\boldsymbol{\theta}_a$ and $\boldsymbol{\theta}_b$, and any q between 0 and 1:

$$q f(\boldsymbol{\theta}_a) + (1 - q) f(\boldsymbol{\theta}_b) \geq f(q\boldsymbol{\theta}_a + (1 - q)\boldsymbol{\theta}_b)$$

This inequality implies that any line segment that connects two points of the function must reside on or above the function itself. Heuristically, this means that whenever we take a small enough step to the right when the gradient is negative or to the left when the gradient is positive, we will head in the direction of the function's minimum.

The formal definition of convexity gives us a precise way to determine whether a function is convex. And we can use this definition to connect the convexity of the average loss $L(\boldsymbol{\theta})$ to the loss function $\ell(\boldsymbol{\theta})$. We have so far in this chapter simplified the representation of $L(\boldsymbol{\theta})$ by not mentioning the data. Recall:

$$L(\boldsymbol{\theta}, \mathbf{X}, \mathbf{y}) = \frac{1}{n} \sum_{i=1}^{n} \ell(\boldsymbol{\theta}, \mathbf{x}_i, y_i)$$

where $\mathbf{X}$ is an $n \times p$ design matrix and $\mathbf{x}_i$ is the ith row of the design matrix, which corresponds to the ith observation in the dataset. This means that the gradient can be expressed as follows:

$$\nabla_\theta L(\boldsymbol{\theta}, \mathbf{X}, \mathbf{y}) = \frac{1}{n} \sum_{i=1}^{n} \nabla_\theta \ell(\boldsymbol{\theta}, \mathbf{x}_i, y_i)$$

If $\ell(\boldsymbol{\theta}, \mathbf{x}_i, y_i)$ is a convex function of $\boldsymbol{\theta}$, then the average loss is also convex. And similarly for the derivative: the derivative of $\ell(\boldsymbol{\theta}, \mathbf{x}_i, y_i)$ is averaged over the data to evaluate the derivative of $L(\boldsymbol{\theta}, \mathbf{X}, \mathbf{y})$. We walk through a proof of the convexity property in the exercises.

Now, with a large amount of data, calculating $\theta^{(t)}$ can be computationally expensive since it involves the average of the gradient $\nabla_\theta \ell$ over all the $(\mathbf{x}_i, y_i)$. We next consider variants of gradient descent that can be computationally faster because they don't average over all of the data.

Variants of Gradient Descent

Two variants of gradient descent, stochastic gradient descent and mini-batch gradient descent, use subsets of the data when computing the gradient of the average loss and are useful for optimization problems with large datasets. A third alternative, Newton's method, assumes the loss function is twice differentiable and uses a quadratic approximation to the loss function, rather than the linear approximation used in gradient descent.

Recall that gradient descent takes steps based on the gradient. At step t, we move from $\boldsymbol{\theta}^{(t)}$ to:

$$\boldsymbol{\theta}^{(t+1)} = \boldsymbol{\theta}^{(t)} - \alpha \cdot \nabla_\theta L(\boldsymbol{\theta}^{(t)}, \mathbf{X}, \mathbf{y})$$

And since $\nabla_\theta L(\boldsymbol{\theta}, \mathbf{X}, \mathbf{y})$ can be expressed as the average gradient of the loss function ℓ, we have:

$$\nabla_\theta L(\boldsymbol{\theta}, \mathbf{X}, \mathbf{y}) = \frac{1}{n} \sum_{i=1}^{n} \nabla_\theta \ell(\boldsymbol{\theta}, \mathbf{x}_i, y_i)$$

This representation of the gradient of the average loss in terms of the average of the gradient of loss at each point in the data shows why the algorithm is also called *batch gradient descent*. Two variants to batch gradient descent use smaller amounts of the data rather than the complete "batch." The first, stochastic gradient descent, uses only one observation in each step of the algorithm.

Stochastic Gradient Descent

Although batch gradient descent can often find an optimal $\boldsymbol{\theta}$ in relatively few iterations, each iteration can take a long time to compute if the dataset contains many observations. To get around this difficulty, stochastic gradient descent approximates the overall gradient by a single, randomly chosen data point. Since this observation is chosen randomly, we expect that using the gradient at randomly chosen observations will, on average, move in the correct direction and so eventually converge to the minimizing parameter.

In short, to conduct stochastic gradient descent, we replace the average gradient with the gradient at a single data point. So, the updated formula is just:

$$\boldsymbol{\theta}^{(t+1)} = \boldsymbol{\theta}^{(t)} - \alpha \cdot \nabla_{\theta} \ell(\boldsymbol{\theta}^{(t)}, \mathbf{x}_i, y_i)$$

In this formula, the ith observations $(\mathbf{x}_i, y_i)$ are chosen randomly from the data. Choosing the points randomly is critical to the success of stochastic gradient descent. If the points are not chosen randomly, the algorithm may produce significantly worse results than batch gradient descent.

We most commonly run stochastic gradient descent by randomly shuffling all of the data points and using each point in its shuffled order until we complete one entire pass through the data. If the algorithm hasn't converged yet, then we reshuffle the points and run another pass through the data. Each *iteration* of stochastic gradient descent looks at one data point; each complete pass through the data is called an *epoch*.

Since stochastic descent only examines a single data point at a time, at times it takes steps away from the minimizer, $\hat{\boldsymbol{\theta}}$, but on average these steps are in the right direction. And since the algorithm computes an update much more quickly than batch gradient descent, it can make significant progress toward the optimal $\hat{\boldsymbol{\theta}}$ by the time batch gradient descent finishes a single update.

Mini-Batch Gradient Descent

As its name suggests, *mini-batch gradient descent* strikes a balance between batch gradient descent and stochastic gradient descent by increasing the number of observations selected at random in each iteration. In mini-batch gradient descent, we average the gradient of the loss function at a few data points instead of at a single point or all the points. We let $\mathcal{B}$ represent the mini-batch of data points that are randomly sampled from the dataset, and we define the algorithm's next step as:

$$\boldsymbol{\theta}^{(t+1)} = \boldsymbol{\theta}^{(t)} - \alpha \cdot \frac{1}{|\mathcal{B}|} \sum_{i \in \mathcal{B}} \nabla_{\theta} \ell(\boldsymbol{\theta}, \mathbf{x}_i, y_i)$$

As with stochastic gradient descent, we perform mini-batch gradient descent by randomly shuffling the data. Then we split the data into consecutive mini-batches and iterate through the batches in sequence. After each epoch, we reshuffle our data and select new mini-batches.

While we have made the distinction between stochastic and mini-batch gradient descent, *stochastic gradient descent* is sometimes used as an umbrella term that encompasses the selection of a mini-batch of any size.

Another common optimization technique is Newton's method.

Newton's Method

Newton's method uses the second derivative to optimize the loss. The basic idea is to approximate the average loss, $L(\boldsymbol{\theta})$, in small neighborhoods of $\boldsymbol{\theta}$, with a quadratic curve rather than a linear approximation. The approximation looks as follows for a small step $\mathbf{s}$:

$$L(\boldsymbol{\theta} + \mathbf{s}) \approx L(\boldsymbol{\theta}) + g(\boldsymbol{\theta})^T \mathbf{s} + \frac{1}{2} \mathbf{s}^T H(\boldsymbol{\theta}) \mathbf{s}$$

where $g(\boldsymbol{\theta}) = \nabla_{\theta} L(\boldsymbol{\theta})$ is the gradient and $H(\boldsymbol{\theta}) = \nabla_{\theta}^2 L(\boldsymbol{\theta})$ is the Hessian of $L(\boldsymbol{\theta})$. More specifically, H is a $p \times p$ matrix of second-order partial derivatives in $\boldsymbol{\theta}$ with i, j elements:

$$H_{i,j} = \frac{\partial^2 \ell}{\partial \theta_i \partial \theta_j}$$

This quadratic approximation to $L(\boldsymbol{\theta} + \mathbf{s})$ has a minimum at $\mathbf{s} = -[H^{-1}(\boldsymbol{\theta})]g(\boldsymbol{\theta})$. (Convexity implies that H is a symmetric square matrix that can be inverted.) Then a step in the algorithm moves from $\boldsymbol{\theta}^{(t)}$ to:

$$\boldsymbol{\theta}^{(t+1)} = \boldsymbol{\theta}^{(t)} + \frac{1}{n} \sum_{i=1}^{n} -[H^{-1}(\boldsymbol{\theta}^{(t)})]g(\boldsymbol{\theta}^{(t)})$$

Figure 20-5 gives the idea behind Newton's method of optimization.

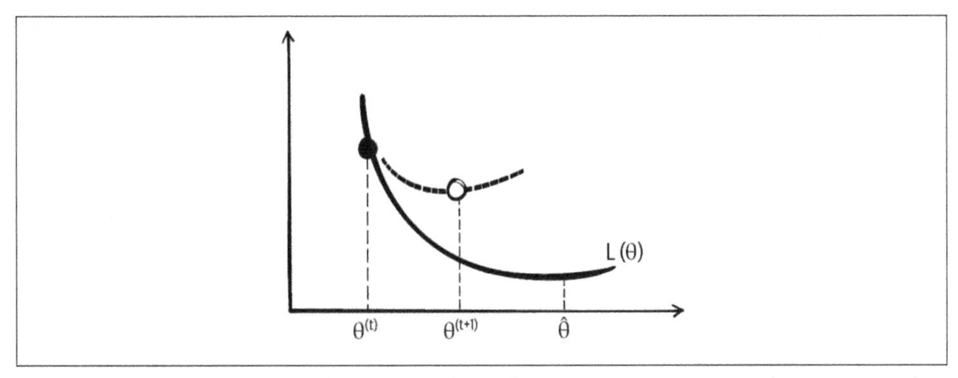

Figure 20-5. Newton's method uses a local quadratic approximation to the curve to take steps toward the minimizing value of a convex, twice-differentiable function

This technique converges quickly if the approximation is accurate and the steps are small. Otherwise, Newton's method can diverge, which often happens if the function is nearly flat in a dimension. When the function is relatively flat, the derivative is near zero and its inverse can be quite large. Large steps can move to $\boldsymbol{\theta}$ that are far from where the approximation is accurate. (Unlike with gradient descent, there is no learning rate that keeps steps small.)

Summary

In this chapter, we introduced several techniques for numerical optimization that take advantage of the shape and smoothness of the loss function in the search for the minimizing parameter values. We first introduced gradient descent, which relies on the differentiability of loss function. Gradient descent, also called batch gradient descent, iteratively improves model parameters until the model achieves minimal loss. Since batch gradient descent is computationally intractable with large datasets, we often instead use stochastic gradient descent to fit models.

Mini-batch gradient descent is most optimal when running on a graphical processing unit (GPU) chip found in some computers. Since computations on these types of hardware can be executed in parallel, using a mini-batch can increase the accuracy of the gradient without increasing computation time. Depending on the memory size of the GPU, the mini-batch size is often set between 10 and 100 observations.

Alternatively, if the loss function is twice differentiable, then Newton's method can converge very quickly, even though it is more expensive to compute one step in the iteration. A hybrid approach is also popular, beginning with gradient descent (of some kind) and then switching the algorithm to Newton's method. This approach can avoid divergence and be faster than gradient descent alone. Typically, the second-order approximation used by Newton's method is more appropriate near the optimum and converges quickly.

Lastly, another option is to set the step size adaptively. Additionally, setting different learning rates for different features can be important if they are of different scale or vary in frequency. For example, word counts can differ a lot across common words and rare words.

The logistic regression model introduced in Chapter 19 is fitted using numerical optimization methods like those described in this chapter. We wrap up with one final case study that uses logistic regression to fit a complex model with thousands of features.

CHAPTER 21

Case Study: Detecting Fake News

Fake news—false information created in order to deceive others—is an important issue because it can harm people. For example, the social media post in Figure 21-1 confidently stated that hand sanitizer doesn't work on coronaviruses. Though factually incorrect, it spread through social media anyway: it was shared nearly 100,000 times and was likely seen by millions of people.

Figure 21-1. A popular post on Twitter from March 2020 falsely claimed that sanitizer doesn't kill coronaviruses

We might wonder whether we can automatically detect fake news without having to read the stories. For this case study, we go through the steps of the data science lifecycle. We start by refining our research question and obtaining a dataset of news articles and labels. Then we wrangle and transform the data. Next, we explore the data to understand its content and devise features to use for modeling. Finally, we build models using logistic regression to predict whether news articles are real or fake, and evaluate their performance.

We've included this case study because it lets us reiterate several important ideas in data science. First, natural language data appear often, and even basic techniques can enable useful analyses. Second, model selection is an important part of data analysis, and in this case study we apply what we've learned about cross-validation, the bias-variance trade-off, and regularization. Finally, even models that perform well on the test set might have inherent limitations when we try to use them in practice, as we will soon see.

Let's start by refining our research question and understanding the scope of our data.

Question and Scope

Our initial research question is: can we automatically detect fake news? To refine this question, we consider the kind of information that we might use to build a model for detecting fake news. If we have hand-classified news stories where people have read each story and determined whether it is fake or not, then our question becomes: can we build a model to accurately predict whether a news story is fake based on its content?

To address this question, we can use the FakeNewsNet data repository as described in Shu et al (*https://arxiv.org/abs/1809.01286*). This repository contains content from news and social media websites, as well as metadata like user engagement metrics. For simplicity, we only look at the dataset's political news articles. This subset of the data includes only articles that were fact-checked by Politifact (*https://www.politifact.com*), a nonpartisan organization with a good reputation. Each article in the dataset has a "real" or "fake" label based on Politifact's evaluation, which we use as the ground truth.

Politifact uses a nonrandom sampling method to select articles to fact-check. According to its website, Politifact's journalists select the "most newsworthy and significant" claims each day. Politifact started in 2007 and the repository was published in 2020, so most of the articles were published between 2007 and 2020.

Summarizing this information, we determine that the target population consists of all political news stories published online in the time period from 2007 to 2020 (we would also want to list the sources of the stories). The access frame is determined by

Politifact's identification of the most newsworthy claims of the day. So the main sources of bias for this data include:

Coverage bias
: The news outlets are limited to those that Politifact monitored, which may miss arcane or short-lived sites.

Selection bias
: The data are limited to articles Politifact decided were interesting enough to fact-check, which means that articles might skew toward ones that are both widely shared and controversial.

Measurement bias
: Whether a story should be labeled "fake" or "real" is determined by one organization (Politifact) and reflects the biases, unintentional or otherwise, that the organization has in its fact-checking methodology.

Drift
: Since we only have articles published between 2007 and 2020, there is likely to be drift in the content. Topics are popularized and faked in rapidly evolving news trends.

We will keep these limitations of the data in mind as we begin to wrangle the data into a form that we can analyze.

Obtaining and Wrangling the Data

Let's get the data into Python using the GitHub page for FakeNewsNet (*https://oreil.ly/0DOHd*). Reading over the repository description and code, we find that the repository doesn't actually store the news articles itself. Instead, running the repository code will scrape news articles from online web pages directly (using techniques we covered in Chapter 14). This presents a challenge: if an article is no longer available online, it likely will be missing from our dataset. Noting this, let's proceed with downloading the data.

The FakeNewsNet code highlights one challenge in reproducible research—online datasets change over time, but it can be difficult (or even illegal) to store and share copies of this data. For example, other parts of the FakeNewsNet dataset use Twitter posts, but the dataset creators would violate Twitter's terms and services if they stored copies of the posts in their repository. When working with data gathered from the web, we suggest documenting the date the data were gathered and reading the terms and services of the data sources carefully.

Running the script to download the Politifact data takes about an hour. After that, we place the datafiles into the *data/politifact* folder. The articles that Politifact labeled as fake and real are in *data/politifact/fake* and *data/politifact/real*. Let's take a look at one of the articles labeled "real":

```
!ls -l data/politifact/real | head -n 5
```

```
total 0
drwxr-xr-x  2 sam  staff  64 Jul 14  2022 politifact100
drwxr-xr-x  3 sam  staff  96 Jul 14  2022 politifact1013
drwxr-xr-x  3 sam  staff  96 Jul 14  2022 politifact1014
drwxr-xr-x  2 sam  staff  64 Jul 14  2022 politifact10185
ls: stdout: Undefined error: 0
```

```
!ls -lh data/politifact/real/politifact1013/
```

```
total 16
-rw-r--r--  1 sam  staff   5.7K Jul 14  2022 news content.json
```

Each article's data is stored in a JSON file named *news content.json*. Let's load the JSON for one article into a Python dictionary (see Chapter 14):

```
import json
from pathlib import Path

article_path = Path('data/politifact/real/politifact1013/news content.json')
article_json = json.loads(article_path.read_text())
```

Here, we've displayed the keys and values in `article_json` as a table:

key	value
url	http://www.senate.gov/legislative/LIS/roll_cal...
text	Roll Call Vote 111th Congress - 1st Session\n\...
images	[http://statse.webtrendslive.com/dcs222dj3ow9j...
top_img	http://www.senate.gov/resources/images/us_sen.ico
keywords	[]
authors	[]
canonical_link	
title	U.S. Senate: U.S. Senate Roll Call Votes 111th...
meta_data	{'viewport': 'width=device-width, initial-scal...
movies	[]
publish_date	None
source	http://www.senate.gov
summary	

There are many fields in the JSON file, but for this analysis we only look at a few that are primarily related to the content of the article: the article's title, text content, URL, and publication date. We create a dataframe where each row represents one article (the granularity in a news story). To do this, we load in each available JSON file as a Python dictionary, and then extract the fields of interest to store as a `pandas DataFrame` named `df_raw`:

```
from pathlib import Path

def df_row(content_json):
    return {
        'url': content_json['url'],
        'text': content_json['text'],
        'title': content_json['title'],
        'publish_date': content_json['publish_date'],
    }

def load_json(folder, label):
    filepath = folder / 'news content.json'
    data = df_row(json.loads(filepath.read_text())) if filepath.exists() else {}
    return {
        **data,
        'label': label,
    }

fakes = Path('data/politifact/fake')
reals = Path('data/politifact/real')

df_raw = pd.DataFrame([load_json(path, 'fake') for path in fakes.iterdir()] +
                      [load_json(path, 'real') for path in reals.iterdir()])

df_raw.head(2)
```

	url	text	title	publish_date	label
0	dailybuzzlive.com/cannibals-arrested-florida/	Police in Vernal Heights, Florida, arrested 3-...	Cannibals Arrested in Florida Claim Eating Hum...	1.62e+09	fake
1	https://web.archive.org/web/20171228192703/htt...	WASHINGTON — Rod Jay Rosenstein, Deputy Attorn...	BREAKING: Trump fires Deputy Attorney General ...	1.45e+09	fake

Exploring this dataframe reveals some issues we'd like to address before we begin the analysis. For example:

- Some articles couldn't be downloaded. When this happened, the `url` column contains `NaN`.
- Some articles don't have text (such as a web page with only video content). We drop these articles from our dataframe.
- The `publish_date` column stores timestamps in Unix format (seconds since the Unix epoch), so we need to convert them to `pandas.Timestamp` objects.

- We are interested in the base URL of a web page. However, the `source` field in the JSON file has many missing values compared to the `url` column, so we must extract the base URL using the full URL in the `url` column. For example, from *dailybuzzlive.com/cannibals-arrested-florida/* we get *dailybuzzlive.com*.
- Some articles were downloaded from an archival website (`web.archive.org`). When this happens, we want to extract the actual base URL from the original by removing the `web.archive.org` prefix.
- We want to concatenate the `title` and `text` columns into a single `content` column that contains all of the text content of the article.

We can tackle these data issues using a combination of `pandas` functions and regular expressions:

```
import re

# [1], [2]
def drop_nans(df):
    return df[~(df['url'].isna() |
                (df['text'].str.strip() == '') |
                (df['title'].str.strip() == ''))]

# [3]
def parse_timestamps(df):
    timestamp = pd.to_datetime(df['publish_date'], unit='s', errors='coerce')
    return df.assign(timestamp=timestamp)

# [4], [5]
archive_prefix_re = re.compile(r'https://web.archive.org/web/\d+/')
site_prefix_re = re.compile(r'(https?://)?(www\.)?')
port_re = re.compile(r':\d+')

def url_basename(url):
    if archive_prefix_re.match(url):
        url = archive_prefix_re.sub('', url)
    site = site_prefix_re.sub('', url).split('/')[0]
    return port_re.sub('', site)

# [6]
def combine_content(df):
    return df.assign(content=df['title'] + ' ' + df['text'])

def subset_df(df):
    return df[['timestamp', 'baseurl', 'content', 'label']]

df = (df_raw
 .pipe(drop_nans)
 .reset_index(drop=True)
 .assign(baseurl=lambda df: df['url'].apply(url_basename))
 .pipe(parse_timestamps)
```

```
    .pipe(combine_content)
    .pipe(subset_df)
)
```

After data wrangling, we end up with the following dataframe named `df`:

```
df.head(2)
```

	timestamp	baseurl	content	label
0	2021-04-05 16:39:51	dailybuzzlive.com	Cannibals Arrested in Florida Claim Eating Hum...	fake
1	2016-01-01 23:17:43	houstonchronicle-tv.com	BREAKING: Trump fires Deputy Attorney General ...	fake

Now that we've loaded and cleaned the data, we can proceed to exploratory data analysis.

Exploring the Data

The dataset of news articles we're exploring is just one part of the larger FakeNewsNet dataset. As such, the original paper doesn't provide detailed information about our subset of data. So, to better understand the data, we must explore it ourselves.

Before starting exploratory data analysis, we apply our standard practice of splitting the data into training and test sets. We perform EDA using only the train set:

```
from sklearn.model_selection import train_test_split

df['label'] = (df['label'] == 'fake').astype(int)

X_train, X_test, y_train, y_test = train_test_split(
    df[['timestamp', 'baseurl', 'content']], df['label'],
    test_size=0.25, random_state=42,
)
```

```
X_train.head(2)
```

	timestamp	baseurl	content
164	2019-01-04 19:25:46	worldnewsdailyreport.com	Chinese lunar rover finds no evidence of Ameri...
28	2016-01-12 21:02:28	occupydemocrats.com	Virginia Republican Wants Schools To Check Chi...

Let's count the number of real and fake articles in the train set:

```
y_train.value_counts()
```

```
label
0    320
1    264
Name: count, dtype: int64
```

Our train set has 584 articles, and there are about 60 more articles labeled `real` than `fake`. Next, we check for missing values in the three fields:

```
X_train.info()
```

```
<class 'pandas.core.frame.DataFrame'>
Index: 584 entries, 164 to 102
Data columns (total 3 columns):
 #   Column     Non-Null Count  Dtype
---  ------     --------------  -----
 0   timestamp  306 non-null    datetime64[ns]
 1   baseurl    584 non-null    object
 2   content    584 non-null    object
dtypes: datetime64[ns](1), object(2)
memory usage: 18.2+ KB
```

Nearly half of the timestamps are null. This feature will limit the dataset if we use it in the analysis. Let's take a closer look at the `baseurl`, which represents the website that published the original article.

Exploring the Publishers

To understand the `baseurl` column, we start by counting the number of articles from each website:

```
X_train['baseurl'].value_counts()
```

```
baseurl
whitehouse.gov               21
abcnews.go.com               20
nytimes.com                  17
                             ..
occupydemocrats.com           1
legis.state.ak.us             1
dailynewsforamericans.com     1
Name: count, Length: 337, dtype: int64
```

Our train set has 584 rows, and we have found that there are 337 unique publishing websites. This means that the dataset includes many publications with only a few articles. A histogram of the number of articles published by each website confirms this:

```
fig = px.histogram(X_train['baseurl'].value_counts(), width=450, height=250,
                   labels={"value": "Number of articles published at a URL"})

fig.update_layout(showlegend=False)
```

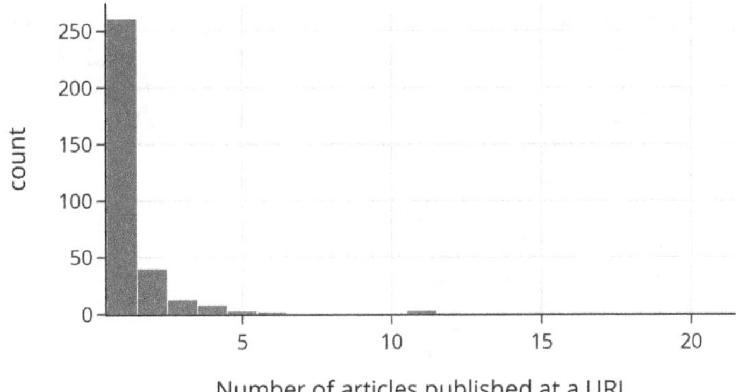

This histogram shows that the vast majority (261 out of 337) of websites have only one article in the train set, and only a few websites have more than five articles in the train set. Nonetheless, it can be informative to identify the websites that published the most fake or real articles. First, we find the websites that published the most fake articles:

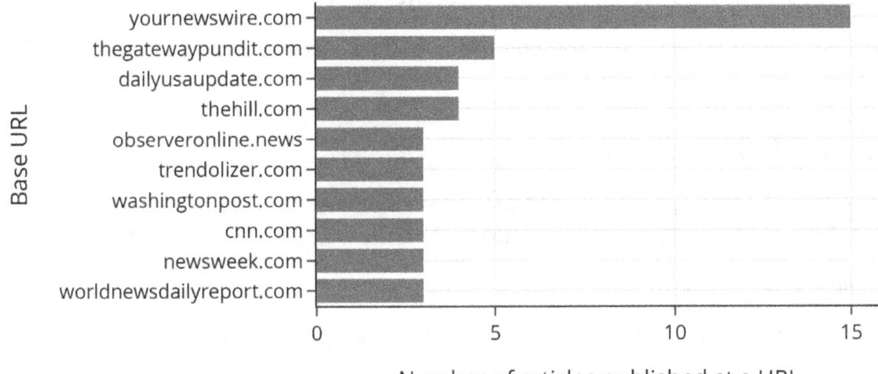

Next, we list the websites that published the greatest number of real articles:

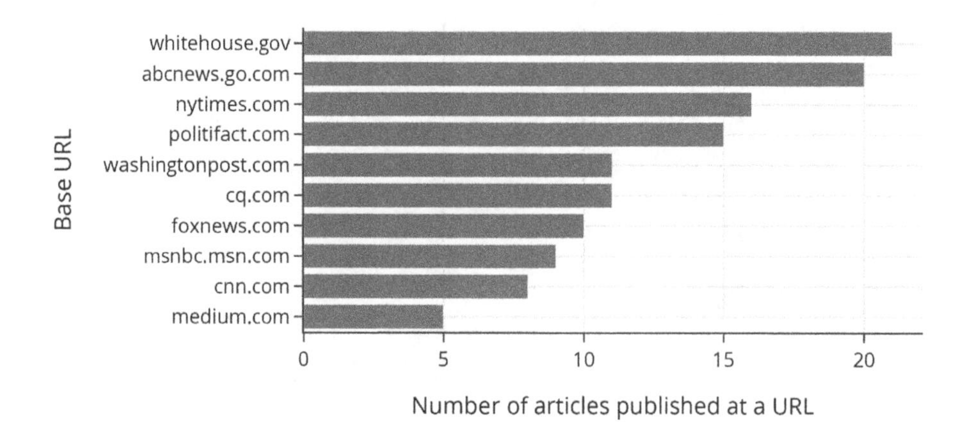

Only `cnn.com` and `washingtonpost.com` appear on both lists. Even without knowing the total number of articles for these sites, we might expect that an article from `yournewswire.com` is more likely to be labeled as `fake`, while an article from `whitehouse.gov` is more likely to be labeled as `real`. That said, we don't expect that using the publishing website to predict article truthfulness would work very well; there are simply too few articles from most of the websites in the dataset.

Next, let's explore the `timestamp` column, which records the publication date of the news articles.

Exploring Publication Date

Plotting the timestamps on a histogram shows that most articles were published after 2000, although there seems to be at least one article published before 1940:

```
fig = px.histogram(
    X_train["timestamp"],
    labels={"value": "Publication year"}, width=550, height=250,
)
fig.update_layout(showlegend=False)
```

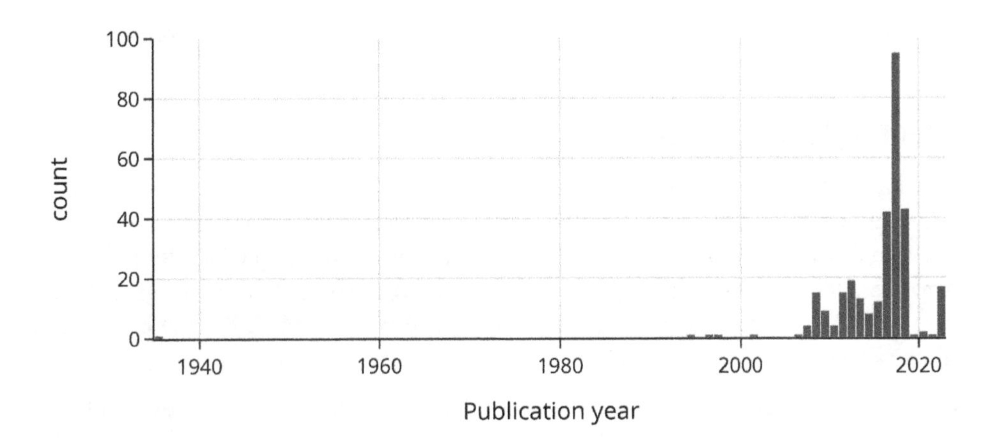

When we take a closer look at the news articles published prior to 2000, we find that the timestamps don't match the actual publication date of the article. These date issues are most likely related to the web scraper collecting inaccurate information from the web pages. We can zoom into the region of the histogram after 2000:

```
fig = px.histogram(
    X_train.loc[X_train["timestamp"] > "2000", "timestamp"],
    labels={"value": "Publication year"}, width=550, height=250,
)
fig.update_layout(showlegend=False)
```

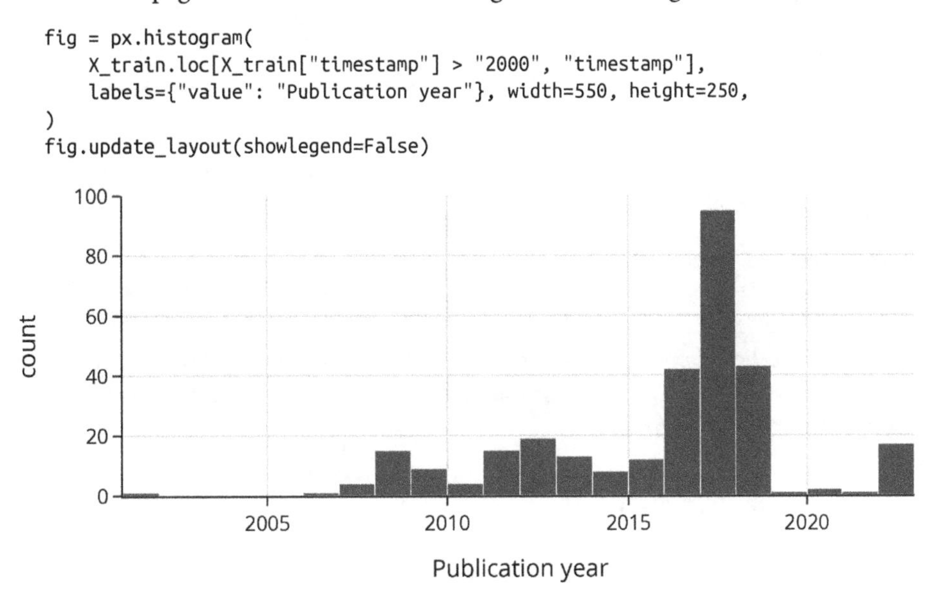

As expected, most of the articles were published between 2007 (the year Politifact was founded) and 2020 (the year the FakeNewsNet repository was published). But we also find that the timestamps are concentrated on the years 2016 to 2018—the year of the controversial 2016 US presidential election and the two years following. This insight is a further caution on the limitation of our analysis to carry over to nonelection years.

Our main aim is to use the text content for classification. We explore some word frequencies next.

Exploring Words in Articles

We'd like to see whether there's a relationship between the words used in the articles and whether the article was labeled as `fake`. One simple way to do this is to look at individual words like *military*, then count how many articles that mentioned "military" were labeled `fake`. For *military* to be useful, the articles that mention it should have a much higher or much lower fraction of fake articles than 45% (the proportion of fake articles in the dataset: 264/584).

We can use our domain knowledge of political topics to pick out a few candidate words to explore:

```
word_features = [
    # names of presidential candidates
    'trump', 'clinton',
    # congress words
    'state', 'vote', 'congress', 'shutdown',

    # other possibly useful words
    'military', 'princ', 'investig', 'antifa',
    'joke', 'homeless', 'swamp', 'cnn', 'the'
]
```

Then we define a function that creates a new feature for each word, where the feature contains `True` if the word appeared in the article and `False` if not:

```
def make_word_features(df, words):
    features = { word: df['content'].str.contains(word) for word in words }
    return pd.DataFrame(features)
```

This is like one-hot encoding for the presence of a word (see Chapter 15). We can use this function to further wrangle our data and create a new dataframe with a feature for each of our chosen words:

```
df_words = make_word_features(X_train, word_features)
df_words["label"] = df["label"]
```

```
df_words.shape
```

```
(584, 16)
```

```
df_words.head(4)
```

	trump	clinton	state	vote	...	swamp	cnn	the	label
164	False	False	True	False	...	False	False	True	1
28	False	False	False	False	...	False	False	True	1
708	False	False	True	True	...	False	False	True	0
193	False	False	False	False	...	False	False	True	1

4 rows × 16 columns

Now we can find the proportion of these articles that were labeled `fake`. We visualize these calculations in the following plots. In the left plot, we mark the proportion of `fake` articles in the entire train set using a dotted line, which helps us understand how informative each word feature is—a highly informative word will have a point that lies far away from the line:

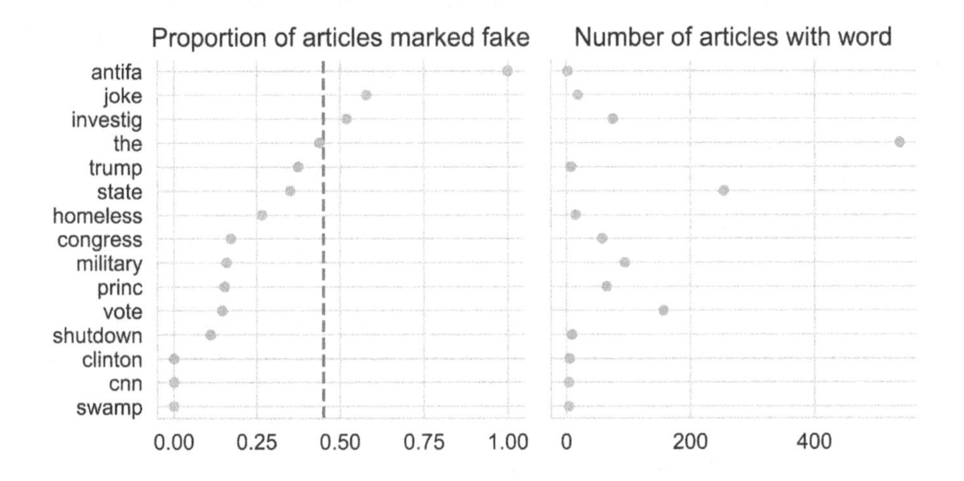

This plot reveals a few interesting considerations for modeling. For example, notice that the word *antifa* is highly predictive—all articles that mention the word *antifa* are labeled `fake`. However, *antifa* only appears in a few articles. On the other hand, the word *the* appears in nearly every article, but is uninformative for distinguishing between `real` and `fake` articles because the proportion of articles with *the* that are fake matches the proportion of fake articles overall. We might instead do better with a word like *vote*, which is predictive and appears in many news articles.

This exploratory analysis brought us understanding of the time frame that our news articles were published in, the broad range of publishing websites captured in the data, and candidate words to use for prediction. Next, we fit models for predicting whether articles are fake or real.

Modeling

Now that we've obtained, cleaned, and explored our data, let's fit models to predict whether articles are real or fake. In this section, we use logistic regression because we have a binary classification problem. We fit three different models that increase in complexity. First, we fit a model that just uses the presence of a single handpicked word in the document as an explanatory feature. Then we fit a model that uses multiple handpicked words. Finally, we fit a model that uses all the words in the train set, vectorized using the tf-idf transform (introduced in Chapter 13). Let's start with the simple single-word model.

A Single-Word Model

Our EDA showed that the word *vote* is related to whether an article is labeled `real` or `fake`. To test this, we fit a logistic regression model using a single binary feature: `1` if the word *vote* appears in the article and `0` if not. We start by defining a function to lowercase the article content:

```
def lowercase(df):
    return df.assign(content=df['content'].str.lower())
```

For our first classifier, we only use the word *vote*:

```
one_word = ['vote']
```

We can chain the `lowercase` function and the function `make_word_features` from our EDA into a `scikit-learn` pipeline. This provides a convenient way to transform and fit data all at once:

```
from sklearn.pipeline import make_pipeline
from sklearn.linear_model import LogisticRegressionCV
from sklearn.preprocessing import FunctionTransformer

model1 = make_pipeline(
    FunctionTransformer(lowercase),
    FunctionTransformer(make_word_features, kw_args={'words': one_word}),
    LogisticRegressionCV(Cs=10, solver='saga', n_jobs=4, max_iter=10000),
)
```

When used, the preceding pipeline converts the characters in the article content to lowercase, creates a dataframe with a binary feature for each word of interest, and fits a logistic regression model on the data using L_2 regularization. Additionally, the `LogisticRegressionCV` function uses cross-validation (fivefold by default) to select the best regularization parameter. (See Chapter 16 for more on regularization and cross-validation.)

Let's use the pipeline to fit the training data:

```
%%time

model1.fit(X_train, y_train)
print(f'{model1.score(X_train, y_train):.1%} accuracy on training set.')
```

```
64.9% accuracy on training set.
CPU times: user 110 ms, sys: 42.7 ms, total: 152 ms
Wall time: 144 ms
```

Overall, the single-word classifier only classifies 65% of articles correctly. We plot the confusion matrix of the classifier on the train set to see what kinds of mistakes it makes:

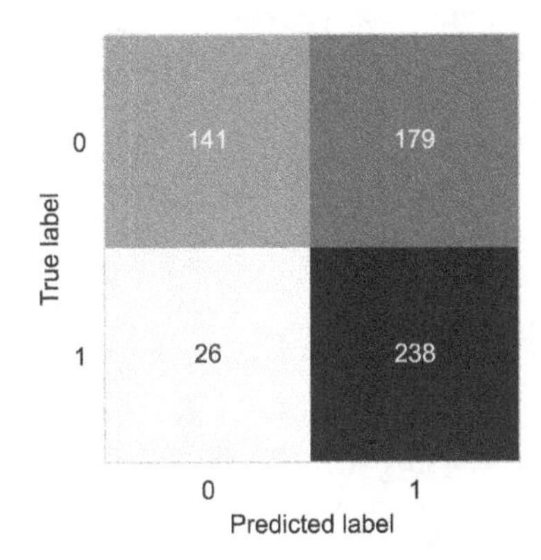

Our model often misclassifies real articles (0) as fake (1). Since this model is simple, we can take a look at the probabilities for the two cases: the word *vote* is in the article or is not:

```
"vote" present: [[0.72 0.28]]
"vote" absent: [[0.48 0.52]]
```

When an article contains the word *vote*, the model gives a high probability of the article being real, and when *vote* is absent, the probability leans slightly toward the article being fake. We encourage readers to verify this for themselves using the definition of the logistic regression model and the fitted coefficients:

```
print(f'Intercept: {log_reg.intercept_[0]:.2f}')
[[coef]] = log_reg.coef_
print(f'"vote" Coefficient: {coef:.2f}')
```

```
Intercept: 0.08
"vote" Coefficient: -1.00
```

As we saw in Chapter 19, the coefficient indicates the size of the change in the odds with a change in the explanatory variable. With a 0-1 variable like the presence or absence of a word in an article, this has a particularly intuitive meaning. For an article with *vote* in it, the odds of being fake decrease by a factor of $\exp(\theta_{vote})$, which is:

```
np.exp(coef)
```

```
0.36836305405149367
```

Remember that in this modeling scenario, a label of `0` corresponds to a real article and a label of `1` corresponds to a fake article. This might seem a bit counterintuitive—we're saying that a "true positive" is when a model correctly predicts a fake article as fake. In binary classification, we typically say a "positive" result is the one with the presence of something unusual. For example, a person who tests positive for an illness would expect to have the illness.

Let's make our model a bit more sophisticated by introducing additional word features.

Multiple-Word Model

We create a model that uses all of the words we examined in our EDA of the train set, except for *the*. Let's fit a model using these 15 features:

```
model2 = make_pipeline(
    FunctionTransformer(lowercase),
    FunctionTransformer(make_word_features, kw_args={'words': word_features}),
    LogisticRegressionCV(Cs=10, solver='saga', n_jobs=4, max_iter=10000),
)
```

```
%%time

model2.fit(X_train, y_train)
print(f'{model2.score(X_train, y_train):.1%} accuracy on training set.')
```

```
74.8% accuracy on training set.
CPU times: user 1.54 s, sys: 59.1 ms, total: 1.6 s
Wall time: 637 ms
```

This model is about 10 percentage points more accurate than the one-word model. It may seem a bit surprising that going from a one-word model to a 15-word model only gains 10 percentage points. The confusion matrix is helpful in teasing out the kinds of errors made:

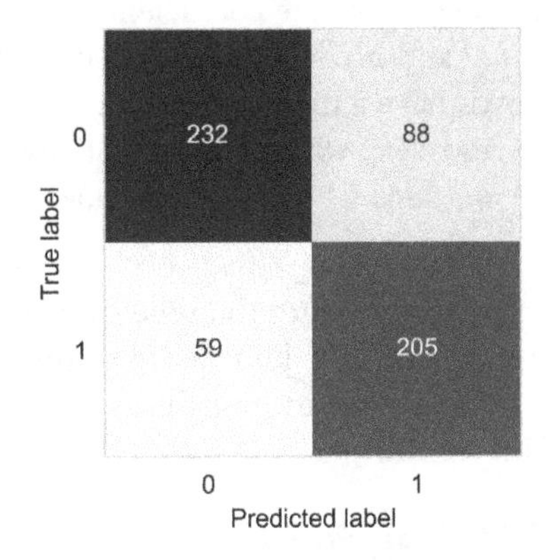

We can see that this classifier does a better job of classifying real articles accurately. However, it makes more mistakes than the simple one-word model when classifying fake article—59 of the fake articles were classified as real. In this scenario, we might be more concerned about misclassifying an article as fake when it is real. So we wish to have a high precision—the ratio of fake articles correctly predicted as fake to articles predicted as fake:

```
model1_precision = 238 / (238 + 179)
model2_precision = 205 / (205 + 88)

[round(num, 2) for num in [model1_precision, model2_precision]]
```

```
[0.57, 0.7]
```

The precision in our larger model is improved, but about 30% of the articles labeled as fake are actually real. Let's take a look at the model's coefficients:

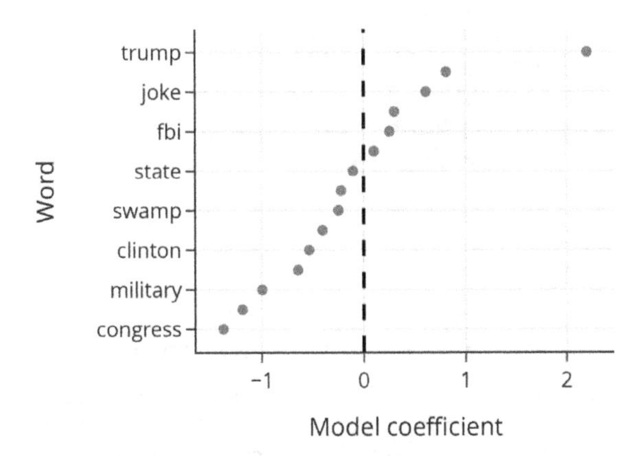

We can make a quick interpretation of the coefficients by looking at their signs. The large positive values on *trump* and *investig* indicate that the model predicts that new articles containing these words have a higher probability of being fake. The reverse is true for words like *congress* and *vote*, which have negative weights. We can use these coefficients to compare the log odds when an article does or does not contain a particular word.

Although this larger model performs better than the simple one-word model, we had to handpick the word features using our knowledge of the news. What if we missed the words that are highly predictive? To address this, we can incorporate all the words in the articles using the tf-idf transform.

Predicting with the tf-idf Transform

For the third and final model, we use the term frequency-inverse document frequency (tf-idf) transform from Chapter 13 to vectorize the entire text of all articles in the train set. Recall that with this transform, an article is converted into a vector with one element for each word that appears in any of the 564 articles. The vector consists of normalized counts of the number of times the word appears in the article normalized by the rareness of the word. The tf-idf puts more weight on words that only appear in a few documents. This means that our classifier uses all the words in the train set's news articles for prediction. As we've done previously when we introduced tf-idf, first we remove stopwords, then we tokenize the words, and then we use the `TfidfVectorizer` from `scikit-learn`:

```
tfidf = TfidfVectorizer(tokenizer=stemming_tokenizer, token_pattern=None)
```

```
from sklearn.compose import make_column_transformer

model3 = make_pipeline(
    FunctionTransformer(lowercase),
    make_column_transformer((tfidf, 'content')),
    LogisticRegressionCV(Cs=10,
                         solver='saga',
                         n_jobs=8,
                         max_iter=1000),
    verbose=True,
)
```

```
%%time

model3.fit(X_train, y_train)
print(f'{model3.score(X_train, y_train):.1%} accuracy on training set.')
```

```
[Pipeline]  (step 1 of 3) Processing functiontransformer, total=   0.0s
[Pipeline] . (step 2 of 3) Processing columntransformer, total=  14.5s
[Pipeline]  (step 3 of 3) Processing logisticregressioncv, total=   6.3s
100.0% accuracy on training set.
```

```
CPU times: user 50.2 s, sys: 508 ms, total: 50.7 s
Wall time: 34.2 s
```

We find that this model achieves 100% accuracy on the train set. We can take a look at the tf-idf transformer to better understand the model. Let's start by finding out how many unique tokens the classifier uses:

```
tfidf = model3.named_steps.columntransformer.named_transformers_.tfidfvectorizer
n_unique_tokens = len(tfidf.vocabulary_.keys())
print(f'{n_unique_tokens} tokens appeared across {len(X_train)} examples.')
```

```
23800 tokens appeared across 584 examples.
```

This means that our classifier has 23,812 features, a large increase from our previous model, which only had 15. Since we can't display that many model weights, we display the 10 most negative and 10 most positive weights:

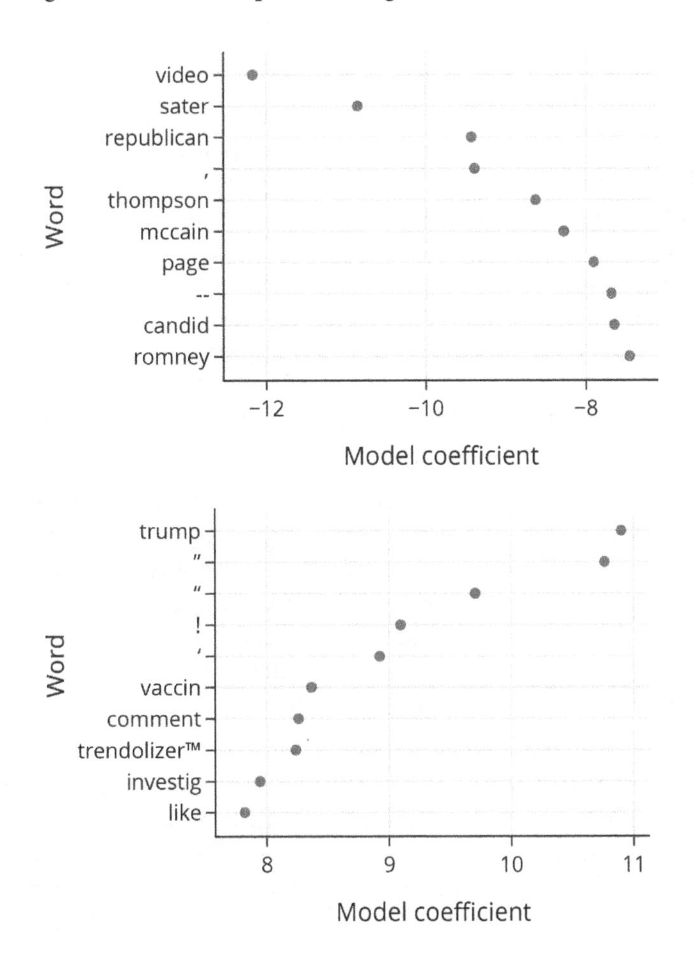

These coefficients show a few quirks about this model. We see that several influential features correspond to punctuation in the original text. It's unclear whether we should clean out the punctuation in the model. On the one hand, punctuation doesn't seem to convey as much meaning as words do. On the other, it seems plausible that, for example, lots of exclamation points in an article could help a model decide whether the article is real or fake. In this case, we've decided to keep punctuation, but curious readers can repeat this analysis after stripping the punctuation out to see how the resulting model is affected.

We conclude by displaying the test set error for all three models:

	test set error
model1	0.61
model2	0.70
model3	0.88

As we might expect, the models became more accurate as we introduced more features. The model that used tf-idf performed much better than the models with binary handpicked word features, but it did not meet the 100% accuracy obtained on the train set. This illustrates a common trade-off in modeling: given enough data, more complex models can often outperform simpler ones, especially in situations like this case study where simpler models have too much model bias to perform well. However, complex models can be more difficult to interpret. For example, our tf-idf model had over 20,000 features, which makes it basically impossible to explain how our model makes its decisions. In addition, the tf-idf model takes much longer to make predictions—it's over 100 times slower compared to model 2. All of these factors need to be considered when deciding which model to use in practice.

In addition, we need to be careful about what our models are useful for. In this case, our models use the content of the news articles for prediction, making them highly dependent on the words that appear in the train set. However, our models will likely not perform as well on future news articles that use words that didn't appear in the train set. For example, our models use the US election candidates' names in 2016 for prediction, but they won't know to incorporate the names of the candidates in 2020 or 2024. To use our models in the longer term, we would need to address this issue of *drift*.

That said, it's surprising that a logistic regression model can perform well with a relatively small amount of feature engineering (tf-idf). We've addressed our original research question: our tf-idf model appears effective for detecting fake news in our dataset, and it could plausibly generalize to other news published in the same time period covered in the training data.

Summary

We're quickly approaching the end of the chapter and thus the end of the book. We started this book by talking about the data science lifecycle. Let's take another look at the lifecycle, in Figure 21-2, to appreciate everything that you've learned.

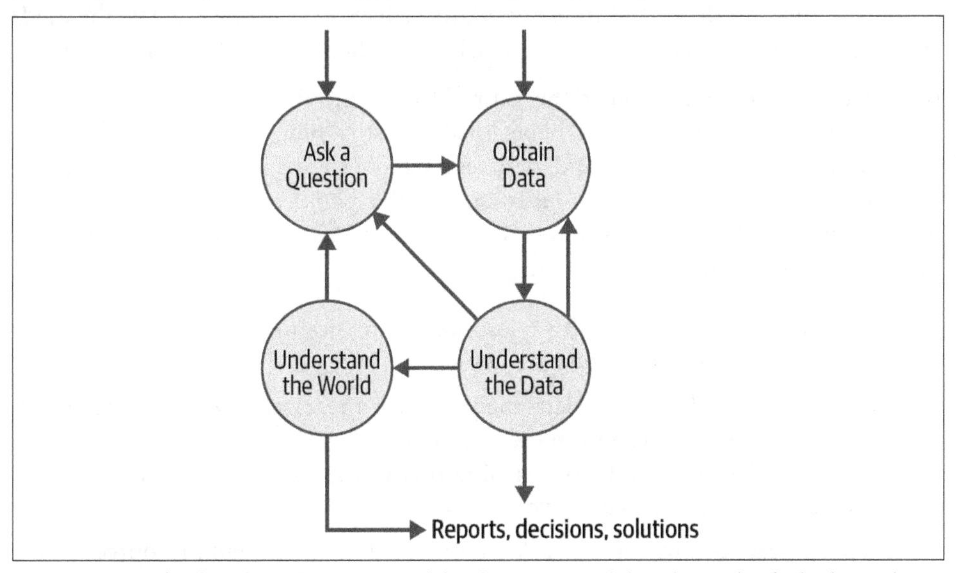

Figure 21-2. The four high-level steps of the data science lifecycle, each of which we dove into throughout this book

This case study stepped through each stage of the data science lifecycle:

1. Many data analyses begin with a research question. The case study we presented in this chapter started by asking whether we can create models to automatically detect fake news.
2. We obtained data by using code found online that scrapes web pages into JSON files. Since the data description was relatively minimal, we needed to clean the data to understand it. This included creating new features to indicate the presence or absence of certain words in the articles.
3. Our initial explorations identified possible words that might be useful for prediction. After fitting simple models and exploring their precision and accuracy, we further transformed the articles using tf-idf to convert each news article into a normalized word vector.
4. We used the vectorized text as features in a logistic model, and we fitted the final model using regularization and cross-validation. Finally, we found the accuracy and precision of the fitted model on the test set.

When we write out the steps in the lifecycle like this, the steps seem to flow smoothly into each other. But reality is messy—as the diagram illustrates, real data analyses jump forward and backward between steps. For example, at the end of our case study, we discovered data cleaning questions that might motivate us to revisit earlier stages of the lifecycle. Although our model was quite accurate, the majority of the training data came from the 2016–2018 time period, so we have to carefully evaluate the model's performance if we want to use it on articles published outside that time frame.

In essence, it's important to keep the entire lifecycle in mind at each stage of a data analysis. As a data scientist, you will be asked to justify your decisions, which means that you need to deeply understand your research question and data. To develop this understanding, the principles and techniques in this book equip you with a foundational set of skills. Going forward into your data science journey, we recommend that you continue to expand your skills by:

- Revisiting a case study from this book. Start by replicating our analysis, then dive deeper into questions that you have about the data.
- Conducting an independent data analysis. Pose a research question you're interested in, find relevant data from the web, and analyze the data to see how well the data matched your expectations. Doing this will give you firsthand experience with the entire data science lifecycle.
- Taking a deep dive into a topic. We've provided many in-depth resources in the Additional Material appendix. Take the resource that seems most interesting to you and learn more about it.

The world needs people like you who can use data to make conclusions, so we sincerely hope that you'll use these skills to help others make effective strategies, better products, and informed decisions.

Additional Material

Collected here are a variety of resources that offer a more in-depth treatment of the larger themes in this book. In addition to recommendations for these topics, we provide resources for several topics that we only lightly touched on. These resources are organized in the order in which the topics appear in the book:

Shumway, Robert, and David Stoffer. *Time Series Analysis and Its Applications*. New York: Springer, 2017.

This book covers how to analyze time-series data, like the Google Flu trends.

Speed, Terry. "Questions, Answers, and Statistics" (*https://oreil.ly/Nw0Rg*) *ICOTS* (1986): 18–28.

Leek, Jeffery and Roger Peng. "What Is the Question?" (*https://doi.org/10.1126/science.aaa6146*) *Science* 347, no. 6228 (February 2015): 1314–1315.

We recommend "Questions, Answers, and Statistics" if you want to learn more about the interplay between questions and data. "What Is the Question?" connects questions with the type of analysis needed.

Lohr, Sharon. *Sampling: Design and Analysis*, 3rd edition. New York: Chapman and Hall, 2021.

More on sampling topics can be found in *Sampling: Design and Analysis*. The book also contains a treatment of the target population, access frame, sampling methods, and sources of bias.

University of California, Berkeley, College of Computing, Data Science, and Society. "HCE Toolkit." Accessed September 15, 2023. *https://oreil.ly/vzkBn*.

Tuskegee University. "National Center for Bioethics in Research and Health Care." Accessed September 15, 2023. *https://oreil.ly/XLsYx*.

These toolkits will help you learn more about the human contexts and ethics of data.

Executive Office of the President. *Big Data: Seizing Opportunities, Preserving Values* (*https://oreil.ly/hTlpq*). May 2014.

This concise White House report provides guidelines and rationale for data privacy.

Ramdas, Aaditya. "Why the Easiest Person to Fool Is Yourself." Accessed September 15, 2023. *https://oreil.ly/dYiKe*.

Ramdas gave a fun, informative talk in our class "Principles and Techniques for Data Science" in fall 2019 on on bias, Simpson's paradox, p-hacking, and related topics. We recommend his slides from the lecture.

Freedman, David et al., *Statistics*, 4th edition. New York: Norton, 2007.

See *Statistics* for an introductory treatment of the urn model, confidence intervals, and hypothesis tests.

Owen, Art B. *Monte Carlo Theory, Methods, and Examples* (*https://artowen.su.domains/mc*). Self-published, 2013.

Owen's online text provides a solid introduction to simulation.

Pitman, Jim. *Probability*. New York: Springer, 1993.

Blitzstein, Joseph K. and Jessica Hwang. *Introduction to Probability*. New York: Chapman and Hall, 2014.

We suggest *Probability* and *Introduction to Probability* for a fuller treatment of probability.

Bickel, Peter J. and Kjell A. Doksum. *Mathematical Statistics: Basic Ideas and Selected Topics Volume I*, 2nd edition. New York: Chapman and Hall, 2015.

You can find a proof that the median minimizes absolute error in *Mathematical Statistics: Basic Ideas and Selected Topics Volume I.*

McKinney, Wes. *Python for Data Analysis*, 3rd edition. Sebastopol, CA: O'Reilly, 2022.

Python for Data Analysis provides in-depth coverage of `pandas`.

Roland, F.D. *The Essence of Databases*. Upper Saddle River, NJ: Prentice Hall, 1998.

W3Schools, Introduction to SQL (*https://w3schools.com/sql/sql_intro.asp*). Accessed September 15, 2023. *https://w3schools.com/sql/sql_intro.asp*.

Kleppmann, Martin. *Designing Data-Intensive Applications*. Sebastopol, CA: O'Reilly, 2017.

The classic *The Essence of Databases* offers a formal introduction to SQL. W3Schools provides SQL basics. *Designing Data-Intensive Applications* surveys and compares different data storage systems, including SQL databases.

Hellerstein, Joseph M. et al. *Principles of Data Wrangling: Practical Techniques for Data Preparation*. Sebastopol, CA: O'Reilly, 2017.

Principles of Data Wrangling: Practical Techniques for Data Preparation is a good resource for data wrangling.

Lohr, "Nonresponse." In *Sampling: Design and Analysis*.

Little, Roderick J. A., and Donald B. Rubin. *Statistical Analysis with Missing Data*. Hoboken, NJ: Wiley, 2019.

For how to handle missing data, see Chapter 8 in *Sampling: Design and Analysis* as well as *Statistical Analysis with Missing Data*.

Tukey, John Wilder. *Exploratory Data Analysis*. Reading, MA: Addison-Wesley, 1977.

Exploratory Data Analysis offers an excellent introduction to EDA.

Silverman, Bernard W. *Density Estimation for Statistics and Data Analysis*. New York: Chapman and Hall, 1998.

The smooth density curve is covered in detail in *Density Estimation for Statistics and Data Analysis*.

Wilke, Claus O. *Fundamentals of Data Visualization*. Sebastopol, CA: O'Reilly, 2019.

See *Fundamentals of Data Visualization* for more on visualization. Our guidelines do not entirely match Wilke's but they come close, and it's helpful to see a variety of opinions on the topic.

Brewer, Cynthia. ColorBrewer2.0. Accessed September 15, 2023. *https://colorbrewer2.org*.

See ColorBrewer2.0 to learn more about color palettes.

Osborne, Christine. "Statistical Calibration: A Review" (*https://doi.org/10.2307/1403690*). *International Statistical Review* 59, no. 3 (Dec 1991): pp. 309–336.

See Osborne for more on calibration.

W3Schools. Python RegEx. Accessed September 15, 2023. *https://w3schools.com/python/python_regex.asp*.

Regular Expressions 101. Accessed September 15, 2023. *https://regex101.com*.

Nield, Thomas. "An Introduction to Regular Expressions." O'Reilly blog. December 13, 2017. *https://oreil.ly/EWuO6*.

Friedl, Jeffrey. *Mastering Regular Expressions*. Sebastopol, CA: O'Reilly, 2006.

You can practice regular expressions with many online resources. We recommend the preceding tutorial, regular expression checker, primer on the topic, and book.

Fox, John. "Collinearity and Its Purported Remedies." In *Applied Regression Analysis and Generalized Linear Models*, 3rd edition. Los Angeles: Sage, 2015.

James, Gareth et al. "Unsupervised Learning." In *An Introduction to Statistical Learning*, 2nd edition. New York: Springer, 2021.

The preceding chapters in *Applied Regression Analysis and Generalized Linear Models* and *An Introduction to Statistical Learning* discuss principal components.

Tompkins, Adrian. "The Beauty of NetCDF". YouTube, April 2, 2021. *https://oreil.ly/3U6Rr*.

"The Beauty of NetCDF" is a helpful video tutorial on how to work with netCDF climate data

Richardson, Leonard. and Sam Ruby. *RESTful Web Services*. Sebastopol, CA: O'Reilly, 2007.

There are many resources on web services. We recommend *RESTful Web Services* for accessible introductory material.

Nolan, Deborah, and Duncan Temple Lang. *XML and Web Technologies for Data Sciences with R*. New York: Springer, 2014.

For more on XML, we recommend *XML and Web Technologies for Data Sciences with R*.

Faraway, Julian J. *Linear Models with Python*. New York: Routledge, 2021.

Fox, *Applied Regression Analysis and Generalized Linear Models*.

James et al. *An Introduction to Statistical Learning*.

Weisberg, Sanford, *Applied Linear Regression* Hoboken, NJ: Wiley, 2005.

The many topics related to modeling, including transformations, one-hot encoding, model-selection, cross-validation, and regularization are covered in several sources. We recommend *Linear Models with Python*, *Applied Regression Analysis and Generalized Linear Models*, *An Introduction to Statistical Learning*, and *Applied Linear Regression*. "The Vector Geometry of Linear Models" in *Applied Regression Analysis and Generalized Linear Models* gives an informative treatment of vector geometry of least squares. "Diagnosing Non-Normality, Nonconstant Error Variance, and Nonlinearity" in *Applied Regression Analysis and Generalized Linear Models* and "Explanation" in *Linear Models with Python* cover the topic of weighted regression.

Perry, Tekla S. "Andrew Ng X-Rays the AI Hype." *IEEE Spectrum*, May 3, 2021.

This IEEE Spectrum interview with Andrew Ng is an interesting read on the gap between the test set and the real world.

James et al. *An Introduction to Statistical Learning*.

"Moving Beyond Linearity" of *An Introduction to Statistical Learning* introduces polynomial regression using orthogonal polynomials.

Chiu, Grace et al. "Bent-Cable Regression Theory and Applications." *Journal of the American Statistical Association* 101, no. 474 (January 1, 2012): pp. 542–553.

For more on broken-stick regression, see "Bent-Cable Regression Theory and Applications."

Rice, John. *Mathematical Statistics and Data Analysis*, 3rd edition. Boston, MA: Cengage, 2007.

A more formal treatment of confidence intervals, prediction intervals, testing, and the bootstrap can be found in *Mathematical Statistics and Data Analysis*.

Wasserstein, Ronald L. and Nicole A. Lazar. "The ASA Statement on *p*-Values: Context, Process, and Purpose". *The American Statistician* 70, no. 2 (2016): pp. 129–133.

Gelman, Andrew and Eric Loken. "The Statistical Crisis in Science." *American Scientist* 102, no. 6 (2014): pp. 460.

"The ASA Statement on p-Values: Context, Process, and Purpose" provides valuable insights into *p*-values. "The Statistical Crisis in Science" addresses p-hacking.

Hettmansperger, Thomas. "Nonparametric Rank Tests." In *International Encyclopedia of Statistical Science*, edited by Miodrag Lovric, 970–972. New York: Springer, 2014.

You can find information about rank tests and other nonparametric statistics in "Nonparametric Rank Tests."

Doerfler, Ron. "The Art of Nomography." January 8, 2008. *https://oreil.ly/twvK5*.

The technique for developing linear models to use in the field is addressed in "The Art of Nomography."

Fox, *Applied Regression Analysis and Generalized Linear Models.*

James et al. *An Introduction to Statistical Learning.*

"Logit and Probit Models for Categorical Response Variables" in *Applied Regression Analysis and Generalized Linear Models* covers the maximum likelihood approach to logistic regression. And "Classification" in *An Introduction to Statistical Learning* covers sensitivity and specificity in more detail.

Wasserman, Larry. "Statistical Decision Theory." In *All of Statistics.* New York: Springer, 2004.

"Statistical Decision Theory" has an in-depth treatment of loss functions and risk.

Segaran, Toby. *Programming Collective Intelligence.* Sebastopol: O'Reilly, 2007.

Programming Collective Intelligence covers the topic of optimization

Bengfort, Benjamin et al. *Applied Text Analysis with Python*. Sebastopol: O'Reilly, 2018.

See *Applied Text Analysis with Python* for more on text analysis.

Data Sources

All of the data analyzed in this book are available on the book's website (*https://learningds.org*) and GitHub repository (*https://github.com/DS-100/textbook*). These datasets are from open repositories and from individuals. We acknowledge them all here, and include, as appropriate, the filename for the data stored in our repository, a description of the resource, a link to the original source, a related publication, and the author(s)/owner(s).

To begin, we provide the sources for the four case studies in the book. Our analysis of the data in these case studies is based on research articles or, in one case, a blog post. We generally follow the line of inquiry in these sources, simplifying the analyses to match the level of the book.

Here are the four case studies:

`seattle_bus_times.csv`
: Mark Hallenbeck of the Washington State Transportation Center (*https://oreil.ly/3hZ_A*) provides the Seattle Transit data. Our analysis is based on "The Waiting Time Paradox, or, Why Is My Bus Always Late?" (*https://oreil.ly/kaQv-*) by Jake VanderPlas.

`aqs_06-067-0010.csv`, `list_of_aqs_sites.csv`, `matched_pa_aqs.csv`, `list_of_purpleair_sensors.json`, *and* `purpleair_AMTS`
: The datasets used in the study of air quality monitors are available from Karoline Barkjohn of the Environmental Protection Agency. These were originally acquired by Barkjohn and collaborators from the US Air Quality System (*https://oreil.ly/Sjku6*) and PurpleAir (*https://www2.purpleair.com*). Our analysis is based on "Development and Application of a United States-Wide Correction for PM 2.5 Data Collected with the PurpleAir Sensor" (*https://oreil.ly/jWuNx*) by Barkjohn, Brett Gantt, and Andrea Clements.

`donkeys.csv`
: Kate Milner collected the data for the Kenyan donkey study on behalf of the UK Donkey Sanctuary. Jonathan Rougier makes the data available in the paranomo package (*https://oreil.ly/oiMNE*) (follow link to download). Our analysis is based on "How to Weigh a Donkey in the Kenyan Countryside" (*https://doi.org/10.1111/j.1740-9713.2014.00768.x*) by Milner and Rougier.

`fake_news.csv`
: The hand-classified fake news data are from "FakeNewsNet: A Data Repository with News Content, Social Context, and Spatiotemporal Information for Studying Fake News on Social Media" (*https://arxiv.org/abs/1809.01286*) by Kai Shu et al.

In addition to these case studies, we use another 20-plus datasets as examples throughout the book. We acknowledge the people and organizations that make these datasets available in the order in which they appear in the book:

`gft.csv`
: The data on the Google Flu Trends is available from Gary King Dataverse (*https://doi.org/10.7910/DVN/24823*) and the plot made from these data is based on "The Parable of Google Flu: Traps in Big Data Analysis" (*https://doi.org/10.1126/science.1248506*) by David Lazer et al.

`WikipediaExp.csv`
: Arnout van de Rijt provides the data for the Wikipedia experiment. These data are analyzed in "Experimental Study of Informal Rewards in Peer Production" (*https://oreil.ly/BDDSV*) by Michael Restivo and van de Rijt.

`co2_mm_mlo.txt`
: The CO_2 concentrations measured at Mauna Loa by the National Oceanic and Atmospheric Administration (NOAA) (*https://noaa.gov*) are available from the Global Monitoring Laboratory (*https://gml.noaa.gov/obop/mlo*).

`pm30.csv`
: We downloaded these air quality measurements for one day and one sensor from the PurpleAir map (*https://www2.purpleair.com*).

`babynames.csv`
: The US Social Security Department (*https://oreil.ly/DBiky*) provides the names from all Social Security card applications.

`DAWN-Data.txt`
: The 2011 DAWN survey (*https://oreil.ly/n8NOQ*) of drug-related emergency room visits is administered by the US Substance Abuse and Mental Health Services Administration (*https://samhsa.gov*).

`businesses.csv`, `inspections.csv`, *and* `violations.csv`
: The data on restaurant inspection scores in San Francisco is from DataSF (*https://datasf.org*).

`akc.csv`
: The data on dog breeds come from Information Is Beautiful's "Best in Show: The Ultimate Data Dog" (*https://oreil.ly/KjIyv*) visualization and was originally acquired from the American Kennel Club (*https://akc.org*).

`sfhousing.csv`
: The housing sale prices for the San Francisco Bay Area were scraped from the *San Francisco Chronicle* (*https://oreil.ly/kaziA*) real estate pages.

`cherryBlossomMen.csv`
: The run times in the annual Cherry Blossom 10-mile run (*https://cherryblossom.org*) were scraped from the race results pages.

`earnings2020.csv`
: The weekly earnings data are made available by the US Bureau of Labor Statistics (*https://oreil.ly/cZG_w*).

`co2_by_country.csv`
: The annual country CO_2 emissions is available from Our World in Data (*https://ourworldindata.org*).

`100m_sprint.csv`
: The times for the 100-meter sprint are from FiveThirtyEight (*https://fivethirtyeight.com/*) and the figure is based on "The Fastest Men in the World Are Still Chasing Usain Bolt" (*https://oreil.ly/ewY7w*) by Josh Planos.

`stateoftheunion1790-2022.txt`
: The State of the Union addresses are compiled from the American Presidency Project (*https://oreil.ly/AnkW8*).

`CDS_ERA5_22-12.nc`
: We collected these data from the Climate Data Store (*https://cds.climate.copernicus.eu*), which is supported by the European Centre for Medium-Range Weather Forecasts (*https://ecmwf.int*).

`world_record_1500m.csv`
: The 1,5000-meter world records come from the Wikipedia page "1,500 Metres World Record Progression" (*https://oreil.ly/2_P4H*).

`the_clash.csv`
: The Clash songs are available at the Spotify Web API (*https://oreil.ly/FYP8B*). The retrieval of the data follows "Exploring the Spotify API in Python" (*https://oreil.ly/mWgYl*) by Steven Morse.

`catalog.xml`
: The XML plant catalog document is from the W3Schools plant catalog (*https://oreil.ly/MNw-G*).

`ECB_EU_exchange.csv`
: The exchange rates are available from the European Central Bank (*https://oreil.ly/Wc61c*).

`mobility.csv`
: These data are available at Opportunity Insights (*https://oreil.ly/W_5KH*), and our example follows "Where Is the Land of Opportunity? The Geography of Intergenerational Mobility in the United States" (*https://doi.org/10.1093/qje/qju022*) by Raj Chetty et al.

`utilities.csv`
: Daniel Kaplan's home energy consumption data is available to download (*https://oreil.ly/YTAsK*) and appear in his first edition of *Statistical Modeling: A Fresh Approach* (self-pub, CreateSpace).

`market-analysis.csv`
: Stan Lipovetsky provides these data, and they correspond to the data in his paper "Regressions Regularized by Correlations" (*https://oreil.ly/UZUJq*).

`crabs.data`
: The crab measurements are from the California Department of Fish and Wildlife (*https://wildlife.ca.gov*), available to download from the Stat Labs Data repository (*https://oreil.ly/mZsQ8*).

`black_spruce.csv`
: Roy Lawrence Rich collected the wind-damaged tree data for his thesis "Large Wind Disturbance in the Boundary Waters Canoe Area Wilderness. Forest Dynamics and Development Changes Associated with the July 4th 1999 Blowdown" (*https://oreil.ly/Pkw8N*). The data are available online in the `alr4` package (*https://oreil.ly/6rPOB*). Our analysis is based on "Logistic Regression" in Weisberg's *Applied Linear Regression*.

Index

Symbols

A

B

C

D

E

F

G

H

I

J

K

L

M

N

O

P

Q

R

S

T

U

V

W

X

Y

Z

About the Authors

Sam Lau is an assistant teaching professor in the Halıcıoğlu Data Science Institute at University of California, San Diego. He has a decade of teaching experience and helped to design and teach flagship data science courses at UC Berkeley and UC San Diego. His research creates novel interfaces for learning and teaching data science, including the popular Pandas Tutor tool, which serves over 40,000 people per year.

Joseph (Joey) Gonzalez is an associate professor in the EECS department at the University of California, Berkeley, and a founding member of the UC Berkeley RISE Lab. His research interests are at the intersection of machine learning and data systems, including dynamic deep neural networks for transfer learning, accelerated deep learning for high-resolution computer vision, and software platforms for autonomous vehicles. Joey is also cofounder of Turi Inc. (formerly GraphLab), which was based on his work on the GraphLab and PowerGraph systems. Turi was recently acquired by Apple Inc.

Deborah (Deb) Nolan is professor emerita of statistics and associate dean for undergraduates in the College of Computing, Data Science, and Society at the University of California, Berkeley, where she held the Zaffaroni Family Chair in Undergraduate Education. Her research has involved the empirical process, high-dimensional modeling, and, more recently, technology in education and reproducible research. Her pedagogical approach connects research, practice, and education, and she is coauthor of four textbooks: *Stat Labs*, *Teaching Statistics*, *Data Science in R*, and *Communicating with Data*.

Colophon

The animal on the cover of *Learning Data Science* is an edible dormouse (*Glis glis*). As you might suspect, these creatures have wound up in human cuisine. The edible dormouse was served grilled as a delicacy in ancient Rome and is still consumed today in Croatia and Slovenia. Edible dormice have squirrel-like bodies with small ears, short legs, large feet, and long, bushy tails. Their front feet have four digits and their hind feet have five. They are predominantly covered in gray to gray-brown fur with white underbellies. Their feet have naked soles that secrete a sticky substance that enables climbing.

These nocturnal creatures spend most of their time in trees. They can be found across Europe and in parts of western and central Asia. While the IUCN categorizes edible dormice as a species of Least Concern, they are threatened by illegal hunting and habitat loss. Many of the animals on O'Reilly covers are endangered; all of them are important to the world. The cover illustration is by Karen Montgomery, based on an antique line engraving from Lydekker's *Royal Natural History*.

Printed in the USA
CPSIA information can be obtained
at www.ICGtesting.com
JSHW061439010424
60353JS00007B/104